Holocene Extinctions

Dedicated to the memory of the Yangtze River dolphin
More should have been done

Holocene Extinctions

EDITED BY

Samuel T. Turvey

Institute of Zoology
Zoological Society of London

OXFORD
UNIVERSITY PRESS

Great Clarendon Street, Oxford, OX2 6DP,
United Kingdom

Oxford University Press is a department of the University of Oxford.
It furthers the University's objective of excellence in research, scholarship,
and education by publishing worldwide. Oxford is a registered trade mark of
Oxford University Press in the UK and in certain other countries

First Edition published in 2009

Published in the United States of America by Oxford University Press
198 Madison Avenue, New York, NY 10016, United States of America

British Library Cataloguing in Publication Data
Data available

Library of Congress Cataloging in Publication Data
Data available

ISBN 978–0–19–953509–5

In the deep discovery of the Subterranean world, a shallow part would satisfie some enquirers... The treasures of time lie high.

Sir Thomas Browne (1605–1682)
Hydriotaphia: Urne-Burial or, A Brief Discourse of the Sepulchrall Urnes Lately Found in Norfolk

Contents

Preface

The scientific community and the wider public have both become increasingly aware that our world is currently experiencing an extinction crisis, a mass extinction of comparable magnitude to the K-T event, the end-Permian event or 'Great Dying', and other rare intervals of hugely elevated extinction from the deep geological past. Although the vast levels of biodiversity loss that are being documented today are uncontroversially recognized to be driven by human actions, present-day species extinctions are only the latest stage in a considerably longer-term sequence of biotic impacts stretching far back into the Quaternary, which follow the prehistoric spread of modern humans out of Africa and into pristine ecosystems around the world. However, large-scale faunal extinctions at the end of the Pleistocene Epoch, which saw the disappearance of much of the world's charismatic continental vertebrate megafauna, also coincided with the major climatically driven environmental changes that accompanied the most recent global shift from glacial to interglacial conditions, and characterizing the role of humans in the end-Pleistocene extinctions has been the subject of intense debate for over a century. In contrast, the subsequent geological epoch, the Holocene—approximately the last 11 500 years from the end of the last glaciation to the present day—has also seen massive levels of extinction that have continued throughout the recent prehistoric and historical eras, but has conversely experienced only relatively minor climatic fluctuations. As such, the Holocene potentially provides a far more useful system in which to investigate the impacts of changing human activities over time on different species and ecosystems. Such research has the potential to provide unique and novel insights into the dynamics of both the prehistoric end-Pleistocene extinctions and also modern-day biodiversity loss, and will also allow us to begin to appreciate the true extent of human-caused extinctions throughout the Late Quaternary.

This book hopes to encourage such research, and to provide a stepping stone towards a better understanding of human impacts both past and present, by presenting an overview of the state of our current knowledge of what we have already lost during the Holocene. Although the majority of research in this field has been conducted into the timing, causation, and magnitude of past human-caused extinctions of mammals and birds, two groups which inevitably again receive a disproportionate amount of attention in this book relative to their overall contribution to global biodiversity, it is also important that human impacts to other major taxonomic groups and also to wider ecosystems across the Holocene are not neglected, and these have also been represented here as fully as possible. In particular, research into changing ecosystems and species losses across the Holocene represents a fertile meeting ground for many academic disciplines, notably zoology, ecology, palaeontology, archaeology, and geography, and stronger interconnections between these potentially disparate fields are also required to lead to the greatest future advances in understanding regional and global human impacts across the recent geological past; it is hoped that this book will be read by researchers and students from all of these different backgrounds.

The compilation and production of this book has been facilitated by a huge number of people. First and foremost I must give my grateful thanks to all of the authors who have provided their time and expertise over the past couple of years to prepare the series of chapters that together provide a unified overview of the broad subject matter of Holocene extinctions. I also wish to thank the long line of academic reviewers who freely gave their invaluable constructive criticism about the various contributions that were sent their way for perusal and comment, and also the many colleagues who gave me many further thoughts and suggestions both about extinctions in the past and about being an editor in the present. In particular, I must give special acknowledgement here to Georgina Mace, who provided me with such great encouragement and support during the early stages of the book's preparation. Finally, my biggest thanks go to the team at Oxford University Press who have seen the book through its development and production from start to finish: in particular, I need to single out Ian Sherman and Stefanie Gehrig, who were there from the beginning and who helped to keep the momentum going at many key moments, and Helen Eaton, who has ably seen the book through to completion in its final stages.

Samuel T. Turvey
London
October 2008

A note on radiocarbon dating conventions

Radiocarbon dating is the most common radiometric dating method for determining the age of subfossil and archaeological samples from the Holocene and Late Pleistocene. This dating method is based on the radioactive decay of the unstable isotope ^{14}C into ^{14}N, which has a half-life of 5568±30 years. Atmospheric ^{14}C is fixed by plants during photosynthesis and constantly incorporated into living organisms, but is not incorporated after death. Measuring the amount of ^{14}C that remains in organic material provides a radiocarbon or ^{14}C age that is usually reported in years before present (years BP), where 'present' corresponds by convention to AD 1950 (so that the year in which the original sample was dated is not needed). However, the production of ^{14}C in the upper atmosphere has varied through time, due to changes in the solar magnetic field. Concentrations of atmospheric ^{14}C have also been influenced by changes in ocean circulation, especially during the Late Glacial. Radiocarbon years are therefore not equivalent to calendar years. To calculate an accurate calendar age, atmospheric ^{14}C fluctuations have to be corrected by means of a calibration curve obtained by comparing raw ^{14}C measurements with true calendar ages provided by independent dating methods (e.g. dendrochronology). The current internationally agreed radiocarbon calibration curve, IntCal04 (Reimer *et al.* 2004), is characterized by a long-term trend with raw ^{14}C ages being significantly younger than calendar ages during most of the last 45 000 years (the temporal limit of radiocarbon dating), and with superimposed abrupt ^{14}C shifts which occurred over centuries to millennia. Although modern radiocarbon studies provide calibrated dates, older studies typically provided only ^{14}C ages, and although efforts have been made herein to minimize potential confusion, many of the chapters in this volume have been forced to include both types of data when reviewing research into different extinction chronologies across the Holocene.

List of contributors

Tim M. Blackburn Institute of Zoology, Zoological Society of London, Regent's Park, London NW1 4RY, UK

Simon Brewer Institut d'Astrophysique et de Géophysique, Université de Liège, Bat. B5c, 17 Allée du Six Août, B-4000 Liège, Belgium

Phillip Cassey School of Biosciences, University of Birmingham, Edgbaston, Birmingham B15 2TT, UK

Ben Collen Institute of Zoology, Zoological Society of London, Regent's Park, London NW1 4RY, UK

Joanne H. Cooper Department of Zoology, Natural History Museum at Tring, Akeman Street, Tring, Hertfordshire HP23 6AP, UK

Nicholas K. Dulvy Centre for Environment, Fisheries and Aquaculture Science, Lowestoft Laboratory, Pakefield Road, Lowestoft, Suffolk NR33 0HT, UK; Department of Biological Sciences, Simon Fraser University, Burnaby, BC V5A 1S5, Canada

Robert R. Dunn Department of Zoology, North Carolina State University, Raleigh, NC 27695, USA

Simon J. Goring BISC, Simon Fraser University, Burnaby, BC V5A 1S5, Canada

Wendell Haag Center for Bottomland Hardwoods Research, USDA Forest Service, 1000 Front Street, Oxford, MI 38655, USA

Tyler S. Kuhn BISC, Simon Fraser University, Burnaby, BC V5A 1S5, Canada

Julie L. Lockwood Ecology, Evolution and Natural Resources, Rutgers University, 14 College Farm Road, New Brunswick, NJ 08901–8551, USA

Anson W. Mackay Environmental Change Research Centre, Department of Geography, University College London, Gower Street, London WC1E 6BT, UK

Rob Marchant York Institute of Tropical Ecosystem Dynamics, Environment Department, University of York, University Road, Heslington, York YO10 5DD, UK

Arne Ø. Mooers BISC and IRMACS, Simon Fraser University, Burnaby, BC V5A 1S5, Canada; Institute for Advanced Study, 14193 Berlin, Germany

Julian D. Olden School of Aquatic and Fishery Sciences, University of Washington, Box 355020, Seattle, WA 98195–5020, USA

John K. Pinnegar Centre for Environment, Fisheries and Aquaculture Science, Lowestoft Laboratory, Pakefield Road, Lowestoft, Suffolk NR33 0HT, UK

John D. Reynolds Department of Biological Sciences, Simon Fraser University, Burnaby, BC V5A 1S5, Canada

R. Paul Scofield Canterbury Museum, Rolleston Avenue, Christchurch 8013, New Zealand

John R. Stewart Department of Palaeontology, Natural History Museum, Cromwell Road, London SW7 5BD, UK

Samuel T. Turvey Institute of Zoology, Zoological Society of London, Regent's Park, London NW1 4RY, UK

Tommy Tyrberg Kimstadsvägen 37, SE-610 20 Kimstad, Sweden

Thompson Webb III Department of Geological Sciences, Brown University, Providence, RI 02912–1846, USA

CHAPTER 1

An introduction to Late Glacial–Holocene environments

Anson W. Mackay

1.1 Introduction

The term Holocene was first coined in the early nineteenth century and was adopted in Bologna by the International Geological Congress in 1885. The Holocene (also known as the Postglacial or the Flandrian) is one of the most easily identified features in palaeoclimate records, and marks the end of the Pleistocene Epoch approximately 11 500 years (11.5 ka) before present (BP). The Holocene is the most recent geological epoch of the Quaternary Period (the past 2.6 million years), which itself is characterized by glacial–interglacial cycles. The Holocene is the Earth's most recent interglacial: a climatically warm interval that separates cooler glacials (or ice ages).

Accurately dating the start of the Holocene has proved challenging. For example, before the widespread use of calibration in radiocarbon (^{14}C) dating, the start of the Holocene was commonly placed at 10 000 ^{14}C years BP. This date undoubtedly persisted so long due to its 'elegant simplicity' (Roberts 1998). However, developments of more accurate chronologies, especially with regard to improved calibration of radiocarbon dates using annually resolved archives (e.g. tree rings, lake sediments, and ice cores) have revealed the start of the Holocene to be *c.*11.5–11.6 calibrated (cal) ka BP. All dates presented in this chapter are calibrated.

1.2 External climate-forcing mechanisms

Factors that drive climate change on Earth but are not in themselves affected by that change are considered external to the climate system. The main natural external factors driving climate on Earth are linked to changes in the Earth's orbit, changes in solar irradiance and volcanic eruptions. The relative importance of each of these factors has changed through time, and trying to disentangle their influences is challenging and often controversial. Understanding the impacts of external forcing mechanisms provides an essential backdrop against which extinctions can be critically assessed. However, these external forcings are by themselves not sufficient to account for the speed and strength of ecosystem responses during the transition from glacial into interglacial environments, which is rapid and often abrupt. That is, while changes in the Earth's orbit force glacial–interglacial cycles, the cycles themselves are not caused by orbital parameters, but rather by the Earth's climate-system feedback mechanisms (Maslin *et al.* 2001). For example, the deglaciation of extensive ice masses requires amplifying factors such as increases in atmospheric greenhouse gases, especially CO_2, changes in the world's ocean currents (thermohaline circulation), and changes in land-surface albedo associated with ice, snow, and vegetation cover. Interactions between external forcings and feedback mechanisms are detailed in, for example, Maslin *et al.* (2001) and Oldfield (2005).

1.2.1 Orbital forcing

During the Quaternary, regular variations in the Earth's orbit (commonly known as Milankovitch cycles) have been the dominant mechanism driving the Earth in and out of intervals of intense glaciation (Imbrie *et al.* 1992) (Fig. 1.1). At a primary level, insolation received from the sun is modified by a combination of three orbital parameters: eccentricity, obliquity, and precession. Eccentricity occurs due to variations in the minor axis of the Earth's orbital ellipse as it goes round the sun, and occurs with a frequency of *c.*100 ka. Obliquity is the term used to describe the tilting of the Earth's axis, which has a frequency of *c.*41 ka. Precession describes the direction of the tilt of the Earth at any given point of its orbit; cycles vary between 23 and 19 ka (mean = 21.7 ka). Changes in orbital parameters have little impact on the total amount of solar radiation received by the Earth from the sun, but they do alter the amount of seasonal solar radiation received at different latitudes. Crucial in this respect is the amount of solar radiation received on continents at northern latitudes, as this determines whether snow and ice can persist through the following summer months. During the Late Glacial–early Holocene, precessional forcing at northern latitudes peaked between *c.*14–10 ka, resulting in warmer summers and colder winters over the northern hemisphere. Since the early Holocene, these levels have declined by approximately 10% relative to recent times (Fig. 1.2a), so that currently the range of temperatures between summer and winter months is less extreme.

1.2.2 Solar variability

The amount of energy emitted from the sun is variable, and changing levels of solar irradiance through the Holocene have had differential, regional impacts on the Earth's climate, primarily through atmospheric interactions in the stratosphere and troposphere (Bradley 2003). A commonly used measure of solar variability is total solar irradiance ($W m^{-2}$). Direct records of past solar activity (extending beyond several centuries) do not exist. Instead, variations in the cosmogenic isotopes of ^{14}C and ^{10}Be in natural archives have been used as proxies of past total solar irradiance (Fig. 1.2b). This is based on the assumption that their production in the Earth's upper atmosphere is inversely proportional to variations in solar radiation; that is, as radiation from the sun decreases, production of cosmogenic isotopes increases. Over the Holocene this has been shown to vary by about 0.4% (in comparison to modern levels). Variations in solar insolation have been implicated in changes in North Atlantic thermohaline circulation (e.g. Björck *et al.* 2001; Bond *et al.* 2001), although the mechanisms of how these processes interact are still uncertain.

1.2.3 Volcanic activity

The third major external forcing to influence Holocene climate is linked to particulate matter and aerosols released into the atmosphere from volcanic eruptions. Large explosive eruptions can result in cooling which lasts several years (Robock 2000). This cooling is caused by the injection of sulphate aerosols into the stratosphere, which cause general radiative cooling of the surface atmospheric layers. However, regional impacts are more complex, and

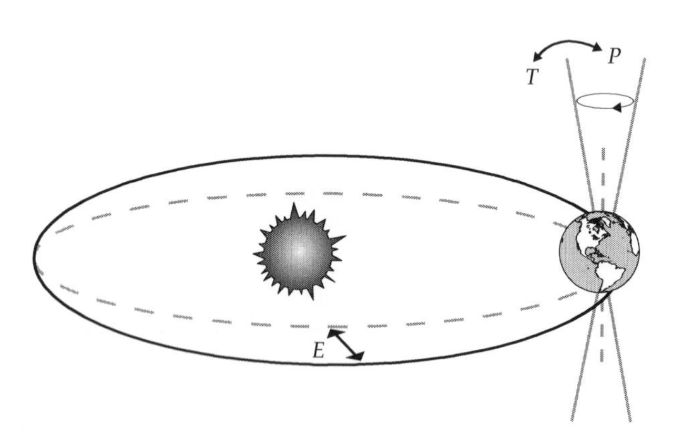

Figure 1.1 Orbital changes of the Earth around the Sun associated with dominant mechanisms of climate forcing. *E*, eccentricity of the Earth's orbit, which is linked to variations in the minor axis of the ellipse; *T*, tilt or obliquity of the Earth's axis; *P*, precession, i.e. the axis tilt direction of the Earth at a given point of its orbit. Source: Rahmstorf and Schellnhuber (2007). Reproduced by permission from Verlag C.H. Beck.

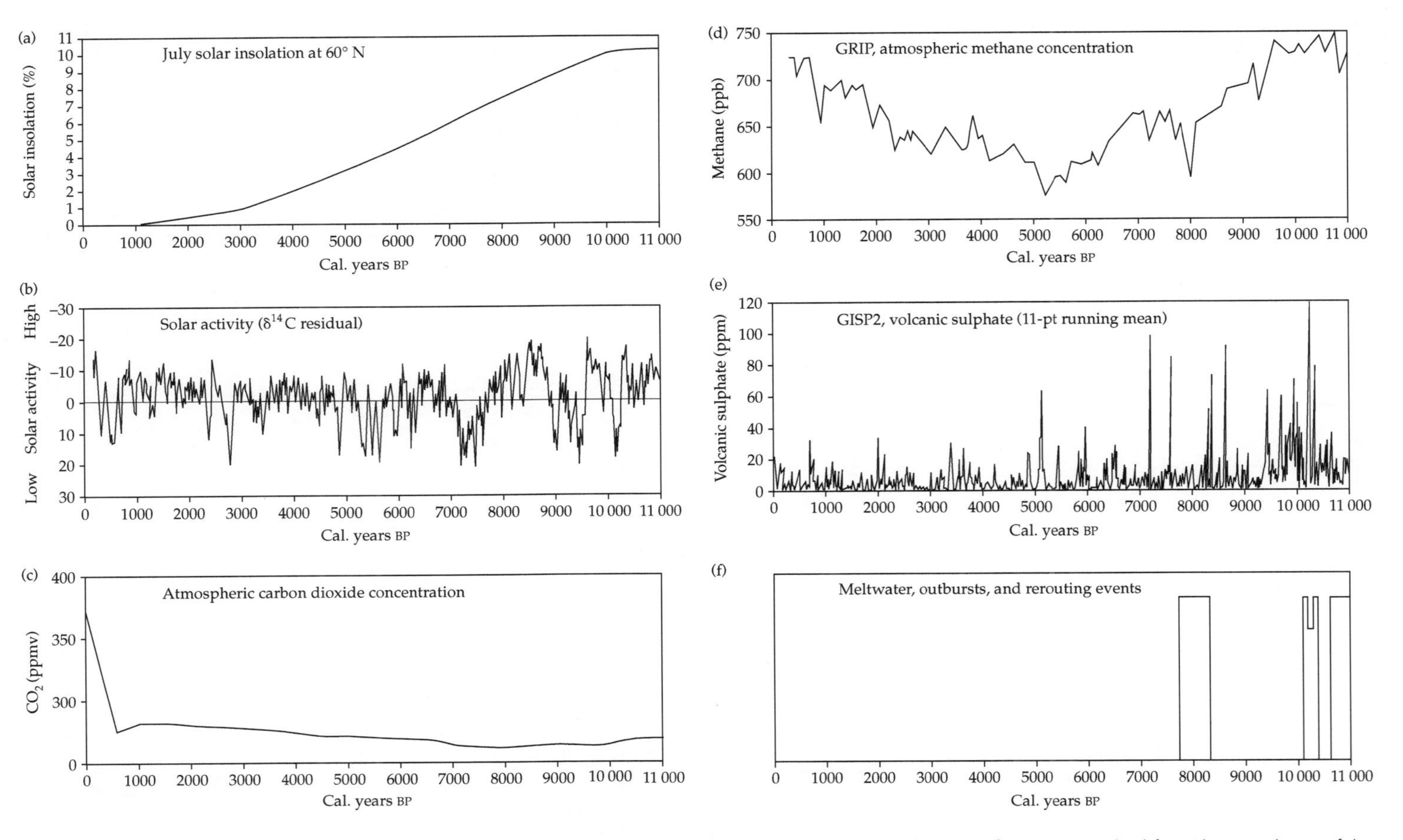

Figure 1.2 Major Holocene climate forcing factors in the northern hemisphere. (a) Percentage insolation anomaly at 60°N (departure from AD 1950 values) for mid June at the top of the atmosphere; (b) solar activity variation estimated from residual ^{14}C variations; (c) atmospheric CO_2 concentrations derived from polar ice cores and Mauna Loa Observatory; (d) concentrations of atmospheric CH_4 found in Greenland Ice Core Project (GRIP) ice core; (e) Greenland Ice Sheet Project 2 (GISP2) volcanic sulphate record (11-pt running mean); (f) early Holocene meltwater outbursts into the North Atlantic from ice-dammed lakes associated with the Laurentide ice sheet. Note that the CH_4 record for the GRIP ice core does not cover the last approximately 200 years; during this interval, atmospheric concentration of CH_4 has increased from approximately 725 parts per billion (ppb) to over 1750 ppb. ppb, parts per billion; ppm, parts per million; ppmv, parts per million by volume. Redrawn and adapted from Nesje *et al.* (2005), which also contains detailed information on sources of data used. Reproduced by permission of the American Geophysical Union.

are dependent on the interactions of aerosol cooling on atmospheric weather patterns. Any one volcanic eruption by itself is unlikely to have had a major climatic impact. However, eruptions occurring in close succession may well have had more significant impacts at the decadal to multi-decadal scale (Bradley 2003). Such clusters occurred during the early Holocene between *c*.10.5 and 9.5 ka BP (Fig. 1.2e). The impact of past volcanic eruptions on human civilization is controversial. There is no dispute that past civilizations have been intricately associated with volcanic regions around the world, but evidence for eruptions invoking civilization collapse is much less certain (Grattan 2006).

Information on past environmental change during the Holocene has the potential to inform questions about contemporary environmental dynamics, most notably how long will warm interglacial climates persist, are current trends in global warming unusual, and can we expect any surprises in the future in the form of increasing instability in the climate system? In order to address these questions, it is necessary to investigate climate forcings of previous interglacials with a view to contextualizing Holocene variability into the future. In this respect two previous interglacials provide much-needed clues: the last interglacial, otherwise known as the Eemian/Marine Isotope Stage (MIS) 5e (*c*.130–116 ka BP ± 1 ka) (although the two are not directly synchronous; Shackleton *et al.* 2003), and MIS11 (*c*.423–362 ka BP) (Droxler *et al.* 2003).

Global mean surface temperatures during the Last Interglacial were at least 2°C higher than the present day (Otto-Bliesner *et al.* 2006), which led to high rates of sea-level rise (average 1.6 m per century; Rohling *et al.* 2008). Palaeontological records demonstrate a relatively stable climate, although proxy records increasingly suggest evidence for millennial-scale variability. The orbital configuration of the Earth during the Last Interglacial was significantly different than during the Holocene. This therefore limits the extent to which we can use palaeontological records from the Last Interglacial as analogues for current and future climate change.

Consequently, there has been growing interest in MIS11 because astronomically driven insolation related to low orbital eccentricity is similar to that during the latter part of the Holocene and to variation predicted for the near future (Loutre and Berger 2003). MIS11 has therefore been used as an analogue from which to predict future climate variability. For example, because the effects of orbital precession are minimized during intervals of low orbital eccentricity, cold summers at high latitude (which occurred at the initiation of the Last Glaciation approximately 116 ka BP) will not occur for at least another 30 ka (IPCC 2007). Moreover, comparisons between the Holocene and MIS11 suggest that natural forcings alone can account for the prediction that Holocene warmth is likely to last for approximately another 50 ka into the future (Loutre and Berger 2003). Nevertheless, whereas previous interglacials can provide us with information on natural variability linked to external forcings and internal feedback mechanisms, the Holocene is still unique due to the unprecedented influence that humans have had on their environment.

1.3 Detecting Late Glacial–Holocene environmental change

Over the Late Glacial–Holocene there is an absence of direct monitoring or documentary records of environmental change, and so natural archives have therefore been exploited. Different archives have their own strengths and weaknesses, but those that are continually deposited offer the best possibility for high-resolution reconstructions at the sub-decadal level or higher. Some of the most exploited archives are those that exhibit annually deposited layers (e.g. varved lake and marine sediments, ice cores, speleothems) or annual growth layers (e.g. tree rings, corals). These archives can provide annually resolved records extending back, in some cases, to at least the Last Glacial Maximum (LGM; *c*.21–22 ka BP).

The distribution of any one particular archive will be geographically restricted. Ice cores, for example, will only be found where conditions are optimal for the build-up of ice layers and their subsequent preservation, for example in the Arctic and Antarctic, but also in temperate and tropical alpine regions. Some archives are more widely distributed than others. Lake sediments, for example, are particularly useful as these are found in most regions across the globe, except for deserts, and even then sediments from palaeo-lakes have often been exploited. Other archives such as tree rings and associated dendroclimatological data owe their significance to some individuals of certain species growing at the margin of their ecological tolerance, which sensitively record even minor fluctuations in climate. This section provides details on annually laminated archives, most notably ice cores and lake and marine sediments. These were selected because they have been particularly important in setting the agenda with regard to the high-resolution reconstruction of Late Glacial–Holocene ecosystems and associated impacts from both natural and anthropogenic events.

Recent studies of ice cores have revealed significant and consistent trends in Late Glacial–Holocene climate variability. Key sites of note include ice cores extracted from Greenland (e.g. Greenland Ice Core Project (GRIP), NorthGRIP, and Greenland Ice Sheet Project 2 (GISP2)) and Antarctica (e.g. Vostok, Dome C/European Project for Ice Coring in Antarctica (EPICA)). The most widely exploited proxies from ice-core layers include greenhouse gases, dust, sulphates (produced by volcanic activity), major ions (such as K^+ and Na^+), and stable isotopes (e.g. $\delta^{18}O$). For example, records of greenhouse gases (e.g. CO_2, CH_4, and N_2O) contained within trapped air in ice cores are important proxies in the debates on early Holocene warming, Holocene cool events, recent global warming, and the processes which control warming during glacial–interglacial cycles (e.g. Raynaud *et al.* 2000). $\delta^{18}O$ records in turn can provide quantitative estimates of past temperature (see below).

Lake and marine sediments are of heterogeneous composition, consisting of both autochthonous and allochthonous components. Sediment accumulation can be influenced by many factors, including (1) seasonal biological production within the water column, (2) transport of sedimentary material either through fluvial or aeolian action, (3) sedimentation processes that allow settling of particles through the water column on to the basin floor, (4) secondary processes such as redeposition, resuspension and focusing, and (5) bioturbation of bottom sediments. Most lake and marine sediments are not annually laminated; specific conditions are required for annual laminations (or varves) to form, including a strong seasonal climate and the prevention of bioturbation of bottom sediments (Zolitschka 2003). The finer temporal resolution provided by annually laminated lake sediments has permitted detailed investigation of many Late Quaternary events, such as the development of agriculture in the Near East during the Late Glacial and Holocene (e.g. Baruch and Bottema 1999).

A large variety of biological proxies have been exploited from sedimentary archives and used to reconstruct past environments. These all require specific properties to enable them to be robust indicators of past change. For example, they need to be produced in large enough quantities so that only relatively small amounts of sediment are needed for analysis. This holds true for pollen, diatoms, and foraminifera, for example. For some other proxies, including beetles and plant macrofossils, larger amounts of sediment material are needed, which usually results in lower-resolution studies. Each proxy needs to be able to preserve in the environment in which it is deposited, and so different proxies all contain properties which aid their preservation. General information is provided below on a few commonly used isotopic and biological proxies that have greatly furthered our understanding of Late Glacial–Holocene environmental change. Several recent comprehensive texts detail other proxies and their uses (e.g. Smol *et al.* 2001; Mackay *et al.* 2003a).

The most widely used geochemical proxies in Holocene studies are stable isotopes of oxygen (^{16}O and ^{18}O), carbon (^{12}C and ^{13}C), and hydrogen (^{1}H and ^{2}H). Interpretations of past environments using stable isotopes are based on the ratios between the isotopes, especially $^{18}O/^{16}O$ and $^{2}H/^{1}H$ in water derived from precipitation, groundwater, rivers, lakes, and oceans. The ratios between stable isotopes are controlled by a large number of factors, especially temperature and evaporative processes which vary over time. For example, during evaporation of water, the heavier isotope (e.g. ^{18}O or ^{2}H) is discriminated against (isotopic fractionation). This results in remaining waters being enriched in heavier isotopes, while the evaporated water vapour is deficient in these isotopes relative to the source. Thus, by studying isotopic ratios in palaeoarchives, past climates (colder/warmer; wetter/drier) can be quantitatively inferred (see Figs 1.3 and 1.5, below). Isotopes are incorporated into lake and marine sediments as either photosynthetically derived precipitates or via biogenic precipitates within organic matter, mollusc shells, diatom frustules, foraminiferan tests, etc. Some of the most significant applications in the use of stable isotopes have been their measurement in annually laminated archives. For example, the $\delta^{18}O$ record from the Greenland GRIP ice cores highlight the abrupt changes in global temperature that characterize the Late Glacial–early Holocene interval (Fig. 1.3).

Pollen analysis is probably the most extensively used palaeoecological technique since pioneering work published by von Post in the beginning of the last century. Pollen grains and spores are useful palaeoenvironmental proxies because they are produced in large numbers and preserve well in anoxic sediments. Interpretation of pollen relies on knowledge of production and dispersal of grains from source vegetation, although rarely do proportions of pollen grains found in sedimentary archives have a linear relationship with past vegetation abundances; careful interpretation is therefore needed. In recent decades there has been substantial progress in the use of pollen to reconstruct population ecologies (e.g. plant migration and plant invasion; Jackson and Overpeck 2000) and community palaeoecology (e.g. past landscape disturbance and biodiversity change) (Seppä and Bennett 2003). Some of the most significant advances in recent years include (1) the improvement and refinement of pollen-based transfer-function models of palaeoclimate (Birks 2003), based on relationships between pollen taxa and modern climatic variables (e.g. Seppä and Birks 2001), and (2) the use of pollen-based palaeoecological records to provide long-term perspectives on conservation and biodiversity (e.g. Willis *et al.* 2007) and on the role of species refugia in temperate and tropical forests (Willis and Whittaker 2000; Colinvaux and Oliveira 2001; Bush 2003).

Diatoms (unicellular eukaryotic algae, Class Bacillariophyceae) are extensively utilized because they can be found in virtually every aquatic environment and they preserve well in most sediments due to their valves being composed of biogenic silica. Diatoms can be identified to the species level, and many species have well-defined niche characteristics, which make them powerful environmental indicators. Diatoms are often used to investigate Holocene climate variability (Mackay *et al.* 2003b), such as past air temperature (Weckström *et al.* 2006). Diatoms are especially useful however for

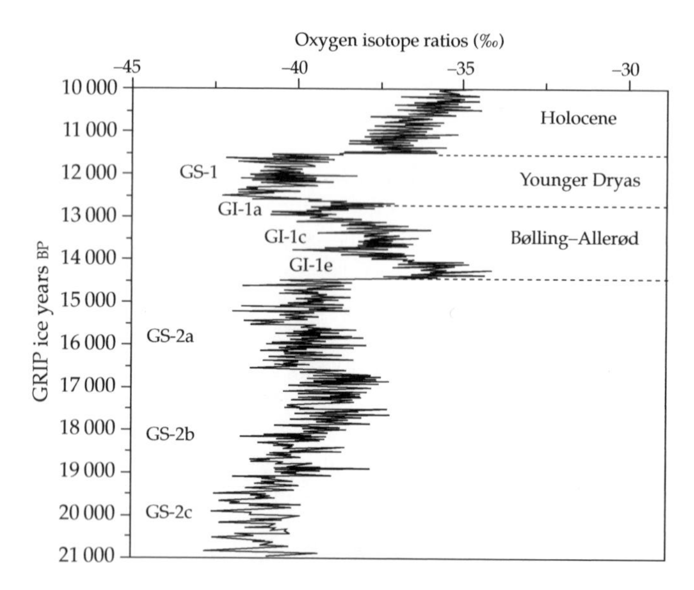

Figure 1.3 $\delta^{18}O$ ‰ GRIP ice core record from Dansgaard *et al.* (1993). The division of the Late Glacial into Greenland stadials (GS) and Greenland interstadials (GI) follows Björck *et al.* (1998). Adapted from Björck *et al.* (1998). ©John Wiley & Sons Limited. Reproduced with permission.

reconstructing the quality of inland and coastal waters which may be impacted by cultural eutrophication. They have therefore played a major role in assessing human impact on freshwater ecosystems throughout the Holocene. For example, Verschuren *et al.* (2002) used diatom analysis of sediments in Lake Victoria to show that eutrophication has resulted in the loss of deep-water oxygen since the 1960s. This is likely to have contributed in part to the extinction of some deep-water endemic cichlids, in addition to impacts linked to the introduction of the Nile perch *Lates niloticus*.

Foraminifera are amoeboid protists found mainly in coastal and marine ecosystems, and their contribution to climate change studies has been significant. Foraminiferan tests preserve in abundance in marine sediments. There are over 10 000 species of foraminifera, which occupy a wide variety of marine environments and ecological niches (Murray 2002), allowing for robust multi-species palaeoenvironmental reconstructions. Indeed, the first empirical transfer function exploited the relationships between marine foraminifera to global sea-surface temperatures and sea-surface salinities (Imbrie and Kipp 1971). The range of palaeoenvironmental studies using foraminifera is extensive. For example, planktonic/benthic ratios within assemblages can provide information on sea-level changes, $\delta^{18}O$ in foraminifera tests has been used as a proxy for changes in Holocene sea-surface temperatures, and $\delta^{13}C$ has been used as a proxy for carbon storage, ocean circulation, and productivity.

One proxy by itself cannot fully account for a complete record of environmental change, and so multi-proxy approaches are increasingly undertaken to reconstruct Late Glacial–Holocene environments (Birks and Birks 2006). Moreover, different proxies may be able to compensate for potential weaknesses in other proxies that have been analysed. Multi-proxy studies are complex and time-consuming, although the rewards in recognizing complimentarities between different proxies are great.

Estimates of past climates can be quantitatively reconstructed either by data (empirical) models or by models based on physical laws (e.g. thermodynamics) of the climate system. Both approaches require substantial numerical processing. Empirical transfer-function methodologies are based on the premise of collecting knowledge on environmental requirements of species represented as fossils in sedimentary records (Birks 2003). Establishing gradients of interest and importance is therefore crucial. Relationships between species and their environment are modelled numerically, based on strong statistical and theoretical bases. Estimates of environmental variables of interest such as summer temperature can then be reconstructed for the past using the fossils present in the relevant archive, based on models developed from the calibration dataset.

Physical models are fundamental tools of climate change science, both for reconstructing past variability and for predicting future changes. They need to be internally consistent, compatible with biophysical laws, perform robustly, and to be data-realistic. If these criteria are satisfied, then the Holocene is an important time frame with which to test models and improve them so that that they can be used, for example, for climate change prediction (Valdes 2003). All models are a simplification of the processes that have actually happened, are happening, or are likely to happen. Models have themselves increased in complexity from rather simple box models (often used to simulate chemical species such as carbon or nitrogen) to general circulation models which are used to understand climate as a three-dimensional process. General circulation models are now frequently coupled with other dynamical models (e.g. ocean circulation models, terrestrial models of vegetation change) but their complexity and computing requirements are enormous. To bridge the gap between these model types, Earth system models of intermediate complexity requiring less computing power are frequently used by palaeoclimatologists, but still attempt to incorporate other biosphere components, such as terrestrial vegetation and land-ice sheets (Valdes 2003). Because these models are quicker to process, they have been used to model past climate at high resolution over many thousands of years. For example, they have made major contributions to assessing sensitivity and stability of North African climate during the early to mid Holocene in relation to concomitant vegetation and landscape changes (Claussen *et al.* 1999).

Climate models demonstrate that elevated summer temperatures at higher latitudes during the early Holocene in part contributed to increased monsoonal intensity throughout many low-latitude regions, resulting in significantly wetter environments. Nowhere has this been more apparent than in the Saharan region of North Africa. Palaeoecological records provide evidence for the widespread existence of freshwater ecosystems, such as Lake Megachad, during the early Holocene. Archaeological records show that the region was extensively populated, while pollen records highlight the persistence of wetlands and associated vegetated landscapes that were likely to have played a significant role in the maintenance of hydrologically wetter conditions. After about *c.*5.5 ka BP, summer monsoons began to fail in the Saharan region, which resulted in increased desertification and which may also have played a role in the extinction of the giant long-horned buffalo *Pelorovis antiquus* (Klein 1994; see also Chapter 2 in this volume). Crucially, however, these hydrological changes cannot be explained by external solar forcing alone. Instead, these models have highlighted the strong role of internal feedback mechanisms, such as the extent of terrestrial vegetation covering land in the region (Claussen *et al.* 1999).

1.4 Late Glacial–early Holocene environments

During the last glacial, ice sheets and glaciers covered approximately 40 million km^2 of the Earth's surface (Anderson *et al.* 2007). The main ice sheets were in Antarctica, North America (the Laurentide and Cordilleran ice sheets), Greenland, and northern Europe (the Scandinavian ice sheet), with smaller ice sheets also present in southern South America. Other notable ice sheets occurred in regions associated with expanded glacier activity; for example, the alpine regions of Europe, Africa, and south-east Asia. At the time of the LGM, ice volume was high and global sea levels were low because water was locked up in ice caps and glaciers. Estimates for the extent of sea-level decline towards the end of the Last Glaciation vary considerably between approximately 120 and 135 m (Bell and Walker 2005). Lowered sea levels resulted in extensive areas of continental shelves being exposed. These provided land bridges both between continents (e.g. Siberia and North America) and between islands on continental shelves (e.g. the Sunda Shelf in south-east Asia). These linkages had significant implications for population expansion and dispersal patterns for many species.

During the LGM temperate and tropical rain forests had greatly restricted geographies (Prentice *et al.* 2000). In Europe, for example, cooler temperatures and increased aridity caused a southwards displacement and reduction of forests. Conversely, tundra, steppe, and grassland communities were much more widespread, extending from the margins of the ice sheets to the coastlines of the Mediterranean. Ice-core evidence shows strong depression of temperatures at high latitudes during the LGM in comparison to the present day; for example, by 21°C in Greenland and 9°C in Antarctica (IPCC 2007). In western Europe, pollen-inferred mean annual temperatures were approximately 15°C cooler than present (Guiot *et al.* 1993). In southern Europe, modelling data suggest that temperatures were between approximately 7 and 10°C cooler during the LGM than the present day. Temperature estimates for lowland tropical regions during the LGM were at least 5°C lower than the present day (Bush *et al.* 2001), although cooling was weaker in the western Pacific Rim (<2°C).

The demise of ice sheets marking the end of the last glacial interval is known as Termination 1. In the last decade, there has been a marked improvement in the resolution of dating of archives used to reconstruct climate change during Termination 1 (Björck *et al.* 1998), and these data show that temperatures fluctuated dramatically by as much as 10°C over intervals of decades, if not years (e.g. Alley 2000). Proxy records highlight the Late Glacial as an interval of significant climate instability in the northern hemisphere. The first sign of a shift to warmer temperatures outside those experienced during the preceding glacial conditions occurred between *c.*14.7 and 14.5 ka BP, with the onset of the Greenland interstadial 1, commonly known as the Bølling–Allerød (GI-1; Björck *et al.* 1998) (Fig. 1.3; Table 1.1). This was followed by an equally rapid shift to cooler temperatures characterizing the Greenland stadial 1 (GS-1; also known

Table 1.1 Chronological history for the Late Glacial based on the Greenland Greenland Ice Core Project (GRIP) ice core. Greenland stadial and interstadial stages are given in ice-core years BP (Björck *et al.* 1998). European population events are derived from Gamble *et al.* (2005), and dates demarking the boundaries between the population events (right-hand column) are also given in GRIP ice-core years BP.

Stage		GRIP ice-core years BP	Continental north-west Europe	Western European population event	Date of boundary between population events (ice-core years BP)
Holocene		11 500	Holocene		
GS-1	GS-1	12 650	Younger Dryas	5	12 900
GI-1	GI-1a	12 900			
	GI-1b	13 150	Allerød	4	14 000
	GI-1c	13 900			
	GI-1d	14 050	Older Dryas		
	GI-1e	14 700	Bølling	3	
GS-2	GS-2a	16 900		2	16 000
	GS-2b	19 500			19 500
	GS-2c	21 200		1	

as the Younger Dryas), which lasted between *c.*12.65 and 11.50 ka BP (Fig. 1.3; Table 1.1). This abrupt cold reversal is linked to an enormous pulse of fresh water being released from retreating ice on the Laurentide ice sheet in North America into the Arctic Ocean (Tarasov and Peltier 2005). This freshwater pulse caused marked changes in the salinity of surface waters of the Arctic Ocean, leading to a slowdown in North Atlantic Deep Water formation. During glacials, these abrupt events occurred approximately every 1500 years and are called Dansgaard–Oeschger events or cycles (Maslin *et al.* 2001). There is increasing evidence to suggest that these events also persist during interglacials, including the Holocene.

1.4.1 Sea-level responses

Late Glacial–early Holocene warming was accompanied by eustatic (globally averaged) increases in sea level, as water once locked up in ice sheets and glaciers was returned to ocean basins. In the millennia following the LGM, sea levels rose by over 120 m (IPCC 2007) but stabilized by *c.*2–3 ka BP. At low latitudes, one of the regions most greatly impacted by sea-level rise was south-east Asia. The once-exposed Sunda Shelf was rapidly inundated in two phases: between *c.*16 and 12.5 ka BP and 11.5 and 8 ka BP. In these lowland habitats there were large declines in both plant and animal populations as shorelines advanced at approximately 40 cm week^{-1}, resulting in the extinction of many species of megafauna from some of the smaller islands (Bush 2003). However, isolation of these islands may have contributed to the persistence in the region of certain other species such as the orang-utan *Pongo pygmaeus*, which became extinct on the south-east Asian mainland (Louys *et al.* 2007).

In the northern hemisphere increasing sea levels resulted in the fragmentation of the vast land area known as Beringia, which encompassed Alaska, the shallow Arctic shelf, and the coastal lowlands of Arctic east Siberia. By *c.*12 ka BP, Wrangel Island in the Arctic Ocean became isolated, resulting in the local persistence of the woolly mammoth *Mammuthus primigenius* well into the Holocene until *c.*4 ka BP (Vartanyan *et al.* 1993; see also Chapter 2 in this volume). Increasing sea levels during the Late Glacial also resulted in the most recent inundation of the intercontinental Bering Land Bridge between North America and Eurasia. Inundation of this region had significant implications for the migration of human populations from continental Eurasia to the Americas (see below). In north-west Europe, inundation of the shallow continental shelf resulted in the loss of large areas of Mesolithic hunting territory, but at the same time created new habitats suitable for shallow-water fishing. New European coastal landscapes were created, which culminated in the isolation of Britain from mainland Europe by *c.*7 ka BP (Shennan *et al.* 2000).

Global changes in relative sea level are also affected by other processes such as vertical land uplift (glacial isostasy) and thermal expansion following melting of heavy ice sheets. The extent of potential uplift can be considerable; for example, the absolute uplift in the Baltic region of Scandinavia was greater than 700 m, while modelled uplift in North America was in excess of 900 m (Bell and Walker 2005). Uplift of land is still occurring today throughout many regions of the North Atlantic that previously supported glacial ice sheets.

1.4.2 Vegetation responses

Plant species responses to climate are individualistic. Palaeoecological studies have shown that during the Late Glacial–early Holocene migrations of vegetation species, populations, and communities were extremely complex, although conceptual models based on ecological niche theory have furthered our understanding of some of these responses in relation to climate change (Jackson and Overpeck 2000). Important factors influencing species migration include not only their individual dispersal characteristics, but also their individual responses to precipitation, seasonal temperature, and other environmental factors (Davis and Shaw 2001; Seppä and Bennett 2003). Through processes of disaggregation and recombination, plant assemblage responses during the Late Glacial–early Holocene therefore formed a changing array of vegetation patterns (Jackson and Overpeck 2000). Moreover, pollen and plant macrofossil records demonstrated that plant associations during the Late Glacial–early Holocene were very different from the associations recognized today. For example, Overpeck *et al.* (1992) showed that in North America major vegetation associations could not be recognized before the early Holocene. Terrestrial plant assemblages can therefore be seen to be highly dynamic, and concepts of so-called climax communities are no longer deemed to be valid.

After the retreat of the ice sheets, trees spread rapidly to occupy new available niches. Migration rates were so fast (10^2–10^3 m year^{-1}) that relatively rare, long-distance dispersal events (leptokurtic dispersal) are likely to have played a prominent role (Clark *et al.* 1998). It has been widely considered that temperate forests were largely restricted to refugia (isolated areas of habitat that retain environmental conditions that were once more widespread, allowing flora and fauna to persist in restricted localities) in the Balkan, Iberian, and Italian peninsulas, and in localized regions of eastern, central, and southwestern Europe during the LGM (Prentice *et al.* 2000). However, palaeoecologists are increasingly questioning whether temperate forests were truly restricted to more southern latitude refugia during this interval, or whether they were actually more extensive in periglacial environments (Clark *et al.* 2001). For example, animals and plants surviving in cryptic northern refugia (Stewart and Lister 2001) may have played a much larger role in the spread of thermophilous species through northern Europe. This is turn is likely to have had a greater impact on mid- to high-latitude biodiversity (Willis and Whittaker 2000) and dispersal patterns of human populations (Gamble *et al.* 2005) than previously considered.

The occurrence of cryptic northern refugia is just one of the questions challenging our understanding of past ecosystems, biodiversity, and speciation (Willis and Whittaker 2000). The role of tropical forest refugia in relation to speciation in the Amazon rainforest has also been subject to considerable debate. In order to account for the high levels of endemicity in the Amazon rainforest, Haffer (1969) proposed the hypothesis that during intervals of full glacial conditions, increased aridity resulted in the Amazon rainforest being restricted to isolated pockets, surrounded by expanded savannah environments. It was speculated that these disjunct, isolated rainforest refugia provided conditions favourable for speciation; new species were then able to expand their ranges on the return of warmer temperatures and interglacial conditions. However, palaeoecological evidence disputes this hypothesis, as forests were shown to have persisted during full glacial conditions despite overall reductions in precipitation (Bush 2003; Augusto de Freitas *et al.* 2001; Colinvaux and Oliveira 2001). The transition from a glacially cooler climate into the warmer Holocene was not accompanied by savanna vegetation succeeding to forest, but of one type of forest succeeding to other different forest types. Cold-adapted forest species can be detected in forest assemblages in Neotropical regions during the LGM. With the onset of the Holocene, some of these species died out, whereas others persisted for many thousands of years (Bush 2003). Overall, the transition into the Holocene is marked by the loss of cold-adapted species, rather than the appearance of species adapted to warmer conditions, as they were continually present.

1.4.3 Human population responses

Several now-extinct hominid species (*Homo habilis/georgicus*, *Homo erectus/ergaster*, and *Homo heidelbergensis*) spread across much of continental and insular Eurasia and evolved into several new species (*Homo antecessor*, *Homo neanderthalensis*, and *Homo floresiensis*) during the early–mid Pleistocene. Late Pleistocene migrations of *Homo sapiens* out of Africa into the Near East, south-east Asia, and Australasia had taken place by *c.*40 ka BP. All of the world's continents other than Antarctica had been colonized by large-scale modern human population migrations by the end of the Pleistocene, and all of these events were associated with similarly large-scale extinctions of megafaunal mammal (and some bird and reptile) species (see also Chapter 2 in this volume).

North and South America have been colonized only since the LGM. Genetic profiling suggests that the most likely origin of these earliest populations was from eastern Siberia (Eshleman *et al.* 2003). These populations crossed into Alaska via the Bering Land Bridge, and spread rapidly through ice-free corridors and via the deglaciated Pacific coastline into the south of the continent (Goebal *et al.* 2008). It is widely believed that this earliest migration of humans into North America represents the Clovis culture, which flourished during the Allerød interstadial between 13.2–13.1 to 12.9–12.8 ka BP (Waters and Stafford 2007). The Clovis culture spread rapidly through North America in less than 300 years (although it did not survive the Younger Dryas interval). However, there is also strong archaeological and genetic evidence for pre-Clovis occupation in both North America (*c.*15 ka BP; Goebal *et al.* 2008) and South America (at Monte Verde) from at least *c.*14.6 ka BP (Dillehay 1999). Further evidence for human presence prior to the Clovis culture in North America between *c.*14.3 and 14.0 ka BP has recently been determined through genetic analysis of human coprolites in southern Oregon (Gilbert *et al.* 2008).

European population numbers during the LGM were low, and archaeological evidence suggests that greatest densities were found in northern Spain, south-west France, eastern European river basins, and the Ukraine (Bailey 2007). With the retreat of ice sheets during the Late Glacial the expansion and dispersal of human populations in Europe was complex, and exhibited non-linear relationships to prevailing climate. Such complexity is probably related to the role of cryptic refugia for subsequent vegetation and animal dispersal throughout northern Europe (Willis and Whittaker 2000; Stewart and Lister 2001). Using a research approach known as dates-as-data, Gamble *et al.* (2005) investigated western European population levels from the Late Glacial through into the early Holocene. This technique is based on the principle that population signals can be determined from analysing the distributions of calibrated radiocarbon dates from archaeological sites. At least five significant population events (1–5) were recognized (Table 1.1), as follows.

1 During the LGM until GS-2b (Table 1.1), Iberia acted as a refugium for human populations, although archaeological evidence highlighted that humans were present in other regions, for example north-central Europe, albeit at low population levels.

2 Populations began to expand throughout western Europe, especially south-west France during GS-2a (*c.*19.5–16 ka BP; Table 1.1). Thus, population increase and expansion started during an interval of cooler rather than ameliorating climate as suggested, for example, by Blockley *et al.* (2000), highlighting in this case the importance of cultural adaptive strategies.

3 Major European population expansion occurred during 16–14 ka BP, along a south-to-north latitudinal gradient. Population numbers may have reached 40 000 individuals (Bocquet-Appel and Demars 2000).
4 An interval of population stasis is recognized to have occurred 14–12.9 ka BP, as settlements re-organized themselves from being dispersed to more nucleated (Gamble *et al.* 2005).
5 Populations contracted during the Younger Dryas (*c.*12.9–11.5 ka BP), although not to the low levels estimated for population events 1 and 2 above. These populations formed the basis for subsequent Holocene/Mesolithic recovery and population growth (Gamble *et al.* 2005).

1.4.4 The onset of agriculture and animal domestication

Agriculture emerged independently in many regions, with some of the earliest records dating from the Late Glacial in the Near East, associated with rapidly fluctuating climates (Fig. 1.4). The role of climate has long been postulated to be a major driver in the development of agriculture (Childe 1952). However, determining the impacts of climate change and associated human cultural development in the Near East with precise resolution has proved more challenging (Sherratt 1997). During the warm Bølling–Allerød (GI-1) interstadial, wild grasses were abundant in the Near East. However, at the onset of the Younger Dryas, a shift to colder, more arid conditions is shown in highly resolved pollen records from annually laminated lake sediments in the Near East; for example, at Lake Huleh in northern Israel (Baruch and Bottema 1999) and Lake Van in eastern Turkey (Wick *et al.* 2003). It is hypothesized that this change to drier and colder climates led human populations to become more closely associated, and to develop strains of cereal crops that were better able to tolerate these new conditions. Whereas cereal crops such as wheat *Triticum* spp. and barley *Hordeum vulgare* were among the first species to be cultivated, the domestication and alteration of other species quickly followed. These included legumes such as peas *Pisum sativum* and lentils *Lens culinaris* in the Near East, rice *Oryza* spp. and soya bean *Glycine max* in the Far East, and squash *Cucurbita* spp. and corn *Zea mays* in the Americas (Bellwood 2005) (Fig. 1.4).

Prior to domestication (*c.*14 ka BP), populations in the Near East had already formed very close associations with certain animals, most notably the wolf *Canis lupus*, from which all modern-day dogs are derived. Domestication of other animals followed during early Holocene including pigs *Sus scrofa domestica*, sheep *Ovis aries*, and goats *Capra aegagrus hircus* by *c.*9 ka BP, and cattle *Bos taurus* by *c.*8 ka BP (Kirch 2005) (Fig. 1.4). Animal domestication in other regions occurred later; for example, the water buffalo *Bubalus bubalis* in the Far East by *c.*7 ka BP, and llamas *Lama glama* and alpacas *Vicugna pacos* in South America by *c.*5 ka BP. Domestication had major, significant influences on global environments. Continent-wide swathes of land have been modified for domestic animals through forest clearance (see Chapter 11 in this volume). Freshwater environments have also been heavily impacted through the introduction of invasive species and modification of water quality (e.g. by nutrient enrichment and subsequent eutrophication) and water quantity (e.g. through abstraction), which has led to catastrophic declines in fish and invertebrate species, sometimes to extinction (Barel *et al.* 1985; Verschuren *et al.* 2002).

1.5 Holocene climate variability

The early to mid Holocene (*c.*9–5 ka BP) was an interval of warmer climate than the present day. This interval is commonly known as the Holocene climatic optimum (or the hypsithermal or altithermal) and was followed by orbitally related cooling which began to occur in the last 4–3 ka BP (Fig. 1.5). However, better-resolved palaeo-records from around the world have identified a much more complex picture than the simple model of early Holocene warming followed by late Holocene cooling. For example, $\delta^{18}O$ records from Greenland ice cores highlight that temperatures during the early stages of the Holocene (approximately the first 1500 years) were highly variable. 'Optimum' temperatures occurred between *c.*8.6 and 4.3 ka BP: thereafter, records demonstrate a general cooling followed by recent global warming (e.g. Johnsen

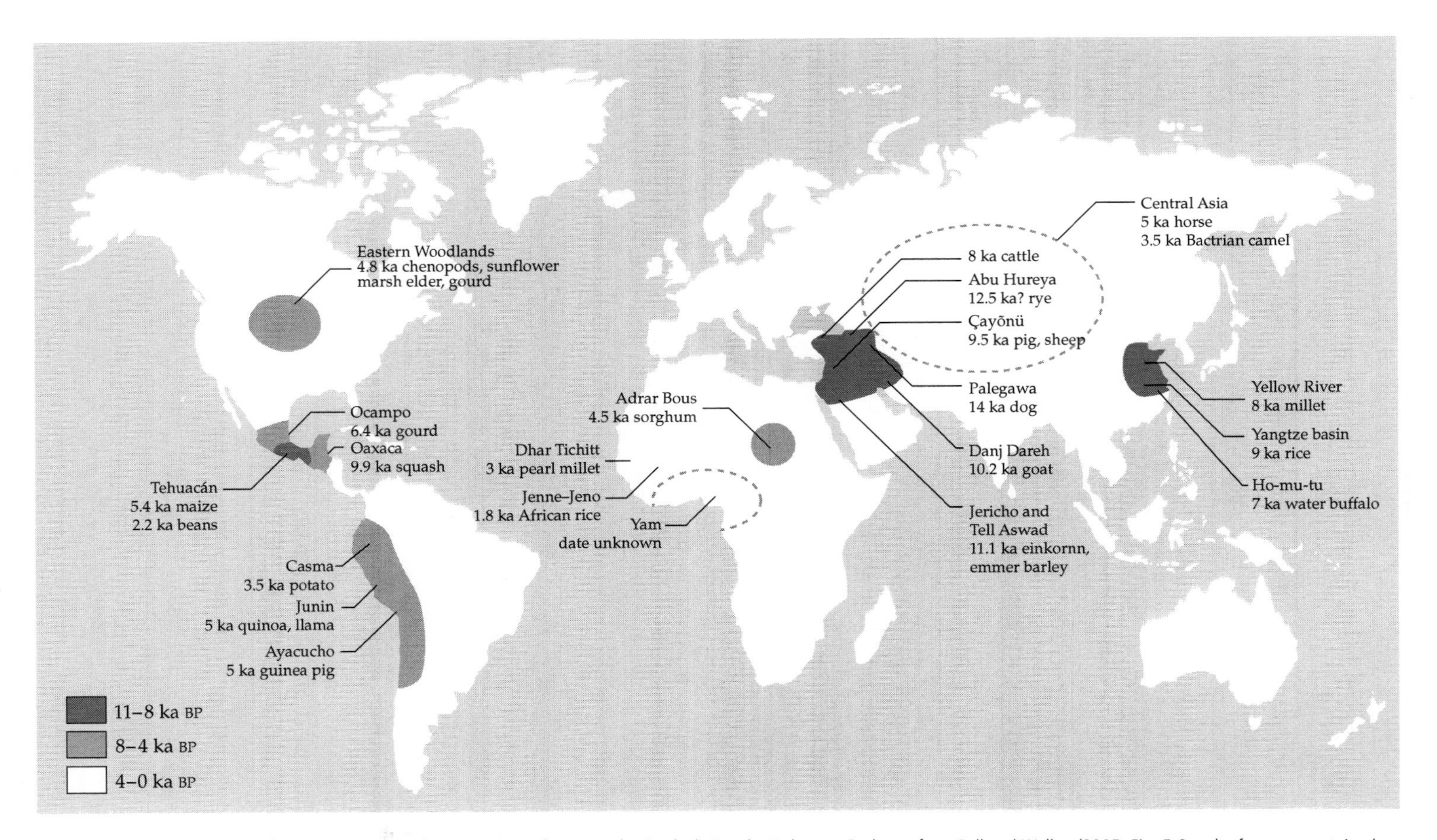

Figure 1.4 Locations and dates for some important domestications of crops and animals during the Holocene. Redrawn from Bell and Walker (2005, Fig. 5.6 and references contained therein).

et al. 2001) (Fig. 1.5). Centennial-scale cool events associated with millennial-scale cycles (i.e. Dansgaard–Oeschger cycles) have been identified worldwide (e.g. Mayewski *et al.* 2004; Nesje *et al.* 2005) (Fig. 1.5). During the early to mid Holocene these cooling events are associated with pulses of freshwater discharge from the Laurentide ice sheet slowing the North Atlantic thermohaline circulation (Teller *et al.* 2002) (Fig. 1.2f), although changes in solar insolation have also been implicated (Bond *et al.* 2001).

Perhaps the most significant Holocene cooling event occurred approximately 8.2 ka BP (Fig. 1.5). Its occurrence is linked to the terminal freshwater pulse of meltwater from the Laurentide ice sheet (Fig. 1.2f). The 8.2 ka event has been recognized from archives throughout the northern hemisphere. However, few records document the impact of this event on human populations, although Neolithic farming communities in the eastern Mediterranean may have undergone a significant reduction at this time (Weninger *et al.* (2006).

Centennial-scale cool events are also observed later in the Holocene, and although their occurrence is still associated with a slowdown in the North Atlantic thermohaline circulation, meltwater pulses originating from North America could not have caused them. Two prominent cooling events can be determined in many natural archives from around the world (6–5 and 3.5–2.5 ka BP) linked to declines in solar insolation and slowdown in thermohaline circulation (Mayewski *et al.* 2004). During both intervals, Na^+ (parts per billion, ppb) and K^+ (ppb) concentrations in the GISP2 Greenland ice core are elevated, indicative of a deepening Icelandic low-pressure zone and a more intense pattern of the central Asian Siberian high-pressure zone respectively. Further evidence of these cool events has been observed from increased concentrations of lithic grains in North Atlantic sediments (Bond *et al.* 2001), from declines in arboreal pollen in the Swiss Alps (Heiri *et al.* 2004), and in $\delta^{18}O$ records in diatom silica from Lake Baikal, south-eastern Siberia (Mackay 2007). Assessing the influence of centennial-scale cool events on past civilizations is also gaining strength, for example in central Europe (Arbogast *et al.* 2006) and in eastern Mediterranean and south Asian societies (Staubwasser and Weiss 2006).

Over the last approximately 1200 years, three distinct phases of climate are apparent in palaeo records: the Medieval Warm Period (*c.*AD 800–1300), the Little Ice Age (sixteenth to eighteenth centuries AD), and recent warming since *c.*AD 1850. However, these terminologies are perhaps rather simplistic and certainly overly restrictive. This is because they assume that 'warm' and 'cold' episodes occurred synchronously across at least the northern hemisphere, whereas high-resolution palaeoclimate records show distinct regional variation (e.g. Nesje *et al.* 2005). The Little Ice Age can be linked to the latest centennial cool event associated with millennial-scale variability. One of the coldest intervals during the Holocene was the Maunder Minimum

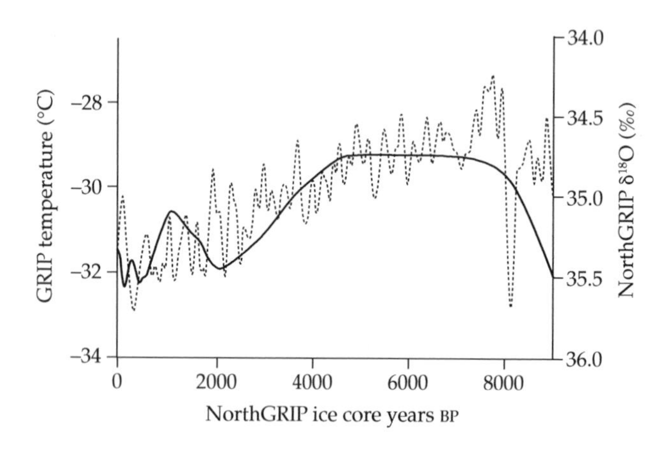

Figure 1.5 $\delta^{18}O$ ‰ NorthGRIP record (thin line) plotted against the long-term temperature trend (thick line) obtained from the NorthGRIP profile. Adapted from Johnsen *et al.* (2001). ©John Wiley & Sons Limited. Reproduced with permission.

(AD 1645–1715), which is characterized by reduced sunspot activity and total solar irradiance. Lozano-Garcia *et al.* (2007) showed that during the Maunder Minimum, increased meridional flow led to higher winter precipitation in the Caribbean, causing an expansion in tropical vegetation. There is still debate as to the actual cause of the Little Ice Age: volcanic eruptions (Crowley 2000), changes in sunspots (Shindell *et al.* 2001), and changes in North Atlantic Ocean circulation (Lund *et al.* 2006) have all been implicated.

In 2000, the term Anthropocene was coined to describe the recent Earth system as being distinct from the rest of the Holocene due to intense human impact, especially industrialization and land use since the latter part of the eighteenth century (Crutzen 2002). Significantly, many of these impacts (including population growth, water use, fertilizer use, species extinction, ozone depletion, etc.) have risen exponentially in the last 50 years. Others have sought to challenge not the concept of the Anthropocene *per se*, but the date of the beginning of major human impact on the Earth. For example, Ruddiman (2003) has recently put forward a set of hypotheses which suggest that human impact on atmospheric CO_2 and CH_4 levels have been significant since *c*.8 ka BP due to deforestation and agriculture. He proposed that such early increases in greenhouse gas concentrations (e.g. see Figs 1.2c and 1.2d) were responsible for delaying natural cooling of the earth. However, this view is controversial, and is not supported by, for example, relatively high concentrations of CO_2 in ice cores during MIS11, which must have been independent of human influence (Siegenthaler *et al.* 2005), or the observation that models do not simulate a glacial inception when run with lower CO_2 concentrations for the latter part of the Holocene (Claussen *et al.* 2005). However, pyrogenic (biomass) burning is also implicated as an important source for, among others, CO_2 since *c*.8 ka BP (Carcaillet *et al.* 2002) and for CH_4 over the last 1000 years (Ferretti *et al.* 2005). There is no doubt therefore that humans have had a major impact on the Earth's ecosystems since the early Holocene: the range of papers covered in this volume are testament to that. Whether human impacts are responsible for delaying the onset of the next glacial interval is still very much under debate.

Summary

Climate has undoubtedly been the major forcing factor for ecosystem change through much of the Late Glacial and Holocene intervals. However, anthropogenic influence on the Earth's ecosystem has increased steadily since the Late Glacial, culminating in the interval now known as the Anthropocene, during which time few if any analogues for past environments exist. It is against this backdrop that subsequent chapters in this book will deal with ecosystem resilience, ecosystem change, and concomitant species extinctions throughout the Late Glacial–Holocene intervals.

CHAPTER 2

In the shadow of the megafauna: prehistoric mammal and bird extinctions across the Holocene

Samuel T. Turvey

2.1 Introduction

The last 500 years of the historical era, an interval that corresponds with the concerted dispersal of European explorers, traders, and colonists around the globe, has witnessed the global disappearance of an alarmingly large number of vertebrate species. Although these extinctions have in many cases been well documented by eye-witness accounts and historical records, and have constituted the primary focus for research into past human impacts on global ecosystems, they represent merely the latest phase of a much longer-term event, comparable in magnitude to prehistoric mass extinctions, that has been underway throughout the Late Quaternary.

The best-studied series of prehistoric Quaternary extinctions is the loss of at least 97 genera of continental megafaunal vertebrates (>44 kg; *sensu* Martin 1984), mostly mammals but also some birds and reptiles, without subsequent ecological replacement during the Late Pleistocene and with very little corresponding extinction of small-bodied vertebrates. The species that became extinct during this interval represent almost two-thirds of Late Pleistocene terrestrial megafaunal genera (Martin and Steadman 1999; Barnosky *et al.* 2004). Continent-wide megafaunal losses occurred earliest in Australia *c.*46 000 years ago (Roberts *et al.* 2001; Miller *et al.* 2005), with the latest well-dated wide-scale series of continental extinctions taking place in North America and northern Eurasia 14 000–10 000 years BP (Stuart 1999; Haynes 2002; MacPhee *et al.* 2002). Human involvement in Pleistocene megafaunal extinction dynamics is now widely accepted by most palaeontologists (although see Firestone *et al.* 2007), as the stepwise nature of the extinctions across different continents correlates with first human arrival in each region. However, the Late Pleistocene is also characterized by major global-scale climatic and associated vegetational shifts during the transition from glacial to interglacial conditions, and the relative importance or possible interactions of prehistoric human activity and natural environmental change in driving this mass extinction event have been debated extensively since the nineteenth century (Grayson 1984).

Huge numbers of continuing prehistoric vertebrate extinctions and large-scale range contractions have also been documented throughout the subsequent Holocene Epoch, across a wide range of body-size classes and taxonomic groups and across both insular and continental regions (Fig. 2.1). Unlike the Late Pleistocene extinctions, these Holocene events all occurred during an interval of modest or minimal climatic variation and under broadly 'modern' environmental and climatic boundary conditions (see Chapter 1 in this volume). Other than the lag in continental vertebrate faunal transitions from cold-adapted Late Glacial species to warm-adapted species several centuries after the end of the Younger Dryas, probably reflecting a delay in the Holocene vegetational succession (Coard and Chamberlain 1999; Hewitt 1999), evidence for direct human involvement in these extinctions and population shifts—through

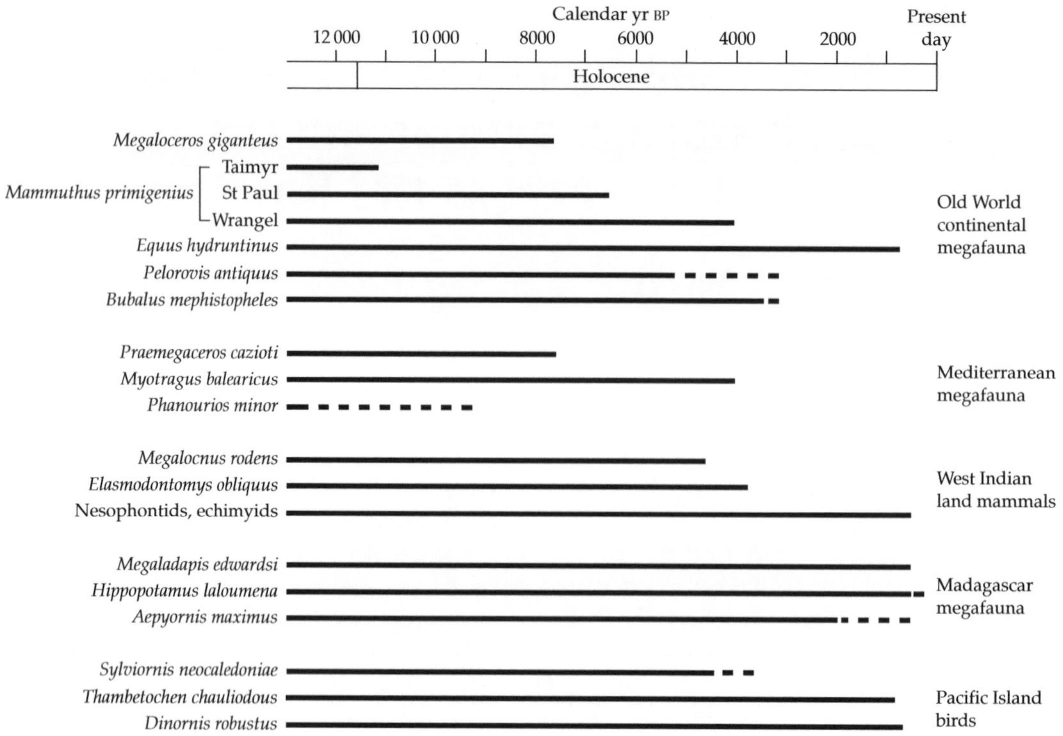

Figure 2.1 Extinction chronologies for representative species on different continents and islands across the Holocene.

both first human arrival events and increased environmental impacts in previously settled regions—is not confounded by other factors and remains relatively undisputed. The Holocene therefore has the potential to act as an ideal study system for investigating the long-term dynamics of anthropogenically mediated extinctions at a global scale. However, the quality of the palaeontological and archaeological records presents ongoing limitations to our ability to resolve specific extinction chronologies, and identify the nature and timing of putative anthropogenic extinction drivers. This impedes our current understanding of the speed and amplitude of the Holocene human-driven extinction wave, and it remains uncertain whether most prehistoric Holocene events occurred as a result of unicausal direct overkill or 'blitzkrieg' (rapid overkill), indirect factors such as habitat destruction through landscape burning and/or fragmentation (gradual 'sitzkrieg'), predation overload, keystone cascades and/or disease, or synergistic effects between multiple factors (Diamond 1989; Martin and Steadman 1999; Barnosky *et al.* 2004). Inclusive overviews of prehistoric Holocene extinctions also remain largely unavailable beyond the regional level, in contrast to the better-studied end-Pleistocene megafaunal extinction event. In order to address this final constraint, data on global patterns of species-level losses in mammals and birds—the two vertebrate groups which have been the subject of the majority of palaeontological research in the Holocene subfossil record—are reviewed here to provide an initial assessment of the patterns of prehistoric human impact across space and time since the end of the last glaciation. It is hoped that this assessment of our current understanding of Holocene extinctions will encourage further attempts to reconstruct the diversity, ecology, and fate of prehuman Quaternary faunas.

2.2 Staggered Pleistocene–Holocene extinctions of continental mammals and birds

The dominant palaeontological paradigm for understanding Late Pleistocene megafaunal losses (e.g. Martin 1984) is of 'eco-catastrophic' (Haynes 2002) rapid and simultaneous extinction of multiple large-bodied taxa at a continental level following first human contact. Although direct radiometric 'last-occurrence' dates from the terminal Pleistocene still remain unavailable for many extinct megafaunal species, it has often been assumed that apparently staggered Quaternary extinction events are artifacts of the Signor–Lipps effect (see Martin and Steadman 1999; see also Chapter 9 in this volume). This interpretation is supported by recognition of a distinct hiatus in radiometric dates at or near the Pleistocene–Holocene boundary in Siberia, which is apparently not related to the true terminal occurrences of megafaunal species but probably reflects taphonomic effects, $\Delta^{14}C$ flux, and biased sampling (MacPhee *et al.* 2002). However, whereas continued research into North American continental megafaunal extinctions has strengthened the evidence for a temporally short, intensive end-Pleistocene 'extinction window' between 11 500 and 10 000 years BP (Barnosky *et al.* 2004; although see Waters and Stafford 2007), last-occurrence data indicate that megafaunal species experienced more complex extinction chronologies in other continental regions.

2.2.1 Europe and northern Asia: staggered megafaunal extinctions

The northern Eurasian fossil record shows distinct phases of megafaunal extinction during the Late Quaternary. European megafaunal species that had been widespread during the warm Last Interglacial (e.g. straight-tusked elephant *Palaeoloxodon antiquus*, hippopotamus *Hippopotamus amphibius*, and rhinoceros *Stepanorhinus* spp.) disappeared by *c.*20 000 years BP during the Last Cold Stage, but cold-adapted tundra/steppe species remained widespread across Europe throughout the main glaciation and disappeared during the Late Glacial between *c.*14 000 and 10 000 years BP. This two-phase extinction model is interpreted as reflecting the sequential restriction of warm-adapted and cold-adapted mammal faunas to habitat refugia during glacial–interglacial climatic cycles, where they would have been increasingly vulnerable to hunting following arrival of modern humans in Europe (Stuart 1999). Different populations of megafaunal species also disappeared at different times across mainland Eurasia, apparently through a complex pattern of range fragmentation. Whereas woolly mammoth *Mammuthus primigenius* disappeared across much of their range around *c.*12 000 years BP, they persisted for over 2000 years on the Taimyr Peninsula and in the north Russian Plain close to the margin of the Fennoscandian ice sheet (Stuart 1999; MacPhee *et al.* 2002; Stuart *et al.* 2002, 2004).

The 'ragged' timing of northern Eurasian continental megafaunal extinctions in contrast to the temporally constrained North American end-Pleistocene event is further emphasized by the survival of one representative of the cold-adapted Last Cold Stage Eurasian megafauna, the giant deer or Irish elk *Megaloceros giganteus*, several thousand years into the Holocene on mainland Eurasia, revealing a more complex pattern of megafaunal extinction chronologies in this region than formerly recognised. The well-studied north-west European Allerød phase *Megaloceros* populations disappeared with the return of open steppe/tundra at the onset of the Younger Dryas severe cold phase around 10 700 years BP, leading to an extrapolated assumption of global extinction at the terminal Pleistocene (Stuart *et al.* 2004). Direct early Holocene radiometric dates of 9225±85 and 9430±65 years BP were reported for *Megaloceros* specimens from the Isle of Man and south-western Scotland by Gonzalez *et al.* (2000), suggesting terminal megafaunal survival on islands at the margin of Eurasia. Although both of these specimens were subsequently re-dated to within the Late Glacial by Stuart *et al.* (2004), these authors presented new accelerator mass spectrometry (AMS) dates demonstrating *Megaloceros* survival in the Urals and western Siberia until around 6900 years BP. The late survival of *Megaloceros* in this region may reflect the ecological requirements imposed by its giant 3.5 m-wide antlers; it has been suggested that the species was constrained to inhabit productive open forest/steppe environments containing willow and birch, which could provide sufficient calcium and phosphate for annual antler growth in adult males (Moen *et al.* 1999) and which persisted in the eastern foothills of the Urals during the early Holocene (Stuart *et al.* 2004). *Megaloceros* had survived previous Quaternary environmentally driven range contractions and expansions, and although the ecological basis for its protracted survival despite the disappearance of other cold-adapted Eurasian 'tundra/steppe' megafaunal species remains unclear, humans may have also played a role in its eventual Holocene extinction. The disappearance of the remnant Asian *Megaloceros* population is associated with the spread of dry steppe environments across the western Siberian refugium around 7000 years BP, which could have forced the species from the Ural foothills into closer contact with Neolithic humans on the West Siberian Plain (Stuart *et al.* 2004).

Another characteristic representative of the northern megafauna, the woolly mammoth, survived even later into the Holocene as dwarfed populations on Wrangel Island (Arctic Sea) and St. Paul Island (Pribilofs, Bering Sea), which were both isolated from respective Eurasian and North American mammoth populations during the end-Pleistocene marine transgression between 13 000 and 12 000 years BP (Vartanyan *et al.* 1993; Guthrie 2004). Other present-day Bering Sea islands (St. Lawrence, St. Matthew, Nunivak) only became separated from Alaska by rising sea levels after 10 000 years BP, subsequent to the extinction of mainland American mammoth populations; other unnamed islands capable of sustaining stranded insular mammoth populations also existed in the Bering Sea during the early Holocene, but have now been inundated (Guthrie 2004). The Wrangel Island population, which had molar teeth about 30% smaller than mainland Siberian mammoths, persisted as recently as 3730±40 years BP (Vartanyan *et al.* 1993), and newly collected mammoth material from St. Paul has been dated to *c.*5700 years BP (Crossen *et al.* 2005). The pattern of relictual survival of insular mammoth populations following widespread continental hunting matches that displayed by Steller's sea cow *Hydrodamalis gigas*. This giant sirenian was distributed in shallow waters around the North Pacific Rim during the Late Pleistocene, but became restricted by historical times to the remote Commander Islands (Bering and Copper Islands) and the western Aleutian Islands (Attu Island and Kiska Island, and possibly also other islands in the Near, Rat, and Andreanof groups) following prehistoric overexploitation across the remainder of its range, and was eventually driven to extinction in these last refuges in the eighteenth century AD by small bands of hunters using primitive pre-industrial hunting technologies (Stejneger 1887; Turvey and Risley 2006; Domning *et al.* 2007).

Vartanyan *et al.* (1993) suggested that the late survival of mammoths on Wrangel Island was associated with local persistence of relict 'tundra/steppe' herbaceous communities consisting of a diverse grass/sedge/forb vegetation mosaic, which also constituted the major vegetational biome associated with mammoth distribution in mainland Siberia during the Late Pleistocene. However, Quaternary pollen cores from St. Paul demonstrate the replacement of tundra/steppe plant species by umbelliferous forb-dominated assemblages characteristic of modern Bering Sea rim vegetation during the early Holocene (Guthrie 2004). The late timing of insular mammoth extinction coincides more closely with the absence of human hunters on either island during the Pleistocene or early Holocene. The earliest archaeological evidence for human occupation of Wrangel Island is around 4300 years BP, close to the last-occurrence date for mammoths on the island. St. Paul remained completely isolated until the late eighteenth century AD, and it is most likely that extirpation of this mammoth population was instead driven by prehistoric reductions in island area caused by continuation of the end-Pleistocene marine transgression before human arrival (Guthrie 2004; Crossen *et al.* 2005).

Further Holocene megafaunal losses before the modern era have also been documented from Europe and northern Asia. Whereas woolly mammoth disappeared in mainland Siberia close to the Pleistocene–Holocene boundary, the globally extinct (although see Shapiro *et al.* 2004) bison *Bison priscus* survived for at least another millennium until 8860±40 years BP in the Taimyr Peninsula, and musk oxen *Ovibos moschatus* also persisted in this region and possibly also southern Siberia until 2700±70 years BP before being extirpated from Eurasia (MacPhee *et al.* 2002). Much later Holocene survival has also been alleged for *B. priscus* and *O. moschatus* in Asia, but these suggestions remain ambiguous and lack robust evidence (MacPhee *et al.* 2002); it remains unclear what, if any, role humans played in the final disappearance of either species. The extinct equid *Equus hydruntinus* is also known from early Holocene deposits and Iron Age sites in Europe and the Middle East (Orlando *et al.* 2006b; Vila 2006). A slender wild equid called 'zebro' (Fig. 2.2, right-hand panel), interpreted as representing a remnant population of *E. hydruntinus*, is recorded from the open grasslands of the Iberian Peninsula in Portuguese manuscripts as recently as AD 1293; this population was apparently driven to extinction by hunting and human encroachment (Antunes 2006). Ancient descriptions of the zebro indicate that it had a black dorsal stripe and muzzle, markings which helped the name become transferred to living African striped equids by Portuguese navigators, missionaries, and chroniclers in the fifteenth century AD (Antunes 2006). Many other large-bodied Eurasian mammal species also experienced extirpations due to human activity across large areas of their former range before the modern era, but have (so far) avoided complete extinction during the Holocene (e.g. lion *Panthera leo*, present in southern Europe until AD 80–100 and the Caucasus region until the tenth century AD; Nowell and Jackson 1996; Sommer and Benecke 2006).

2.2.2 North America and Australia: limited prehistoric extinctions

Similar patterns of terrestrial vertebrate declines and extinctions are shown by the relatively well-studied Holocene palaeontological, zooarchaeological, and historical records of both North America and Australia. The primary faunal trends documented for both of these continental regions across the Holocene are population depressions and local extirpations of many mammal and bird species. Many of these events were driven by overexploitation by prehistoric hunters even under relatively low human population densities, and which have continued up to the present as human pressures

Figure 2.2 Possible contemporary depictions of now-extinct Holocene megafaunal mammals. Left: diminutive adult elephantid on a frieze in an Eighteenth Dynasty pharaonic tomb (c.1475–1445 BC), which may depict a late-surviving Mediterranean species of *Elephas*. From Rosen (1994); reprinted by permission from Macmillan Publishers Ltd: *Nature 369*: 364, ©1994. Right: Folius 120 from the *Apocalipse de Lorvão*, showing a saint riding an equid that may be a European wild ass or zebro *Equus hydruntinus*. From Antunes (2006).

have escalated (Morrison *et al.* 2007), although other local extinctions reflect climatic changes across the Holocene (e.g. Grayson 2000, 2005). The most significant and chronostratigraphically earliest faunal changes in Holocene midden sequences consistently show reductions of both terrestrial and coastal species most attractive to hunters or otherwise most sensitive to hunting pressure; for example, *Branta*, *Anser*, and *Chen* geese in the Californian Emery Shell Mound between 2600 and 700 years BP (Broughton 2004), artiodactyls in the eastern Great Basin and northern plateau of Utah between 1400 and 600 years BP (Janetski 1997), and mountain sheep *Ovis canadensis* in high-elevation archaeological sites in the Californian White Mountains from 3500 years BP onwards (Grayson 2001). The dramatic abundances of artiodactyls and waterfowl reported by early European explorers in western North America probably reflect recent population irruptions following the large-scale sixteenth–seventeenth century AD population crash of native Amerindian hunters after European arrival.

Early explorers and collectors in Australia, where widespread indigenous hunting continued until more recently than in North America, instead noted very low population densities of a wide range of both large-bodied and smaller-bodied species, including emus, terrestrial and arboreal macropods (*Dendrolagus*, *Macropus*, *Setonix*, and *Thylogale*), rat-kangaroos, bettongs, koalas, possums, and bandicoots, many of which were unknown to the first European settlers. Later historical records report diachronically parallel large-scale population increases for all of these species during the mid–late nineteenth century (Johnson and Wroe 2003; Johnson 2006). Although European scrub clearance for sheep and cattle pasture may have played a part, many species (notably arboreal mammals) are ecologically unlikely to have been favoured by this anthropogenic habitat modification and others became abundant before extensive land clearing began, and it is likely that heavy Aboriginal hunting had previously been holding all of these populations well below the environmental carrying capacity (Johnson 2006).

However, both North America and Australia experienced very few complete species extinctions during the Holocene before the modern historical era, in marked contrast to the major losses seen during both the Late Pleistocene megafaunal extinction events on each continent and the subsequent series of well-documented recent and ongoing vertebrate extinctions in both regions. Only three North American terrestrial vertebrate species are known to have become globally extinct during this period: the giant pika *Ochotona whartoni*, the flightless diving duck *Chendytes lawi*, and the small turkey *Meleagris crassipes* (Morejohn 1976; Rea 1980; Mead and Grady 1996; Jones *et al.* 2008b); other supposed Holocene-era extinctions such as the large marten *Martes nobilis* have now been discounted (Youngman and Schueler 1991). Other than the Hunter Island penguin *Tasidyptes hunteri*, which may have survived into the early historical era (Van Tets and O'Connor 1983; Meredith 1991), and the cormorant *Microcarbo serventyorum*, about which little remains known (Van Tets 1994), the only recorded prehistoric Holocene mammal and bird extinctions from Australia are the extirpations of the mainland populations of thylacine or Tasmanian tiger *Thylacinus cynocephalus*, Tasmanian devil *Sarcophilus harrisii*, and Tasmanian native hen *Gallinula mortierii*. All three of these species persisted into the recent historical period in Tasmania, and were widespread across the continent during the Pleistocene but may have experienced major mainland range reductions by the start of the Holocene (Johnson and Wroe 2003; Johnson 2006). The native hen persisted until 4670±90 years BP in south-western Victoria (Baird 1991) and the youngest mainland thylacine specimen has been dated directly to 3280±90 years BP (Partridge 1967; although see also Chapter 10 in this volume). Tasmanian devils may have survived considerably later, with the youngest mainland devil remains dated indirectly to 430 years BP (Archer and Baynes 1972), although the association of the devil specimen with the dated material has been questioned (Johnson 2006). The introduction of the dingo *Canis lupus dingo* to Australia by around 3500 years BP is frequently blamed for the mainland extinction of all three species, through either direct predation or competition (e.g. Baird 1991, Corbett 1995). However, this hypothesis has been questioned by Johnson and Wroe (2003) and

Johnson (2006), who instead suggested that these regional extinctions were driven by greater anthropogenic environmental exploitation associated with the Late Holocene Aboriginal cultural 'intensification' that occurred after 5000 years BP. This event involved a complex of socio-economic changes including artistic and technological innovations (e.g. new tool types and tool-making culture), utilization of a much wider range of plant and animal species, more intensive and permanent settlement patterns including occupation of marginal habitats, and increasing human population densities. Both dingoes and mainland Aboriginal cultural changes failed to reach Tasmania, making it difficult to distinguish between the relative impacts of these different putative extinction drivers.

2.2.3 South America, Africa, and south-east Asia: an unknown quantity

The magnitude and timing of prehistoric Holocene mammal and bird extinctions in other continental regions remains poorly understood. Direct radiometric dating and other dates derived from deposits containing extinct species have frequently been interpreted as evidence for the survival of several members of the South American megafauna (e.g. horses, camelids, ground sloths) into the early or mid Holocene (Table 2.1). These supposed Holocene survival records have been accepted at face value as the basis for developing new models of megafaunal extinction in some modern reviews (e.g. de Vivo and Carmignotto 2004). However, recent critical radiometric investigations, including re-dating of supposed mid-Holocene megafaunal specimens (e.g. Araujo *et al.* 2004), have failed to find evidence that any South American megafaunal species survived long beyond the Pleistocene–Holocene boundary. Steadman *et al.* (2005) considered that no South American ground sloth material has been reliably dated to within the last 10000 years BP, with the youngest ground sloth dung deposits in South America being only centuries younger than deposits from North America. Considerable further research is required to clarify temporal trends and dynamics of Late Quaternary South American mammal extinctions, and in particular to re-analyse anomalous Holocene radiometric dates to assess the possibility of contamination (see also Stuart 1991 for review of similar poorly supported Holocene radiometric dates for several North American megafaunal species, e.g. *Mammut americanum*). The only now-extinct mammal species that probably survived until the mid–late Holocene in continental South America are the giant vampire bat *Desmodus draculae* (tooth size 25% larger than the living vampire bat *Desmodus rotundus*), for which a canine provisionally assigned to this species has been reported from historical-era deposits from Argentina (Pardiñas and Tonni 2000); and the approximately 15kg (Prevosti and Vizcaíno 2006) canid *Dusicyon avus*, which has been reported from several mid–Late Holocene archaeological sites (e.g. Tonni and Politis 1982; Berman and Tonni 1987; Mazzanti and Quintana 1997). Relatively little also remains known about patterns of Holocene exploitation of extant Neotropical bird and mammal species, although recent studies indicate that several species have been locally extirpated across areas of their former range through prehistoric human activity, indicating that it is difficult or impossible to determine pre-human species-level compositions of Neotropical vertebrate communities under conditions free from human influence (Steadman *et al.* 2003).

Late Quaternary megafaunal extinctions were markedly less severe in Africa and south-east Asia, and these two biogeographic regions still retain a wide range of large-bodied mammal species, including all living terrestrial vertebrates over 1000kg; this is interpreted as reflecting the long period of megafaunal coevolution with primitive hunting strategies employed by behaviourally non-modern *Homo sapiens* in Africa from *c.*195000 years ago (McDougall *et al.* 2005) and by earlier hominin species in both regions (e.g. Martin 1984; Martin and Steadman 1999; Barnosky *et al.* 2004). These extinction events may also have been staggered across the Pleistocene–Holocene boundary, although robust radiometric or stratigraphic data are again generally unavailable for reconstructing megafaunal extinction chronologies from either region, and critical radiometric investigations are required to properly evaluate this possibility; the taxonomic validity of several apparently extinct species also remains unclear (e.g. Peters *et al.* 1994).

Table 2.1 Holocene last-occurrence dates proposed for extinct South American megafaunal and large-bodied mammal species.

Species	Locality	Last-occurrence date (years BP)	Dating method	Reference
Dasypodidae				
Propaopus cf. *sulcatus*	Toca de Serrote do Artur, Brazil	8490±120–6890±60	^{14}C in charcoal from upper and lower layers of blackish silt bed containing megafaunal remains	Faure *et al.* (1999)
Glyptodontidae				
Doedicurus clavicaudatus	Arroyo Seco 2, Argentina	6555±160	Direct ^{14}C date	Borrero *et al.* (1998)
Glyptodon clavipes	Toca de Serrote do Artur, Brazil	8490±120–6890±60	^{14}C in charcoal from upper and lower layers of blackish silt bed containing megafaunal remains	Faure *et al.* (1999)
Hoplophorus euphractus	Toca de Serrote do Artur, Brazil	8490±120–6890±60	^{14}C in charcoal from upper and lower layers of blackish silt bed containing megafaunal remains	Faure *et al.* (1999)[1]
Megatheriidae				
Megatherium americanum	Arroyo Seco 2, Argentina	7320±50	Direct AMS date	Borrero *et al.* (1998)
Mylodontidae				
Glossotherium sp.	El Cautivo, Ecuador	8680±80	AMS date of stratigraphically associated mollusc shell	Ficcarelli *et al.* (2003)[2]
Scelidodon chiliensis	Pampa de los Fósiles, Peru	8910±200	Direct AMS date	Hubbe *et al.* (2007)
Scelidotheriidae				
Catonyx cuvieri	Lagoa Santa, Brazil	9580±200	^{14}C in charcoal	Neves *et al.* (2004)
Felidae				
Smilodon populator	Lagoa Santa, Brazil	9130±150	Direct AMS date	Auler *et al.* (2006)
Gomphotheriidae				
Haplomastodon chimborazi	El Cautivo, Ecuador	8680±80	AMS date of stratigraphically associated mollusc shell	Ficcarelli *et al.* (2003)[2]
Equidae				
Equus neogeus	Toca de Serrote do Artur, Brazil	8490±120–6890±60	^{14}C in charcoal from upper and lower layers of blackish silt bed containing megafaunal remains	Faure *et al.* (1999)
	Arroyo Seco 2, Argentina	8890±90	Direct AMS date	Borrero *et al.* (1998)
Equus santaelenae	El Cautivo, Ecuador	8680±80	AMS date of stratigraphically associated mollusc shell	Ficcarelli *et al.* (2003)[2]
Camelidae				
Palaeolama major	Toca de Serrote do Artur, Brazil	8490±120–6890±60	^{14}C in charcoal from upper and lower layers of blackish silt bed containing megafaunal remains	Faure *et al.* (1999)
Toxodontidae				
Toxodon platensis	Abismo Ponta de Flecha	5000±1600 (enamel), 6700±1300 (dentine)	ESR in teeth	Baffa *et al.* (2000)

AMS, accelerator mass spectrometry; ESR, electron spin resonance.

[1]There is some uncertainty over whether this species occurs in the blackish silt bed (Faure *et al.* 1999).

[2]Older collections from this region, apparently representing the same fossiliferous level, also contain *Eremotherium laurillardi* or *Eremotherium rusconii, Scelidotherium reyesi, Glossotherium tropicorum, Chlamytherium occidentale, Palaeolama aequatorialis, Odocoileus salinae, Neochoerus sirasakae, Dusicyon sechurae, Protocyon orcesi*, and *Smilodon* sp. (Ficcarelli *et al.* 2003), but the mixed faunas represented in these deposits are suggestive of reworking.

The only well-supported prehistoric Holocene megafaunal extinction event in Africa is that of the giant long-horned buffalo *Pelorovis antiquus*, now regarded as a valid species distinct from the extant buffalo *Syncerus caffer* (Klein 1994). *Pelorovis* is present in Pleistocene fossil assemblages from eastern and southern Africa, but was restricted by the mid Holocene to relict North African populations in the Maghreb and Sahara. It is relatively common in Holocene rock art from this region, with two probable representations dating as recently as the third millennium BP. The youngest well-supported radiometric date for the species is 4715±295 years BP (Gautier and Muzzolini 1991). Klein (1984) suggested that the extinction of this species was a natural event caused by increasing aridification and contraction of North African savanna environments at the end of the early Holocene moist phase, but Gautier and Muzzolini (1991) suggested that the primary extinction driver was instead probably competition for food and water with domestic cattle, which were widespread across the region by the mid Holocene. The pre-European spread of cattle associated with the Bantu pastoralist expansion also impacted other native ungulates in eastern and southern Africa and led to prehistoric population declines in several large mammals, notably the bluebuck *Hippotragus leucophaeus* and bontebok *Damaliscus dorcas dorcas*; the former species is common in early–mid Holocene sites in the Cape Zone antedating the regional introduction of stock around 2000 years ago, but was rare by the time of European arrival and became extinct shortly afterwards (Klein 1974).

Indirect Holocene radiometric dates from archaeological sites for other extinct South African megafaunal species are tantalizing but require further verification. The extinct giant hartebeest *Megalotragus priscus* occurs in a stratigraphic horizon at Wonderwerk Cave dated between 8000±80 and 5970±70 years BP; the extinct equid *Equus capensis* is also present at this site in an older Pleistocene–Holocene boundary level dated between 10 200±90 and 9130±90 years BP (Beaumont 1990). Holocene survival of Bond's springbok *Antidorcas bondi* is suggested by an associated charcoal date of 7570±120 years BP at Kruger's Cave, where other mixed assemblages reportedly show good concordance between the dates of different components (Brown and Verhagen 1985). Klein (1984) also reported the global extinction of three Maghreb/Sahara megafaunal savanna species during the mid Holocene (5000–4000 years BP): Thomas' camel *Camelus thomasi*, the North African giant deer *Megaloceros* [now *Megaceroides*] *algericus*, and possibly the Atlantic gazelle *Gazella atlantica*. However, *C. thomasi* and *M. algericus* are now interpreted as having died out in North Africa during the Late Pleistocene, as they are not reliably recorded from Holocene deposits or depicted in Holocene rock art (Muzzolini 1986; Peters 1992), and *C. thomasi* may represent the direct ancestor of the living one-humped camel *Camelus dromedarius* (Peters 1998), so that its disappearance only represents a pseudo-extinction event (cf. early historical-era extinctions of aurochs *Bos primigenius* and tarpan *Equus ferus* in Eurasia). The only other reliably documented Holocene African mammal extinction dating from before the recent historical era is that of the shrew *Crocidura balsamifera*, which is known from embalmed and mummified individuals found in Egyptian tombs, and was apparently endemic to swamps and gallery forests along the Nile that have now been completely destroyed (Hutterer 1994). Prehistoric Holocene degradation of wetland habitats in north-east Africa and south-west Asia may also have been responsible for the extinction of a wetland bird species, the Bennu heron *Ardea bennuides*, which is last recorded from the Arabian Peninsula around 4500 years BP (Hoch 1977).

The southern Chinese fossil record displays progressively fewer megafaunal extinctions through the Quaternary, suggesting that early Pleistocene extinctions were more significant than in other regions (Louys *et al.* 2007). However, Holocene last-occurrence dates have been suggested in the recent literature for a surprisingly large number of megafaunal mammals that disappeared during the Late Pleistocene elsewhere in Eurasia. Ma and Tang (1992) and Tong and Liu (2004) reported dates of 7815±385 years BP for the hyaena *Crocuta* [*crocuta*] *ultima* (the taxonomic status of which remains unclear; Kurtén 1956; Rohland *et al.* 2005), 7380±100 years BP for *Mammuthus primigenius*, 4235 years BP for the giant tapir *Megatapirus augustus*, 4100 years BP for the elephantid *Stegodon orientalis*, 3630±90 years BP for the woolly rhinoceros *Coelodonta antiquitatis*, and 3140±110 years BP for the short-horned water buffalo *Bubalus mephistopheles*. Although *B. mephistopheles* has also been reported from a fifteenth–twelfth century BC Chinese Shang Dynasty archaeological site (Teilhard de Chardin and Young 1936), further fieldwork and critical radiometric studies are required to clarify these other extinction chronologies, and they are considered here to be questionable. Other species such as the orang-utan *Pongo pygmaeus* also disappeared from mainland south-east Asia during the Holocene, but persist on islands on the Sunda Shelf (Louys *et al.* 2007).

2.3 Holocene extinctions of global island faunas

Whereas prehistoric Holocene extinctions of continental mammals and birds were relatively minor in comparison to species losses during the substantial Late Pleistocene megafaunal extinctions, insular faunas have experienced massive-scale extinction events of varying complexity over the past few thousand years and into the historical period, as a direct result of the colonization of most of the world's major oceanic and oceanic-type islands (i.e. those possessing ecologically unbalanced and markedly endemic faunas) by prehistoric seafarers and settlers during this interval. All of the island systems to be colonized by modern humans during the Late Pleistocene or Pleistocene–Holocene boundary (Japan, south-east Asia, New Guinea and Solomon Islands, California Channel Islands, western Mediterranean islands) also experienced substantial extinctions following first human arrival (Martin 1984; Martin and Steadman 1999), and in some cases also suffered further losses during the Holocene caused by changes in the intensity and quality of anthropogenic environmental impacts. Direct human involvement in these events is strongly supported by the lack of temporal synchrony between Late Quaternary insular extinctions, even between different island groups within the same insular system (e.g. Mediterranean, Pacific), which typically match the sequential chronology of human colonization of these geographic regions rather than that of major environmental changes. The palaeontological and archaeological records have provided further extensive evidence of a range of anthropogenic impacts on island faunas and ecosystems (e.g. overhunting of now-extinct species, exotic introductions, and habitat alteration, degradation, and destruction), and prehistoric insular extinction events have been widely used as indirect corroborative evidence for human involvement in earlier continental Pleistocene extinctions for which fewer data are available on causative drivers (Martin 1984; Martin and Steadman 1999). However, most extinct island species are still typically known only from undated pre-human Late Quaternary deposits, and their extinction dynamics and chronologies must be inferred in large part from known dates for first human arrival and subsequent anthropogenic environmental impacts in different island systems (often by analogy with modern-era insular extinctions observed during the recent historical period), on the general assumption that the great majority of now-extinct species survived until the arrival of humans (see also Chapter 10 in this volume); this carries the danger of circularity for developing general models of insular extinction drivers.

Extinctions of insular mammals, birds, and other taxa (together with human-aided colonizations; see also Chapter 12 in this volume) have served to obscure many of the faunistic differences between different island groups (e.g. Palmer *et al.* 1999), and warn against the development of general biogeographic principles based on modern-day insular data in the absence of a fuller understanding of the pre-human diversity and ecology of these systems. In particular, the pre-human faunas of many oceanic-type island systems contained endemic megafaunal mammals, birds, and/or reptiles, which are all now extinct except for the giant tortoises of the Galápagos Islands and Aldabra (Burness *et al.* 2001). Many formerly widespread island species became regionally but not globally extinct across large parts of their range during the Holocene. Large numbers of species disappeared from islands but survived on neighbouring continental regions; for example, the Mascarene Islands lost their native populations of greater flamingo *Phoenicopterus roseus*, reed cormorant *Phalacrocorax africanus*, dugong *Dugong dugon*, and turtles *Chelonia mydas* and *Eretmochelys imbricata* following human arrival (Cheke and Hume 2008), and at least 50 of the continental European bird species present in Pleistocene and/or Holocene deposits from the islands of the western Mediterranean (Mallorca, Menorca, Eivissa, Corsica, Sardinia and Tavolara) became extinct on at least one island, with 22 now entirely restricted to neighbouring mainland regions (Alcover *et al.* 1999b). Both species and higher-order taxa still present on island systems (e.g. Pacific island birds) typically display restricted 'pseudo-endemic' distributions that lack any obvious evolutionary, geographical, or ecological basis, and instead reflect substantial range contractions caused by prehistoric anthropogenically driven extirpation from other islands in their pre-human ranges (Steadman 2006a; Fig. 2.3). In particular, many formerly widespread procellariiforms were extirpated across wide geographic regions by the introduction of rats by both prehistoric settlers and modern-era European explorers, but although a number of species now persist only on remote predator-free offshore islands representing a tiny fraction of their pre-human ranges, there have been relatively few global extinctions in this group (see Chapter 7 in this volume). More fundamentally in terms of global biodiversity loss, huge numbers of insular vertebrate species, representing both geologically recent island radiations and ancient relict lineages, experienced complete extinction following human colonization of different island systems during the Holocene. The major insular regions impacted by humans over the past 10000 years BP are detailed below.

2.3.1 Mediterranean islands

This intercontinental mega-archipelago possesses an excellent Late Quaternary vertebrate fossil record containing numerous extinct endemic species from faunistically distinct assemblages across the western, central, and eastern Mediterranean regions (Alcover *et al.* 1998, 1999b), including a

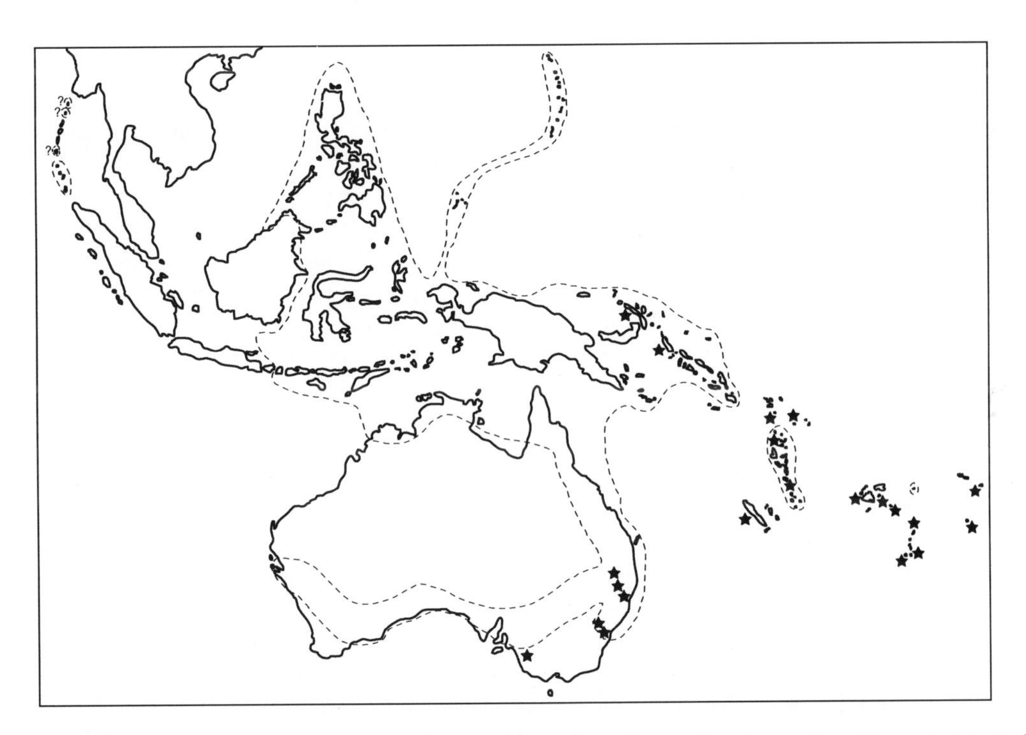

Figure 2.3 Modern-day and prehistoric distribution of megapodes (Aves: Megapodiidae, Sylviornithidae) across the Indo-Pacific region, demonstrating a substantial global range contraction caused by prehistoric human-caused extinctions on many islands. Modern-day distribution indicated by dotted line, fossil finds indicated by stars. After Jones *et al.* (1995) and Steadman (2006b).

substantial number of large-bodied and even megafaunal mammals displaying convergent insular adaptations such as dwarfing and reduction of distal limb elements. These include several species of hippopotami, dwarf proboscideans and cervids mainly from the eastern Mediterranean, the Balearic cave goat *Myotragus balearicus*, five genera of otters, and a canid (Table 2.2). The earliest reliable evidence for anatomically modern human presence in the insular Mediterranean (as opposed to earlier Pleistocene hominin species, which have also been reported from Sardinia; Alcover *et al.* 1998) varies between different island groups across the region; several islands (e.g. Cyprus, the Cyclades, Corsica and Sardinia) were first colonized around the Pleistocene–Holocene boundary or the beginning of the Holocene (>9000 years BP), but other large islands (Crete, Malta) lack evidence of human presence before *c.*7500 years BP, and the earliest reliable *terminus ante quem* for human arrival in the Gymnesic Islands is 3770±30 years BP (Alcover *et al.* 1999b; Ramis and Alcover 2005; Broodbank 2006). The Holocene mammal fauna of the insular Mediterranean also includes a range of continental Eurasian mammals that were introduced before the historical era, including not only rodents and domestic species but also insectivores, cervids, and several carnivore groups.

It remains extremely unclear whether the majority of the Mediterranean's endemic large-bodied mammals persisted into the Holocene or even the terminal Pleistocene, as limited collagen preservation under local environmental conditions has meant that direct radiometric last-occurrence dates remain unavailable for developing meaningful extinction chronologies for nearly all of these species. The Mediterranean region experienced a dynamic series of pre-human extinction events during the Plio-Pleistocene with varying impacts on different island groups, including both regional faunal turnovers and extinctions lacking subsequent faunal replacement, associated partly with changing sea levels and inter-island distances (Alcover *et al.* 1999b). It is likely that the end-Pleistocene marine transgression could have driven the extinctions of megafaunal species that required large island areas to support viable populations (e.g. elephants from Delos, Kithnos, Milos, Naxos and Serifos) in the absence of human hunting (cf. the St. Paul mammoth population). Data on Quaternary stratigraphical provenance are also lacking for many Mediterranean megafaunal species, raising the possibility that several taxa may in fact represent diachronic components of evolving lineages or sequential faunal replacements rather than temporarily sympatric elements of the region's end-Pleistocene biota.

The only robust data indicating the survival of any Mediterranean megafaunal mammal species into the Holocene are a series of direct Holocene radiometric dates for *M. balearicus*, with a last-occurrence date of 4785±40 years BP (Ramis and Alcover 2005), a recently published direct radiometric date of 6970±80 years BP (supported by uranium-series dating) for the Sardinian giant deer *Praemegaceros cazioti* (Benzi *et al.* 2007), and a date of 9240±550 years BP for the dwarf hippopotamus *Phanourios minor* on Cyprus (Simmons 1999). Although the late survival of *M. balearicus* reflects the late colonization of the Gymnesic Islands during the mid Holocene, there remains no evidence for human interaction with the species (Bover and Alcover 2003); V-trimmed horn cores initially interpreted as proof of human manipulation associated with domestication (to avoid damage to corralled or stabled animals) have also been reported from pre-human sites and are now recognized as evidence of natural osteophagic behaviour in the species (Bover and Ramis 2005). More controversially, direct bone dates have been reported in the older literature for undescribed elephants from 7090±680 and 4390±600 years BP on Tilos (Dodecanese, Greece), and from 9160±240 years BP on Kithnos (Cyclades, Greece), suggesting that some dwarf elephants may also have survived for millennia beyond human colonization of the Mediterranean and into the Aegean Bronze Age (Honea 1975; Symeonidis *et al.* 1973; Bachmayer and Symeonidis 1975; Bachmayer *et al.* 1976). It has also been suggested that a diminutive adult elephantid included in a frieze painted on an Eighteenth Dynasty pharaonic tomb (*c.*1475–1445 BC) may depict a late-surviving Mediterranean species of *Elephas* (Rosen 1994; Masseti 2001; Fig. 2.2, left-hand panel). However, it has proved impossible to verify these radiometric dates independently (Adrian Lister, personal communication 2007), and

Table 2.2 Late Pleistocene Mediterranean megafaunal and large-bodied mammal species. There is little direct radiometric evidence to suggest that most of these species occurred during the Holocene or terminal Pleistocene, but post-Pleistocene arrival of modern humans across most of the insular Mediterranean raises the possibility that the regional megafauna may have persisted into the Holocene. Dubious last-occurrence dates are indicated with asterisks.

Species	Occurrence	Direct radiometric last-occurrence date (years BP)		Reference
		End Pleistocene	Holocene	
Canidae				
Cynotherium sardous	Corsica, Sardinia	11 350±100		Ramis and Alcover (2005)
Mustelidae				
Algarolutra majori	Corsica, Sardinia			
Lutra euxena	Malta			
Lutrogale cretensis	Crete			
Megalenhydris barbaricina	Sardinia			
Sardolutra ichnusae	Sardinia			
Viverridae				
Genetta plesictoides	Cyprus			
Elephantidae				
Elephas chaniensis	Crete			
Elephas creutzbergi	Crete			
Elephas cypriotes	Cyprus			
Elephas sp. A	Tilos (Dodecanese)		4390±600*	Symeonidis *et al.* (1973); Bachmayer and Symeonidis (1975); Bachmayer *et al.* (1976)
Elephas sp. B	Delos, Kithnos, Milos, Naxos, Serifos (Cyclades)		9160±240*	Honea (1975)
Hippopotamidae				
Hippopotamus creutzbergi	Crete	12 135±485*		Reese *et al.* (1996)
Hippopotamus melitensis	Malta			
Phanourios minor	Cyprus		9240±550	Simmons (1999)
Cervidae				
Candiacervus cretensis	Crete			
Candiacervus major	Crete			
Candiacervus rethymensis	Crete			
Candiacervus ropalophorus	Crete			
Candiacervus sp. A	Crete			
Candiacervus sp. B	Crete			
Candiacervus cerigensis	Karpathos, Kasos			
Candiacervus pigadiensis	Karpathos, Kasos			
Cervus dorothensis	Crete			
Praemegaceros cazioti	Corsica, Sardinia		6970±80	Benzi *et al.* (2007)
Cervidae gen. et sp. A	Malta			
Cervidae gen. et sp. B	Amorgos (Cyclades)			
Bovidae				
Myotragus balearicus	Mallorca, Menorca, Cabrera		4785±40	Ramis and Alcover (2005)

further radiometric analysis is required to support the possibility of late survival of Mediterranean elephantids into the Holocene. The only well-documented potential evidence for direct hunting of Mediterranean megafauna by early settlers is from the Akrotiri *Aetokremnos* rockshelter on Cyprus, where burnt *Phanourios minor* and *Elephas cypriotes* remains have been reported from early human occupation horizons (Simmons 1999); however, this site may alternatively represent a natural fossil assemblage (Ramis and Alcover 2005). Undated fossil genet material tentatively assigned to the extinct species *Genetta plesictoides* is also known from Akrotiri *Aetokremnos*, but this site apparently spans the glacial–interglacial transition, and genet material is only known from the oldest stratigraphical unit (Simmons 1999).

The extinction chronologies of relatively few extinct endemic insular Mediterranean bird or small mammal species have been investigated radiometrically. Direct mid-Holocene radiometric last-occurrence dates are available for the semi-flightless Eivissa rail *Rallus eivissensis* and the extinct Balearic micromammals *Asoriculus hidalgoi* and *Eliomys morpheus*, corresponding with first human arrival on these islands, with the dated rail material from a stratigraphically associated assemblage also containing other regionally or globally extinct bird species (*Anser* sp. nov., *Grus grus*, *Haliaeetus albicilla*, *Pyrrhocorax pyrrhocorax*) (Alcover *et al.* 1999b; McMinn *et al.* 2005; Ramis and Alcover 2005; Bover and Alcover 2008). Other extinct small-bodied species are known to have persisted considerably beyond first human arrival in the insular Mediterranean. The latest stratigraphic occurrences of the extinct shrew *Asoriculus corsicanus*, field mouse *Rhagamys orthodon*, vole *Tyrrhenicola henseli*, and pika *Prolagus sardus* in owl-accumulated small vertebrate bone assemblages from northern Corsica are constrained between 1960±120 and 610±120 years BP, an interval which corresponds with both the introduction of the black rat *Rattus rattus* and a rapid intensification of agricultural activities and major vegetation clearance in the region (Vigne and Valladas 1996). There is considerable evidence for direct prehistoric human consumption of *P. sardus* during the early Holocene, and radiometric dates of modified *Prolagus* bones have been used to determine the earliest *terminus ante quem* for human presence in Corsica and Sardinia (Ramis and Alcover 2005), but historical eye-witness reports suggest that this species may have survived on the small Sardinian island of Tavolara as recently as AD 1774 (Alcover *et al.* 1999b; MacPhee and Flemming 1999). Other bird and micromammal species are known only from mid or Late Pleistocene deposits, but may be inferred uncontroversially to have persisted into the Holocene. Many of the other extinct endemic birds present in the pre-human Quaternary Mediterranean insular avifauna were raptors (*Aquila*, *Gyps*), strigiforms (*Athene*, *Bubo*, *Tyto*), or large-bodied waterfowl (*Anser*, *Cygnus*), which would have been particularly vulnerable to human hunting or indirect ecological impacts by early settlers (Alcover *et al.* 1999b).

2.3.2 West Indies

The tectonically complex terrestrial landmasses of the Caribbean basin contained a diverse pre-human Quaternary land mammal fauna, comprising over 100 endemic species of sloths, monkeys, and primitive insectivores, and a series of insular radiations of four families of small- and large-bodied rodents. Some of the sloths (*Megalocnus*) and the giant heptaxodontid rodent *Amblyrhiza inundata* from the Anguilla Bank probably exceeded 100 kg in size, whereas other sloths (*Neocnus*) were smaller than living arboreal species (McFarlane *et al.* 1998; MacPhee *et al.* 2000). West Indian land mammals have experienced the most severe extinctions of any Holocene mammal faunas, and today only two species of large solenodontid insectivore (*Solenodon cubanus* and *Solenodon paradoxus*) and approximately 10 species of capromyid rodents in the genera *Capromys*, *Geocapromys*, *Mesocapromys*, *Mysateles*, and *Plagiodontia* survive in the region. West Indian bats and aquatic mammals have also been impacted heavily during the Late Quaternary: over 30 chiropteran species have experienced extirpation of insular Caribbean populations, and nine species from Cuba, Hispaniola, and Puerto Rico have become globally extinct (Morgan 2001; Suárez and Díaz-Franco 2003; Mancina and García-Rivera 2005; Suárez 2005); and the Caribbean monk seal

Monachus tropicalis, which was last documented in the 1950s, is one of only two seal species to have disappeared as a result of human activity. The West Indian Quaternary avifauna has experienced numerous island-scale extirpations (Pregill and Olson 1981), and has also suffered almost 70 global species-level extinctions (of both regional endemics and species also recorded from the Pleistocene of continental America), including several species of giant owl (*Ornimegalonyx*, *Tyto*), a condor (*Gymnogyps*), a teratorn (*Teratornis*), several other large-bodied or giant raptors (*Amplibuteo*, *Gigantohierax*, *Titanohierax*), a crane (*Grus*), storks (*Ciconia*, *Mycteria*), an unusual flightless ibis (*Xenicibis*), and numerous parrots and rails. Other Caribbean taxa such as reptiles have also experienced either global extinctions or local extirpations of island populations during the Late Quaternary (e.g. Pregill 1981; Vélez-Juarbe and Miller 2007).

Although much remains to be learned about early human colonization of the Caribbean, radiometric dating of Archaic (pre-ceramic) archaeological sites and elevated levels of charcoal particles in sediment cores (interpreted as evidence of extensive anthropogenic forest burning soon after first contact) suggests that humans arrived in the Greater Antilles (with the apparent exception of Jamaica) around 6000 years BP, and in the Lesser Antilles around 4000 years BP (Burney *et al.* 1994; Wilson 1997). The region experienced further waves of human migration, population expansion, and cultural and socio-economic change during the period before first European arrival in AD 1492. This was associated with increasing anthropogenic pressures on terrestrial environments, as hunter/fisher/gatherer communities with subsistence economies primarily based on marine resources were replaced by fully agricultural populations who progressively colonized the interior highlands of the larger islands (Keegan 1992; Rouse 1992; Wilson 1997; Turvey *et al.* 2007a).

Other than for a few now-extinct birds and rodents that were collected as live specimens during the nineteenth and twentieth centuries, well-constrained last-occurrence dates are still unavailable for most extinct West Indian species, and the taxonomic status of several West Indian mammal groups (notably Cuban capromyid rodents and Lesser Antillean rice rats) needs to be clarified before meaningful extinction models can be developed for these species (see also Chapter 10 in this volume). Earlier models of West Indian extinctions suggested that many species disappeared as a result of shifts from xeric to more mesic habitats at the Pleistocene–Holocene boundary (Pregill and Olson 1981), and Morgan (2001) has suggested that most West Indian bat extirpations and extinctions may have been driven by changes in cave microclimates, or the disappearance of large cave systems through flooding from rising sea levels or erosional collapse, during this climatic transition. Uranium-series disequilibrium dates support a non-anthropogenic Late Pleistocene extinction for the megafaunal rodent *Amblyrhiza inundata*, probably caused by inundation of the Anguilla Bank at the end of the last glaciation (McFarlane *et al.* 1998). Pre-Holocene extinctions have also been postulated for other species including the Jamaican giant rodent *Clidomys osborni*, the Puerto Rican rodent *Puertoricomys corozalus*, and the Cuban monkey *Paralouatta varonai*, on the basis of the heavy fossilization of all known specimens and their absence from well-studied Late Quaternary deposits (Williams and Koopman 1951; Morgan and Wilkins 2003; MacPhee and Meldrum 2006). However, these hypotheses remain unsupported by adequate data. Instead, despite problems with collagen degradation under moist, humid subtropical environmental conditions (e.g. Turvey *et al.* 2007a), considerable recent progress in generating robust radiometric last-occurrence dates for a range of extinct land mammal species suggests that most West Indian mammals disappeared during two waves of extinction during the Holocene; it is likely that bird extinctions, which remain much more poorly constrained, followed a broadly similar pattern.

Most of the nesophontid island-shrews and several small- and medium-sized rodents (*Brotomys voratus*, *Heteropsomys insulans*, *Isolobodon portoricensis*, *Megalomys audreyae*) are now known on the basis of direct radiometric dates to have survived until around the time of European arrival in the West Indies, although none of these species has yet been shown to have survived well beyond first European contact and into the sixteenth century (Flemming and MacPhee 1999; MacPhee and Flemming 1999; MacPhee *et al.* 1999; McFarlane *et al.* 2000; Turvey *et al.* 2007a). Many additional extinct small- and medium-sized rodent and insectivore species, as well as the Jamaican monkey *Xenothrix mcgregori*, are stratigraphically associated with introduced *Rattus* material in superficial cave sediments, and so can similarly be interpreted with reasonable confidence to have survived into the European historical era (MacPhee and Flemming 1999). As well as *Isolobodon portoricensis* (which was introduced from Hispaniola by Amerindians across the islands of the Puerto Rican shelf and St. Croix), *Nesophontes edithae*, and *Heteropsomys insulans*, many undated smaller extinct land mammals and birds are also abundantly represented in ceramic-age archaeological sites, notably all of the undescribed oryzomyine rice rats of the Lesser Antilles and large *Nesotrochis* rails (Wetmore 1918; Quitmyer 2003; Newsom and Wing 2004). Although they were all exploited for food by Amerindians, the abundance and continued persistence of most species into culturally late horizons (dated to as recently as 595±30 years BP for a hearth containing the undescribed rice rat from Saba; Hoogland 1996) makes it very likely that they also survived until European arrival. It is probable that the extinctions of these species were driven by interactions (predation/competition) with *Rattus rattus*, which reached the Caribbean by the early 1500s, although the subsequent deliberate introduction of mongoose and massive-scale forest clearance for sugarcane and other crops could also have been a key factor in some native mammal and bird extinctions.

Several large-bodied land mammals, including the Hispaniolan monkey *Antillothrix bernensis*, the heptaxodontid rodent *Quemisia gravis*, and several sloths (e.g. *Neocnus comes*, *Paulocnus petrifactus*), have been reported from pre-Columbian archaeological sites in the older literature (e.g. Miller 1929; Hooijer 1963), but no published stratigraphic descriptions are available to indicate that extinct mammal material was genuinely associated with archaeological horizons, so these records cannot be used to support hypotheses of Late Holocene survival or Amerindian exploitation of these species. Recent research has not revealed any bones of large-bodied mammals displaying evidence of human modification or in an unquestionable archaeological context (excluding monk seals and manatees in coastal archaeological sites); evidence for pre-Columbian exploitation of large-bodied terrestrial species therefore remains minimal. However, recent direct radiometric studies on Cuban and Hispaniolan sloths have demonstrated the protracted survival of megafaunal mammals following first human arrival in the Greater Antilles (Steadman *et al.* 2005); the youngest known date for identified sloth material is 4190±40 years BP for the giant Cuban sloth *Megalocnus rodens*, approximately a millennium younger than the island's oldest dated archaeological site (5140±170 years BP) (MacPhee *et al.* 2007), and unidentified sloth material from Haiti has been directly dated to 3755±175 years BP (MacPhee *et al.* 2000). Other large-bodied mammals also apparently persisted for millennia beyond Amerindian colonization, as charcoal stratigraphically associated with the approximately 13 kg Puerto Rican rodent *Elasmodontomys obliquus* has been dated to 3512±28 years BP (Turvey *et al.* 2007a). At least some large mammals therefore appear to have become extinct in protracted sitzkrieg-style events rather than blitzkrieg-style overkill following initial Amerindian colonization. Although no large-bodied extinct West Indian birds have been adequately dated or are present in robust late stratigraphic or archaeological associations, the legend of the so-called chickcharnie on Andros Island, Bahamas, is frequently interpreted as reflecting the historical-era survival of the giant (1 m tall) flightless barn owl *Tyto pollens*. It is possible that the substantial reliance on marine resources shown in pre-ceramic archaeological sites may have limited the potential for massive exploitation and resultant rapid extinction of terrestrial vertebrate faunas by early Amerindian colonists on larger West Indian islands (Turvey *et al.* 2007a).

2.3.3 Madagascar

Human arrival in Madagascar is associated with the extinction of all of the island's megafauna, which included the gigantic elephant birds (*Aepyornis*) and related ratites (*Mullerornis*), giant tortoises (*Dipsochelys*), three species of pygmy hippopotamus (*Hippopotamus*), a leopard-sized fossa (*Cryptoprocta*), and at least 17 species of lemurs weighing ≥10 kg (ranging up to approximately 200 kg for the giant terrestrial lemur *Archaeoindris*; Simons 1997), comprising approximately one-third of the island's lemur species. Analysis of fossil spore densities of the coprophilous fungus *Sporormiella* from sediment cores across Madagascar suggests that the megafauna was apparently most abundant in semiarid coastal wooded savannas (Burney *et al.* 2003). In contrast to human-driven extinctions in the Mediterranean and West Indies, relatively few smaller-bodied (<10 kg) species disappeared during this extinction event; together with two enigmatic aardvark-like mammals (*Plesiorycteropus*), only two Malagasy rodent species (*Hypogeomys australis* and an undescribed species of *Nesomys*; Goodman *et al.* 2003) and a recently described species of shrew tenrec (*Microgale macpheei*; Goodman *et al.* 2007) are known to have become extinct.

Human arrival on Madagascar around 2300 years BP is indicated by abundant associated palynological and sedimentological evidence for human-caused landscape transformation that continued and escalated over subsequent centuries, including a decrease in palm pollen and an increase in pollen spectra of grasses and ruderal herbs, reflecting the progressive replacement of native forests by a mosaic of grassland, succulent bushland, and dry woodland; the presence of prehistorically introduced plants (e.g. *Cannabis*); a large spike in charcoal particles in lake and bog sediments; and culturally induced eutrophication of lakes driven by increased run-off from burnt landscapes (Burney 1999; Burney *et al.* 2004). Although this ongoing phase of anthropogenic environmental modification was preceded by a period of natural climatic aridity beginning approximately four millennia ago, which shifted Madagascar's vegetation balance to a more dry-adapted palm savanna mosaic between 3500 and 2500 years BP, extensive radiocarbon dating studies have now demonstrated the survival of nearly all of the island's megafaunal species until human arrival (Burney *et al.* 2004).

Whereas there is minimal evidence for direct human exploitation of extinct megafauna in the Mediterranean and West Indies, the earliest date for human presence in Madagascar is a radius of the extinct 'sloth lemur' *Palaeopropithecus ingens* dated to 2325±43 years BP with cut marks suggesting flesh removal by a sharp object (Burney *et al.* 2004), and human-modified bones, teeth, and eggs from a range of other extinct large-bodied mammals and birds (*Aepyornis, Hippopotamus, Daubentonia robusta, Pachylemur insignis*) indicative of both butchery and postmortem utilization (Perez *et al.* 2005) provide substantial evidence for direct exploitation of megafauna by proto-Malagasy settlers. Burney (1999) has proposed a sitzkrieg extinction scenario for the Malagasy megafauna, with a synergistic role for direct human hunting and landscape modification by anthropogenic fire and forest clearance, combined with natural climatic desiccation and loss of forage and water to introduced livestock. However, a drastic decrease in *Sporormiella* spores at 1720±40 years BP in sediment cores indicates that human arrival was followed almost immediately by a precipitous decline in megafaunal biomass, which predates both the sedimentological charcoal spike and the subsequent introduction of cattle dated from archaeological evidence and resurgence in *Sporormiella* (Burney *et al.* 2003). This suggests that the primary driver of megafaunal declines may have been prehistoric overhunting, and that changes in regional fire ecology may have resulted from an increased build-up of plant biomass in wooded savannas in the absence of cropping regimes from native herbivorous megafauna (a mechanism also proposed to explain landscape-level vegetation changes in Australia following prehistoric human arrival; Flannery 1994).

However, despite early regional extirpation or population reduction of Madagascar's megafauna, direct radiometric last-occurrence dates indicate the protracted persistence of many species after this initial period of decline. Notably, very recent dates of 213±40 and 99±36 years BP for *Hippopotamus laloumena*, 510±80 years BP for the 40–55 kg 'sloth lemur' *Palaeopropithecus ingens*, and 630±50 years BP for the 75+ kg 'koala lemur' *Megaladapis edwardsi* support the survival of these species into the European colonial period (Simons 1997; Burney *et al.* 2004). Several other megafaunal species also survived for at least a millennium or so after first human arrival, including *Aepyornis maximus, Dipsochelys abrupta*, and the giant lemurs *Archaeolemur* sp. (cf. *edwardsi*) and *Daubentonia robusta* (Burney *et al.*

2004). These dates support local traditions and legends of fantastic animals, and even rituals associated with their killing, that have been collected since European arrival and may be interpreted as relatively recent cultural memories of extinct megafauna. Brief, almost certainly second-hand descriptions of the *mangarsahoc* or *mangarotsaoka* (? = native hippopotamus), *vouron patra* or *vorom patrana* (? = *Aepyornis maximus*), *tretretretre* or *tratratratra* (? = *Megaladapis edwardsi* or *Palaeopropithecus ingens*), and *antamba* (? = *Cryptoprocta spelea*) were given by de Flacourt (1658; see Simons 1997; Burney and Ramilisonina 1998; MacPhee and Flemming 1999; Goodman *et al.* 2004; Kay 2004); either aepyornithids or the extinct Madagascar hawk-eagle *Stephanoaetus mahery* may have inspired Arabian legends of the *roc* (Goodman 1994); and putative reports of hippopotami and archaeolemurids have continued into the nineteenth and even twentieth centuries (Burney and Ramilisonina 1998).

It is likely that the delayed extinction of Madagascar's megafauna reflects the protracted, gradual pattern of human colonization and population growth across the island indicated by palynological and archaeological data. The oldest evidence for human arrival and environmental impact in Madagascar is in the south-west, several centuries to almost a millennium before the first evidence for their appearance elsewhere; however, this semiarid region may never have supported a large human population, and late radiometric dates for *Megaladapis* and *Palaeopropithecus* indicate co-occurrence of humans and megafauna for several centuries after first contact (Burney 1999). *Sporormiella* and charcoal spike sediment core signatures indicate that human-caused megafaunal extinctions and associated landscape transformation took several centuries to spread up the fairly accessible west coast to wetter climates, first appearing in the central highlands about 1400 years BP, and with remaining eastern and north-western low-elevation humid forests and higher elevation areas being penetrated relatively slowly; humans were absent from humid interior lowland sites in the north-west until 700 years BP, but high human population densities associated with deforestation, erosion, and eutrophication were achieved rapidly in this region (Burney 1999; Burney *et al.* 2004). The latest dates for megafauna come from inland sites, suggesting that relictual populations survived longest in remote and inaccessible regions of the interior, but this may be an artifact of incomplete sampling and a continuing relatively limited number of radiometric dates. However, although anthropogenic landscape transformation occurred relatively slowly following first human arrival in Madagascar, escalating human impacts continue to drive increasingly massive-scale environmental problems and species extinctions across the island today.

2.3.4 Pacific islands

Almost all of the islands of the tropical Pacific have been inhabited during the Late Holocene. Humans arrived in New Guinea and Near Oceania (the Bismarck Archipelago and the main chain of the Solomon Islands) in the Late Pleistocene as early as 33000 years BP, but it was not until 3600–3000 years BP that a pottery-making people known as the Lapita Cultural Complex spread from Taiwan, the Philippines and the Greater Sundas through the Maluku Islands across Near Oceania and into Remote Oceania (eastern Melanesia and all of Polynesia and Micronesia) (Steadman 2006b). The timing and pattern of this massive-scale human dispersal event remains the subject of considerable debate; the central South Pacific (Fiji, Tonga, Samoa) was colonized by about 2800 years ago, and accumulating evidence suggests that continued expansion into eastern Polynesia (e.g. Cooks, Societies, Marquesas, Hawai'i) may not have taken place until after AD 800, with the most remote islands (Easter Island, New Zealand) not colonized until around AD 1200 (Holdaway and Jacomb 2000; Hunt and Lipo 2006).

Other than the well-studied Quaternary fossil record for New Zealand, which has been the subject of intense scientific interest since the nineteenth century (Worthy and Holdaway 2002), prehuman faunas remained almost unknown from across the insular Pacific until recent decades. The region's only indigenous mammals are bats and murid rodents, which decrease in diversity eastward from Near to Remote Oceania. Although these two groups do not have a substantially wider distribution or species richness in the recent fossil record, increasing numbers of extinct Quaternary representatives of both groups continue to be documented from across the Indo-Pacific region, and it is likely that several more extinct species remain to be discovered. Extensive palaeontological and zooarchaeological research has now led to the published description of avifaunas from 224 Pleistocene and Holocene sites on 65 islands, representing 23 island groups distributed across Oceania (Steadman 2006b; excluding New Zealand and Hawai'i), and it is increasingly apparent that insular avifaunas experienced a massive-scale wave of extinctions associated with prehistoric human dispersal across the Pacific before the well-documented series of avian extinctions associated with the later historical-era arrival of Europeans and their commensal mammalian predators in the region. No widespread family of Pacific land-birds has been spared human-caused reductions in diversity and distribution, and losses of land-birds have been greatest for megapodes (Fig. 2.3), parrots, pigeons and rails. Notably, every island in the tropical Pacific with a good prehistoric record of birds, even small flat islands such as Wake and Laysan, has yielded bones of one to four endemic species of flightless rail in the genera *Amaurornis, Cabalus, Capellirallus, Diaphorapteryx, Fulica, Gallinula, Gallirallus, Gymnocrex, Pareudiastes, Poliolimnas, Porphyrio, Porzana, Rallina,* and/or *Vitirallus*, representing one of the most remarkable examples of avian evolutionary radiation (Steadman 1995, 2006b; Worthy and Holdaway 2002). Whereas fewer than 20 endemic Oceanic rail species survive today (excluding widely distributed 'tramp' species such as *Gallirallus philippensis* that show little geographical variation across or beyond the region), Steadman (1995) estimated that flightless rails alone may have accounted for as many as 2000 species of birds (approximately 20% of extant avian species diversity, and more than any other family of birds) until human arrival in the insular Pacific (see further discussion in Chapter 10 in this volume). Seabirds, especially shearwaters and petrels but also albatrosses, frigatebirds, boobies, terns, and gulls, have also experienced severe losses across the region (Steadman 2006b; see also Chapter 7 in this volume). Recognizing the magnitude and pattern of these extinctions highlights the need to consider the prehistoric record in order to develop a meaningful understanding of the ecological properties of insular faunas; for example, in contrast to anthropogenically disturbed modern Oceanic bird communities, few volant bird species were formerly naturally endemic to only one or two islands, many islands formerly supported two or three species within a single genus, and several widespread genera have now been extirpated from wide areas of Polynesia (Steadman 2006b).

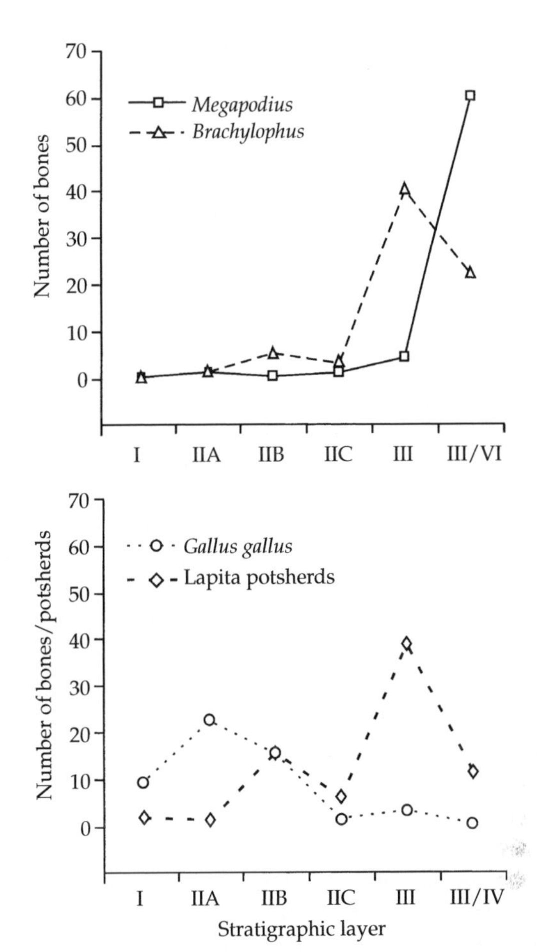

Figure 2.4 Stratigraphic distribution of bones of the extinct megapode *Megapodius alimentum*, the extinct iguana *Brachylophus gibbonsi*, introduced chicken *Gallus gallus*, and decorated (Lapita) potsherds between Layer III/IV (lowest cultural stratum) and Layer I (uppermost stratum) in Unit 10, Tongoleleka site, Lifuka, Tonga. From Steadman *et al.* (2002); ©2002 National Academy of Sciences, USA.

Much more unusual, highly endemic Holocene faunas also occurred on several Oceanic islands before human arrival, representing both ancient lineages and recent adaptive radiations. As well as extinct flightless waterfowl, rails, adzebills (predatory gruiforms), owlet-nightjars, and xenicids (basal passerines), New Zealand's Quaternary avifauna also formerly included 10 species of moa, representing two families of dinornithoid ratites that occupied a variety of large herbivore niches in the absence of native mammalian competitors and displayed extreme female-biased sexual size dimorphism; females of the largest species (*Dinornis novaezealandiae* and *Dinornis robustus*) weighed up to 240 kg and measured about 2 m tall at the back (Worthy and Holdaway 2002; Bunce *et al.* 2003). The Hawai'ian avifauna was originally characterized by radiations of endemic drepanidine finches and flightless ibises, and extraordinary goose-like ducks or *moa-nalo* (*Chelychelynechen, Ptaiochen, Thambetochen*) that were probably ecologically analogous to giant tortoises (Olson and James 1991). The larger islands of Melanesia and the central South Pacific also contained several megafaunal birds and reptiles, notably a huge (30–40 kg) megapode-like bird (*Sylviornis*) on New Caledonia, a dodo-sized flightless pigeon (*Natunaornis*) on Fiji, giant horned cryptodiran turtles (*Meiolania*) on New Caledonia and Lord Howe Island, a radiation of terrestrial mekosuchine crocodiles (*Mekosuchus, Volia*) on Vanuatu, New Caledonia, and Fiji, and giant iguanas (*Brachylophus, Lapitiguana*) on Fiji and Tonga.

Relatively little is known about the magnitude and duration of prehistoric Quaternary avian extinctions in the large, topographically complex islands of Near Oceania, although the available data suggest a protracted series of losses with many now-extinct species persisting well into the Holocene (Steadman *et al.* 1999). In contrast, increasing numbers of well-dated archaeological sites across Remote Oceania indicate that prehistoric human arrival on these islands was frequently associated with the almost instantaneous disappearance of a range of endemic species, even on the largest islands such as the South Island of New Zealand, sometimes over a time interval equal to or less than the statistical error associated with radiometric dating and therefore too rapid to be resolved by this technique (Holdaway and Jacomb 2000; Steadman *et al.* 2002b; Steadman and Martin 2003; Fig. 2.4). The subfossil record reveals acute losses of species across a range of body size classes, probably in response to both direct human overexploitation of large-bodied species, and predation of the eggs and nestlings of terrestrial and/or ground- and burrow-nesting species by introduced Pacific rats *Rattus exulans* and other exotic mammalian predators (Holdaway 1999b; Cassey 2001; Worthy and Holdaway 2002). In contrast to the pattern apparently shown in the West Indies, archaeological data indicate that native birds and other terrestrial species initially constituted a substantial proportion of the diet of the first settlers, often more than half of all vertebrate food, although this overexploitation declined drastically following avian population crashes and extinctions (Steadman 1995). For example, the presence of moa remains was formerly used to characterize the earliest (so-called Moa-Hunter) period of human occupation in New Zealand, with later sites lacking evidence of moa exploitation and instead characterized by greater reliance on marine and other resources (Duff 1950; Holdaway and Jacomb 2000). Analysis of subfossil material from associated archaeological and natural sites in New Zealand demonstrates that now-extinct bird species were more selectively and intensively targeted for food than surviving species by early Polynesian hunters (Duncan *et al.* 2002). Extensive deforestation also followed Polynesian arrival on many Pacific islands (Rolett and Diamond 2004), although the extent to which prehistoric habitat modification drove regional extinctions remains unclear (Blackburn *et al.* 2004; Didham *et al.* 2005). Many extinctions and extirpations in other insular groups (e.g. reptiles) also correlate with prehistoric human arrival across the Pacific, but evidence of direct predation is generally lacking except for some large-bodied species such as the extinct Tongan iguana *Brachylophus gibbonsi* (Steadman *et al.* 2002b; Steadman 2006b).

2.4 Conclusions: explaining Holocene species losses

In total, 523 bird species and 255 mammal species recorded from the subfossil or historical records are currently suspected to have become extinct during the Holocene (see Chapter 3 and 4 in this volume; excluding contentious large-bodied and megafaunal Holocene survivors from South America, China, and the insular Mediterranean), and this number is certain to increase as further research is carried out into both insular and continental Quaternary faunas. Only an extremely small proportion of Holocene species or population losses can even questionably be interpreted as non-anthropogenic events. For example, Steadman *et al.* (1991) suggested that of the 29–34 vertebrate populations or species lost in the Galápagos during the Holocene, only three (land iguana *Conolophus subcristatus*, gecko *Phyllodactylus* sp., and rice rat *Nesoryzomys* sp. on Rábida) are not known from the period of human occupation from either the historical record or 'modern' radiometric dates, and even these possible pre-human losses may reflect poor sampling rather than background extinctions. Similarly, although several authors have suggested that the Martinique giant rice rat *Megalomys desmarestii* was wiped out by the eruption of Mount Pelée in 1902 (e.g. Allen 1942; Balouet and Alibert 1990; Flannery and Schouten 2001), it is more likely that this extinction event on an island with a long volcanic history was instead primarily driven by anthropogenic impacts, notably the introduction of the Indian mongoose *Herpestes javanicus* a few years earlier in 1889 (Horst *et al.* 2001). The disappearance of the St. Paul mammoth population represents one of the few Holocene vertebrate extinction events that was probably mediated by natural environmental change rather than anthropogenic activity. However, the stochastic disappearance of this mammoth population would have had little relevance to global mammoth extinction if the species had not already died out across the rest of its range. However, it remains possible that other species-level extinctions of insular vertebrates may have been driven by similar sea-level changes during the early Holocene, such as the flooding of the shallow lagoon flats around Rodrigues, which were twice the size of the current island, around 7000 years ago (Cheke and Hume 2008).

Considerable further research is still required to clarify both the pattern (extinction chronologies) and process (causative drivers) of extinctions during the Holocene. Limited historical or radiometric records may provide little meaningful information about the final disappearance of a species or the first appearance of different possible extinction drivers, posing major challenges when attempting to infer cause and effect, and many perceived regional differences undoubtedly reflect major variation in the quality and resolution of available data on regional Holocene chronologies (see also Chapter 10 in this volume). However, differing levels of prehistoric human impacts are evident across different geographical regions and taxonomic or ecological groups, notably between the protracted and staggered population declines shown by faunas on continental systems and some larger islands (e.g. Madagascar, Greater Antilles, Near Oceania) during the Holocene, and the massive-scale and apparently rapid extinction events both on other island systems during the Holocene and across continental regions during the Late Pleistocene.

A series of intrinsic biological factors that correlate with increased risk of extinction were shared by many of the bird and mammal species that have become extinct during the Holocene. Modern and well-dated prehistoric faunas show that large-bodied vertebrates are typically the first species to become extinct following human arrival or subsequent cultural change on both continents and islands (e.g. Duncan and Blackburn 2004), either through selective targeting by hunters or because size-dependent scaling of ecological and life-history traits (e.g. population size, reproductive rate) increases their vulnerability to anthropogenic disturbance (Cardillo *et al.* 2005). For example, whereas the growth periods of living birds are apomorphically shortened to less than a year, New Zealand's moa frequently display annual growth marks in their cortical bone (Fig. 2.5), indicating that they took several years (in some cases almost a decade) to reach skeletal maturity (Turvey *et al.* 2005). Reproductive maturity was therefore extremely delayed relative to all extant birds and population recruitment would have been very slow, leaving them highly vulnerable to early human colonists. The absence of native mammalian predators in most oceanic-type island systems is also associated with the evolution of a suite of characteristic adaptations in insular birds and mammals,

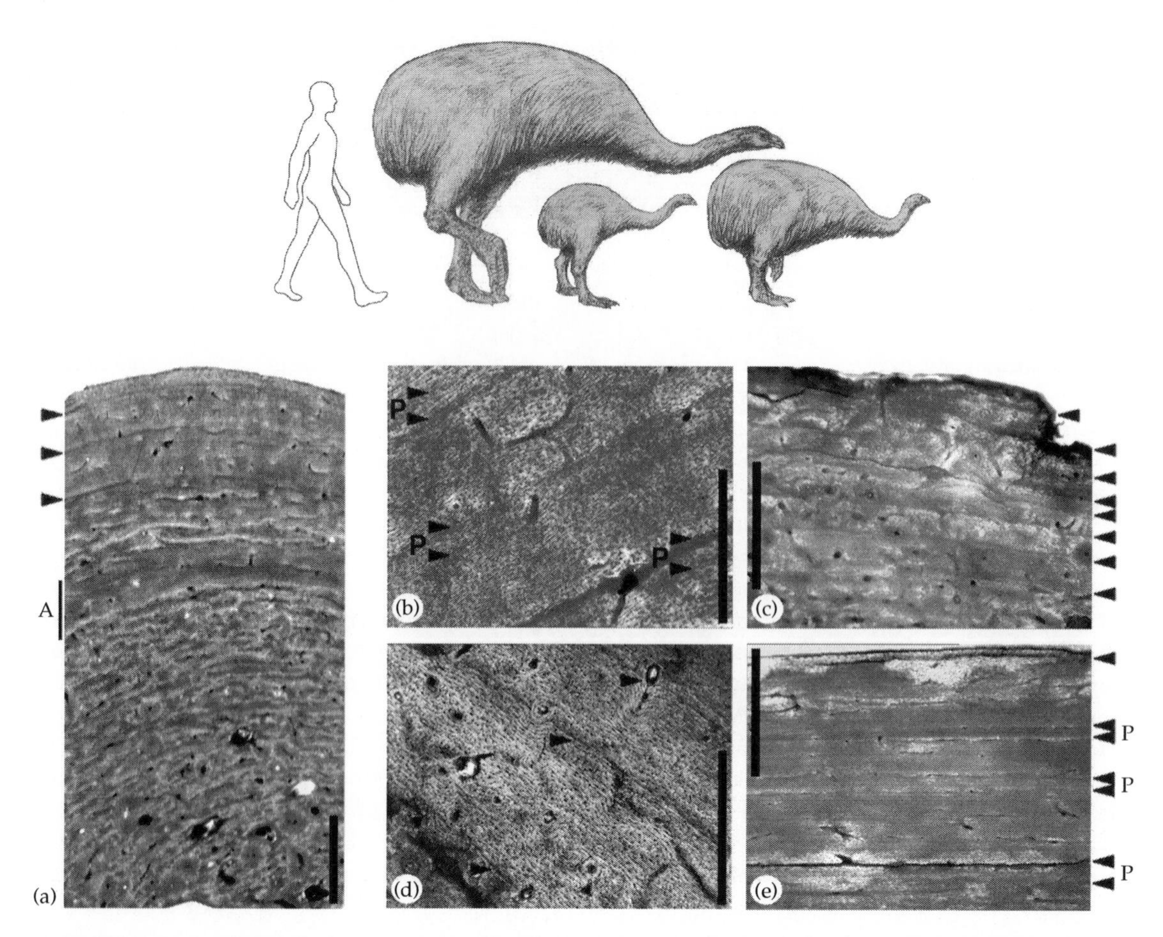

Figure 2.5 Top: reconstructions of dinornithid and emeid moa species, with a 1.8 m-tall person for scale. Left to right: female *Dinornis novaezealandiae* (Dinornithidae), *Megalapteryx didinus* (Emeidae), and *Pachyornis elephantopus* (Emeidae). Images courtesy of Colin Edgerley and New Zealand Geographic. Bottom: transverse histological thin sections through adult moa long bones, showing annual cortical growth marks. The bone periphery is at the top. (a, b) *Megalapteryx didinus* tibiotarsus: (a) cortical cross-section, showing lamellar annulus (A) in the mid-cortex and three lines of arrested growth in a second annulus in the outer cortex (arrows); (b) close-up of three paired lines of arrested growth (LAGs). (c) *Anomalopteryx didiformis* tibiotarsus, showing eight single LAGs. (d) *M. didinus* tibiotarsus, showing three single LAGs. (e) *A. didiformis* tibiotarsus, showing four single or paired (P) LAGs. Scale bars: 1 mm (a, c), 0.5 mm (b, d). From Turvey *et al.* (2005).

notably 'approachability' or naïve behaviours lacking defensive or escape responses, flightlessness in insular birds (and analogous adaptations in other groups, such as semi-terrestriality in the West Indian primate *Paralouatta varonai*; MacPhee and Meldrum 2006), and the development of strongly *K*-selected life-history strategies compared to closely related continental taxa (e.g. New Zealand takahe *Porphyrio hochstetteri* compared with the cosmopolitan purple swamphen *P. porphyrio*; see Lee and Jamieson 2001); these would have left insular species highly vulnerable to extinction following the arrival of humans and exotic mammalian predators. The tendency for protracted life histories in insular species emphasizes the fact that reproductive rate irrespective of body size is a key biological trait directly related to risk of extinction, with 'bradyfauna' rather than megafauna representing the most vulnerable class of species (Johnson 2002). A striking example is provided by the differing status of otherwise ecologically similar insectivorous mammals in the West Indies and Madagascar. Over 80% of West Indian Quaternary insectivore species have become extinct and the two surviving species of *Solenodon* are both classified as Endangered by IUCN (2008), whereas Madagascar has lost only one of its 29 species of tenrecs (Goodman *et al.* 2007). Although solenodons have only two litters of one or two young each year, relatively long gestation and lactation periods and protracted lifespans characteristic of island species (Symonds 2005), tenrecs have extremely fast life-histories with up to 32 young per litter (the largest known mammalian litters) and associated low adult survivorship, possibly as an evolutionary response to climatic unpredictability as well as the presence of native mammalian predators in Madagascar (Dewar and Richard 2007).

Probability of extinction is also strongly influenced by extrinsic environmental and cultural variables, which further help to explain the observed pattern of species losses across the Holocene. The only species with slow reproductive rates to survive the series of Late Quaternary megafaunal extinction events were arboreal, nocturnal, and/or lived in closed habitats, factors that would all have reduced their exposure to direct interaction with humans (Johnson 2002). Steadman and Martin (2003) qualitatively reviewed the extrinsic factors affecting speed and extent of anthropogenic extinction of birds on islands across the Pacific after human arrival, to develop the ABC (abiotic, biotic, cultural) model of extinction risk. Probability of survival increases on larger islands with nutrient-poor soils unsuitable for cultivation, reliably wet forests that are difficult to burn, and steep, rugged karst limestone (e.g. *makatea* limestone) or knife-edge volcanic bedrock that provides refugia for native species and deters deforestation and human access, and where human populations are temporary (e.g. seasonal hunting or fishing outpost), restricted to coastal regions, and practise hunter/fisher/gatherer rather than horticultural subsistence. Overexploitation of specific native species generally decreases on less isolated islands with high levels of marine and terrestrial biodiversity, where human populations are not dependent upon a single resource sink, although high renewable resource levels also support greater human population growth. Bird species with drab plumage, a bad taste, and short, curved bones of little use for tool-making would also be expected to experience less persecution. The importance of prehistoric cultural variation in determining the rate and magnitude of Holocene extinctions is illustrated by the rapid extinction of all species of moa, which not only had extremely protracted life histories but were also preferentially hunted—and wastefully overexploited—by Maori colonists (Trotter and McCulloch 1984; Duncan *et al.* 2002), resulting in their disappearance within 100 years of Polynesian settlement of New Zealand (Holdaway and Jacomb 2000). Conversely, the significance of overhunting as a major factor in megafaunal extinctions remains disputed for other island systems (e.g. Madagascar, West Indies) because archaeological evidence suggests that early colonists were fishermen, herders, and agriculturalists rather than big game hunters (in contrast to New Zealand's early Moa-Hunter culture), and although hunting does not need to have been a culturally important practice to play a significant part in the extinction of species with a reduced capacity for rapid population recruitment, less intensive exploitation may have resulted in the more protracted megafaunal declines observed in these regions.

Acknowledgements

I am extremely grateful to Ken Aplin, John Borg, Cyrian Broodbank, Anthony Cheke, Kristofer Helgen, Victoria Herridge, Thomas Higham, Mike Hoffmann, Robert Ingle, Adrian Lister, Ross MacPhee, Colin McEwan, Claire Mellish, Gary Morgan, George Theodorou, and Tim Wacher for invaluable discussion about extinction patterns and processes, and for providing crucial information on poorly known faunas. Support was provided by a NERC Postdoctoral Fellowship.

CHAPTER 3

Holocene mammal extinctions

Samuel T. Turvey

This table presents last-occurrence dates for 255 extinct mammal species that are currently known or suspected to have died out during the Holocene. This includes species known to have survived into the historical period, and those represented in Late Quaternary subfossil deposits known or believed to be Holocene in age, or from island systems which humans did not reach until the Holocene. The table excludes several large-bodied mammal species from South America, China and the insular Mediterranean for which Holocene survival has been suggested based on questionable radiometric dates (see Chapter 2 in this volume), and also species for which wild progenitors are extinct but domestic derivates still survive (e.g. *Bos primigenius*; see Gentry *et al.* 2004 for further discussion). Only the youngest known occurrence dates are given for each species, or for sites from which the species has been reported. Taxonomic order and validity of historical-era species follow Wilson and Reeder (2005). Calibrated dates are given as either AD/BC (if the date range falls at least partly within the past 2000 years) or BP. An asterisk indicates that calibrated dates have not previously been published, and have been calculated for this chapter using OxCal 4.0 (Bronk Ramsey 1995, 2001). In radiometric dating, the 1 σ age range corresponds to 68.2% probability and the 2 σ age range corresponds to 95.4% probability.

Scientific name	Common name	Range	Last occurrence						Reference
			Historical record	Radiometric date			Archaeological or stratigraphic association	Other estimate (see list in footnotes; see also Notes)	
				^{14}C age (years BP), ±1 σ	Calibrated date, ±2 σ	Direct date?			
DIDELPHIMORPHIA									
Didelphidae									
Cryptonanus ignitus	Red-bellied gracile mouse opossum	Argentina	1962						Díaz *et al.* (2002)
DASYUROMORPHIA									
Thylacinidae†									
Thylacinus cynocephalus	Thylacine, Tasmanian tiger	Tasmania	1936						Rounsevell and Mooney in Strahan (1995); Guiler (1985)
		Mainland Australia	(1830s–1840s?)	3280±90	3816–3336 BP*	Yes			Partridge (1967); Paddle (2000)
Thylacinus sp. ?nov.		New Guinea		9200±200	10 490–10 210 BP	No			van Deusen (1963); Menzies (1977); Note 1
PERAMELEMORPHIA									
Thylacomyidae									
Macrotis leucura	Lesser bilby	Australia	1950s						Burbidge *et al.* (1988)
Chaeropodidae†									
Chaeropus ecaudatus	Pig-footed bandicoot	Australia	1950s						Burbidge *et al.* (1988)
Peramelidae									
Perameles eremiana	Desert bandicoot	Australia	1960s						Burbidge *et al.* (1988)
'*Peroryctes*' sp. nov.		Pulau Kobroor (Aru Islands)		9400±50	11 055–10 432 BP	No			Aplin and Pasveer (2005); K. Aplin (personal communication)
Peramelidae gen. et sp. nov.		Halmahera (Maluku)		5170±100 to 3410±70	6190–5663 BP to 3836–3479 BP*	No			Flannery *et al.* (1995); Note 2
DIPROTODONTIA									
Pseudocheiridae									
Petauroides ayamaruensis		New Guinea		6900±80	7891–7546 BP	No			Pasveer (1998); Aplin *et al.* (1999)

Petauridae								
Dactylopsila kambuayai		New Guinea		6900±80	7891–7546 BP	No		Pasveer (1998); Aplin *et al.* (1999)
Potoroidae								
Bettongia pusilla	Nullarbor dwarf bettong	Australia					2, 4	McNamara (1997); Johnson (2006)
Caloprymnus campestris	Desert rat-kangaroo	Australia	1935 (1980s?)					Carr and Robinson (1997)
Potorous platyops	Broad-faced potoroo	Australia	1875					Kitchener in Strahan (1995)
Macropodidae								
Lagorchestes asomatus	Central hare-wallaby	Australia	*c.*1960					Burbidge *et al.* (1988)
Lagorchestes leporides	Eastern hare-wallaby	Australia	1890					Strahan in Strahan (1995)
Macropus greyi	Toolache wallaby	Australia	1939 (1972?)					Robinson and Young (1983); Smith in Strahan (1995)
Onychogalea lunata	Crescent nailtail wallaby	Australia	1956					Burbidge in Strahan (1995)
Thylogale christenseni		New Guinea		3250±80	3688–3335 BP*	No		Hope (1981); Hope *et al.* (1993); Flannery (1995b)
Thylogale sp. nov.		New Guinea		3250±80	3688–3335 BP*	No		Hope (1981); Hope *et al.* (1993); Flannery (1995b)
AFROSORICIDA								
Tenrecidae								
Microgale macpheei		Madagascar		2480±40	2739–2359 BP*	No		Goodman *et al.* (2007)
BIBYMALAGASIA†								
Plesiorycteropidae†								
Plesiorycteropus germainepetterae		Madagascar		Note 3			1	
Plesiorycteropus madagascariensis		Madagascar		Note 3			1	
PROBOSCIDEA								
Elephantidae								
Mammuthus primigenius	Woolly mammoth	Wrangel Island (Arctic Sea)		3730±40	4141–3987 BP	Yes		Vartanyan *et al.* (1993)

Table *Continued*

Scientific name	Common name	Range	Last occurrence						Reference
			Historical record	Radiometric date			Archaeological or stratigraphic association	Other estimate (see list in footnotes; see also Notes)	
				^{14}C age (years BP), ±1 σ	Calibrated date, ±2 σ	Direct date?			
		St. Paul Island (Bering Sea)		*c.*5700	*c.*6500 BP	Yes			Crossen *et al.* (2005)
		Taimyr Peninsula (Siberia)		9670±60	11 200–11 040 BP	Yes			MacPhee *et al.* (2002)
Elephas cypriotes		Cyprus		9420±550	12 649–9249 BP	No			Simmons (1999); Ramis and Alcover (2005)
SIRENIA									
Dugongidae									
Hydrodamalis gigas	Steller's sea cow	Bering and Copper Island (Bering Sea)	1768						Stejneger (1887)
		Attu Island (Aleutians)	post-1750?						Domning *et al.* (2007)
PILOSA									
Megalonychidae									
Acratocnus antillensis		Cuba						1	
Acratocnus odontrigonus		Puerto Rico, Antigua	3330±50	3688±3450	BP	NO		1	Steadman *et al.* (1984)
Acratocnus simorhynchus		Hispaniola						1	
Acratocnus ye		Hispaniola		Note 4				1, Note 5	
Paulocnus petrifactus		Curaçao (Caribbean)						1	
Neocnus comes		Hispaniola		4391±42; Note 4	4840–5260 BP	Yes		Note 5	Steadman *et al.* (2005)
Neocnus dousman		Hispaniola		Note 4				1, Note 5	
Neocnus gliriformis		Cuba						1	
Neocnus major		Cuba						1	
Neocnus toupiti		Hispaniola		Note 4				1, Note 5	

Megalocnus rodens		Cuba		4190±40	4580–4840 BP	Yes		MacPhee *et al.* (2007)
Megalocnus zile		Hispaniola					1	
Parocnus browni		Cuba		4960±280	6350–4950 BP	Yes		Steadman *et al.* (2005)
Parocnus serus		Hispaniola		Note 4			1	
Galerocnus jaimezi		Cuba					1	
Paramiocnus riveroi		Cuba					1	
PRIMATES								
Lemuridae								
Pachylemur insignis		Madagascar		1220±50	AD 715–985	Yes		Burney *et al.* (2004)
Pachylemur jullyi		Madagascar					1	
Megaladapidae†								
Megaladapis madagascariensis		Madagascar		Note 6			1	
Megaladapis grandidieri		Madagascar		Note 6			1	
Megaladapis edwardsi	Koala lemur; ?tretretretre	Madagascar	pre-1658?	630±50	AD 1280–1420	Yes		MacPhee and Flemming (1999); Burney *et al.* (2004)
Palaeopropithecidae†								
Archaeoindris fontoynonti		Madagascar		2291±55	2362–2149 BP	Yes		Burney *et al.* (2004)
Babakotia radofilai		Madagascar		4400±60	5290–4840 BP	Yes		Burney *et al.* (2004)
Mesopropithecus globiceps		Madagascar		1694±40	AD 245–429	Yes		Burney *et al.* (2004)
Mesopropithecus pithecoides		Madagascar		1410±40	AD 570–679	Yes		Burney *et al.* (2004)
Mesopropithecus dolichobrachion		Madagascar					1	
Palaeopropithecus ingens	Sloth lemur; ?tretretretre	Madagascar	pre-1658?	510±80	AD 1300–1620	Yes	4	Simons (1997); Burney *et al.* (2004)
Palaeopropithecus maximus		Madagascar					1	
Palaeopropithecus sp. nov.		Madagascar					1	Gommery *et al.* (2004)
Archaeolemuridae†								
Archaeolemur edwardsi		Madagascar		2060±70; Notes 7, 9	332 BC–AD 110	Yes	4?	Burney and Ramilisonina (1998); Burney *et al.* (2004)

Table *Continued*

Scientific name	Common name	Range	Last occurrence						Reference
			Historical record	Radiometric date			Archaeological or stratigraphic association	Other estimate (see list in footnotes; see also Notes)	
				^{14}C age (years BP), ±1 σ	Calibrated date, ±2 σ	Direct date?			
Archaeolemur majori		Madagascar		1650±50; Notes 8, 9	AD 370–575	Yes		4?	Burney and Ramilisonina (1998); Burney *et al.* (2004)
Hadropithecus stenognathus		Madagascar		1413±80	AD 444–772	Yes		4?	Burney and Ramilisonina (1998); Burney *et al.* (2004)
Daubentoniidae									
Daubentonia robusta		Madagascar		1065±40	AD 891–1027	Yes			Burney *et al.* (2004)
Cebidae									
Antillothrix bernensis		Hispaniola		3850±135	4789–3872 BP*	No		Note 5	Rímoli (1977)
Paralouatta varonai		Cuba						1, 5	MacPhee and Meldrum (2006)
Xenothrix mcgregori		Jamaica	1769?					3	MacPhee and Fleagle (1991)
Atelidae									
Caipora bambuiorum		Brazil						2, 5	Cartelle and Hartwig (1996); Hartwig and Cartelle (1996)
Protopithecus brasiliensis		Brazil						2, 5	Cartelle and Hartwig (1996); Hartwig and Cartelle (1996)
LAGOMORPHA									
Ochotonidae									
Ochotona whartoni		Eastern North America		8670±220	10 251–9140 BP*	Yes			Mead and Grady (1996)
Prolagus sardus	Sardinian pika	Corsica		1960±120–610±120	322 BC–AD 283 to AD 1224–1459	No			Vigne and Valladas (1996)
		Tavolara (Sardinia)	1774						Cetti (1777)
SORICOMORPHA									
Nesophontidae[†]									
Nesophontes edithae		Puerto Rico		990±24	AD 991–1153	Yes			Turvey *et al.* (2007a)

		Virgin Islands					Cinnamon Bay, St. John: AD 1000–1490		Quitmyer (2003)
Nesophontes hypomicrus		Hispaniola		790±50	AD 1175–1295	Yes		3	MacPhee *et al.* (1999)
Nesophontes major		Cuba		7864±96	8993–8453 BP	No		3?	Jiménez Vázquez *et al.* (2005)
Nesophontes micrus		Cuba		590±50	AD 1295–1430	Yes		3	MacPhee *et al.* (1999)
Nesophontes paramicrus		Hispaniola		680±50	AD 1265–1400	Yes		3	MacPhee *et al.* (1999)
Nesophontes zamicrus		Hispaniola		590±50	AD 1295–1430	No		3?	MacPhee *et al.* (1999)
Nesophontes sp. nov. A		Cayman Brac						1, 3	
Nesophontes sp. nov. B		Grand Cayman						1, 3	
Solenodontidae									
Solenodon arredondoi		Cuba						1	
Solenodon marcanoi		Hispaniola						1, 3	
Soricidae									
Crocidura balsamifera		Egypt		2400±140	2771–2121 BP*	Yes			Hutterer (1994)
Asoriculus corsicanus		Corsica		1960±120–610±120	322 BC–AD 283 to AD 1224–1459	No			Vigne and Valladas (1996)
Asoriculus hidalgoi		Mallorca, Menorca, Cabrera (Mediterranean)		4280±50	4979–4639 BP	Yes			Ramis and Alcover (2005)
Asoriculus similis		Sardinia						1	
CHIROPTERA									
Pteropodidae									
Pteropus allenorum	Small Samoan flying fox	Upolu (Samoa)	1856						Helgen *et al.* (2009)
Pteropus brunneus	Dusky flying fox	Percy Island (Australia)	1874						Conder in Strahan (1995)
Pteropus coxi	Large Samoan flying fox	Samoa	1840						Helgen *et al.* (2009)
Pteropus pilosus	Large Palau flying fox	Palau	pre-1874						Flannery (1995a)
Pteropus subniger	Lesser Mascarene flying fox	Mauritius, Réunion (Mascarenes)	1870s?						Cheke and Hume (2008)
Pteropus tokudae	Guam flying fox	Guam	1974?						Flannery (1995a)

Table *Continued*

Scientific name	Common name	Range	Last occurrence						Reference
			Historical record	Radiometric date			Archaeological or stratigraphic association	Other estimate (see list in footnotes; see also Notes)	
				^{14}C age (years BP), ±1σ	Calibrated date, ±2σ	Direct date?			
Nyctimene sanctacrucis	Nendö tube-nosed fruit bat	Nendö Island (Solomon Islands)	1907?						MacPhee and Flemming (1999)
Mystacinidae									
Mystacina robusta	Greater short-tailed bat	New Zealand	1967					6a, c-d	Lloyd (2001)
Phyllostomidae									
Desmodus draculae	Giant vampire bat	Argentina	pre-1820?	290±40	AD 1482–1795*	No			Waterton (1825); Pardiñas and Tonni (2000)
Desmodus puntajudensis		Cuba		7864±96; Note 10	8993–8453 BP	No			Jiménez Vázquez *et al.* (2005)
Desmodus stocki		San Miguel Island (California Channel Islands)					Note 11	5	Guthrie (1980)
Cubanycteris silvai		Cuba						1	
Phyllonycteris major		Puerto Rico, Antigua		3330±50	3688–3450 BP*	No		1	Steadman *et al.* (1984)
Artibeus anthonyi		Cuba		7864±96	8993–8453 BP	No			Jiménez Vázquez *et al.* (2005)
Phyllops silvai		Cuba						1	
Phyllops vetus		Cuba						1	
Mormoopidae									
Mormoops magna		Cuba		7864±96	8993–8453 BP	No			Jiménez Vázquez *et al.* (2005)
Pteronotus pristinus		Cuba						1	
Pteronotus sp. nov.		Hispaniola							Morgan (2001)
Vespertilionidae									
Myotis insularum	Samoan myotis	Samoa	1860s						Helgen *et al.* (2008)
Nyctophilus howensis	Lord Howe long-eared bat	Lord Howe Island (Australia)	1887						Etheridge (1889); McKean (1973)

Pharotis imogene	New Guinea big-eared bat	New Guinea	1890					6a	Flannery (1995b)
Vespertilionidae gen. et sp. indet.		Maui (Hawai'i)		7750±500	9905–7656 BP*	No			James *et al.* (1987); Ziegler (2002)
Family indet.									
Boryptera alba		Réunion (Mascarenes)	1801					7	Cheke and Hume (2008)
CARNIVORA									
Eupleridae									
Cryptoprocta spelea	Giant fossa	Madagascar	pre-1658?					1, 4	Goodman *et al.* (2004)
Canidae									
Dusicyon australis	Falkland Islands wolf, warrah	Falkland Islands	1876						Allen (1942)
Dusicyon avus		Argentina		4865±65	5746–5335 BP*	No			Mazzanti and Quintana (1997)
Otariidae									
Zalophus japonicus	Japanese sea lion	Sea of Japan	1951						Rice (1988)
Phocidae									
Monachus tropicalis	Caribbean monk seal	Caribbean	1952					6d	Adam and Garcia (2003)
Mustelidae									
Neovison macrodon	Sea mink	Gulf of Maine (North America)	1860 (1894?)						Allen (1942); Mead *et al.* (2000)
PERISSODACTYLA									
Equidae (Note 12)									
Equus hydruntinus	European wild ass, zebro	Spain, Portugal	1293				Recorded from Iron Age sites in Europe and Iran		Antunes (2006); Orlando *et al.* (2006b); Vila (2006)
ARTIODACTYLA									
Hippopotamidae									
Hippopotamus laloumena		Madagascar	(1976?)	213±40, 99±36; Notes 13, 14	AD 1639–1950	Yes		4	Burney and Ramilisonina (1998); Burney *et al.* (2004)
Hippopotamus lemerlei		Madagascar	(1976?)	1740±50; Note 13	AD 155–415	Yes		4	Burney and Ramilisonina (1998); Burney *et al.* (2004)

Table *Continued*

Scientific name	Common name	Range	Last occurrence						Reference
			Historical record	Radiometric date			Archaeological or stratigraphic association	Other estimate (see list in footnotes; see also Notes)	
				^{14}C age (years BP), ±1σ	Calibrated date, ±2σ	Direct date?			
Hippopotamus madagascariensis		Madagascar	(1976?)	Note 13				1, 4	Burney and Ramilisonina (1998)
Phanourios minor		Cyprus		9420±550	12 649–9249 BP	Yes			Simmons (1999); Ramis and Alcover (2005)
Cervidae									
Megaloceros giganteus	Irish elk, giant deer	Western Siberia		6816±35	7725–7585 BP	Yes			Stuart *et al.* (2004)
Praemegaceros cazioti		Corsica, Sardinia		6790±80	7650–7530 BP	Yes			Benzi *et al.* (2007)
Rucervus schomburgki	Schomburgk's deer	Thailand	1938					6b, d	Schroering (1995); Duckworth *et al.* (1999)
Bovidae									
Antidorcas bondi	Bond's springbok	South Africa		7570±100	8557–8180 BP*	No			Brown and Verhagen (1985)
Eudorcas rufina	Red gazelle	Algeria	pre-1894						de Smet and Smith (2001); Note 15
Bison priscus	Primitive bison	Siberia		8860±40	10 160–9710 BP	Yes			MacPhee *et al.* (2002)
Bubalus cebuensis		Cebu Island (Philippines)						5	Croft *et al.* (2006)
Bubalus mephistopheles	Short-horned water buffalo	China		Note 16		?	Anyang: 1400–1100 BC		Teilhard de Chardin and Young (1936)
Pelorovis antiquus	Giant long-horned buffalo	North Africa		4715±295; Note 17	6180–4645 BP*	Yes?	Two probable rock art representations from first millennium BC		Gautier and Muzzolini (1991); Klein (1994)
Myotragus balearicus	Balearic cave goat	Mallorca		5720±60	6669–6349 BP	Yes			Ramis and Alcover (2005)
		Menorca		5060±40	5919–5709 BP	Yes			Ramis and Alcover (2005)
		Cabrera		4785±40	5599–5329 BP	Yes			Ramis and Alcover (2005)
Hippotragus leucophaeus	Bluebuck	South Africa	1800						Klein (1974)

Megalotragus priscus		South Africa		8000±80	9078–8604 BP*	No			Beaumont (1990)
CETACEA									
Lipotidae[†]									
Lipotes vexillifer	Yangtze River dolphin, baiji	Yangtze and Qiantang Rivers (China)	pre-2006; Note 18					6a-b	Turvey *et al.* (2007b)
RODENTIA									
Gliridae									
Eliomys morpheus		Mallorca, Menorca, Cabrera (Mediterranean)		5890±35	6789–6639 BP	Yes			Bover and Alcover (2008)
Eliomys wiedincetensis		Malta						1	
Nesomyidae									
Hypogeomys australis		Madagascar		1536±35	AD 428–618	Yes			Burney *et al.* (2004)
Nesomys sp. nov.		Madagascar						1, 6c	Goodman *et al.* (2003)
Cricetidae									
Microtus melitensis		Malta						1	
Microtus sp. nov.		Santa Rosa Island (California Channel Islands)						5	Guthrie (1993)
Tyrrhenicola henseli		Corsica, Sardinia		1960±120–610±120	322 BC–AD 283 to AD 1224–1459	No			Vigne and Valladas (1996)
Neotoma anthonyi	Anthony's woodrat	Todos Santos Island (Mexican Pacific)	1926					6a-b	Cortés-Calva *et al.* (2001a)
Neotoma bunkeri	Bunker's woodrat	Coronados Island (Gulf of California)	1931					6a-b	Smith *et al.* (1993); Álvarez-Castañeda and Ortega-Rubio (2003)
Neotoma martinensis	San Martin Island woodrat	San Martin Island (Mexican Pacific)	1950s					6a-b	Cortés-Calva *et al.* (2001b)
Peromyscus anayapahensis		West Anacapa Island (California Channel Islands)						5	Guthrie (1993)
Peromyscus nesodytes		San Miguel and Santa Rosa Islands (California Channel Islands)					8000 BP; Note 19		Guthrie (1993)

Table *Continued*

Scientific name	Common name	Range	Last occurrence						Reference
			Historical record	Radiometric date			Archaeological or stratigraphic association	Other estimate (see list in footnotes; see also Notes)	
				^{14}C age (years BP), $\pm 1\sigma$	Calibrated date, $\pm 2\sigma$	Direct date?			
Peromyscus pembertoni	Pemberton's deer mouse	San Pedro Nolasco Island (Gulf of California)	1931						Álvarez-Castañeda and Cortés-Calva (2003); Álvarez-Castañeda and Ortega-Rubio (2003)
Juscelinomys candango	Brasilia burrowing mouse, Candango mouse	Brazil	1960						Moojen (1965); Hershkovitz (1998)
"*Ekbletomys hypenemus*"		Antigua, Barbuda (Caribbean)						1, 3	Note 20
Megalomys audreyae		Barbuda (Caribbean)		750±50	AD 1173–1385*	Yes		1, 3	MacPhee and Flemming (1999); Note 20
Megalomys curazensis		Aruba, Curaçao (Caribbean)						1, 5	Hooijer (1966); McFarlane and Lundberg (2002a)
Megalomys desmarestii	Martinique giant rice rat	Martinique (Caribbean)	*c.*1897						Allen (1942)
Megalomys luciae	St. Lucia giant rice rat	St. Lucia (Caribbean)	pre-1881						Allen (1942)
Megaoryzomys curioi	Curio's giant rice rat	Santa Cruz (Galápagos)		210±55	AD 1520–1950	Yes		3	Steadman *et al.* (1991)
Megaoryzomys sp. nov.	Isabela giant rice rat	Isabela (Galápagos)		265±200	AD 1515–1950	Yes			Steadman *et al.* (1991)
Nesoryzomys darwini	Darwin's Galápagos rice rat	Santa Cruz (Galápagos)	pre-1940						Dowler *et al.* (2000)
Nesoryzomys indefessus	Indefatigable Galápagos rice rat	Santa Cruz (Galápagos)	1934					6b	Dowler *et al.* (2000)
Nesoryzomys sp. nov. A		Baltra, Santa Cruz (Galápagos)						1	Steadman *et al.* (1991)
Nesoryzomys sp. nov. B		Isabela (Galápagos)						1, 3	Steadman *et al.* (1991)
Nesoryzomys sp. nov. C		Isabela (Galápagos)						1, 3	Steadman *et al.* (1991)
Nesoryzomys sp. nov. D		Rábida (Galápagos)		5700±70	6721–6316 BP	Yes			Steadman *et al.* (1991)

Noronhomys vespuccii	Vespucci's rice rat	Fernando de Noronha Island (Brazil)	1503			Carleton and Olson (1999)
Oligoryzomys victus	St. Vincent pygmy rice rat	St. Vincent (Caribbean)	1892		6a	Allen (1942)
Oryzomys antillarum	Jamaican rice rat	Jamaica	1877			Allen (1942)
Oryzomys nelsoni	Nelson's rice rat	Tres Marías Island (Mexican Pacific)	1897			Wilson (1991)
Thomasomys sp. nov.		Bonaire (Caribbean)			1, 5	Hooijer (1966); McFarlane and Lundberg (2002b)
Oryzomyini gen. et sp. indet. A		Anguilla, St. Martin, Tintamarre (Caribbean)		Shoal Bay East, Anguilla: AD 940–1320		Pregill *et al.* (1994); Crock (2000); Note 21
Oryzomyini gen. et sp. indet. B		Barbados (Caribbean)	1847?			Schomburgk (1848); Marsh (1985); Note 21
Oryzomyini gen. et sp. indet. C		Grenada (Caribbean)		Pearls: 37 BC–AD 533*		Pregill *et al.* (1994); Haviser (1997); Note 21
Oryzomyini gen. et sp. indet. D		Grenada (Caribbean)		Pearls: 37 BC–AD 533*		Pregill *et al.* (1994); Haviser (1997); Note 21
Oryzomyini gen. et sp. indet. E		Carriacou (Caribbean)		Grand Bay: AD 390–1280		LeFebvre (2007); S. Fitzpatrick (personal communication); Note 21
Oryzomyini gen. et sp. indet. F		Guadeloupe (Caribbean)		Morel: AD 21–881*		Pregill *et al.* (1994); Haviser (1997); Note 20
Oryzomyini gen. et sp. indet. G		La Desirade (Caribbean)		Petite Rivière: AD 600–1400		Pregill *et al.* (1994); de Waal (1996); Note 21
Oryzomyini gen. et sp. indet. H		Marie Galante (Caribbean)		Taliseronde: AD 350–665*		Pregill *et al.* (1994); Haviser (1997); Note 21
Oryzomyini gen. et sp. indet. I		Montserrat (Caribbean)		Trants: 774 BC–AD 622		Pregill *et al.* (1994); Petersen (1996); Note 21
Oryzomyini gen. et sp. indet. J		Montserrat (Caribbean)		Trants: 774 BC–AD 622		Pregill *et al.* (1994); Petersen (1996); Note 21
Oryzomyini gen. et sp. indet. K		Nevis, St. Eustatius, St. Kitts (Caribbean)		Sulphur Ghaut, Nevis: AD 900–1200		Pregill *et al.* (1994); Newsom and Wing (2004); Note 21
Oryzomyini gen. et sp. indet. L		Saba (Caribbean)		Kelbey's Ridge II: AD 1290–1400		Pregill *et al.* (1994); Hoogland (1996); Note 21

Table *Continued*

Scientific name	Common name	Range	Last occurrence						Reference
			Historical record	Radiometric date			Archaeological or stratigraphic association	Other estimate (see list in footnotes; see also Notes)	
				^{14}C age (years BP), ±1 σ	Calibrated date, ±2 σ	Direct date?			
Oryzomyini gen. et sp. indet. M		Bonaire (Caribbean)		Note 22				1	McFarlane and Lundberg (2002b)
Muridae									
Meriones malatestae		Lampedusa (Mediterranean)						1	
Mesocricetus rathgeberi		Armathia (Mediterranean)						1, 3	Pieper (1984)
Leimacomys buettneri	Groove-toothed forest mouse, Togo mouse	Togo	1890					4, 6a,d	Decher and Abedi-Lartey (2002)
Apodemus sp. nov.		Naxos (Mediterranean)						1	Dermitzakis and Sondaar (1978)
Rhagamys orthodon		Corsica, Sardinia		1960±120–610±120	322 BC–AD 283 to AD 1224–1459	No			Vigne and Valladas (1996)
Lenomys sp. nov.		Sulawesi						5	Musser and Holden (1991); Note 23
Papagomys theodorverhoeveni	Verhoeven's giant tree rat	Flores		3550±525	5450–2725 BP*	No		6c	Musser (1981)
Rattus macleari	Maclear's rat	Christmas Island (Indian Ocean)	1903						Pickering and Norris (1996)
Rattus nativitatis	Bulldog rat	Christmas Island (Indian Ocean)	1903						Pickering and Norris (1996)
Rattus sanila		New Ireland		1630±30	AD 347–535*	No			Flannery and White (1991); Gosden and Robertson (1991)
Rattus sp. nov. A		Manus Island (New Guinea)					Pamwak: found throughout Quaternary stratigraphic sequence		Flannery (1995a)

Rattus sp. nov. B		Timor					Present to surface layer in all excavated deposits		Musser (1981); O'Connor and Aplin (2007); K. Aplin and K. Helgen (personal communication)
Solomys spriggsarum		Buka Island (New Guinea)		7900±110 to 6670±80	9014–8455 BP to 7663–7431 BP*	No			Flannery and Wickler (1990); Note 24
Melomys spechti		Buka Island (New Guinea)		7900±110 to 6670±80	9014–8455 BP to 7663–7431 BP*	No			Flannery and Wickler (1990)
Melomys/ *Pogonomelomys* sp. nov. A		Timor					Present to surface layer in all excavated deposits		Musser (1981); O'Connor and Aplin (2007); K. Aplin and K. Helgen (personal communication)
Melomys/ *Pogonomelomys* sp. nov. B		Timor					Present to surface layer in all excavated deposits		Musser (1981); O'Connor and Aplin (2007); K. Aplin and K. Helgen (personal communication)
Spelaeomys florensis	Flores cave rat	Flores		3550±525	5450–2725 BP*	No			Musser (1981)
Coryphomys buehleri		Timor					Present in Late Quaternary deposits, but less abundant in late Holocene horizons	3	Hooijer (1965); Musser (1981); O'Connor and Aplin (2007); K. Aplin and K. Helgen (personal communication)
Coryphomys sp. nov.		Timor					Present in Late Quaternary deposits, but less abundant in late Holocene horizons		O'Connor and Aplin (2007); K. Aplin and K. Helgen (personal communication)
Conilurus albipes	White-footed rabbit-rat	Australia	1845						Johnson (2006)
Leporillus apicalis	Lesser stick-nest rat	Australia	1933 (1970?)						Robinson in Strahan (1995)
Pseudomys gouldii	Gould's mouse	Australia	1857						Johnson (2006)
Pseudomys glaucus	Blue-grey mouse	Australia	1956					2	Dickman in Van Dyck and Strahan (2008)
Notomys amplus	Short-tailed hopping mouse	Australia	pre-1896						Dixon in Strahan (1995)

Table *Continued*

Scientific name	Common name	Range	Last occurrence						Reference
			Historical record	Radiometric date			Archaeological or stratigraphic association	Other estimate (see list in footnotes; see also Notes)	
				^{14}C age (years BP), ±1 σ	Calibrated date, ±2 σ	Direct date?			
Notomys longicaudatus	Long-tailed hopping mouse	Australia	1901						Dixon in Strahan (1995)
Notomys macrotis	Big-eared hopping mouse	Australia	1843						Dixon in Strahan (1995)
Notomys mordax	Darling Downs hopping mouse	Australia	1840s						Watts in Strahan (1995)
Notomys robustus	Great hopping mouse, broad-cheeked hopping-mouse	Australia							Medlin in Van Dyck and Strahan (2008)
Malpaisomys insularis	Volcano rat	Fuerteventura, La Graciosa, Lanzarote (Canary Islands)		1070±50	AD 784–1116*	No	Malpais de Arena, Fuerteventura: c.800 BP	3	Hutterer *et al.* (1988); Boye *et al.* (1992)
Canariomys bravoi		Tenerife (Canary Islands)		2305±40	2359–2155 BP	Yes			Bocherens *et al.* (2006)
Canariomys tamarani		Gran Canaria (Canary Islands)					El Hormiguero: ? AD 220; unnamed cultural level: 130 BP	3	Michaux *et al.* (1996)
Mus minotaurus		Crete						1	
Mus sp. nov.		Karpathos (Mediterranean)						1	Dermitzakis and Sondaar (1978)
Muridae gen. nov. 1, sp. nov. A		Timor					Present in Late Quaternary deposits, but less abundant in late Holocene horizons		Musser (1981); O'Connor and Aplin (2007); K. Aplin and K. Helgen (personal communication)

Muridae gen. nov. 1, sp. nov. B		Timor					As for Muridae gen. nov. 1, sp. nov. A		Musser (1981); O'Connor and Aplin (2007); K. Aplin and K. Helgen (personal communication)
Muridae gen. nov. 1, sp. nov. C		Timor					As for Muridae gen. nov. 1, sp. nov. A		Musser (1981); O'Connor and Aplin (2007); K. Aplin and K. Helgen (personal communication)
Muridae gen. nov. 2, sp. nov. A		Timor					As for Muridae gen. nov. 1, sp. nov. A		Musser (1981); O'Connor and Aplin (2007); K. Aplin and K. Helgen (personal communication)
Muridae gen. nov. 2, sp. nov. B		Timor					As for Muridae gen. nov. 1, sp. nov. A		Musser (1981); O'Connor and Aplin (2007); K. Aplin and K. Helgen (personal communication)
Muridae gen. nov. 2, sp. nov. C		Timor					As for Muridae gen. nov. 1, sp. nov. A		Musser (1981); O'Connor and Aplin (2007); K. Aplin and K. Helgen (personal communication)
Muridae gen. nov. 3, sp. nov.		Timor					As for Muridae gen. nov. 1, sp. nov. A		Musser (1981); O'Connor and Aplin (2007); K. Aplin and K. Helgen (personal communication)
Muridae gen. nov. 4, sp. nov. A		Morotai Island (Maluku)		5530±70	6464–6194 BP*	No			Flannery *et al.* (1999); K. Aplin (personal communication)
Muridae gen. nov. 4, sp. nov. B		Morotai Island (Maluku)		5530±70	6464–6194 BP*	No			Flannery *et al.* (1999); K. Aplin (personal communication)
Abrocomidae									
Cuscomys oblativa	Inca tomb rat	Peru					Machu Picchu: AD 1450–1532	6d	Emmons (1999)
Chinchillidae									
Lagostomus crassus	Peruvian plains viscacha	Peru	pre-1910						Thomas (1910)
Echimyidae									
Boromys offella		Cuba		7864±96	8993–8453 BP	No		3	Jiménez Vázquez *et al.* (2005)
Boromys torrei		Cuba		7864±96	8993–8453 BP	No		3	Jiménez Vázquez *et al.* (2005)
Brotomys contractus		Hispaniola						1	
Brotomys voratus		Hispaniola	1536–1546	430±60	AD 1410–1640	Yes		3	Miller (1929); McFarlane *et al.* (2000)
Heteropsomys insulans		Puerto Rico		1219±26	AD 694–887	Yes			Turvey *et al.* (2007a)
Puertoricomys corozalus		Puerto Rico						1, 5	Williams and Koopman (1951)

Table *Continued*

Scientific name	Common name	Range	Last occurrence						Reference
			Historical record	Radiometric date			Archaeological or stratigraphic association	Other estimate (see list in footnotes; see also Notes)	
				^{14}C age (years BP), ±1 σ	Calibrated date, ±2 σ	Direct date?			
Capromyidae									
Capromys antiquus		Cuba						1	Note 25
Capromys arredondoi		Cuba						1	Note 25
Capromys latus		Cuba						1	Note 25
Capromys pappus		Cuba						1	Note 25
Capromys robustus		Cuba						1	Note 25
Capromys sp. nov.		Grand Cayman, Little Cayman, Cayman Brac	1586	375±60	AD 1439–1643*	No		3	Morgan (1994); G. Morgan, personal communication
Geocapromys columbianus		Cuba						1, 3	Note 25
Geocapromys pleistocenicus		Cuba		7864±96	8993–8453 BP	No			Jiménez Vázquez *et al.* (2005); Note 23
Geocapromys thoracatus	Little Swan Island hutia	Little Swan Island (Caribbean)	1950s						Morgan (1989a)
Geocapromys sp. nov. A		Cayman Brac						1, 3?	
Geocapromys sp. nov. B		Grand Cayman						1, 3	
Mesocapromys barbouri		Cuba						1	Note 25
Mesocapromys beatrizae		Cuba						1	Note 25
Mesocapromys delicatus		Cuba						1	Note 25
Mesocapromys gracilis		Cuba						1	Note 25
Mesocapromys kraglievichi		Cuba						1	Note 25

Mesocapromys minimus		Cuba						1	Note 25
Mesocapromys sanfelipensis	Little earth hutia	Juan García Key (San Felipe Keys, Cuba)	1978					6a	Frías *et al.* (1988); Berovides Alvarez and Comas González (1991); Meier (2004)
Mesocapromys silvai		Cuba						1	Note 25
Mysateles jaumei		Cuba						1	Note 25
Hexolobodon phenax		Hispaniola						1, 3?	
Hexolobodontinae gen. et sp. nov.		Hispaniola		3755±175	4785–3641 BP*	No			Woods (1989); Note 25
Plagiodontia araeum		Hispaniola						1	
Plagiodontia ipnaeum		Hispaniola	?1536–1546					1, 3, 4?, 6c	Miller (1929); Woods (1989)
Rhizoplagiodonta lemkei		Hispaniola		3755±175	4785–3641 BP*	No		3	Woods (1989)
Isolobodon montanus		Hispaniola						1, 3	
Isolobodon portoricensis	Puerto Rican hutia	Hispaniola	?1536–1546	710±50	AD 1234–1389	Yes	Puerto Real: AD 1503–1578	4?, 6a,c	Miller (1929); Woods (1989); Reitz and McEwan (1995); McFarlane *et al.* (2000)
		Puerto Rico	1800s (1970s?)	620±60	AD 1280–1425	Yes		6a,c-d	Wetmore (1922b); Raffaele (1979); Flemming and MacPhee (1999)
		Mona Island		380±60	AD 1480–1655	Yes		6d	Nieves-Rivera and McFarlane (2001)
		Virgin Islands					Cinnamon Bay, St. John: AD 1490		Quitmyer (2003)
Heptaxodontidae†									
Clidomys osborni		Jamaica						1, 5	Morgan and Wilkins (2003)
Elasmodontomys obliquus		Puerto Rico		3512±28	3862–3700 BP	No			Turvey *et al.* (2007a)
Quemisia gravis		Hispaniola	?1536–1546					1, 4?	Miller (1929)
Xaymaca fulvopulvis		Jamaica						1	
Family indet.									
Tainotherium valei		Puerto Rico						1	
Rodentia? gen. et sp. nov.		Jamaica						1	MacPhee and Flemming (2003)

Table Continued

Other estimates column: indirect last-occurrence date estimates (only listed for species known only from Pleistocene or undated Late Quaternary deposits).

1. First human arrival in region during Holocene (only noted if no other evidence exists for Holocene survival).
2. Associated with still-extant Holocene regional mammal fauna (only noted for regions with pre-Holocene human arrival).
3. Associated with anthropogenically introduced mammal species; for further information see MacPhee and Flemming (1999).
4. Local names or traditions exist (or are recorded in the historical literature) that may refer to extinct species, suggesting relatively late survival (only noted for species otherwise known only from fossil material).
5. May have died out before the Holocene.
6. Considered to be extant or possibly extant by (a) IUCN (2008), (b) MacPhee and Flemming (1999), (c) Alcover *et al.* (1998), or (d) cited reference.
7. No specimens; known only from historical records.

Notes

1. This specimen was later assigned an age of between 6000 and 9000 years BP (Plane 1976).
2. Subfossil *Dorcopsis* wallaby material from Halmahera, dated indirectly to 2540±70–1870±80 years BP (Flannery *et al.* 1995), is now interpreted as the extant *D. mulleri mysoliae*, and representing a prehistoric introduction to Halmahera and Gebe from Misool (Flannery *et al.* 1999).
3. *Plesiorycteropus* sp. material has been directly dated to 2154±40 years BP (Burney *et al.* 2004).
4. Unidentified sloth material from Trou Woch Sa Wo Cave, Haiti, possibly referable to this species, has been directly dated to 3715±50 years BP (MacPhee *et al.* 2000).
5. Woods (1989; see MacPhee *et al.* 2000) reported that this species may have co-occurred in superficial stratigraphic levels with *Rattus* and *Mus* at Trouing Jérémie #5, but this is considered unlikely.
6. *Megaladapis* sp. material has been directly dated to 1591±60 and 1815±60 years BP (Burney *et al.* 2004).
7. Dated specimen identified as *Archaeolemur* cf. *edwardsi* (Burney *et al.* 2004).
8. Specimen identified as *Archaeolemur* cf. *majori* has been directly dated to 1370±40 years BP (Burney *et al.* 2004).
9. *Archaeolemur* sp. skeletal material has been directly dated to 1020±50 years BP; subfossil coprolite identified as belonging to cf. *Archaeolemur* sp. has been directly dated to 830±60 years BP (Burney *et al.* 2004).
10. Unpublished Holocene radiometric dates for *Desmodus puntajudensis* are also mentioned by Suárez (2005).
11. Guthrie (1980) reported that *Desmodus stocki* material was recovered from a cave deposit level dating to 5000–2500 years BP, but later considered that this stratigraphical interpretation was incorrect, and that all remains were recovered from below the level of 10 700 years BP (Guthrie 1993).
12. Eurasian caballoid horses have been extremely oversplit, and although extinct putative Late Quaternary caballoid taxa correspond in some cases with subfossil material subsequently dated to the Holocene, we place these in *Equus caballus sensu lato* (cf. MacPhee *et al.* 2002).
13. Unidentified Malagasy hippopotamus material has been directly dated to 980±200 years BP (AD 660–1400) (Burney *et al.* 2004). There are several possible eyewitness accounts of Malagasy hippopotami from the nineteenth and early twentieth centuries, the latest dating from 1976 (Burney and Ramilisonina 1998).
14. Burney *et al.* (2004) have suggested that the dated specimen may be the skull of a modern African hippopotamus (*Hippopotamus amphibius*).
15. Two other gazelle species, *Gazella bilkis* and *Gazella saudiya*, are currently recognized as having become extinct during the recent historical period by IUCN (2008), but the taxonomic status of North African and Middle Eastern gazelles remains the subject of considerable confusion, and doubts remain over the validity of both species (T. Wacher, personal communication).
16. An unsupported radiometric date of 3140±110 years BP is also available for this species (Tong and Liu 2004; see Chapter 2 in this volume). Note added in proof: new Holocene radiometric dates for *Bubalus mephistopheles* have recently been published by Yang *et al.* (2008).
17. A more recent direct apatite date of 2350 years BP is considered dubious (Gautier and Muzzolini 1991).
18. More recent unconfirmed sightings were reported until 2007 (Turvey *et al.* 2007b).
19. Walker (1980) reported material of *Peromyscus nesodytes* from a stratified archaeological midden on San Miguel Island dating to 400 BC–AD 300, but Guthrie (1993) reported that this material came from an area that had been extensively disturbed by grave robbers, resulting in mixing of materials between levels.

20. '*Ekbletomys hypenemus*' remains a *nomen nudum*, as the name has never been formally published (see Ray 1962). However, examination of the *Megalomys audreyae*-type material and associated rice rat fossil material from Barbuda at the Natural History Museum, London, UK, suggests that two distinct oryzomyin species were formerly present on this island (S.T. Turvey, personal observation).
21. The species diversity and taxonomic status of the extinct rice rats present in Lesser Antillean zooarchaeological collections remains extremely poorly understood, and populations recorded from different islands or adjacent island groups are treated separately here.
22. Based on fluorine relative dating results, accelerator mass spectrometry (AMS) ^{14}C dating is expected to yield a mid-to-late Holocene age for this deposit (McFarlane and Lundberg 2002b).
23. The occurrence of this species on Sulawesi is indicated only by undated subfossil samples (Musser and Holden 1991).
24. *Solomys salamonis*, *Uromys imperator*, and *Uromys porculus*, three closely related large-bodied murids from the Solomon Islands that are known only from specimens collected during the late nineteenth century, have also sometimes been considered to be extinct (Groves and Flannery 1994; Flannery 1995a). However, rodent survey efforts in the archipelago have been insufficient to demonstrate that these species have definitely died out (K. Helgen, personal communication).
25. The taxonomic validity of most extinct capromyine and heptaxodontine capromyids remains uncertain, and it is likely that many of these species are synonymous. However, little taxonomic revision of this group has been carried out, and so they are provisionally listed as distinct species pending further review. The initial taxonomic revisions of Díaz-Franco (2001) and Woods and Kilpatrick in Wilson and Reeder (2005) are followed herein.

Note added in proof: new Holocene radiometric dates for *Bubalus mephistopheles* have recently been published by Yang *et al.* (2008).

CHAPTER 4

Holocene avian extinctions

Tommy Tyrberg

This chapter presents last-occurrence dates for 523 extinct bird species that are currently known or believed to have died out during the Holocene. This includes species known to have survived into the historical period, and those represented in Late Quaternary subfossil deposits known or believed to be Holocene in age, or from island systems which humans did not reach until the Holocene. It also interprets fossil Holocene taxa that have been designated 'cf.' in relation to other living or extinct species as provisionally being distinct at the species level. Only the youngest known occurrence dates are given for each species, or for sites from which the species has been reported. Calibrated dates are given as either AD/BC (if the date range falls at least partly within the past 2000 years) or BP. An asterisk indicates that calibrated dates have not previously been published, and have been calculated for this chapter using OxCal 4.0 (Bronk Ramsey 1995, 2001).

Scientific name	Common name	Range	Last occurrence				Archaeological or stratigraphic association	Other estimate (see list in footnotes; see also Notes)	Reference
			Historical record	Radiometric date					
				^{14}C age (years BP), ±1 σ	Calibrated date, ±2 σ	Direct date?			
Aepyornithidae†									
Aepyornis gracilis		Madagascar	–					1, Note 1	Monnier (1913); Brodkorb (1963b)
Aepyornis hildebrandti		Madagascar	–					1, Note 1	Burckhardt (1893); Brodkorb (1963b)
Aepyornis maximus		Madagascar	–	1830±60	AD 55–343*	Yes		1, Note 1	Geoffroy Saint-Hilaire (1851); Brodkorb (1963b)
Aepyornis medius		Madagascar	–					1, Note 1	Milne-Edwards and Grandidier (1866); Brodkorb (1963b)
Mullerornis agilis		Madagascar	–					1, Note 1	Milne-Edwards and Grandidier (1894); Brodkorb (1963b)
Mullerornis betsilei		Madagascar	–					1, Note 1	Milne-Edwards and Grandidier (1894); Brodkorb (1963b)
Mullerornis rudis		Madagascar	–					1, Note 1	Milne-Edwards and Grandidier (1894); Brodkorb (1963b)
Dinornithidae†									
Dinornis novaezealandiae	Large bush moa, North Island giant moa	North Island (New Zealand)	–	2913±69	3317–2867 BP*	Yes	Occurs in Maori middens, post c. AD 1280	8	Bunce *et al.* (2003); Huynen *et al.* (2003); Worthy *et al.* (2005)
Dinornis robustus	South Island giant moa	South Island (New Zealand)	–	530±67	AD 1289–1469*	Yes	Occurs in Maori middens, post c. AD 1280	8	Cooper *et al.* (2001); Worthy and Holdaway (2002); Bunce *et al.* (2003); Huynen *et al.* (2003); Worthy *et al.* (2005)
Emeidae†									
Anomalopteryx didiformis	Little bush moa	North Island+South Island (New Zealand)	–	4735±72	5591–5319 BP*	Yes	Occurs in Maori middens, post c. AD 1280	8	Worthy and Holdaway (2002); Tennyson and Martinson (2006)
Emeus crassus	Eastern moa	South Island (New Zealand)	–		AD 620–790	Yes	Occurs in Maori middens, post c. AD 1280	8	Cooper *et al.* (2001); Worthy and Holdaway (2002); Tennyson and Martinson (2006)

Eurapteryx curtus	Coastal moa	North Island (New Zealand)	–	1114±50	AD 780–1018*	Yes	Occurs in Maori middens, post *c.* AD 1280	8	Worthy and Holdaway (2002); Tennyson and Martinson (2006)
Euryapteryx gravis	Stout-legged moa	North Island+South Island (New Zealand)	–	363±40	AD 1448–1635*	Yes	Occurs in Maori middens, post *c.* AD 1280	8	Worthy (2005b)
Megalapteryx didinus	Upland moa	South Island (New Zealand)	–	628±83	AD 1252–1440*	Yes	Occurs in Maori middens, post *c.* AD 1280	8	Worthy (1990); Worthy and Holdaway (2002); Tennyson and Martinson (2006)
Pachyornis australis	Crested moa	South Island (New Zealand)	–					1	Oliver (1949); Worthy (1989b, 1990); Worthy and Holdaway (2002); Tennyson and Martinson (2006)
Pachyornis elephantopus	Heavy-footed moa	South Island (New Zealand)	–				Occurs in Maori middens, post *c.* AD 1280	8	Worthy and Holdaway (2002); Tennyson and Martinson (2006)
Pachyornis geranoides	Mantell's moa	North Island (New Zealand)	–	624±58	AD 1278–1415*	No	Occurs in Maori middens, post *c.* AD 1280	8	Worthy and Holdaway (2002); Worthy (2005b); Tennyson and Martinson (2006)
Casuariidae									
Dromaius ater	King Island emu	King Island (Australia)	1802						Greenway (1967); BirdLife International (2000)
Dromaius baudinianus	Kangaroo Island emu	Kangaroo Island (Australia)	*c.*1827						Parker (1984); BirdLife International (2000)
Apterygidae									
Apteryx undescribed species	Eastern kiwi	South Island (New Zealand)	–					1	Holdaway and Worthy (1997); Worthy (1998a, 1998b); Worthy and Holdaway (2002)
Anatidae									
cf. *Dendrocygna* undescribed species	Polynesian whistling-duck	Aitutaki (Cook Islands)	–				Ureia site, *c.*1000 Cal BP	8	Allen and Steadman (1990); Steadman (1991, 2006b)
Cygnus [*atratus*] *sumnerensis*	Black swan	North Island+South Island+Chatham Islands (New Zealand)	–	792±77	AD 1059–1401	Yes	Occurs in Moriori middens, post *c.* AD 1350	8	Millener (1999); Worthy and Holdaway (2002); Tennyson and Martinson (2006)
Cygnus equitum		Malta+Sicily	–					1, 5, Note 2	Bate (1916); Northcote (1988, 1992); Pavia (2000)

Table *continued*

Scientific name	Common name	Range	Last occurrence				Archaeological or stratigraphic association	Other estimate (see list in footnotes; see also Notes)	Reference
			Historical record	Radiometric date					
				^{14}C age (years BP), ±1σ	Calibrated date, ±2σ	Direct date?			
Cygnus falconeri		Malta+Sicily	–					1, 5, Note 2	Parker (1865, 1869); Northcote (1982b, 1992); Pavia (2000)
Anser undescribed species aff. *erythropus*		Eivissa	–				Es Pouàs	1	Seguí and Alcover (1999)
Branta hylobadistes	Nene-nui	Maui	–	1050±50	AD 883–1151*	Yes			Olson and James (1984, 1991); James *et al.* (1987); Paxinos *et al.* (2002)
Branta undescribed species aff. *hylobadistes*	Medium Kauai goose	Kauai	–				Maha'ulepu Cave Unit IV–VI, <6000 BP	1	Olson and James (1982, 1984, 1991); Burney *et al.* (2001)
Branta undescribed species aff. *hylobadistes*		Oahu	–	770±70	AD 1046–1380*	No		8	Olson and James (1982, 1984, 1991)
Branta very large undescribed species		Hawai'i	–	510±60	AD 1296–1485*	Yes	Umi'i Manu Cave		Olson and James (1982, 1984, 1991); Giffin (2003)
Cnemiornis gracilis	North Island goose	North Island (New Zealand)	–				Occurs in Maori middens, post *c.* AD 1280	8	Worthy and Holdaway (2002); Tennyson and Martinson (2006)
Cnemiornis calcitrans	South Island goose	South Island (New Zealand)	–				Occurs in Maori middens, post *c.* AD 1280	8	Worthy and Holdaway (2002); Tennyson and Martinson (2006)
Pachyanas chathamica	Chatham Island duck	Chatham Islands	–	1529±57	AD 448–657	Yes	Occurs in Moriori middens, post *c.* AD 1350	8	Oliver (1955); Millener (1999)
Neochen barbadiana		Barbados	–				Ragged Point, Barbados	1	Brodkorb (1965)
Alopochen kervazoi	Réunion Island sheldgoose	Réunion	1672						Cowles (1994); Mourer-Chauviré *et al.* (1999)

Alopochen mauritianus	Mauritian shelduck	Mauritius	1693						Cheke (1987)
Alopochen sirabensis		Madagascar	–	1380±90	AD 437–875*	Yes	Beloha		Andrews (1897); Goodman and Rakotozafy (1997)
Centrornis majori	Madagascar sheldgoose	Madagascar	–				Sirabé	1	Andrews (1897)
Tadorna cristata	Crested shelduck	East Asia	1964					6	Greenway (1967); BirdLife International (2000)
Tadorna undescribed species	Chatham Island shelduck	Chatham Islands	–	1534±62	AD 455–612	Yes	Occurs in Moriori middens, post c. AD 1350	8	Millener (1999)
aff. *Tadorna* undescribed species		Kauai	–				Maha'ulepu Cave Unit IV–VI, <6000 BP	1	Burney *et al.* (2001)
Chenonetta finschi	Finsch's duck	North Island + South Island (New Zealand)	–	305±70	AD 1442–1953*	Yes	Occurs in Maori middens, post c. AD 1280	8	Holdaway (1999b); Holdaway *et al.* (2002b); Worthy and Olson (2002)
Anas marecula	Amsterdam Island duck	Amsterdam Island	–					1	Martinez (1987); Olson and Jouventin (1996); Worthy and Jouventin (1999)
Anas oustaleti	Marianas mallard	Guam	1981						Steadman (2006b)
Anas theodori	Mauritian duck	Mauritius + Réunion	1696						Cheke (1987); Mourer-Chauviré *et al.* (1999)
Anas undescribed species	Macquarie Island teal	Macquarie Island	1821?						Worthy and Holdaway (2002); Tennyson and Martinson (2006)
Anas undescribed species		St. Paul Island	1793					7	Olson and Jouventin (1996)
Anas undescribed species		Viti Levu (Fiji)	–					1	Worthy (2003)
aff. *Anas* undescribed species		Kauai	–				Maha'ulepu Cave Unit IV–VI, <6000 BP		Burney *et al.* (2001)
Chelychelynechen quassus	Kauai moa-nalo, turtle-jawed moa-nalo	Kauai	–					1	Olson and James (1982, 1984, 1991)
Ptaiochen pau	Small-billed moa-nalo	Maui	–					1	Olson and James (1984, 1991)

Table *continued*

Scientific name	Common name	Range	Last occurrence				Archaeological or stratigraphic association	Other estimate (see list in footnotes; see also Notes)	Reference
			Historical record	Radiometric date					
				^{14}C age (years BP), ±1 σ	Calibrated date, ±2 σ	Direct date?			
Thambetochen chauliodous	Maui Nui moa-nalo	Molokai+Maui+Lanai	–	770±70	AD 1057–1375	No	Molohai heiau	3, 8	Olson and Wetmore (1976); Kolb (1994)
Thambetochen xanion	Oahu moa-nalo	Oahu	–		AD 440–639	Yes	Sinkhole 4907-D+K2062, 'Ewa plain	8	Olson and James (1982, 1984, 1991); James (1987); Athens *et al.* (2002)
Anatidae 'supernumerary Oahu goose'		Oahu	–					1	Olson and James (1982, 1984, 1991)
Camptorhynchus labradorius	Labrador duck	Eastern North America	1875 (1878?)						Greenway (1967)
Rhodonessa caryophyllacea	Pink-headed duck	India, Bangladesh, Bhutan, Myanmar	1949					6	Greenway (1967); BirdLife International (2000)
Malacorhynchos scarletti	Scarlett's duck	North Island+South Island+Chatham Islands (New Zealand)	–				Occurs in Maori middens, post *c.* AD 1280	8	Olson (1977a)
Chendytes lawi	Law's diving goose	Channel Islands and California coast	–	2910±40	2720–2350 BP	No	Little Sycamore, California	8	Miller (1925); Morejohn (1976); Guthrie (1993); Jones *et al.* (2008b)
Mergus australis	New Zealand merganser, southern merganser	North Island+South Island (New Zealand) Chatham Islands, Auckland Island	1902					8, Note 3	Greenway (1967); Kear and Scarlett (1970); Millener (1999); BirdLife International (2000); Worthy and Holdaway (2002); Tennyson and Martinson (2006)
Oxyura vantetsi	New Zealand stiff-tailed duck	North Island+South Island (New Zealand)	–				Occurs in Maori middens, post *c.* AD 1280	8, Note 4	Scofield *et al.* (2003); Worthy (2005a)
Biziura delautouri	New Zealand musk duck	North Island+South Island (New Zealand)	–				Occurs in Maori middens, post *c.* AD 1280	8	Worthy and Holdaway (2002); Tennyson and Martinson (2006)

Anatidae undescribed species	Marianas flightless duck	Rota (Marianas)	–	930±70	AD 990–1260	Noo	Payapai Cave, Layer II	3, 8	Steadman (1999, 2006b)
Megapodidae									
Megapodius alimentum	Consumed megapode	Lakeba+Aiwa Levu (Lau Group)+Mago+ Ha'apai group+ Tongatapu+ 'Eua (Tonga)	–	2740±50	2948–2757 BP*	Yes		3, 8	Steadman (1989b, 2006b); Steadman *et al.* (2002a)
Megapodius amissus	Lost megapode	Viti Levu (Fiji)	1926?				Udit Cave, Voli Voli Cave	1, 4?, Notes 5, 6	Worthy (2000)
Megapodius molistructor	Pile-builder megapode	Grande Terre+Isle des Pins (New Caledonia)+Ha'apai Group+Tongatapu (Tonga)	–	1750±70	AD 86–428*	No	Pindai	3, 8, Note 6	Balouet and Olson (1989); Steadman (2006b)
Megapodius undescribed species		Ofu (Samoa)	–				Toaga Site, 1900–2800 BP	8	Steadman (1990, 1994, 2006b)
Megapodius undescribed species A	Large Bismarcks megapode	New Ireland	–				Balof Rockshelter, c.10 000–14 000 BP	2, 8	Steadman *et al.* (1999); Steadman (2006b)
Megapodius undescribed species B	Large Solomon Islands megapode	Buka (Solomons)	–				Kilu Cave, Layer I, <11,000 BP	2, 8	Steadman (2006b)
Megapodius undescribed species C	Large Vanuatu megapode	Efate (Vanuatu)	–				Arapus site, <2500 BP	8	Steadman (2006b)
Megapodius undescribed species D	New Caledonia megapode	Grande Terre (New Caledonia)	–				Lapita site, c.2900 BP	8	Steadman (1997b, 2006b)
Megapodius undescribed species E	Loyalty megapode	Lifu+Mare (Loyalty Islands)	–					8	Steadman (2006b)
Megapodius undescribed species F	Small-footed megapode	'Eua (Tonga)	–				Several archaeological sites, <3000 BP	3, 8	Steadman (2006b)
Megapodidae undescribed species	Stout Tongan megapode	Tongatapu (Tonga)	–				Ha'ateiho site, Layer 3, <2800 BP	8	Steadman (2006b)
Megavitiornis altirostris	Deep-billed megapode	Viti Levu+Naigani (Fiji)	–				Naigani Island, Site VL21/5, <3000 BP	8	Worthy *et al.* (1999); Worthy (2000)

Table *continued*

Scientific name	Common name	Range	Last occurrence				Archaeological or stratigraphic association	Other estimate (see list in footnotes; see also Notes)	Reference
			Historical record	Radiometric date					
				^{14}C age (years BP), ±1σ	Calibrated date, ±2σ	Direct date?			
Sylviornithidae†									
Sylviornis neocaledoniae	Giant flightless megapode, du	Grande Terre+Isle des Pins (New Caledonia)	–	3470±210	4407–3269 BP*	Yes	Known from several sites; the youngest is possibly WBR001 (Nessadiou), *c.*2800 BP	3, 4, 8, Note 7	Poplin (1980); Poplin *et al.* (1983); Poplin and Mourer-Chauviré (1985); Steadman (1997b, 2006b); Mourer-Chauviré and Balouet (2005)
Phasianidae									
Coturnix gomerae	Canaries quail	Gomera+El Hierro+Tenerife+Fuerteventura (Canaries)	–					8	Jaume *et al.* (1993); Rando and Perera (1994); Rando and Lopez (1996); Rando *et al.* (1997); Castillo *et al.* (2001)
Coturnix novaezelandiae	New Zealand quail	North Island+South Island (New Zealand)	1875						Worthy and Holdaway (2002); Tennyson and Martinson (2006)
Coturnix undescribed species		Madeira+Porto Santo	–	–				1	Pieper (1985)
Ophrysia superciliosa	Himalayan quail	India	1868 (1877?)					6	Greenway (1967); BirdLife International (2000)
Argusianus bipunctatus	Double-banded argus	South-east Asia?	1871						BirdLife International (2000)
Spheniscidae									
Eudyptes undescribed species	Chatham crested penguin	Chatham Islands	–				Occurs in Moriori middens, post *c.* AD 1350	3, 8	Tennyson and Millener (1994)
Tasidyptes hunteri	Hunter Island penguin	Australia	–	760±70	AD 1050–1392*	No		8	Van Tets and O'Connor (1983); Meredith (1991)
Megadyptes waitaha		South Island + Stewart Island (New Zealand)					Occurs in Maori middens, post *c.* AD 1280		Boessenkool *et al.* (2009)

Podicipedidae									
Tachybaptus rufolavatus	Alaotra grebe	Madagascar	1985					6	BirdLife International (2000)
Podilymbus gigas	Atitlán grebe	Lake Atitlán (Guatemala)	1986						BirdLife International (2000)
Podiceps andinus	Colombian grebe	Colombia	1977						BirdLife International (2000)
Procellaridae									
Pterodroma caribbaea	Jamaica petrel	Jamaica	1879					6	BirdLife International (2000)
Pterodroma jugabilis	Oahu petrel, gracile petrel	Oahu+Hawai'i	–					1	Olson and James (1991)
Pterodroma rupinarum	Large St. Helena petrel	St. Helena	–					1	Olson (1975)
Pterodroma undescribed species	Bourne's petrel	Rodrigues	*c.*1726						Cheke (1987); Cheke and Hume (2008)
Pterodroma undescribed species		Chatham Islands	–	4935±73	5893–5489 BP*	Yes	Taupeka dunes		Tennyson and Millener (1994); Millener (1999); Worthy and Holdaway (2002)
Pterodroma undescribed species		Norfolk Island	–				Cemetery Beach, Nepean Island	1, 3	Meredith (1991)
Pterodroma undescribed species	'Eua petrel	'Eua (Tonga)	–					1	Steadman (2006b)
Bulweria bifax	Small St. Helena petrel	St. Helena	–					1	Olson (1975)
Pseudobulweria undescribed species		Taravai and Agakauiti Islands, Mangareva	-				Onema, *c.* AD 1000–1200	8	Worthy and Tennyson (2004); Kirch (2007)
Puffinus holeae		Fuerteventura+ Lanzarote	–					1, 8, Note 8	Walker *et al.* (1990)
Puffinus olsoni	Lava shearwater	Fuerteventura+ Lanzarote	–	1265±25	AD 1060–1220	Yes	Cueva de la Moscas, Fuerteventura	8	McMinn *et al.* (1990); Alcover and McMinn (1995); Rando and Alcover (2007)
Puffinus parvus		Bermuda						8	Shufeldt (1916); Olson (2004)
Puffinus spelaeus	Scarlett's shearwater, cave shearwater	South Island (New Zealand)	–		AD 1556–1858	Yes			Holdaway and Worthy (1994)
Puffinus undescribed small species	–	Mallorca+Minorca	–					1	Seguí (1998); Alcover *et al.* (1999a)
Puffinus undescribed species	'Eua shearwater	'Eua (Tonga)	–					1	Steadman (2006b)

Table *continued*

Scientific name	Common name	Range	Last occurrence				Archaeological or stratigraphic association	Other estimate (see list in footnotes; see also Notes)	Reference
			Historical record	Radiometric date					
				^{14}C age (years BP), $\pm 1\sigma$	Calibrated date, $\pm 2\sigma$	Direct date?			
Procellaridae undescribed genus and species	Easter Island petrel	Easter Island	–		AD 1000–1430	Yes	Ahu Naunau at Anakena	8	Steadman *et al.* (1994); Steadman (2006b)
Hydrobatidae									
Oceanodroma macrodactyla	Guadalupe storm-petrel	Isla Guadalupe	1911					6	Greenway (1967); BirdLife International (2000)
Phalacrocoracidae									
Microcarbo serventyorum		Western Australia	–						Van Tets (1994)
Phalacrocorax perspicillatus	Pallas's cormorant	Bering Island	1850						Greenway (1967)
Ardeidae									
Ardea bennuides	Bennu heron	United Arab Emirates	–				Umm-an-Nar, *c.*2500 BC	8	Hoch (1977)
Nyctanassa carcinocatactes	Crab-eating night-heron	Bermuda	1623						Olson and Wingate (2006)
Nycticorax duboisi	Réunion night-heron	Réunion	Probable report 1671/1672						Cowles (1994); Mourer-Chauviré *et al.* (1999)
Nycticorax kalavikai	Niue night-heron	Niue	–				Anakuli Cave, Hakupu, 3500–4500 BP	1	Steadman *et al.* (2000); Steadman (2006b)
Nycticorax olsoni		Ascension	Probable report 1502						Olson (1977c); Ashmole and Ashmole (1997); Bourne *et al.* (2003)
Nycticorax mauritianus	Mauritius night-heron	Mauritius	Probable report 1693						Günther and Newton (1879); Cheke (1987)

Nycticorax megacephalus	Rodrigues night-heron	Rodrigues	1726						Cheke (1987)
Nycticorax undescribed species A	Solomon Islands night-heron	Buka (Solomons)	–				Kilu Cave, Layer I, <11 000 BP	2, 8	Steadman (2006b)
Nycticorax undescribed species B	Tonga night-heron	Ha'apai group+'Eua (Tonga)	–				Tongoleleka site, 2700–2850 BP	8	Steadman (2006b)
Nycticorax undescribed species C	Cook Islands night-heron	Mangaia (Cook Islands)	–		AD 1390–1470	No	Te Ana Manuku Rockshelter (MAN-84)	8	Steadman and Kirch (1990); Steadman (2006b)
Ixobrychus novaezelandiae	New Zealand little bittern	North Island+South Island+Chatham Islands (New Zealand)	1870						Potts (1871); Worthy and Holdaway (2002)
cf. Ardeidae undescribed species		Easter Island	–		AD 1000–1430	No	Ahu Naunau at Anakena	8	Steadman *et al.* (1994); Steadman (2006b)
Ciconiidae									
Mycteria wetmorei		Cuba+southern North America	–					1, Note 9	Howard (1935); Olson (1991); Suárez and Olson (2003b)
Ciconia maltha	Asphalt stork	Cuba+southern North America+South America?	–					1, Note 9	Suárez and Olson (2003b); Agnolin (2006)
Ciconia undescribed small species		Cuba	–				Las Breas de San Felipe	1	Suárez and Olson (2003b)
Threskiornithidae									
Apteribis brevis	Maui highland apteribis	Maui	–	1850±270	538 BC–AD 687*	No	Puu Naio Cave, E24	1	Olson and James (1991)
Apteribis glenos	Molokai apteribis	Molokai	–					1	Olson and Wetmore (1976); Olson and James (1982, 1991)
Apteribis undescribed species	Maui lowland apteribis	Maui	–					1	Olson and James (1984, 1991)
Threskiornis solitarius	Réunion solitaire, Réunion flightless ibis	Réunion	1705						Mourer-Chauviré and Moutou (1987); Mourer-Chauviré *et al.* (1995a, 1995b, 1999)
Xenicibis xympithecus	Club-winged ibis	Jamaica	–	2145±220	787 BC–AD 320*	No	Long Mile Cave	1	Olson and Steadman (1977, 1979); Morgan (1993); McFarlane *et al.* (2002)

Table *continued*

Scientific name	Common name	Range	Last occurrence				Archaeological or stratigraphic association	Other estimate (see list in footnotes; see also Notes)	Reference
			Historical record	Radiometric date					
				^{14}C age (years BP), ±1 σ	Calibrated date, ±2 σ	Direct date?			
Cathartidae									
Gymnogyps varonai		Cuba	–					1	Arredondo (1971); Arredondo and Olson (1976); Suárez (2000a); Suárez and Emslie (2003)
Teratornithidae†									
Teratornis olsoni		Cuba	–					1	Arredondo and Arredondo (2002a)
Accipitridae									
Gyps melitensis		Corsica+Sardinia+ Malta and mainland Southern Europe	–	14 260±60 BP	17 314–16 876 BP	Yes	Castiglione 3, Fracture PL	1, 8, Note 10	Lydekker (1890); Tyrberg (1998); Louchart (2002); Pereira *et al.* (2006)
Circus dossenus		Oahu+Molokai	–				Moomomi Dunes, Molokai, Barbers Point Oahu	1	Olson and James (1991)
Circus eylesi	Eyles's harrier	North Island+South Island (New Zealand)	–				Occurs in Maori middens, post *c.* AD 1280	8	Worthy and Holdaway (2002); Tennyson and Martinson (2006)
Gigantohierax suarezi		Cuba	–					1	Arredondo and Arredondo (2002b)
Accipiter efficax	Powerful goshawk	Grande Terre (New Caledonia)	–	1750±70	86–428 AD*	No	Pindai		Balouet and Olson (1989)
Accipiter quartus	Fourth goshawk	Grande Terre (New Caledonia)	–	1750±70	86–428 AD*	No	Pindai		Balouet and Olson (1989)
Amplibuteo woodwardi		Cuba+southern North America	–				Cueva de Sandoval	1, Note 9	Suárez (2004a)

Aquila nipaloides		Corsica+Sardinia	–				Grotta di Corbeddu (Sardinia), *c.*8000–15 000 BP, and Castiglione 3 Locus PL (Corsica), 9000–17 000 BP	1	Louchart *et al.* (2005)
Aquila undescribed large species		Madagascar	–				Ampasambazimba 1000–8000 BP	1	Goodman and Rakotozafy (1995)
Aquila undescribed small species		Madagascar	–				Lamboharana 1200–3500 BP	1	Goodman and Rakotozafy (1995)
Hieraaetus moorei	Haast's eagle	South Island (New Zealand)	–	2096±72	359 BC–AD 52*	Yes	Occurs in Maori middens, post *c.* AD 1280	8	Worthy and Holdaway (2002); Bunce *et al.* (2005); Tennyson and Martinson (2006)
Stephanoaetus mahery		Madagascar	–				Ampasambazimba 1000–8000 BP	1	Goodman (1994)
Titanohierax gloveralleni		New Providence, Little Exuma (Bahamas)+ Grand Cayman (Cayman Islands)	–					1	Wetmore (1937b); Olson and Hilgartner (1982); Morgan (1994)
Titanohierax undescribed species		Hispaniola	–					1	Olson and Hilgartner (1982)
Buteogallus borrasi		Cuba	–					1	Arredondo (1970); Arredondo and Olson (1976); Suárez (2001); Suárez and Olson (2007)
Falconidae									
Caracara creightoni	Bahamas caracara	New Providence+ Abaco (Bahamas)+ Cuba+Grand Cayman (Cayman Islands)	–				Sawmill Sink, Abaco, 1040–3820 BP	8, Note 11	Brodkorb (1959); Olson and Hilgartner (1982); Morgan (1994); Suárez and Olson (2001b); Steadman *et al.* (2007)
Caracara lutosa	Guadalupe caracara	Isla Guadalupe	1900						Greenway (1967); BirdLife International (2000)
Caracara undescribed species		Jamaica	–				Jackson's Bay Caves	1	Suárez and Olson (2001b); McFarlane *et al.* (2002)

Table *continued*

Scientific name	Common name	Range	Last occurrence				Archaeological or stratigraphic association	Other estimate (see list in footnotes; see also Notes)	Reference
			Historical record	Radiometric date					
				^{14}C age (years BP), ±1σ	Calibrated date, ±2σ	Direct date?			
Milvago alexandri		Hispaniola	–				Cave at St. Michel de l'Atalaye	1	Olson (1976a)
Milvago carbo		Cuba	–				Las Breas de San Felipe	1	Suárez and Olson (2003a)
Polyborus latebrosus	Puerto Rican caracara	Puerto Rico	–					1	Wetmore (1920, 1922b)
Falco duboisi	Réunion kestrel	Réunion	Probable report 1671/2						Cheke (1987); Cowles (1994); Mourer-Chauviré *et al.* (1999)
Falco kurochkini		Cuba	–					1	Suárez and Olson (2001a); Suárez (2004b)
Falconidae undescribed small species		Réunion	Probable report 1671/2						Mourer-Chauviré *et al.* (1999)
Gruidae									
Grus cubensis		Cuba	–				Caverna de Pio Domingo	1	Fischer (1968); Fischer and Stephan (1971a)
Grus melitensis		Malta+Sicily	–					1, 5, Note 2	Lydekker (1890); Northcote (1982a, 1984, 1992); Pavia (2000)
Grus primigenia		Western Europe+ Mallorca	–				Glastonbury, pre-Roman Iron Age	8	Harrison and Cowles (1977); Northcote and Mourer-Chauvire (1985, 1988); von den Driesch (1999)
Grus undescribed flightless species		Miyako (Ryukyus)	–	25800±900	–	No	Pinza-Abu Cave	2, 8	Matsuoka (2000)
Rallidae									
Dryolimnas augusti	Réunion rail	Réunion	Probable report 1671/2						Mourer-Chauviré *et al.* (1999)

Dryolimnas undescribed species	Sauzier's wood-rail	Mauritius	–					1	Hume and Prŷs-Jones (2005); Cheke and Hume (2008)
Aramides gutturalis	Red-throated wood-rail	Peru	pre-1843						BirdLife International (2000)
Gallirallus dieffenbachi	Dieffenbach's rail, mehoriki	Chatham Islands	1840	677±60	AD 1283–1515	Yes			Greenway (1967); Millener (1999)
Gallirallus epulare	Nuku Hiva rail	Nuku Hiva (Marquesas)	–				Ha'atuatua Site	8	Steadman (2006b); Kirchman and Steadman (2007)
Gallirallus ernstmayri	New Ireland rail	New Ireland	–				Panakiwuk and Balof Rockshelters, Matenkupkum and Matenbek Caves, youngest record <2000 BP	8	Steadman *et al.* (1999); Kirchman and Steadman (2006b); Steadman (2006b)
Gallirallus gracilitibia	Ua Huka rail	Ua Huka (Marquesas)	–				Hane Dune site (MUH-1), Phase I–II, AD 300–1300	8	Steadman (2006b); Kirchman and Steadman (2007)
Gallirallus huiatua	Niue rail	Niue	–				Anakuli Cave, Hakupu, 3500–4500 BP	1	Steadman *et al.* (2000)
Gallirallus lafresnayanus	New Caledonian rail	Grande Terre+Isle des Pins (New Caledonia)	1890 (1984?)					6	BirdLife International (2000); Steadman (2006b)
Gallirallus macquariensis	Macquarie Island rail	Macquarie Island	1880						Worthy and Holdaway (2002); Tennyson and Martinson (2006)
Gallirallus pacificus	Tahiti rail	Tahiti (Society Islands)	1777						Greenway (1967); BirdLife International (2000)
Gallirallus pendiculentus	Tinian rail	Tinian (Marianas)	–	1880±50	AD 20–245*	Yes	Railhunter Rockshelter, Layer I–III	8	Steadman (1999, 2006b); Kirchman and Steadman (2006b)
Gallirallus pisonii	Aguiguan rail	Aguiguan (Marianas)	–	1780±70	AD 80–420	No	Pisonia Rockshelter, Layer III–V	8	Steadman (1999, 2006b); Kirchman and Steadman (2006b)
Gallirallus poecilopterus	Bar-winged rail	Viti Levu+Ovalau (Fiji)	1890 (1970?)						Steadman (2006b)
Gallirallus ripleyi	Ripley's rail	Mangaia (Cook Islands)	–				Te Ana Manuku Rockshelter (MAN-84), AD 1390–1470 cal	3, 8	Steadman (1987, 2006b); Steadman and Kirch (1990)

Table *continued*

Scientific name	Common name	Range	Last occurrence				Archaeological or stratigraphic association	Other estimate (see list in footnotes; see also Notes)	Reference
			Historical record	Radiometric date					
				^{14}C age (years BP), ±1σ	Calibrated date, ±2σ	Direct date?			
Gallirallus roletti	Tahuata rail	Tahuata (Marquesas)	–				Hanamiai Site, Phase I–III, 550–1000 BP	8	Steadman and Rolett (1996); Steadman (2006b); Kirchman and Steadman (2007)
Gallirallus sharpei	Sharpe's rail	?	pre-1893					6	Olson (1986b); BirdLife International (2000)
Gallirallus storrsolsoni	Huahine rail	Huahine (Society Islands)	–		AD 700–1150	No	Fa'ahia site	8	Kirchman and Steadman (2006a); Steadman (2006b)
Gallirallus temptatus	Rota rail	Rota (Marianas)	–				Mochong archaeological site, <2500 BP	8	Steadman (1999, 2006b); Kirchman and Steadman (2006b)
Gallirallus vekamatolu	'Eua rail	'Eua (Tonga)	–				Anatú site, Layer II–III, >3000 BP	1	Kirchman and Steadman (2005); Steadman (2006b)
Gallirallus wakensis	Wake rail	Wake	1942						Steadman (2006b)
Gallirallus undescribed flightless species		Miyako (Ryukyus)	–	25 800±900	–	No	Pinza-Abu Cave	2, 5	Matsuoka (2000)
Gallirallus undescribed species B	Buka rail	Buka (Solomons)	–				Kilu Cave, Layer I and II, <11 000 BP	2	Steadman (2006b)
Gallirallus undescribed species C	Toga rail	Toga (Vanuatu)	–					1	Steadman (2006b)
Gallirallus undescribed species D	Lakeba rail	Lakeba (Lau group)	–				Quara ni Puqa rock-shelter, <2800 BP	3, 8	Steadman (2006b)
Gallirallus undescribed species E		Lifuka (Tonga)	–				Tongoleleka site, 2700–2850 BP	8	Steadman (2006b)
Gallirallus undescribed species F	Ha'afeva rail	Ha'afeva (Tonga)	–				Mele Havea site, 2700–2850 BP	8	Steadman (2006b)
Gallirallus undescribed species J	Hiva Oa rail	Hiva Oa (Marquesas)	–				Hanatekua Rockshelter	8	Steadman (2006b); Kirchman and Steadman (2007)
Gallirallus undescribed species O	Saipan rail	Saipan (Marianas)	–				Chalan Piao archaeological site, 3400 BP	8	Steadman (2006b)

Gallirallus undescribed species	Norfolk Island rail	Norfolk Island	–				Cemetery Beach, Nepean Island	1, 3	Meredith (1991)
Gallirallus undescribed species		Vava'u (Tonga)	1793					7	Olson (2006)
Mundia elpenor	Ascension flightless crake	Ascension	1656						Ashmole (1963); Olson (1973); Bourne *et al.* (2003)
Nesotrochis debooyi	Antillean cave-rail	Puerto Rico+Virgin Islands	–				Midden site Richmond estate, Chistiansted, St. Croix	3, 8, Note 12	Wetmore (1918, 1922b, 1927, 1937a); Olson (1977b)
Nesotrochis picapicensis	Cuban cave-rail	Cuba	Possible report 1625	4190±60	4853–4537 BP*	No		4, 8	Fischer and Stephan (1971b); Olson (1974, 1977b); Jimenez Vazquez (1997, 2001)
Nesotrochis steganinos	Haitian cave-rail	Hispaniola	–				Cave at St. Michel de l'Atalaye	1	Olson (1974, 1977b)
Rallus eivissensis		Eivissa	–	6,130±80 BP	7249–6789 BP*	Yes	Es Pouás	1	Alcover *et al.* (2005); McMinn *et al.* (2005)
Rallus undescribed species		Fernando de Noronha	–					1	Olson (1982b); Carleton and Olson (1999)
Rallus undescribed species	Abaco flightless rail	Abaco (Bahamas)	–				Sawmill Sink, Owl Roost deposit	1	Steadman *et al.* (2007)
Vitirallus watlingi	Viti Levu rail	Viti Levu (Fiji)	–					1	Worthy (2004b)
Atlantisia podarces	St. Helena crake	St. Helena	–					1	Wetmore (1963); Olson (1975)
Cabalus modestus	Chatham rail, Hutton's rail, mâtirakahu	Chatham Islands	1893	1270±120	AD 344–882	Yes			Greenway (1967); Trewick (1997); Millener (1999); Worthy and Holdaway (2002); Tennyson and Martinson (2006)
Capellirallus karamu	Snipe rail	North Island (New Zealand)	–				Occurs in Maori middens, post c. AD 1280	8	Falla (1954); Worthy and Holdaway (2002); Tennyson and Martinson (2006)
Diaphorapteryx hawkinsi	Forbes's rail, Hawkins's rail, mehonui	Chatham Islands	–	1860±150	152 BC–AD 550	Yes	Occurs in Moriori middens, post c. AD 1350	4, 8, Note 13	Trewick (1997); Millener (1999); Cooper and Tennyson (2004)
Porzana astrictocarpus	St. Helena rail	St. Helena	–					1	Olson (1973, 1975)
Porzana keplerorum	Kepler's crake	Maui	–				Auwahi Cave	1	Olson and James (1984, 1991); James *et al.* (1987)

Table *continued*

Scientific name	Common name	Range	Last occurrence				Archaeological or stratigraphic association	Other estimate (see list in footnotes; see also Notes)	Reference
			Historical record	Radiometric date					
				^{14}C age (years BP), ±1 σ	Calibrated date, ±2 σ	Direct date?			
Porzana menehune	Small Molokai rail	Molokai	–				Ilio Point and Moomomi dunes	1	Olson and James (1991)
Porzana monasa	Kosrae crake	Kosrae	1828						Greenway (1967); Slikas *et al.* (2002)
Porzana nigra	Tahiti crake, Miller's rail	Tahiti	1784					Note 14	BirdLife International (2000)
Porzana palmeri	Laysan crake	Laysan	1944						Slikas *et al.* (2002)
Porzana ralphorum	Medium Oahu rail	Oahu	–				Kuliouou shelter, <1500 BP	8	Olson and James (1991)
Porzana rua	Mangaian crake	Mangaia (Cook Islands)	–				Te Ana Manuku Rockshelter (MAN-84), AD 1390–1470 cal	3, 8	Steadman (1987, 2006b); Steadman and Kirch (1990)
Porzana sandwichensis	Hawai'ian crake	Hawai'i	1884 (1893?)						Olson and James (1991); BirdLife International (2000); Slikas *et al.* (2002)
Porzana severnsi	Severn's crake	Maui	–				Auwahi Cave	1	Olson and James (1982, 1984, 1991)
Porzana ziegleri	Small Oahu rail	Oahu	–		AD 650–869	Yes	Sinkhole 4907-D1, 'Ewa plain	1	Olson and James (1982, 1984, 1991); James (1987); Athens *et al.* (2002)
Porzana large undescribed species	Large Hawai'i rail	Hawai'i						8	Olson and James (1982, 1991)
Porzana small undescribed species	Small Hawai'i rail	Hawai'i		920±60	AD 1014–1238*	No	Pohakuloa, MPRC Site 10269	8	Olson and James (1991)
Porzana undescribed species 1	Medium Kauai rail	Kauai	–				Maha'ulepu Cave Unit IV–VI, <6000 BP	1	Olson and James (1982, 1984, 1991); Burney *et al.* (2001)
Porzana undescribed species 2	Large Kauai rail	Kauai	–					1	Olson and James (1991)

Porzana undescribed species	Medium Maui rail	Maui	–					1	Olson and James (1982, 1984, 1991)
Porzana undescribed species A	Malakula crake	Malakula	–				Yalu and Navaprah sites, <2900 BP	8	Steadman (2006b)
Porzana undescribed species B	Aiwa crake	Aiwa Levu (Lau Group)	–				Aiwa Levu Rockshelter 1, >2400 BP	8	Steadman (2006b)
Porzana undescribed species C	Small Mangaian crake	Mangaia (Cook Islands)	–				Te Ana Manuku Rockshelter (MAN-84), AD 1390–1470 cal	8	Steadman and Kirch (1990); Steadman (2006b)
Porzana undescribed species D	Nuku Hiva crake	Nuku Hiva (Marquesas)	–				Ha'atuatua site	8	Steadman (2006b)
Porzana undescribed species E		Ua Huka (Marquesas)	–				Hane site (MUH-1), Phase I–II, AD 300–1300	8	Steadman (2006b)
Porzana undescribed species F		Ua Huka (Marquesas)	–				Hane site (MUH-1), Phase I–II, AD 300–1300	8	Steadman (2006b)
Porzana undescribed species G	Easter Island crake	Easter Island	–		AD 1000–1430	No	Ahu Naunau at Anakena	8	Steadman *et al.* (1994); Steadman (2006b)
Porzana undescribed species H	Rota crake	Rota (Marianas)	–				Alaguan Rockshelter and As Matmos Cliffside Cave	3, 8	Steadman (1999, 2006b)
Porzana undescribed species I	Aguiguan crake	Aguiguan (Marianas)	–	540±60	AD 1290–1450	No	Pisonia Rockshelter, Layer I–V	8	Steadman (1999, 2006b)
Porzana undescribed species J	Tinian crake	Tinian (Marianas)	–	2420±60	2708–2346 BP*	Yes	Railhunter Rockshelter, Layer I–III	8	Steadman (1999, 2006b)
cf. *Porzana* undescribed species		Lisianski Island	1828					7	Olson and Ziegler (1995)
Gallinula hodgenorum	Hodgen's waterhen	North Island+South Island (New Zealand)	–				Parewanui, Lower Rangitikei River, *c.* AD 1769	8	Olson (1986a); Cassels *et al.* (1988); Worthy and Holdaway (2002); Tennyson and Martinson (2006)
Gallinula nesiotis	Tristan moorhen	Tristan da Cunha	1873						Greenway (1967); BirdLife International (2000)
cf. *Gallinula* sp.		Viti Levu (Fiji)	–					1	Worthy (2004b)

Table *continued*

Scientific name	Common name	Range	Last occurrence				Archaeological or stratigraphic association	Other estimate (see list in footnotes; see also Notes)	Reference
			Historical record	Radiometric date					
				^{14}C age (years BP), ±1σ	Calibrated date, ±2σ	Direct date?			
'Hovacrex' roberti		Madagascar	–				Antsirabe	1	Andrews (1897); Olson (1977b)
Pareudiastes pacifica	Samoan gallinule	Savai'i	1873 (1987?)					6	BirdLife International (2000)
Pareudiastes undescribed species	Buka gallinule	Buka (Solomons)	–				Kilu Cave, Layer I, <11 000 BP	8	Steadman (2006b)
Porphyrio albus	Lord Howe swamphen	Lord Howe Island	1790						Greenway (1967); BirdLife International (2000)
Porphyrio coerulescens	'Oiseau bleu'	Réunion	*c.*1734						Mourer-Chauviré *et al.* (1999)
Porphyrio kukwiedei	Divine swamphen, New Caledonia gallinule	Grande Terre+Isle des Pins (New Caledonia)	1860?	1750±70	AD 86–428*	No	Pindai	3, 4	Balouet and Olson (1989)
Porphyrio mantelli	North Island takahe, moho, mohoau	North Island (New Zealand)	1894				Occurs in Maori middens, post *c.* AD 1280	8	Phillipps (1959); Worthy and Holdaway (2002); Tennyson and Martinson (2006)
Porphyrio mcnabi	Huahine swamphen	Huahine	–		AD 700–1150	No	Fa'ahia site	8	Kirchman and Steadman (2006a); Steadman (2006b)
Porphyrio paepae	Marquesan swamphen	Hiva Oa+Tahuata (Marquesas)	1902?				Hanamiai Site, Phase III, Tahuata AD 1300–1450	8, Note 15	Steadman (1988, 2006b); Steadman and Rolett (1996)
Porphyrio undescribed species A	New Ireland swamphen	New Ireland	–				Panakiwuk and Balof Rockshelters and Matenkupkum Cave, youngest record <1600 BP	8	Steadman *et al.* (1999); Steadman (2006b)
Porphyrio undescribed species B	Buka swamphen	Buka (Solomons)	–				Kilu Cave, Layer I and II, <11 000 BP	2, 8	Steadman (2006b)
Porphyrio undescribed species C	Mangaia swamphen	Mangaia (Cook Islands)	–				Tangatatau rockshelter (MAN 44), <700 BP	8	Steadman and Kirch (1990); Steadman (2006b)

Porphyrio undescribed species E	Mariana swamphen	Rota+Tinian (Marianas)	–	930±70	AD 990–1260	No	Payapai Cave	3, 8	Steadman (1999, 2006b)
Fulica chathamensis	Chatham Island coot	Chatham Islands	–	2278±70	2651–2069 BP*	Yes	Occurs in Moriori middens, post *c.* AD 1350	8	Millener (1999)
Fulica newtoni	Mascarene coot	Mauritius+Réunion	Probable report 1671/1672 and possible report 1693						Cheke (1987); Mourer-Chauviré *et al.* (1999)
Fulica prisca	New Zealand coot	North Island+South Island (New Zealand)	–				Occurs in Maori middens, post *c.* AD 1280	8	Worthy and Holdaway (2002); Tennyson and Martinson (2006)
Fulica undescribed species		Oahu	–						James (1987); Hearty *et al.* (2005)
'*Fulica*' *podagrica*		Barbados	–				Ragged Point, Barbados	1	Brodkorb (1965); Olson (1974, 1977b)
Aphanapteryx bonasia	Red rail	Mauritius	1693						Olson (1977b); Cheke (1987)
Aphanapteryx leguati	Rodrigues rail	Rodrigues	1726						Olson (1977b); Cheke (1987)
Rallidae undescribed species		Madeira	–					1	Pieper (1985)
Rallidae undescribed species		Madeira	–					1	Pieper (1985)
Rallidae undescribed species		Madeira	–					1	Pieper (1985)
Rallidae undescribed species		Easter Island	–		AD 1000–1430	No	Ahu Naunau at Anakena	8	Steadman *et al.* (1994); Steadman (2006b)
Rhynochetidae									
Rhynochetos orarius	Coastal kagu	Grande Terre+Isle des Pins (New Caledonia)	–	1750±70	AD 86–428*	No	Pindai, Kanumera		Balouet and Olson (1989)
Aptornithidae†									
Aptornis defossor	South Island adzebill	South Island (New Zealand)	–				Occurs in Maori middens, post *c.* AD 1280	8	Worthy and Holdaway (2002); Tennyson and Martinson (2006)

Table *continued*

Scientific name	Common name	Range	Last occurrence				Archaeological or stratigraphic association	Other estimate (see list in footnotes; see also Notes)	Reference
			Historical record	Radiometric date					
				^{14}C age (years BP), $\pm 1\sigma$	Calibrated date, $\pm 2\sigma$	Direct date?			
Aptornis otidiformis	North Island adzebill	North Island (New Zealand)	–				Occurs in Maori middens, post *c.* AD 1280	8	Worthy and Holdaway (2002); Tennyson and Martinson (2006)
Haematopodidae									
Haematopus meadewaldoi	Canary Islands oystercatcher	Fuerteventura+ Lanzarote	1913 (1940?)						Collar *et al.* (1994); BirdLife International (2000)
Burhinidae									
Burhinus [*bistriatus*] *nanus*		New Providence (Bahamas)	–					1, Note 16	Brodkorb (1959); Olson and Hilgartner (1982)
Charadriidae									
Vanellus macropterus	Javanese lapwing	Java	1940					6	Greenway (1967)
Vanellus madagascariensis	Malagasy lapwing	Madagascar	–				Ampoza, Lamboharana	1	Goodman (1996)
Scolopacidae									
Prosobonia cancellata	Christmas sandpiper	Kirimati	1778						Greenway (1967)
Prosobonia ellisi	Ellis's sandpiper, white-winged sandpiper	Mo'orea (Society Islands)	1777						BirdLife International (2000)
Prosobonia leucoptera	White-winged sandpiper, Tahitian sandpiper	Tahiti (Society Islands)	1773						Greenway (1967); BirdLife International (2000)
Prosobonia aff. *parvirostris*		Ua Huka (Marquesas)	–				Hane site (MUH-1), Phase I–II, AD 300–1300	8	Steadman (2006b)
Prosobonia undescribed species A	Cook Islands sandpiper	Mangaia (Cook Islands)	–				Te Ana Manuku Rockshelter (MAN-84), AD 1390–1470 cal	8	Steadman and Kirch (1990); Steadman (2006b)

Prosobonia undescribed species B	Pitcairn sandpiper, Henderson sandpiper	Henderson	–				Several archaeological sites, AD 1000–1600	8	Weisler (1994); Wragg and Weisler (1994); Wragg (1995); Steadman (2006b)
Scolopax anthonyi	Puerto Rican woodcock	Puerto Rico	–					1	Wetmore (1920, 1922b); Olson (1976b)
Coenocorypha barrierensis	North Island snipe	North Island (New Zealand)	1870						Oliver (1955); Worthy and Holdaway (2002); Tennyson and Martinson (2006)
Coenocorypha chathamica	Forbes's snipe	Chatham Islands	–				Occurs in Moriori middens, post *c.* AD 1350	8	Worthy and Holdaway (2002); Tennyson and Martinson (2006)
Coenocorypha iredalei	South Island snipe	South Island (New Zealand)	1964						Worthy and Holdaway (2002); Tennyson and Martinson (2006)
Coenocorypha miratropica	Fijian snipe	Viti Levu (Fiji)	–				Vatuma Cave	1	Worthy (2003)
Coenocorypha undescribed species	Norfolk Island snipe	Norfolk Island	–				Cemetery Beach dunes	3	Meredith (1991)
Coenocorypha? undescribed species	New Caledonian snipe	Grande Terre (New Caledonia)	–				Gilles	1	Balouet and Olson (1989)
Gallinago undescribed species		Little Exuma (Bahamas)+Cuba+Cayman Brac (Cayman Islands)	–					1	Wetmore (1937b); Olson and Hilgartner (1982); Morgan (1994); Suárez (2004b)
Laridae									
Larus utunui	Society Islands gull	Huahine	–		AD 700–1150	No	Fa'ahia site	8	Steadman (2002a)
Larus undescribed species	Kauai gull	Kauai	–				Maha'ulepu Cave, Unit IV–VI, <6000 BP	1	Burney *et al.* (2001)
Alcidae									
Pinguinus impennis	Great auk	North Atlantic	1844 (1848?)						Grieve (1885); Gaskell (2000)
Fratercula dowi		Channel Islands	–	11 890±75	13 947–13 571 BP*	Yes		5	Guthrie *et al.* (2002); Guthrie (2005)

Table *continued*

Scientific name	Common name	Range	Last occurrence				Archaeological or stratigraphic association	Other estimate (see list in footnotes; see also Notes)	Reference
			Historical record	Radiometric date					
				^{14}C age (years BP), ±1σ	Calibrated date, ±2σ	Direct date?			
Columbidae									
'Raperia' godmanae	Lord Howe pigeon	Lord Howe Island	*c.*1790						Greenway (1967)
Columba jouyi	Ryukyu wood-pigeon	Okinawa+Daito Islands (Ryukyus)	1936						Greenway (1967); Matsuoka *et al.* (2002)
Columba melitensis		Malta	–					1	Lydekker (1891)
Columba versicolor	Bonin wood-pigeon	Nakondo Shima+Chichi-jima (Bonin Islands)	1889						Greenway (1967); BirdLife International (2000)
Nesoenas duboisi	Réunion pigeon	Réunion	Probably extinct by 1705						Mourer-Chauviré *et al.* (1999)
Macropygia arevarevauupa	Society Islands cuckoo-dove	Huahine (Society Islands)	–		AD 700–1150	No	Fa'ahia site	8	Steadman (1992, 2006b)
Macropygia heana	Marquesan cuckoo-dove	Nuku Hiva+Ua Huka (Marquesas)	–				Hane, Ua Huka, Phase I–II (AD 300–1200)	8	Steadman (1992, 2006b)
Microgoura meeki	Meek's ground pigeon, Choiseul pigeon	Choiseul	1904						BirdLife International (2000); Mayr and Diamond (2001)
Natunaornis gigoura	Fiji flightless pigeon	Viti Levu (Fiji)	–					1	Worthy (2001)
Pezophaps solitaria	Solitaire, Rodrigues solitaire	Rodrigues	1754 (1761?)						Cheke (1987)
Raphus cucullatus	Dodo	Mauritius	1662						Cheke (1987, 2006)
Geotrygon larva	Puerto Rican quail-dove	Puerto Rico	–				Midden at Mesa hill, Mayaguez	8	Wetmore (1920, 1922b, 1923)
Caloenas canacorum	Kanaka pigeon	Grande Terre (New Caledonia)+Ha'apai Group (Tonga)	–	1750±70	AD 86–428*	No	Pindai		Balouet and Olson (1989); Steadman (2006b)

Gallicolumba ferruginea	Tanna ground-dove	Tanna (Vanuatu)	1774						Greenway (1967)
Gallicolumba leonpascoi	Henderson ground-dove	Henderson	–				Several archaeological sites, AD 1000–1600	8	Worthy and Wragg (2003)
Gallicolumba longitarsus	Long-tarsus ground-dove	Grande Terre (New Caledonia)	–	1750±70	AD 86–428*	No	Pindai		Balouet and Olson (1989)
Gallicolumba nui	Immense ground-dove	Mangaia (Cook Islands)+Huahine (Society Islands)+ Mangareva (Gambier Islands)+Ua Huka+ Hiva Oa+Tahuata (Marquesas)	–				Te Ana Manuku Rockshelter (MAN-84), Mangaia (AD 1390–1470 cal)	8	Steadman and Kirch (1990); Steadman (1992, 2006b); Steadman and Rolett (1996); Steadman and Justice (1998)
Gallicolumba salamonis	Thick-billed ground-dove	San Cristobal+Ramos (Solomons)	1927						BirdLife International (2000)
Gallicolumba undescribed species	Large Marianas ground-dove	Rota (Marianas)	–	400±60	AD 1420–1640	No	As Matmos Cliffside Cave	3, 8	Steadman (1999, 2006b)
Gallicolumba? norfolciensis	Norfolk Island ground-dove	Norfolk Island	1790						Schodde *et al.* (1983)
Didunculus placopedetes	Tongan tooth-billed pigeon	Ha'apai group+ Tongatapu+'Eua (Tonga)	–				Several archaeological sites, 2700–2850 BP	8	Steadman (2006a, 2006b)
Ptilinopus mercierii	Red-moustached fruit-dove	Nuku Hiva, Hiva Oa (Marquesas)	1922					Note 17	BirdLife International (2000)
Ptilinopus undescribed species	Tubuai fruit-dove	Tubuai	–				Atiahara site, AD 1000–1400	8	Steadman (2006b)
Alectroenas nitidissima	Mauritius blue pigeon	Mauritius	1826						Cheke (1987)
Alectroenas rodericana	Rodrigues pigeon	Rodrigues	1708						Cheke (1987)
cf. *Alectroenas* undescribed species		Réunion	Probably extinct by 1705						Mourer-Chauviré *et al.* (1999)
Ducula david	David's imperial-pigeon	'Uvea (Loyalty group)	–				Utuleve (WF-U-MU 21A) site, Bed VI, 2000–2500 BP	8	Balouet and Olson (1987); Steadman (2006b)

Table *continued*

Scientific name	Common name	Range	Last occurrence				Archaeological or stratigraphic association	Other estimate (see list in footnotes; see also Notes)	Reference
			Historical record	Radiometric date					
				^{14}C age (years BP), ±1σ	Calibrated date, ±2σ	Direct date?			
Ducula harrisoni	Henderson Island imperial-pigeon	Henderson	–				Several archaeological sites, AD 1000–1600	8	Wragg and Weisler (1994); Wragg and Worthy (2006)
Ducula lakeba	Lau imperial-pigeon	Viti Levu (Fiji)+Lakeba+Aiwa Levu (Lau Group)	–					8	Worthy (2001); Steadman (2006b)
Ducula undescribed species	Tongan imperial-pigeon	Ha'apai group+'Eua (Tonga)	–				Several archaeological sites, 2700–2850 BP	8	Steadman (2006b)
Ducula undescribed species		Taravai (Mangareva)	–				Onemea site *c.* AD 1000–1200	8	Worthy and Tennyson (2004); Kirch (2007)
Ectopistes migratorius	Passenger pigeon	North America	1914						Greenway (1967)
Hemiphaga spadicea	Norfolk Island pigeon	Norfolk Island	1839						Schodde *et al.* (1983)
Bountyphaps obsoleta	Henderson archaic pigeon	Henderson	–				Several archaeological sites, AD 1000–1600	8	Worthy and Wragg (2008)
Columbidae undescribed genus and species A	Small-winged ground-pigeon	Buka (Solomons)	–				Kilu Cave, Layer I and II, <11 000 BP	2, 8	Steadman (2006b)
Columbidae undescribed genus and species B	Kilu ground-pigeon	Buka (Solomons)	–				Kilu Cave, Layer I and II, <11 000 BP	2, 8	Steadman (2006b)

Columbidae undescribed genus C	Royal Tongan pigeon	Ha'apai group+ Tongatapu+'Eua (Tonga)	–				Several archaeological sites, 2700–2850 BP	8	Steadman (2006b)
Psittacidae									
Vini sinotoi	Sinoto's lorikeet	Mangaia (Cook Islands)+Huahine (Society Islands)+ Nukuhiva+Ua Huka+Hiva Oa+ Tahuata (Marquesas)	–	1000±250	AD 559–1418*	No	Te Ana Manuku Rockshelter (MAN-84), Mangaia (AD 1390–1470 cal)	8	Steadman and Zarriello (1987); Steadman and Kirch (1990); Steadman and Rolett (1996); Steadman (2006b)
Vini vidivici	Conquered lorikeet	Mangaia (Cook Islands)+Huahine (Society Islands)+ Nukuhiva+Ua Huka+Hiva Oa+ Tahuata (Marquesas)	–	1000±250	AD 559–1418*	No	Te Ana Manuku Rockshelter (MAN-84), Mangaia (AD 1390–1470 cal)	8	Steadman and Zarriello (1987); Steadman and Kirch (1990); Steadman and Rolett (1996); Steadman (2006b)
Charmosyna diadema	New Caledonian lorikeet	Grande Terre (New Caledonia)	1859 (1976?)					6	BirdLife International (2000)
Conuropsis carolinensis	Carolina parakeet	Southeastern North America	1918						Greenway (1967)
Cacatua aff. *galerita* undescribed species	New Ireland cockatoo	New Ireland	–				Balof Rockshelter, c.1000–5000 BP	8	Steadman *et al.* (1999); Steadman (2006b)
Cacatua undescribed species B	New Caledonian cockatoo	Grande Terre (New Caledonia)	–				Lapita site, c.2900 BP	8	Steadman (1997b, 2006b)
Nestor productus	Norfolk Island kaka	Norfolk Island	1851						Schodde *et al.* (1983)
Nestor undescribed species	Chatham Island kaka	Chatham Islands	–				Occurs in Moriori middens, post c. AD 1350	8	Millener (1999)
Eclectus infectus	Oceanic eclectus parrot	Malakula (Vanuatu)+ Ha'apai group+'Eua (Tonga)	–				Found at a number of archaeological sites, c.2700–2850 BP	8, Note 18	Steadman (2005, 2006b)
Eclectus cf. *infectus*		Vava'u (Tonga)	1793					7	Olson (2006)
Lophopsittacus mauritianus	Broad-billed parrot, raven parrot	Mauritius	1673						Cheke (1987); Hume (2007)

Table *continued*

Scientific name	Common name	Range	Last occurrence				Archaeological or stratigraphic association	Other estimate (see list in footnotes; see also Notes)	Reference
			Historical record	Radiometric date					
				^{14}C age (years BP), ±1σ	Calibrated date, ±2σ	Direct date?			
Mascarinus mascarinus	Mascarene parrot	Réunion	c.1785						Mourer-Chauviré *et al.* (1999); Hume and Prŷs-Jones (2005); Hume (2007)
Necropsittacus rodericanus	Rodrigues parrot	Rodrigues	1761						Cheke (1987); Hume (2007)
Psephotus pulcherrimus	Paradise parrot	Australia	1927						BirdLife International (2000)
Cyanoramphus ulietanus	Ra'iatea parakeet	Ra'iatea (Society Islands)	1774						Greenway (1967); BirdLife International (2000)
Cyanoramphus zealandicus	Black-fronted parakeet	Tahiti (Society Islands)	1844						Greenway (1967); BirdLife International (2000)
Psittacula bensoni	Mauritius grey parrot; Thirioux's grey parrot	Mauritius	1764						Cheke (1987); Hume (2007)
Psittacula cf. *bensoni*	Réunion grey parrot	Réunion	1732						Hume (2007)
Psittacula exsul	Newton's parakeet, Rodrigues parakeet	Rodrigues	1875						Cheke (1987); Hume (2007)
Psittacula wardi	Seychelles parakeet	Seychelles (Mahé+ Silhouette+ Praslin)	1907						BirdLife International (2000); Hume (2007)
Anodorhynchus glaucus	Glaucous macaw	South America	1951 (1960?)					6	BirdLife International (2000)
Ara atwoodi	Dominican green-and-yellow macaw	Dominica	1791					7	Greenway (1967); BirdLife International (2000); Williams and Steadman (2001)
Ara autochthones	St. Croix macaw	St. Croix, Puerto Rico	–					8	Wetmore (1937a); Williams and Steadman (2001)
Ara erythrocephala	Jamaican green-and-yellow macaw	Jamaica	1847					7	Greenway (1967); BirdLife International (2000); Williams and Steadman (2001)

Ara gossei	Jamaican red macaw	Jamaica	*c.*1765					7	Greenway (1967); BirdLife International (2000); Williams and Steadman (2001)
Ara guadeloupensis	Lesser Antillean macaw	Guadeloupe+ Martinique	1742					7	Greenway (1967); BirdLife International (2000); Williams and Steadman (2001)
Ara tricolor	Cuban macaw	Cuba	1864 (1876?)						Greenway (1967); Williams and Steadman (2001)
Ara undescribed small species	Monserrat macaw	Montserrat	–				Trants archaeological site, <2500 BP	8	Williams and Steadman (2001)
Aratinga labati	Guadeloupe parakeet	Guadeloupe	1750					7	Greenway (1967); BirdLife International (2000)
Aratinga undescribed species	Barbudan parakeet	Barbuda	–				Barbuda I and II palaeontological sites	8	Pregill *et al.* (1994); Williams and Steadman (2001)
Amazona martinicana	Martinique parrot	Martinique	1742					7	Greenway (1967); BirdLife International (2000)
Amazona violacea	Guadaloupe parrot	Guadeloupe	1742					7	BirdLife International (2000); Williams and Steadman (2001)
Amazona cf. *violacea*	Guadaloupe parrot?	Marie Galante	–				Folle Anse archaeological site, <2500 BP	8	Williams and Steadman (2001)
Amazona undescribed large species	Turks and Caicos parrot	Grand Turk Island	–	1280±60	AD 650–885	No	Coralie archaeological site	8	Carlson (1999); Williams and Steadman (2001)
Amazona undescribed species	Monserrat parrot	Montserrat	–				Trants archaeological site, <2500 BP	8	Reis and Steadman (1999); Williams and Steadman (2001)
Amazona undescribed species	Grenada parrot	Grenada	1667					7	Butler (1992); Williams and Steadman (2001)
'*Necropsittacus*' *borbonicus*	Réunion red and green parakeet	Réunion	1671						Hume (2007)
Psittacidae undescribed species		Rota (Marianas)	–	930±70	AD 990–1260	No	Payapai Cave, Layer II	3, 8	Steadman (1999, 2006b)

Table *continued*

Scientific name	Common name	Range	Last occurrence				Archaeological or stratigraphic association	Other estimate (see list in footnotes; see also Notes)	Reference
			Historical record	Radiometric date					
				^{14}C age (years BP), ±1σ	Calibrated date, ±2σ	Direct date?			
cf. Psittacidae undescribed species 1		Easter Island	–		AD 1000–1430	No	Ahu Naunau at Anakena	8	Steadman *et al.* (1994); Steadman (2006b)
cf. Psittacidae undescribed species 2		Easter Island	–		AD 1000–1430	No	Ahu Naunau at Anakena	8	Steadman *et al.* (1994); Steadman (2006b)
Cuculidae									
Coua berthae	Bertha's coua	Madagascar	–					1	Goodman and Ravoavy (1993)
Coua delalandei	Snail-eating coua	Madagascar	1834						Greenway (1967); Collar *et al.* (1994); BirdLife International (2000)
Coua primaeva		Madagascar	–	1980±60	166 BC–AD 134*	Yes	Tsiandroina	1	Milne-Edwards and Grandidier (1895); Goodman and Rakotozafy (1997)
Nannococcyx psix	St. Helena cuckoo	St. Helena	–						Olson (1975)
Tytonidae									
Tyto cavatica	Puerto Rican barn owl	Puerto Rico	–					4	Wetmore (1920, 1922b)
Tyto melitensis		Malta	–					1	Lydekker (1891)
Tyto neddi		Barbuda	–				Rat Pocket, Two Foot Bay	1	Steadman and Hilgartner (1999)
Tyto noeli	Noel's barn owl	Cuba	–	17 406±161	21 056–20 159 BP*	Yes	Cueva el Abrón	1	Arredondo (1972a); Arredondo and Olson (1976); Suárez and Diaz-Franco (2003)
Tyto ostologa	Hispaniolan barn owl	Hispaniola	–					1	Wetmore (1922a)
Tyto pollens	Bahaman barn owl	New Providence+ Great Exuma+ Andros? (Bahamas)	–					1	Wetmore (1937b); Brodkorb (1959)
Tyto riveroi	Rivero's barn owl	Cuba	–					1	Arredondo (1972b); Arredondo and Olson (1976)

Tyto undescribed species A		New Ireland	–				Matenkupkum Cave, >10 000 BP	2	Steadman *et al.* (1999); Steadman (2006b)
Tyto undescribed small species	Cuban barn owl	Cuba	–	17 406±161	21 056–20 159 BP*	No	Cueva el Abrón	1	Suárez and Diaz-Franco (2003); Suárez (2004b)
Tyto undescribed species		Madeira	–					1	Jaume *et al.* (1993)
Tyto? letocarti	Letocart's owl	Grande Terre (New Caledonia)	–				Gilles	1	Balouet and Olson (1989)
Strigidae									
Otus siaoensis	Siau scops owl	Siau	1866					6	BirdLife Internstional (2000)
Otus undescribed species		Madeira	–					1	Pieper (1985)
Gymnoglaux undescribed species		Cuba	–				Cueva de Paredones	1	Louchart (2005)
Mascarenotus murivorus	Rodrigues owl	Rodrigues	1726						Cheke (1987)
Mascarenotus grucheti	Réunion owl	Réunion	–					1	Mourer-Chauviré *et al.* (1994, 1999)
Mascarenotus sauzieri	Mauritius owl	Mauritius	1800–1850						Cheke (1987)
Ornimegalonyx oteroi	Cuban giant owl	Cuba	–					1	Arredondo (1958); Arredondo and Olson (1976)
Ornimegalonyx species 2		Cuba	–					1, Note 19	Arredondo (1982); Louchart (2005)
Bubo insularis		Corsica+Sardinia	–	8225±80	9416–9015 BP*	No			Mourer-Chauviré and Weesie (1986); Vigne *et al.* (1997); Louchart (2002)
Bubo osvaldoi		Cuba	–					1	Arredondo and Olson (1994)
Grallistrix auceps	Long-legged Kauai owl	Kauai	–	2328±60	2694–2152 BP*	Yes	Maha'ulepu Cave, Unit VI. *c.* AD 1425–1660	8	Olson and James (1982, 1984, 1991)
Grallistrix erdmani	Long-legged Maui owl	Maui	–		AD 1057–1440	No	Molohai heiau	3, 8	Olson and James (1984, 1991); James *et al.* (1987); Kolb (1994)
Grallistrix geleches	Long-legged Molokai owl	Molokai	–				Moomomi and Ilio Point Dunes	1	Olson and James (1982, 1984, 1991)
Grallistrix orion	Long-legged Oahu owl	Oahu	–				Barbers Point	1	Olson and James (1982, 1984, 1991)
Pulsatrix arredondoi		Cuba	–					1	Brodkorb (1959)

Table *continued*

Scientific name	Common name	Range	Last occurrence				Archaeological or stratigraphic association	Other estimate (see list in footnotes; see also Notes)	Reference
			Historical record	Radiometric date					
				^{14}C age (years BP), ±1σ	Calibrated date, ±2σ	Direct date?			
Sceloglaux albifacies	Laughing owl	North Island+South Island (New Zealand)	1914 (1930?)						Worthy and Holdaway (2002); Tennyson and Martinson (2006)
Athene angelis		Corsica	–				Castiglione 3 Locus PL (Corsica) 9000–17 000 BP	1	Mourer-Chauviré *et al.* (1997); Louchart (2002)
Athene cretensis	Cretan little owl	Crete+Armathia	–					1	Weesie (1982, 1988); Pieper (1984)
Athene undescribed small species		Puerto Rico	–					1	Pregill and Olson (1981); Olson (1985b)
Athene undescribed species		Euboia and adjoining Greek mainland	–				Southeast Gate, New Halos: Hellenistic period, 220–260 BC	8	Prummel (2005)
Asio priscus		Channel Islands	–	12 020±270	14 861–13 344 BP*	No	San Miguel Island, Site V-12 (Cuyler locality)	5	Howard (1964); Guthrie (1993, 2005)
Strigidae undescribed small species	–	Eivissa	–					1	Florit *et al.* (1989)
Aegothelidae									
Aegotheles novaezealandiae	New Zealand owlet-nightjar	North Island+South Island (New Zealand)	–	954±60	AD 987–1214*	Yes	Occurs in Maori middens, post *c.* AD 1280	8	Holdaway *et al.* (2002a)
Caprimulgidae									
Siphonorhis americana	Jamaican pauraque	Jamaica	1859						Greenway (1967); BirdLife International (2000)
Siphonorhis daiquiri	Cuban pauraque	Cuba	–	7864±96	8993–8457 BP*	No	Cuevas Blancas		Olson (1985a); Suárez (2000b); Jiménez-Vázquez *et al.* (2005)

Apodidae						
Collocalia manuoi	Mangaia swiftlet	Mangaia (Cook Islands)	–	Te Ana Manuku Rockshelter (MAN-84), AD 1390–1470 cal	8	Steadman (2002b, 2006b)
Tachornis uranoceles	Puerto Rican palm swift	Puerto Rico	–	Blackbone Cave	5	Olson (1982a)
Trochilidae						
Chlorostilbon bracei	Brace's emerald	New Providence (Bahamas)	1877			Graves and Olson (1987)
Chlorostilbon elegans	Gould's emerald	Jamaica or Bahamas?	1860			Weller (1999)
Heliangelus zusii	Bogota sunangel	Colombia	1909		6	BirdLife International (2000)
Eriocnemis godini	Turquoise-throated puffleg	Colombia-Ecuador	pre-1900 (1976?)		6	BirdLife International (2000)
Alcedinidae						
Halcyon miyakoensis	Miyako kingfisher	Miyako (Sakishima) (Ryukyus)	1887			Greenway (1967)
Brachypteracidae						
Brachypteracias langrandi		Madagascar	–		1	Goodman (2000)
Upupidae						
Upupa antaios	St. Helena hoopoe	St. Helena	–		1	Olson (1975)
Bucerotidae						
Aceros undescribed species	New Caledonian hornbill	Lifu (Loyalty Islands)	–		8	Steadman (2006b)
Rhinocryptidae						
Merulaxis stresemanni	Stresemann's bristlethroat	Brazil	1945			BirdLife International (2000)
Scytalopus undescribed species		Cuba	–	Cuevas Blancas, 7864±96 BP	6b	Olson and Kurochkin (1987); Jiménez-Vázquez *et al.* (2005)
Acanthisittidae						
Dendroscansor decurvirostris	Long-billed wren	South Island (New Zealand)	–		1	Millener and Worthy (1991)

Table *continued*

Scientific name	Common name	Range	Last occurrence				Archaeological or stratigraphic association	Other estimate (see list in footnotes; see also Notes)	Reference
			Historical record	Radiometric date					
				^{14}C age (years BP), ±1 σ	Calibrated date, ±2 σ	Direct date?			
Pachyplicas yaldwyni	Stout-legged wren	North Island+South Island (New Zealand)	–				Occurs in Maori middens, post *c.* AD 1280	8, Note 20	Millener (1988); Worthy and Holdaway (2002); Tennyson and Martinson (2006)
Traversia lyalli	Lyall's wren	North Island+South Island (New Zealand)	1895						Worthy and Holdaway (2002); Tennyson and Martinson (2006)
Xenicus longipes	Bush wren	North Island+South Island (New Zealand)	1972						Worthy and Holdaway (2002); Tennyson and Martinson (2006)
Hirundinidae									
Eurychelidon serintarae	White-eyed river martin	Thailand	1978 (1986?)					6	BirdLife International (2000)
Campephagidae									
cf. *Lalage* sp.	Tongan triller	Tongatapu+'Eua (Tonga)	–				Ha'ateiho site, Layer 3, <2800 BP	8	Steadman (2006b)
Pycnonotidae									
Hypsipetes undescribed species	Rodrigues merle	Rodrigues	–					1	Cheke (1987); Cheke and Hume (2008)
Muscicapidae									
Myadestes myadestinus	Kama'o	Kauai	1983 (1989?)					6	Greenway (1967); BirdLife International (2000)
Myadestes woahensis	Amaui	Oahu	1825						Greenway (1967); BirdLife International (2000)
Myadestes undescribed species aff. *lanaiensis*		Maui	–					1	James and Olson (1991)
Turdidae									
Zoothera terrestris	Bonin Island thrush	Ogasawara Shoto (Bonin Islands)	1828						Greenway (1967); BirdLife International (2000)

Turdus ravidus	Grand Cayman thrush	Grand Cayman (Cayman Islands)	1938			Johnston (1969); BirdLife International (2000)
Meridiocichla salotti		Corsica+Sardinia+ Crete	–		1	Louchart (2004)
Turdidae undescribed species		Madeira	–		1	Pieper (1985)
Timaliidae						
Timaliinae undescribed genus and species	Rodrigues babbler	Rodrigues	–		1	Cheke (1987); Cheke and Hume (2008)
Sylviidae						
cf. *Cettia* sp.	Tongan warbler	'Eua (Tonga)	–	Anatú site, Layer II–III, >3000 BP	8	Steadman (2006b)
Acrocephalus undescribed species	Majuro warbler	Majuro (Marshall islands)	–		8	Steadman (2006b)
Nesillas aldabrana	Aldabra warbler	Aldabra	1983			Hambler *et al.* (1985)
Bowdleria rufescens	Chatham Island fernbird	Chatham Islands	1892			Greenway (1967); BirdLife International (2000); Worthy and Holdaway (2002); Tennyson and Martinson (2006)
Pardalotidae						
Gerygone insularis	Lord Howe Island flycatcher	Lord Howe Island	1918			Greenway (1967); BirdLife International (2000)
Monarchidae						
Myiagra freycinet	Guam flycatcher	Rota+Guam, (Marianas)	1983			Collar *et al.* (1994); Steadman (1999, 2006b); BirdLife International (2000)
Pomarea fluxa	Eiao monarch	Eiao (Marquesas)	1977			Cibois *et al.* (2004)
Pomarea mira	Ua Pou monarch	Ua Pou (Marquesas)	1985			Cibois *et al.* (2004)
Pomarea nukuhivae	Nukuhiva monarch	Nuku Hiva (Marquesas)	1940 (1975?)			Cibois *et al.* (2004)
Pomarea pomarea	Maupiti monarch	Maupiti	1823			BirdLife International (2000)
Zosteropidae						
Zosterops strenuus	Robust white-eye	Lord Howe Island	1913			Greenway (1967); BirdLife International (2000)
Zosteropidae undescribed species		Tongatapu (Tonga)	–	Ha'ateiho site, Layer 3, <2800 BP	8	Steadman (2006b)

Table *continued*

Scientific name	Common name	Range	Last occurrence				Archaeological or stratigraphic association	Other estimate (see list in footnotes; see also Notes)	Reference
			Historical record	Radiometric date					
				^{14}C age (years BP), ±1σ	Calibrated date, ±2σ	Direct date?			
Zosteropidae undescribed species	'Eua white-eye	'Eua (Tonga)	–				Anatú site, Layer II–III, >3000 BP	8	Steadman (2006b)
Meliphagidae									
Chaetoptila angustipluma	Kioea	Hawai'i	1859						Greenway (1967); BirdLife International (2000); James and Olson (1991)
Chaetoptila undescribed species aff. *angustipluma*	Oahu kioea	Oahu	–		AD 890–1040	Yes	Site CLST B.EU-46, 'Ewa Plain		Athens *et al.* (2002)
Chaetoptila undescribed species aff. *angustipluma*		Maui	–		AD 1057–1440	No	Molohai heiau	3, 8	James and Olson (1991); Kolb (1994)
cf. *Chaetoptila* undescribed species	Narrow-billed kioea	Maui	–					1	James and Olson (1991)
Moho apicalis	Oahu O'o	Oahu	1837						Greenway (1967); BirdLife International (2000)
Moho bishopi	Bishop's O'o	Molokai+Maui	1904 (1981?)						Greenway (1967); BirdLife International (2000)
Moho braccatus	Kauai O'o	Kauai	1985 (1987?)						BirdLife International (2000)
Moho nobilis	Hawaii O'o	Hawai'i	1934 (1957?)						Greenway (1967); BirdLife International (2000); Giffin (2003)
Moho undescribed species		Maui	–					1	James and Olson (1991)
Anthornis melanocephala	Chatham Island bellbird, kômako	Chatham Islands	1906						Greenway (1967); BirdLife International (2000); Worthy and Holdaway (2002); Tennyson and Martinson (2006)

Emberizidae						
Emberiza alcoveri	Long-legged bunting	Tenerife	–	Cueva del Viento	1	Rando *et al.* (1999)
Melopyrrha latirostris		Cayman Brac (Cayman Islands)	–		1	Steadman and Morgan (1985); Morgan (1994)
Pedinorhis stirpsarcana		Puerto Rico	–	Blackbone Cave, Cueva del Perro	1	Olson and McKitrick (1982); Neas and Jenkinson (1986)
Parulidae						
Vermivora bachmanni	Bachmann's warbler	North America	1962 (1988?)		6	Hamel (1986)
Leucopeza semperi	Semper's warbler	St. Lucia	1961 (1995?)		6	BirdLife International (2000)
Fringillidae						
Hemignathus lucidus	Nukupuu	Kauai+Maui+Oahu	1998			Greenway (1967); BirdLife International (2000)
Hemignathus vorpalus		Hawai'i	–	Petrel Cave, Pu'u Wa'awa'a, <3000 BP	1	Giffin (2003); James and Olson (2003)
Dysmodrepanis munroi	Lanai hookbill	Lanai	1913 (1918?)			James *et al.* (1989); BirdLife International (2000)
Paroreomyza flammea	Kakawahie	Molokai	1963			Greenway (1967); BirdLife International (2000)
Loxops sagittirostris	Greater amakihi	Hawai'i	1901			Greenway (1967); BirdLife International (2000)
Melamprosops phaeosoma	Po'ouli	Maui	2004		6	BirdLife International (2000)
Orthiospiza howarthi	Highland finch	Maui	–	Lua Manu, Puu Makua and Auwahi Caves	1	James and Olson (1991); James (2004)
Psittirostra kona	Kona grosbeak	Hawai'i	1894			Greenway (1967); BirdLife International (2000)
Rhodacanthis flaviceps	Lesser koa-finch	Hawai'i	1891			Greenway (1967); BirdLife International (2000)
Rhodacanthis forfex		Kauai+Maui	–	Makauwahi Cave, Kauai, Puu Naio Cave, Maui, <8000 BP	1	James and Olson (2005)

Table *continued*

Scientific name	Common name	Range	Last occurrence				Archaeological or stratigraphic association	Other estimate (see list in footnotes; see also Notes)	Reference
			Historical record	Radiometric date					
				^{14}C age (years BP), ±1 σ	Calibrated date, ±2 σ	Direct date?			
Rhodacanthis litotes		Oahu+Maui					Sinkhole 50-Oa-B6–22, Ewa Plain, Oahu, Puu Naio Cave, Maui, <8000 BP	1	James and Olson (2005)
Rhodacanthis palmeri	Greater koa-finch	Hawai'i	1896						Greenway (1967); BirdLife International (2000)
Telespiza persecutrix		Kauai+Oahu	–				Maha'ulepu Cave, Kauai, Unit VI. *c.* AD 1425–1660	8	James and Olson (1991)
Telespiza ypsilon	Maui nui finch	Molokai+Maui	–				Puu Naio Cave, <8000 BP	1	Olson and James (1982, 1984); James and Olson (1991)
Telespiza undescribed species aff. *ypsilon*		Maui	–				Lua Lepo Cave	1	James and Olson (1991)
Telespiza undescribed species	'Owl Cave telespiza'	Hawai'i	–				Owl Cave, Umi'i Manu Cave, <5000 BP	1	Giffin (2003); James (2004)
Vangulifer mirandus		Maui	–				Puu Naio and Lua Lepo Caves, <8000 BP	1	James *et al.* (1987); James and Olson (1991)
Vangulifer neophasis	Thin-billed finch	Maui	–				Puu Naio Cave, <8000 BP	1	James *et al.* (1987); James and Olson (1991)
Xestospiza conica	Cone-billed finch	Kauai	–				Maha'ulepu Cave, Kauai, Unit V, *c.*4380 BP–AD 1240	1	Olson and James (1982, 1984); James and Olson (1991)
Xestospiza fastigialis	Ridge-billed finch	Oahu+Molokai+Maui	–				Several sites, <5500 BP	3	James and Olson (1991)
Loxioides kikuchi		Kauai	–				Maha'ulepu Cave, <(2000) BP	1	James and Olson (2006)

Loxioides undescribed species		Hawai'i	–		1	James (2004)
Himatione freethii	Laysan honeyeater, Laysan apapane	Laysan	1923			Rothschild (1893–1900); BirdLife Internstional (2000)
Aidemedia chascax		Oahu	–	Site-50-Oa-B6–100b, Barbers Point	1	James and Olson (1991)
Aidemedia lutetiae		Molokai+Maui	–	Puu Naio, Lua Lepo and Puu Makua Caves, <8000 BP	1	Olson and James (1982, 1984); James and Olson (1991)
Aidemedia zanclops	Sickle-billed gaper	Oahu	–	Site-50-Oa-B6–78, Barbers Point	1	James and Olson (1991)
Akialoa ellisiana	Oahu akialoa	Oahu	1837 (1940?)			Greenway (1967); BirdLife International (2000)
Akialoa lanaiensis	Maui Nui akialoa	Lanai	1892			Greenway (1967); BirdLife International (2000)
Akialoa cf. *lanaiensis*		Maui	–	Pu'u Naio Cave, Lua Lepo Cave, <8000 BP	1	James (2004)
Akialoa obscura	Lesser akialoa	Hawai'i	1895 (1940?)			Greenway (1967); BirdLife International (2000)
Akialoa stejnegeri	Kauai akialoa	Kauai	1969			Greenway (1967); BirdLife International (2000)
Akialoa upupirostris	Hoopoe-like sicklebill	Kauai+Oahu	–	Maha'ulepu Cave, Unit IV–VI, <6000 BP	1	Olson and James (1982, 1984, 1991); James (1987); James and Olson (1991)
Akialoa aff. *upupirostris*		Maui	–		1	James (2004)
Akialoa large undescribed species		Hawai'i	–	Umi'i Manu Cave, Pu'u Wa'awa'a, <5000 BP	1	Giffin (2003); James (2004)
Akialoa undescribed species aff. *lanaiensis*		Molokai	–		1	James and Olson (1991)
Chloridops regiskongi	King Kong finch	Oahu	–	Site 50-Oa-B6–22, Barbers Point	1	James and Olson (1991)
Chloridops wahi	Oahu grosbeak, Wahi grosbeak	Kauai+Oahu+Maui	–	Maha'ulepu Cave, Kauai, Unit VI, c. AD 1425–1660	8	Olson and James (1982, 1984, 1991); James (1987, 2004); Hearty *et al.* (2005)

Table *continued*

Scientific name	Common name	Range	Last occurrence				Archaeological or stratigraphic association	Other estimate (see list in footnotes; see also Notes)	Reference
			Historical record	Radiometric date					
				^{14}C age (years BP), ±1σ	Calibrated date, ±2σ	Direct date?			
Chloridops undescribed small species		Maui	–				Puu Naio Cave, <8000 BP	1	James and Olson (1991)
Ciridops anna	Ula-ai-hawane	Hawai'i	1892						Greenway (1967); BirdLife International (2000)
Ciridops cf. *anna*		Molokai	–				Moomi Dunes	1	Olson and James (1982, 1991)
Ciridops tenax	Stout-legged finch	Kauai	–				Site K-2, Makawehi Dunes	1	Olson and James (1982, 1984); James and Olson (1991)
Ciridops undescribed species		Oahu	–				Barbers Point	1	Olson and James (1982); James (1987, 2004); James and Olson (1991)
Drepanis funerea	Black mamo	Molokai+Maui	1907						Greenway (1967); James and Olson (1991); BirdLife International (2000); James (2004)
Drepanis pacifica	Hawaii mamo	Hawai'i	1899 (1960?)						Greenway (1967); BirdLife International (2000); Giffin (2003)
Carduelis triasi	Trias's greenfinch	La Palma	–					1	Alcover and Florit (1987)
Chaunoproctus ferreorostris	Bonin grosbeak	Chichi-jima (Bonin Islands)	1828 (1890?)					1	Greenway (1967)
Icteridae									
Sturnella undescribed species		New Providence+Little Exuma (Bahamas)	–					1	Wętmore (1937b); Olson and Hilgartner (1982)
Quiscalus palustris	Slender-billed grackle	Mexico	1910						BirdLife International (2000)
Dolichonyx kruegeri		Cuba	–				Caverna de Pio Domingo	1	Fischer and Stephan (1971b)

Estrildidae									
Erythrura undescribed species	Marianas parrot-finch	Rota (Marianas)	–	400±60	AD 1420–1640	No	Crevice north of Payapai Cave and As Matmos Cliffside Cave	3, 8	Steadman (1999, 2006b)
Passeridae									
Foudia belloni	Réunion fody	Réunion	Probable report 1671/2						Mourer-Chauviré *et al.* (1999); Cheke and Hume (2008)
Sturnidae									
Aplonis corvina	Kosrae starling	Kosrae	1828						Greenway (1967)
Aplonis diluvialis	Huahine starling	Huahine (Society Islands)	–		AD 700–1150	No	Fa'ahia site	8	Steadman (1989a, 2006b)
Aplonis fusca	Norfolk Island starling	Norfolk Island	1923						Schodde *et al.* (1983)
		Lord Howe Island	1918						BirdLife International (2000)
Aplonis mavornata	Mauke starling, mysterious starling	Ma'uke (Cook Islands)	1825						Olson (1986c)
Aplonis undescribed species	Erromango starling	Erromango (Vanuatu)	–				Ponamla site, <2900 BP	8	Steadman (2006b)
Fregilupus varius	Réunion starling	Réunion	1838						Mourer-Chauviré *et al.* (1999)
Necropsar rodericanus	Rodrigues starling	Rodrigues	1726						Cheke (1987)
Callaeidae									
Callaeas cinerea	South Island kokako	South Island (New Zealand)	1967						Worthy and Holdaway (2002); Tennyson and Martinson (2006)
Heteralocha acutirostris	Huia	North Island (New Zealand)	1907 (1925?)						Worthy and Holdaway (2002); Tennyson and Martinson (2006)
Turnagridae†									
Turnagra capensis	South Island piopio	South Island (New Zealand)	1905 (1947?)						Worthy and Holdaway (2002); Tennyson and Martinson (2006)
Turnagra tanagra	North Island piopio	North Island (New Zealand)	1902						Worthy and Holdaway (2002); Tennyson and Martinson (2006)

Table *continued*

Scientific name	Common name	Range	Last occurrence				Archaeological or stratigraphic association	Other estimate (see list in footnotes; see also Notes)	Reference
			Historical record	Radiometric date					
				^{14}C age (years BP), ±1σ	Calibrated date, ±2σ	Direct date?			
Corvidae									
Corvus antipodum	New Zealand raven	North Island+South Island (New Zealand)	–				Occurs in Maori middens, post c. AD 1280	8	Worthy and Holdaway (2002); Gill (2003); Tennyson and Martinson (2006)
Corvus impluviatus	Deep-billed crow	Oahu	–				Site 50-Oa-B6–139, Barbers Point	8	Olson and James (1982, 1984); James and Olson (1991)
Corvus moriorum	Chatham Island raven	Chatham Islands	–	3410±150	4084–3358 BP*	Yes	Occurs in Moriori middens, post c. AD 1350	8	Millener (1999); Gill (2003)
Corvus pumilis	Puerto Rican crow	Puerto Rico+Virgin Islands	–					4, 8	Wetmore (1920, 1922b, 1937a)
Corvus viriosus	Slender-billed crow	Oahu+Molokai	–				Site 50-Oa-B6–139, Barbers Point	1	Olson and James (1982, 1984); James and Olson (1991)
Corvus large species	New Ireland crow	New Ireland	–				Balof Rockshelter, 1000–5000 BP	8	Steadman *et al.* (1999); Steadman (2006b)
Corvus undescribed species aff. *hawaiiensis*		Maui						1	James *et al.* (1987); James and Olson (1991)
Corvus undescribed species aff. *hawaiiensis*		Hawai'i	–					1, Note 21	Giffin (2003); James and Olson (2003)
Passeriformes incertae sedis									
'*Turdus*' *ulietensis*	Ra'iatea thrush	Ra'iatea (Society Islands)	1774					7	Greenway (1967)
aff. *Carduelis* undescribed species		Madeira	–					1	Pieper (1985)

Passeriformes undescribed slender-billed species	Kauai	–	Maha'ulepu Cave, Unit IV–VI, <6000 BP	1	Burney *et al.* (2001)
Passeriformes undescribed very small species	Kauai	–	Maha'ulepu Cave, Unit IV–VI, <6000 BP	1	Burney *et al.* (2001)
Aves incertae sedis					
'Aquila' simurgh	Crete	–	Liko Cave	1	Weesie (1988); Louchart *et al.* (2005)

Other estimates column: indirect last-occurrence date estimates (only listed for species known only from Pleistocene or undated Late Quaternary deposits).

1. First human arrival in region during Holocene (only noted if no other evidence exists for Holocene survival).
2. Associated with still-extant Holocene regional avifauna (only noted for regions with pre-Holocene human arrival).
3. Associated with anthropogenically introduced mammal or bird species.
4. Local names or traditions exist (or are recorded in the historical literature) that may refer to extinct species, suggesting relatively late survival (only noted for species otherwise known only from fossil material).
5. May have died out before the Holocene.
6. Considered to be extant or possibly extant by (a) IUCN (2006) or (b) cited reference.
7. No specimens; known only from historical records.
8. Associated with archaeological remains.

Notes

1. Aepyornithid bones and, especially, eggshell are fairly common in archaeological contexts on Madagascar but are usually not determinable to species level. The same is true for anecdotal evidence that aepyornithids ('vorompatra') survived in southern Madagascar until the seventeenth century. The systematics of the Aepyornithidae is also badly in need of revision. The treatment here follows Brodkorb (1963b).
2. This species probably became extinct on Sicily well before the end of the Pleistocene when Sicily became connected to the mainland; whether it survived into the Holocene on Malta is uncertain since none of the find sites are datable.
3. Survived into historical times (AD 1902) only on Auckland Island. Extinct on North Island post-AD 1280, South Island post-AD 1500, and on Chatham Islands post-AD 1350. The Chatham and Auckland forms may be distinct species or subspecies.
4. The occurrence of this species in the South Island is uncertain since it is based on a single bone.
5. Possibly identical with the 'sasa' mentioned by Wood and Wetmore (1926).
6. Subfossil megapode material from Rockshelter 1 on Aiwa Levu (Lau Group, Fiji), dated to <2700 years BP, has been tentatively identified as either *Megapodius amissus* or *Megapodius molistructor* (Worthy 2001; Steadman 2006b).
7. The forms on New Caledonia and Isle des Pins are possibly separate species (Mourer-Chauviré and Balouet 2005).
8. Also found in Late Pleistocene sites in mainland Portugal and Spain.
9. Known from the latest Pleistocene on the North American mainland. May have survived longer on Cuba.
10. Probably extinct before the end of the Pleistocene on the European mainland. Youngest known record is from Corsica.
11. The Grand Cayman form may actually be identical with the undescribed species found on Jamaica (Suárez and Olson 2001b).

Table *continued*

12. Possibly introduced by humans to the Virgin Islands from Puerto Rico. If the reported association with *Gallus gallus* is correct, it must have survived into post-Columbian times, and possibly into the recent historical past: Wetmore (1927) considered that stories of a bird called the 'carrao' (the modern name for the limpkin *Aramus guaruana*) that could be captured using dogs in Puerto Rico prior to 1912 may have referred to this species, and Nichols (1943) was told that 'flightless waterhens' that could be killed with sticks had been common on Virgin Gorda into the early twentieth century, before the introduction of mongoose to the island.
13. The detailed information on this species given by the Moriori in the late nineteenth century suggests that it survived until relatively recently, perhaps the early 1800s.
14. This species has been considered by some authors to be conspecific with *Porzana tabuensis* (see Lysaght 1956; Walters 1988).
15. The 1902 record refers to a painting by Gauguin (*Le Sorcier d'Hiva Oa*) which almost certainly depicts this species (see also Chapter 10 in this volume).
16. There are also fossil records of *Burhinus* in Cuba and on Grand Turk Island. Whether these are the same as the extant form on Hispaniola is not clear at present.
17. Also known as a subfossil from Ua Huka (unknown subspecies).
18. If this species is identical with the *Eclectus* parrot illustrated on Vavau by the Malaspina Expedition (Olson 2006), it survived at least until AD 1793. However, in the absence of any parrot fossils from Vava'u, this is only a plausible conjecture.
19. Only two species in this genus are now considered valid (Suárez and Jimenez *fide* Louchart 2005).
20. Includes *Pachyplicas jagmi.*
21. Two extinct *Corvus* species may actually be present on Hawai'i, one thin-billed and one thick-billed (Giffin 2003).

CHAPTER 5

Past and future patterns of freshwater mussel extinctions in North America during the Holocene

Wendell R. Haag

5.1 Introduction

Humans have had profound impacts on the ecology of North America both before and since colonization by Europeans. Modern-day human impacts extend to nearly every type of habitat, but evidence for pre-Columbian human impacts is limited almost exclusively to terrestrial ecosystems. In pre-Columbian times, human activities, especially burning and agriculture, transformed significant areas of North America (Delcourt and Delcourt 2004; Mann 2005; see also Chapter 11 in this volume) and, in some cases, even short-term, small-scale agriculture resulted in persistent ecosystem changes (Briggs *et al.* 2006). The linkage between current-day land-use practices and freshwater ecosystem integrity is clear and central to some of the most pressing contemporary conservation issues (e.g. Diaz and Rosenberg 1995; Malakoff 1998), but this linkage has not been shown widely for pre-Columbian human land use. Recent studies in the Americas and Europe have shown that prehistoric Holocene human activities, including low-intensity agriculture, caused long-lasting and sometimes drastic changes in productivity, faunal composition, and water chemistry of small lakes and ponds (Douglas *et al.* 2004; Ekdahl *et al.* 2004; Miettinen *et al.* 2005). Although these studies demonstrate the potential for early human impacts to aquatic habitats, the impact of pre-Columbian humans on the diverse riverine ecosystems of North America remains virtually unknown. In rivers of the south-eastern and central USA the relative abundance of the freshwater mussel genus *Epioblasma* declined steadily over the last 5000 years prior to European colonization but most steeply after the advent of maize agriculture (Peacock *et al.* 2005), suggesting that even larger freshwater ecosystems might have experienced impacts from pre-Columbian human activities.

In the past 100 years, North American freshwater mussels (order Unionoida) have experienced one of the highest rates of extinction of any group of organisms (Neves *et al.* 1997). North America is home to the most diverse freshwater mussel fauna on Earth, including about 300 species; in contrast, there are about 85 species in Africa and 11 species in Europe (Graf and Cummings 2007). Highest mussel diversity and endemism occurs in the eastern half of the continent, especially in the large streams of the central and south-eastern USA. Mussels have a number of ecological attributes that render them especially vulnerable to changes in aquatic habitats. First, as mostly sedentary filter feeders, mussels are directly impacted by changes in the landscape that affect water and substrate quality (Brim Box and Mossa 1999; Arbuckle and Downing 2002). Second, because many species are long-lived (>50 years) and have low recruitment rates (Haag 2002), they may not be able to sustain high adult mortality and may be slow to recolonize disturbed habitats. Finally, mussels have a complex life history in which larvae (glochidia) are obligate parasites on fishes for a brief period before becoming free-living juveniles. Host use varies among mussel species from generalists that use a wide range of fish species to specialists that are able to complete metamorphosis to the juvenile stage on only a few closely related fish species (Haag and Warren 1997). Long-term survival of mussel populations is therefore dependent not only on the presence of suitable mussel habitat but also on the distribution and abundance of host fishes.

Freshwater mussels offer an excellent opportunity to study long-term patterns of extinction and assemblage change in response to human impacts. Spurred by conservation concern about mussels, intensive field surveys over the last 30 years have made possible a detailed accounting of recent species losses and declines at multiple scales across much of North America (e.g. Metcalfe-Smith *et al.* 1998; Brim Box and Williams 2000; Warren and Haag 2005). In addition, a large body of historical and pre-Columbian material exists that allows assessment of trends prior to recent surveys. Because of the beauty and easy preservation of their shells (Fig. 5.1), mussels were a particular fascination of nineteenth- and early twentieth-century naturalists who collected extensively and solicited specimens from other collectors throughout the country. Many of these large private collections survived and ultimately formed the nuclei of most major institutional research collections. Furthermore, the economic importance of mussels for pearl harvest and shell button manufacturing, both of which peaked in the early 1900s, spurred systematic surveys of many river systems before major human impacts

Figure 5.1 Some extinct North American mussel species. From top (left to right): *Lampsilis binominata*, *Pleurobema marshalli*, *Quadrula stapes*, *Epioblasma biemarginata*, *Epioblasma flexuosa* (male), *Epioblasma flexuosa* (female), *Epioblasma haysiana*, and *E. haysiana* (shell interior). Photographs © Richard T. Bryant.

to these streams had occurred. Finally, extensive shell middens resulting from harvest of bivalves by aboriginal peoples occur along many streams, providing a record of mussel abundance and diversity spanning a period of over 4000 years prior to European contact. Together, these sources provide an extraordinarily comprehensive record of distribution and in some cases abundance throughout the Holocene that exists for few organisms in general and is unprecedented for invertebrates.

In this chapter, I present a systematic treatment of freshwater mussel extinctions in North America throughout the Holocene. First, I evaluate how human activities from pre-Columbian times to the present day have affected the North American mussel fauna and identify the major causes and timing of mussel extinctions. Second, I examine ecological and life-history attributes that have rendered freshwater mussel species more or less vulnerable to extinction. Finally, I examine current patterns of species imperilment and make a prognosis about how additional mussel extinctions will occur in the future.

5.2 Pre-Columbian human impacts on mussel communities

Since at least 5000 years BP, humans in North America have used freshwater mussels extensively. Aboriginal peoples used mussel shells for production of jewelry and implements, and as a source of tempering material for pottery, but harvested mussels primarily for food (Parmalee and Klippel 1974). Shell middens associated with human habitation sites occur commonly along water bodies ranging in size from small streams to the Mississippi River (e.g. Klippel *et al.* 1978; Theler 1987a) but are particularly numerous and extensive along the large rivers of the central and south-eastern USA (Fig. 5.2). Hundreds of middens are found throughout the 1050 km length of the Tennessee River (e.g. Parmalee *et al.* 1982; Hughes and Parmalee 1999); along the middle portion of the river in Alabama, 'the banks of the river are lined with shell middens up to 18–20 ft. (5.5–6.1 m) in depth' (Webb and DeJarnette 1942). Similarly, at least 48 shell deposits are known within a four-county area along the Green River in Kentucky, and several of these deposits each cover nearly 1 ha (Morey and Crothers 1998). Although harvest of mussels was likely periodic and may have coincided with periods of scarcity of other resources (Lyman 1984; Theler and Boszhardt 2006), pre-Columbian people doubtlessly exerted tremendous local pressure on mussel populations.

To illustrate the magnitude of aboriginal harvest pressure exerted on mussel populations I estimated the number of mussels contained in three adjacent shell middens along the Green River, using data from Morey and Crothers (1998). Three large shell middens (Haynes, Carlston Annis, and DeWeese) along an 8 km reach of river each ranged from 0.6 to 1.0 ha in area, 2.2 to 2.6 m in depth, and 5300 to 5800 m^3 in volume. I assumed an average volume of 5550 m^3 for each midden and estimated that 20% of the material in the midden was shell, based on the average representation of shell reported across the stratigraphic profile of one of the excavated features (Haynes). In the laboratory, I counted the number of disarticulated shell valves contained in 0.001 m^3 (1 litre; $\bar{x} = 82$ valves $\pm$ 2.9 SE, three replicate counts); shells were of small to medium-sized individuals representing a diverse array of species characteristic of dense, main-channel mussel assemblages in the region. I divided the mean number of valves/0.001 m^3 by 2 (to convert number of disarticulated valves to number of individual bivalves), then multiplied by a density of 0.2 (20% of midden material; see above) to yield an average of 8.2 individuals/0.001 m^3 of midden material. Extrapolating this shell density to the combined volume of material in the middens yields an estimate of over 135 million mussels contained in all three middens. Radiocarbon dates from one of the middens (Haynes) indicated that shells were deposited over a time period of approximately 560 years (see Morey and Crothers 1998). Applying this time frame to all three middens indicates that on average 242 000 mussels were harvested from this 8 km reach of river each year.

Apart from direct mortality due to harvest, pre-Columbian human populations could have impacted mussel resources indirectly if land-use activities altered aquatic habitats. Beginning about 5000 years BP, the growing human population of North America resulted in increased rates of land

Figure 5.2 Prehistoric Native American freshwater mussel shell middens along the Tennessee River, Colbert County, Alabama, USA (feature ct27, excavated 1936). Top: east profile of mound. Bottom: detail of interior of mound showing shell material; for scale, note trowel at bottom of photo. Photos courtesy Tennessee Valley Authority and University of Alabama Museums.

clearance and disturbance associated with agriculture, acquisition of fuel wood, and burning for game management (Delcourt 1987a; Johannessen 1993). These disturbances intensified with the adoption of large-scale maize agriculture beginning about 1000 years BP (Lopinot 1992). Fields reaching tens to hundreds of hectares in size became a common feature across the landscape (Peacock 1998), and supported large settlements organized around an agricultural surplus (Peebles 1978; Mann 2005). During this time, indicators of anthropogenic disturbance such as charcoal influx and sedimentation rates increased markedly (Chapman *et al.* 1982), showing widespread intensification of land use and soil erosion (Delcourt 1987a, 1997; Steponaitis 1986). The impact of these land-use changes on riverine ecosystems is poorly known, but even low-intensity agriculture associated with small, transient settlements was sufficient in some cases to cause dramatic and persistent changes in diatom and rotifer assemblages in small lakes (Ekdahl *et al.* 2004). The large human population of North America prior to European contact had profound impacts on the terrestrial landscape, and it is highly plausible if not probable that these impacts also resulted in changes to aquatic communities.

5.2.1 Evidence for pre-Columbian mussel extinctions

Despite high harvest pressure and the potential effects of human land-use practices on aquatic habitats, no extinctions of mussel species have been documented in prehistory. Zooarchaeological research over nearly the past 100 years has resulted in examination of millions of shells from pre-Columbian middens across North America. To date, only a single species identified in these middens (*Fusconaia apalachicola*) was not subsequently documented by naturalists early in the historical period (Bogan 1990). *F. apalachicola* was found at sites along the Apalachicola River, Florida, ranging in age from 1500 to 650 years BP (Williams and Fradkin 1999), but the timing of its extinction is not known. This species continued to occur at multiple sites into late pre-Columbian times (at least 650 years BP). However, this region was impacted heavily by humans by the early 1800s, and little scientific collecting occurred prior to those impacts, suggesting that this species may have gone extinct during early phases of European settlement (Williams and Fradkin 1999). Other than *F. apalachicola*, all species recorded from pre-Columbian middens throughout North America survived until at least the late 1800s, supporting the conclusion that no North American species were driven to extinction by harvest pressures or human-mediated stream alterations prior to European colonization.

Similarly, there is little evidence that pre-Columbian human activities resulted in local extinctions of species. I compiled data on species presence/absence for 15 stream reaches in the central and south-eastern USA for which comprehensive pre-Columbian and historical data were available. At all sites, a high percentage of species present in the pre-Columbian fauna persisted until the historical period (Table 5.1). Further, pairwise distance matrices of sites (Euclidean distance) based on species presence/absence were highly associated between the pre-Columbian and historical periods (randomized Mantel test, 1000 permutations: Mantel $r = 0.9094$, $P < 0.001$), showing that local patterns of species composition changed little over an extended time period prior to major, modern impacts to streams.

At all stream reaches, historical surveys recorded more species than were found in pre-Columbian middens (Table 5.1). A high degree of consistency in species missing from middens suggests that the higher richness of historical surveys is due to collection and preservation bias associated with midden samples. The genera *Leptodea, Lampsilis, Lasmigona, Potamilus, Truncilla*, and *Villosa* were documented historically but were absent in middens at over half of the sites; *Leptodea fragilis* was documented at 14 sites historically, but occurred in middens at only two sites. Other species that were consistently underrepresented in middens relative to historical samples included *Alasmidonta marginata, Cumberlandia monodonta, Megalonaias nervosa, Pyganodon grandis*, and *Utterbackia imbecillis*. Aboriginal people concentrated harvest efforts in high-density mussel beds characteristic of main-channel river habitats (Matteson 1960; Peacock 2000a), resulting in the consistent absence in middens of species characteristic of lentic backwaters and sloughs (e.g. *L. fragilis, Potamilus* spp., *Pyganodon grandis, U. imbecillis*) or specialized habitats (e.g. *C. monodonta*). Thin-shelled species can be underrepresented in middens due to poor preservation or destruction of these specimens during archaeological recovery (Klippel *et al.* 1978; Morey and Crothers 1998). Thin-shelled species include those characteristic of lentic habitats, compounding their rarity in middens, but also riverine species (e.g. *Alasmidonta* spp., *Lampsilis* spp., *Lasmigona* spp., *Leptodea leptodon, Villosa* spp.). Finally, harvesters consistently avoided both very large species such as *Megalonaias nervosa* that are less palatable and occur in deeper habitats (Ortmann 1909; Parmalee 1956) and very small species that are difficult to collect (e.g. *Truncilla* spp., *Villosa fabalis*). *M. nervosa* occurs sporadically in archaeological contexts, but specimens are often modified for use as tools (Theler 1991). In contrast to aboriginal harvesters, early naturalists were interested in generating comprehensive species lists from a wide variety of habitats regardless of food value or the efficiency and ease of procurement.

Unlike the consistent absence of certain species in middens, species that were present in middens but not detected by historical surveys were not consistent across sites in Table 5.1. With the exception of *Epioblasma stewardsoni*, which was present in middens but not detected historically at two sites, there was no overlap among sites in species not detected historically. Species not detected historically but present in middens included mussels that

Table 5.1 Persistence of pre-Columbian mussel assemblages into the historical period in 15 streams in the central and south-eastern USA. The pre-Columbian fauna represents the number of species recovered from aboriginal human shell middens adjacent to the streams. The historical fauna represents the number of species documented in stream reaches before major twentieth-century human impacts to these streams. Species persisting to the historical period represents the percentage of species recorded in pre-Columbian assemblages that were also documented by historical collections.

Stream reach	Number of species		Species persisting to historical period (%)	Sources
	Pre-Columbian fauna	Historical fauna		
Big Black River, Hinds Co., MS	20	29	90	Hartfield and Rummel (1985); Peacock and James (2002)
Clinch River, Roane/Anderson/Knox Co., TN	39	52	92	Ortmann (1918); Johnson (1978); Parmalee and Bogan (1986)
Cumberland River, Smith/Trousdale Co., TN	37	44	89	Wilson and Clark (1914); Parmalee *et al.* (1980)
Duck River, Bedford/Marshall/Maury Co., TN	28	43	96	Ortmann (1924); Isom and Yokley (1968); Parmalee and Klippel (1986)
Ohio River, upper river, WV/PA	30	39	90	Johnson (1978); Taylor (1989)
Illinois River, Fulton Co., IL	28	40	100	Starrett (1971); Warren (1995)
Illinois River, Pike Co., IL	30	40	90	Matteson (1959); Starrett (1971)
Mississippi River, near Prairie Du Chien, WI	28	39	100	Havlik and Stansbery (1978); Theler (1987a)
Pomme de Terre River, Hickory Co., MO	16	24	100	Klippel *et al.* (1978)
South Fork Holston River, Sullivan Co., TN	28	38	93	Ortmann (1918); Parmalee and Polhemus (2004)
Tennessee River, Loudon/Knox Co., TN	38	57	100	Ortmann (1918); Hughes and Parmalee (1999)
Tennessee River, Muscle Shoals, Colbert/Lauderdale Co., AL	62	68	94	Ortmann (1925); Stansbery (1964); Hughes and Parmalee (1999); Garner and McGregor (2001)
Tombigbee River, Lowndes/Clay Co., MS	33	35	97	van der Schalie (1981); Robison (1983)
Tombigbee River, Pickens/Greene Co., AL	25	31	96	van der Schalie (1981); Robison (1983); Williams *et al.* (1992)
Wabash River, lower river, IL/IN	35	43	94	Parmalee (1969); Cummings *et al.* (1988)
Average (±SE)			95 (±1)	

remain widespread currently (e.g. *Elliptio dilatata, Lampsilis teres, Ptychobranchus fasciolaris, Villosa iris*) as well as those that are now extinct (e.g. *E. stewardsoni*) or imperiled (e.g. *Pegias fabula, Pleurobema clava, Quadrula intermedia*), and collectively these species include a representative cross-section of sensitivity to modern-day human impacts. Across all sites, 68% of species present in middens but not detected historically are currently of conservation concern, similar to 72% of North American species overall (Williams *et al.* 1993), showing that species absent from historical samples are not disproportionately composed of species highly sensitive to habitat alteration. In the late 1800s and early 1900s, the difficulty of travel and lack of modern sampling gear (e.g. SCUBA) and methods (e.g. Smith 2006) precluded the exhaustive sampling effort necessary to detect most or all mussel species in an assemblage. In contrast, midden samples are composed of thousands of shells collected over many years resulting in a more complete representation of species richness in main channel habitats, even

Table 5.2 Freshwater mussel community composition at different times during prehistory in the eastern USA. Sample date was obtained either from estimates given in the source or by taking the midpoint of the reported cultural period (e.g. late Woodland; see Peacock *et al.* 2005). Within a site, sample sizes were standardized using rarefaction (1000 randomizations; Gotelli and Entsminger 2001) to interpolate estimated species richness and evenness of larger samples based on a sample size equal to the smallest sample at the site; values for the smallest samples have no 95% confidence intervals (95% CI) because they are observed values. Richness (including confidence intervals) is rounded to the nearest whole number. Evenness gives the probability that two randomly sampled individuals from the assemblage represent two different species (Gotelli and Entsminger 2001). Correlation coefficients (Pearson's) are for the association of individual species abundances between samples.

Site[1]	Sample	Date (years BP)	Species richness (±95% CI)	Evenness (Hurlbert's PIE, ±95% CI)	Correlation coefficient (*r*)	
					Sample 2	Sample 3
Clinch River	1	1697	35 (±32–37)	0.913 (±0.910–0.916)	0.906	–
	2	750	34	0.937	–	–
Cumberland River A	1	852	33 (±31–35)	0.905 (±0.901–0.909)	0.959	–
	2	652	40	0.925	–	–
Cumberland River B	1	3502	30	0.874	0.987	–
	2	3027	29 (±28–31)	0.874 (±0.872–0.876)	–	–
Green River	1	4850	33	0.886	0.950	–
	2	4520	31 (±30–31)	0.887 (±0.886–0.887)	–	–
Platte River	1	3407	16 (±16–16)	0.538 (±0.535–0.541)	0.992	–
	2	1407	16	0.518	–	–
Tennessee River A	1	1750	30	0.882	0.952	0.937
	2	1250	28 (±25–30)	0.808 (±0.790–0.825)	–	0.943
	3	750	22 (±20–24)	0.835 (±0.823–0.846)	–	–
Tennessee River B	1	3750	21	0.704	0.983	0.984
	2	2250	22 (±21–22)	0.744 (±0.738–0.749)	–	0.954
	3	1500	23 (±21–24)	0.75 (±0.739–0.761)	–	–
Tennessee River C	1	3750	33 (±30–36)	0.866 (±0.853–0.879)	0.958	–
	2	1500	33	0.765	–	–

[1]Sources and localities: Clinch River (Roane Co., TN; Parmalee and Bogan 1986); Cumberland River A (Davidson Co., TN; Peacock 2000b); Cumberland River B (Jackson Co., TN; Breitburg 1986); Green River (Butler Co., KY; Morey and Crothers 1998); Platte River (Grant Co., WI; Theler 1987b); Tennessee River A (Meigs/Rhea Co., TN; Parmalee *et al.* 1982); Tennessee River B (Jackson Co., AL; Warren 1975); Tennessee River C (Lauderdale Co., AL; Parmalee 1994)

though species characteristic of other habitats not sampled by aboriginal harvesters are underrepresented or absent. The lack of consistent patterns of species present during pre-Columbian times but not detected in historical surveys strongly suggests that these discrepancies are due to random sampling error and provide no evidence of systematic patterns of local extinctions in prehistory.

5.2.2 Pre-Columbian changes in mussel assemblage composition

Even if human activities did not result in range-wide or local extinctions, heavy harvest pressure or impacts to streams due to land use should be expected to cause detectable shifts in mussel assemblage composition. I compared mussel assemblage composition between pairs of temporally successive samples representing assemblages harvested by pre-Columbian people from the same stream site at different times (eight sites, 18 time periods; Table 5.2). These pairs of samples represent a range of time periods including prior, during, and after adoption of maize agriculture. Species richness differed among time periods at only three sites, but decreased at two and increased at one. Assemblage evenness differed among time periods at six sites but the magnitude of change was small and evenness decreased at three sites and increased at three others. Further, evenness of most samples was high (>0.75), showing that throughout pre-Columbian times these were extremely diverse assemblages that were not dominated by one or a few species. Finally, individual species abundances were highly correlated between all temporally successive pairs of samples ($P > 0.90$) which, along with lack of consistent trends in richness or evenness, shows that assemblage structure was remarkably similar over time.

The only evidence of significant changes in mussel assemblages during prehistory is a steady decline in relative abundance of *Epioblasma* spp. occurring over the 5000 years preceding European settlement (Fig. 5.3; Peacock *et al.* 2005). Apart from avoiding very large or very small species, aboriginal people probably harvested mussels within high-density mussel beds indiscriminate of species (Matteson 1960; Peacock 2000a). It is therefore unlikely that the disproportionate decline of *Epioblasma* was caused by preferential harvest of these species. Over the past 100 years, *Epioblasma* has suffered the most severe decline of any mussel genus (Johnson 1978), suggesting that these species are especially intolerant of human impacts to streams. It is unknown whether the decline of *Epioblasma* in prehistory is also a result of accelerating human environmental impacts or due to other long-term changes unrelated to human activities. However, the rate of decline increased after the widespread adoption of maize agriculture about 1000 years BP (Peacock *et al.* 2005), suggesting that intensification of anthropogenic disturbance associated with a rapidly expanding human population (e.g. Delcourt 1987a, 1997) began to exert substantial pressures on riverine ecosystems late in prehistory.

Because data on pre-Columbian mussel assemblages are derived almost exclusively from shell middens harvested by humans, it is impossible to determine whether assemblages changed during the early phases of pre-Columbian human settlement of North America. However, data spanning nearly 5000 years of human occupation show that mussel assemblages changed little even in the face of intense harvest and during periods of major human-induced changes on the landscape. Although humans intensively exploited mussel communities as a food resource, harvesters likely operated under an optimal foraging model in which harvest location shifted as local stocks became depleted but well before they were extirpated completely (e.g. Raab 1992). Frequent shifts in the location of harvest and the presence of large mussel populations throughout an interconnected river network would have allowed recovery of mussel numbers after local depletion. It remains unclear to what extent mussels were affected by habitat deterioration due to pre-Columbian human land-use activities, but at most these impacts resulted in only subtle changes in assemblage composition (e.g. declines of *Epioblasma*) and did not result in species extinctions or consistent, widespread

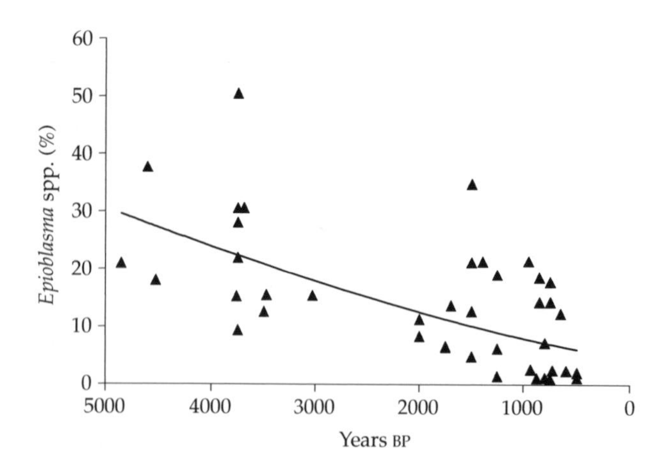

Figure 5.3 Relationship between time and relative abundance of *Epioblasma* spp. in the eastern USA (arcsine [relative abundance] = 12.119–0.0043 time, $r^2 = 0.369$, $P < 0.0001$). From Peacock *et al.* (2005).

patterns of local extinctions. Overall, the mussel fauna of North America was intact at the time of European settlement.

5.3 Recent human impacts on mussel communities

5.3.1 European settlement until 1924

European settlers began to significantly impact the aquatic communities of North America by at least the early 1800s. Harvest of mussels for pearls represented the first large-scale and widespread historical impact to freshwater mussels. Pearling was widespread over much of the continent by 1860, and several areas experienced pearl rushes during which large numbers of prospectors converged on a stream following the discovery of pearls (Anthony and Downing 2001). Because gem-quality pearls occur in a small percentage of individuals, pearling was a highly wasteful endeavour that required harvest of large numbers of mussels in order to realize even a modest return. By the early 1900s, pearling had resulted in a reduction of mussel abundance in many streams. In 1911, piles of shells discarded by pearlers, each as large as 3 tons, were present along much of the Cumberland River (≈845 km) in Kentucky and Tennessee, and larger mussels most likely to have pearls were absent from mussel beds in some areas (Wilson and Clark 1914). Pearling also resulted in impacts to stream habitats in some areas. Pearlers were known to work 'a plow drawn by a strong team' through shoals in the Clinch River, Tennessee, to expose buried mussels (Böpple and Coker 1912).

Harvest of mussels for pearls paled in comparison to harvests by the shell-button industry. Use of mussel shells for manufacture of buttons and other mother-of-pearl items began on a large scale in 1892 and soon grew into a multi-million dollar industry, peaking in 1916 but lasting until the mid-1960s (Anthony and Downing 2001). The pearl-button industry encompassed at least 20 states along the large rivers of the central and south-eastern USA and was therefore overlaid precisely on the region of highest mussel diversity. Similar to harvests for food by pre-Columbian settlements, the pearl button fishery resulted in the harvest of staggering numbers of mussels (Fig. 5.4). In 3 years, over 9000 tons of shells were taken from a single mussel bed less than 0.75 km^2 in size in the Mississippi River in Illinois; this harvest represented over 100 million animals (Smith 1898 in Anthony and Downing 2001). Harvests of similar magnitude were reported throughout the region and eventually mussel resources of many streams were seriously overexploited from a commercial fishery perspective (Anthony and Downing 2001).

In addition to intense harvest pressures, mussel populations suffered from a variety of other insults associated with a rapidly expanding human population and an increasingly industrialized society. With an absence of environmental regulation of any kind, by the early 1900s mussel populations had been reduced or eliminated completely in stream reaches throughout the USA by chronic and severe point-source pollution, dams and other stream channel modifications, and massive erosion and sedimentation (Bogan 1993; Neves *et al.* 1997). Between 1900 and 1920 mussel life was eliminated almost completely in nearly 200 km of the Illinois River due to discharge of raw sewage from Chicago and other cities (Starrett 1971). The extinction of *F. apalachicola* in the Apalachicola River system is probably attributable to discharge of industrial effluents combined with sedimentation caused by widespread hillside clearing for cotton production after the Civil War (Williams and Fradkin 1999). These impacts to the Apalachicola system demonstratively resulted in the elimination of many other mussel species from large portions of the system by the early 1900s (Brim Box and Williams 2000).

Despite intense harvest pressures and a wide variety of serious insults to streams, no extinctions of freshwater mussels were documented by contemporary observers prior to 1924. Although commercial harvest resulted in massive reduction in mussel abundance in many streams, of the 50 most important commercial species (Anthony and Downing 2001), 38% are currently of conservation concern, which is about half the rate of imperilment for the North American fauna as a whole (72%; Williams *et al.* 1993). This result can be interpreted in two non-mutually exclusive ways: (1) commercially exploited species in general are less sensitive to human impacts or (2) commercial harvest had

Figure 5.4 Commercial harvest of freshwater mussels in the early twentieth century. Top: barges loaded with shells in Arkansas (from Coker 1919). Bottom: mussel shells at a button factory on the upper Mississippi River (photo courtesy US Fish and Wildlife Service).

little effect in propelling species into a sustained downward spiral of abundance. Regardless, even intense unregulated harvest did not result in species extinctions. Similarly, although extensive stream reaches were practically defaunated by a wide array of insults to stream habitats during this period, resulting in extinction of local populations, species richness of the North American fauna as a whole remained largely unchanged.

Mussels remained remarkably abundant and diverse in many areas, with large streams supporting the highest diversity of mussel species. Species composition across much of the eastern USA remained similar to pre-Columbian assemblages (Table 5.1) and numerous authors of the time describe dense aggregations of mussels that in some cases extended for many stream kilometres (e.g. Wilson and Clark 1914; Ortmann 1924, 1926; Clench 1926). However, these same authors reported localized declines in mussel populations due to harvest, pollution, and habitat destruction. In the Green River, Kentucky, in 1925, at one site '...most of the Unionidae have been killed here by pearl-hunters', while at other sites mussels '...were extremely abundant' or were 'so thick that they touched one another' (Ortmann 1926). In the Cumberland River, Kentucky and Tennessee, in 1910–1911, 'in spite of the great number of mussels taken out [for the button industry], the river as a whole...does not show any marked depletion except in one or two restricted localities' (Wilson and Clark 1914). In the Tennessee River system in 1912–1915 with regard to mussel abundance and diversity '...conditions are fair, in some parts splendid; but there are already polluted streams, in which the fauna is gone' (Ortmann 1918).

These descriptions exemplify the state of the North American mussel fauna and the freshwater landscape in general in the early decades of the twentieth century. Stream systems at this time were a mosaic of disturbed and relatively undisturbed reaches. Similar to pre-Columbian harvest, harvest for pearls and buttons shifted location frequently when local mussel abundance fell below commercially exploitable levels (Coker 1919; Neves *et al.* 1997; Anthony and Downing 2001). Similarly, some impacts such as sedimentation due to logging and mining were also shifting in nature. Because stream systems remained largely contiguous and were fragmented by few permanent physical barriers, the enormous reproductive potential contained within dense mussel populations in undisturbed reaches allowed repopulation of impacted areas following abatement of impacts. In at least some instances, mussel beds depleted by harvest but subsequently abandoned recovered to commercially exploitable levels in several years (Coker 1919). Despite intense harvest pressure and a variety of severe insults to stream habitats, the scattered and shifting nature of these impacts coupled with the interconnectivity of stream systems allowed the North American mussel fauna as a whole to survive intact well into the twentieth century.

5.3.2 Systematic habitat destruction 1924–1984

The building of dams...also has a deteriorating effect upon mussel life, and...surely will increase in the future (Ortmann 1918)

Despite his prophetic words, the early mussel biologist Arnold E. Ortmann could not likely have imagined the scale and rapidity of dam construction in North America that commenced in earnest shortly after these words were written. Although many dams were built before 1924 and dam construction continues, the end points of 1924–1984 encompass the most intensive period of large dam construction in the USA and are symbolic in the context of freshwater mussel extinctions. As one of the largest dams in the world at the time of its completion in 1924, Wilson Dam on the Tennessee River in Alabama simultaneously drowned much of Muscle Shoals, the most diverse site for freshwater bivalves on the planet (≈70 species; Stansbery 1964; Garner and McGregor 2001), and ushered in the age of large dams in North America. The next 60 years witnessed a frenzy of dam building and stream channelization for the ostensible purposes of flood control, hydroelectric power generation, navigation, water storage, and recreation. The Tennessee-Tombigbee Waterway in Alabama and Mississippi was completed in 1984, and included construction of 10 locks and dams and 377 km of navigation channel (Tennessee-Tombigbee Waterway Development Authority 2007). This project effectively destroyed the Tombigbee River, which represented the last unpolluted, free-flowing large stream reach in the Mobile Basin, one of the most diverse stream systems in North America (Abell *et al.* 2000), and to date has resulted in the extinction of three endemic mussel species (*Pleurobema curtum*, *P. marshalli*, and *Quadrula stapes*). Whereas the necessity and cost-benefit ratios of earlier dam and channelization projects varied widely, the Tennessee-Tombigbee Waterway eclipsed any single previous project in terms of cost, dubious need, misrepresented justification, and environmental destruction (Stine 1993), and was the grand finale of the golden age of large dam building and stream channelization in the USA.

The result of this frenzy of dam building was to eliminate most free-flowing large rivers and many small and medium-sized rivers in the USA (Benke 1990; Dynesius and Nilsson 1994). In their natural state, even the largest rivers had extensive gravel and sand bars that created shallow shoals at times of low water. The Ohio River could be crossed seasonally by wading at Cincinnati, Ohio (Fig. 5.5), and Muscle Shoals on the Tennessee River was a shallow, 85 km complex of islands, shoals, and rocky reefs (Garner and McGregor 2001) that blocked river traffic at low water. This type of shallow, shoal habitat was lost completely from most large rivers after impoundment. Stream reaches not directly impounded but located downstream from large dams were fundamentally modified by dam releases having highly altered flow, temperature, and oxygen regimes (Miller *et al.* 1984; Layzer *et al.* 1993). During this period, four of the most diverse rivers in the world, from a freshwater mussel perspective (Tennessee, Cumberland, Ohio, and Coosa), were transformed into a series of reservoirs and regulated reaches with little or no free-flowing main-channel habitat remaining. In addition, most of the large tributaries in these systems were

Figure 5.5 The Ohio River at Cincinnati, Ohio (approximately 1888), before impoundment. Photo was taken at low water, showing presence of shallow shoals and gravel bars. Note people on gravel bars in the distance for scale. From the collections of The Public Library of Cincinnati and Hamilton County, and the Cincinnati Historical Society Library.

impounded. In the Tennessee River drainage alone there are 53 major dams (defined as impounding >40 ha): nine on the main channel and the remainder on tributaries (Etnier and Starnes 1993).

This systematic destruction of large-stream habitat resulted in the first wave of mussel species extinctions, beginning in the 1930s (Table 5.3). The exact timing of extinction is difficult to determine for any species (e.g. Diamond 1987; Reed 1996) and can be especially difficult for freshwater mussels because relict individuals of some species can survive for more than 30 years in radically altered habitats that no longer support viable populations (Parmalee and Klippel 1982; Ahlstedt and McDonough 1993). A species can be considered functionally extinct when reports of its existence cease or when all populations are no longer viable and extinction becomes inevitable (Holdaway 1999b; DeLord 2007). Here, I define functional extinction of mussels as occurring when all suitable habitat has been destroyed or when a species becomes so rare that the chances of finding an individual, or of the species reproducing, becomes essentially zero. For the remainder of this chapter I discuss the time of functional extinction of species but refer to this as simply extinction.

The first wave of mussel extinctions was composed primarily of obligate large-river species. Unaltered main-channel habitat in the Ohio and Tennessee Rivers was completely eliminated between 1924 and 1944, and species endemic to these habitats were likely the first to become extinct (e.g. *Epioblasma flexuosa, E. f. florentina, E. personata, E. propinqua*; Table 5.3). Due to the difficulty of accessing and sampling large rivers, the relative dearth of mussel biologists during this period, and the rapid pace of dam construction, the temporal and spatial sequence of the disappearance of many of these species is poorly known; rather, these species were simply never seen again after impoundment of large stream habitat was complete. For example, *E. flexuosa* was last documented with certainty in 1900 from the Ohio River (Stansbery 1970), but because early twentieth-century collecting in the middle and lower reaches of this river was sporadic and restricted to only a few localities, the species probably persisted here until complete impoundment of the river. For other species the timing of extinction can be determined with more precision. By the mid-1940s, all known habitat for *E. lewisii* had been impounded or altered by dam release except

Table 5.3 Species of freshwater mussels in North America that became extinct in the twentieth century. Time of extinction is the probable time of functional extinction (see text). Several taxa of uncertain taxonomic status (e.g. *Alasmidonta robusta, Pleurobema* spp., *Quadrula tuberosa*; see text) are omitted from this list.

Species	Time of extinction	Cause of extinction[1]	Last known occurrence
Alasmidonta mccordi	1964	1	Coosa River, AL
Alasmidonta wrightiana	1930s	2,3	Ochlockonee River, FL
Elliptio nigella	1950s	2	Coolewahee Creek, GA
Epioblasma arcaeformis	1940s	1	Holston River, TN
Epioblasma biemarginata	1970	2	Elk River, TN
Epioblasma flexuosa	1920s–1930s	1	Ohio River, KY
Epioblasma florentina curtisi	1990s	3	Little Black River, MO
Epioblasma florentina florentina	1940s	1	Holston River, TN
Epioblasma haysiana	1970	2	Clinch River, VA
Epioblasma lenior	1967	1	Stones River, TN
Epioblasma lewisii	1950	1	Cumberland River, KY
Epioblasma metastriata	1980s	2	Conasauga River, GA
Epioblasma othcaloogensis	1970s	2	Conasauga River, GA
Epioblasma personata	1920s–1930s	1	Tennessee River, AL
Epioblasma propinqua	1936	1	Clinch River, TN
Epioblasma sampsoni	1930s–1940s	2	Wabash River, IL/IN
Epioblasma stewardsoni	1940s	1	Holston River, TN
Epioblasma torulosa gubernaculum	1980s	2	Clinch River, TN
Epioblasma torulosa torulosa	1970s	1	Kanawha River, WV
Epioblasma turgidula	1976	1	Duck River, TN
Lampsilis binominata	1970s	3	Flint River, GA
Pleurobema curtum	1990s	2	East Fork Tombigbee River, MS
Pleurobema marshalli	1984	1	Tombigbee River, AL/MS
Quadrula couchiana	Early 1900s	3	Rio Grande, TX
Quadrula stapes	1980s	2	Sipsey River, AL
Quincuncina mitchelli	1970s	3	Rivers in central Texas

[1]Causes of extinction: 1, direct loss of all habitat by stream impoundment or channelization; 2, indirect effects of fragmentation due to habitat destruction; 3, small original range and non-impoundment related habitat degradation.

for the Caney Fork River, Tennessee (Cumberland River system), and the upper Cumberland River, Kentucky. The species persisted in both of these stream reaches until 1948 and 1950 respectively (Neel and Allen 1964; Layzer *et al.* 1993), when dam construction on these streams eliminated all remaining habitat for the species. Similarly, by the 1970s, *Pleurobema marshalli* persisted only in a single free-flowing reach of the Tombigbee River in Alabama and Mississippi which was destroyed by completion of the Tennessee-Tombigbee Waterway in 1984, resulting in the extinction of the species (Haag 2004a).

Mussels were eliminated in many river reaches due to the abrupt and profound transformation of shallow, riverine habitat into deep, still, and often hypolimnetic reservoirs designed primarily for floodwater storage or hydroelectric generation, or by severely altered hydrological conditions downstream of such reservoirs. These transformations created habitats to which few or no mussel species could adapt, resulting in near total loss of the mussel fauna (Miller *et al.* 1984; Layzer *et al.* 1993). However, dams designed primarily for navigation, such as those on the Alabama, lower Cumberland, Mississippi, Ohio, and Tennessee Rivers, eliminated shallow shoal habitats but created run-of-the-river impoundments that retain some riverine characteristics, especially in tailwater reaches downstream of dams. Relative to original river conditions, tailwater reaches below navigation dams have greatly increased depth (≈6–15 m) and altered daily hydrographs (Garner and McGregor 2001; Freeman *et al.* 2005) and can experience periods of low dissolved oxygen (<5 mg/l; Voightlander and Poppe 1989) sufficient to cause mussel mortality (Johnson 2001). However, unlike other reservoir habitats, tailwater reaches have flow sufficient to keep gravel and sand substrates silt-free, providing habitat for a variety of riverine organisms (Voightlander and Poppe 1989). In these impounded riverine habitats, only a portion of the original mussel fauna was eliminated while other species were able to maintain recruiting populations.

Navigation dams affected species selectively and the likelihood of persisting in impounded riverine habitats was not simply a function of pre-impoundment abundance. In the Tennessee River, some of the most abundant species both in pre-Columbian and historical times were eliminated by impoundment (e.g. *Epioblasma torulosa torulosa*, *E. biemarginata*, *Dromus dromas*) while other species that were rare before impoundment persisted or even increased (e.g. *Fusconaia ebena*, *Megalonaias nervosa*; Ortmann 1918; Morrison 1942; Garner and McGregor 2001). This selective loss of species was caused in large part by differences in fish host use among species and not by inter-species differences in habitat requirements of the mussels themselves.

Mussel species that adapted to impounded riverine habitats are either host-generalists or specialize in the use of host fishes that also could adapt to these habitats. For example, in the Tennessee River, all but one host-specialist mussel species that survived as a reproducing population after impoundment use catfishes, freshwater drum, skipjack herring, sunfishes (including black basses), gar, or sauger (Table 5.4). These fishes all thrive in the run-of-the-river reservoirs of the Tennessee River (Etnier and Starnes 1993). The single exception, *Ptychobranchus fasciolaris*, is extremely rare in the impounded Tennessee River (Garner and McGregor 2001). In contrast, 89% of mussel species that did not adapt to impoundment of the Tennessee River used darters, riverine minnows, sturgeon, or rock bass as hosts (Table 5.4). These fishes were eliminated or greatly reduced in impounded rivers, even in riverine tailwater reaches, largely as a result of the loss of shallow shoal habitat (Voightlander and Poppe 1989; Etnier and Starnes 1993; Freeman *et al.* 2005). The logperch *Percina caprodes*, a darter that serves as primary host for several mussel species, including species of *Epioblasma*, persists in some sections of the impounded Tennessee River, but likely leaves the river in spring to spawn in shoal habitat of small tributary streams (Etnier and Starnes 1993). Such a migration, coinciding with the period of glochidial release by these mussel species, would effectively render these fishes unavailable as hosts, despite their continued seasonal presence in the reservoirs. Because small stream fishes such as darters and minnows have short lifespans (<5 years), impoundment resulted in abrupt changes in the fish assemblage, eliminating hosts for a large number of mussel species.

Even after loss of their fish hosts precluded recruitment, many long-lived mussel species continued to persist in impounded riverine habitats as aging, relict populations. *Pleurobema cordatum* was a dominant component of main channel mussel assemblages in the Tennessee River but has realized little recruitment subsequent to impoundment of the river. In 1957 the mean age of *P. cordatum* was 22 years, but increased to 49 years by 1993 (Scruggs 1960; Ahlstedt and McDonough 1995–1996); in both cases, these estimates show that most individuals recruited just before or just after construction of dams in this section of the river in 1940 and 1942. Diverse, relict assemblages of species intolerant of impoundment (e.g. *Cyprogenia stegaria*, *Dromus dromas*, *Epioblasma* spp., *Obovaria retusa*) persisted in the Cumberland River, Tennessee, for at least 25 years after dam construction (Parmalee *et al.* 1980; Parmalee and Klippel 1982) and in several sections of the Tennessee River, Alabama and Tennessee (Ahlstedt and McDonough 1993; Garner and McGregor 2001). Similarly, *Epioblasma torulosa torulosa* continued to be harvested by mussel fisherman in the lower Ohio River for 30 years after impoundment (Parmalee 1967). These relict populations were composed exclusively of older individuals that had

Table 5.4 Host fish use by the mussel fauna of the Tennessee River near Muscle Shoals, Alabama. Impoundment-tolerant species are defined here as those that were able to maintain reproducing populations in impounded riverine habitat remaining in dam tailwaters in this section of the river (see text). Intolerant species are those that were eliminated completely soon after impoundment, or those for which little or no recruitment occurred after impoundment even though adults may persist for many years. Species occurrence and impoundment tolerance was assessed from Garner and McGregor (2001) and Ortmann (1925). Host use was determined from a large body of published and unpublished literature (for an introduction to this literature, see Watters 1994, and Mussel/host database, The Ohio State University Museum of Biological Diversity, Division of Mollusks, www.biosci.ohio-state.edu/~molluscs/OSUM2/index.htm). For some species with unknown hosts, host use was inferred based on information for congeners. Species with no host information or information for congeners were excluded from the table, but these included both tolerant (e.g. *Obliquaria reflexa*) and intolerant (e.g. *Hemistena lata*) species.

Mussel species	Primary host fish use
Impoundment-tolerant species	
Amblema plicata	Generalist
Cyclonaias tuberculata	Catfishes (*Ictalurus*, *Pylodictis olivaris*)
Ellipsaria lineolata	Freshwater drum (*Aplodinotus grunniens*)
Elliptio crassidens	Skipjack herring (*Alosa chrysochloris*)
Fusconaia ebena	Skipjack herring
Lampsilis abrupta	Black basses (*Micropterus*)
Lampsilis ovata	Black basses
Lampsilis teres	Gar (Lepisosteidae)
Leptodea fragilis	Freshwater drum
Ligumia recta	Sauger (*Sander canadense*), black basses
Megalonaias nervosa	Generalist
Plethobasus cyphyus	Sauger
Potamilus alatus	Freshwater drum
Ptychobranchus fasciolaris	Darters (*Ammocrypta*, *Etheostoma*, or *Percina*)
Quadrula pustulosa	Catfishes
Quadrula quadrula	Catfishes
Toxolasma (2 spp.)	Sunfishes (Centrarchidae)
Tritogonia verrucosa	Catfishes
Truncilla (2 spp.)	Freshwater drum
Villosa vanuxemensis	Sculpins (*Cottus* spp.), sunfishes
Impoundment-intolerant species	
Cyprogenia stegaria	Darters
Dromus dromas	Darters
Elliptio dilatata	Darters
Epioblasma (10 spp.)	Darters
Fusconaia barnesiana	Riverine minnows (e.g. *Cyprinella*, *Erimystax*, *Nocomis*, or *Notropis*)
Fusconaia cor	Riverine minnows
Fusconaia cuneolus	Riverine minnows
Lampsilis fasciola	Black basses, rock bass (*Ambloplites rupestris*)
Lasmigona costata	Generalist
Lemiox rimosus	Darters
Leptodea leptodon	Freshwater drum
Lexingtonia dollabelloides	Riverine minnows
Medionidus conradicus	Darters
Obovaria olivaria	Sturgeon (*Scaphirhynchus*)
Obovaria retusa	Darters

Table 5.4 *Continued*

Mussel species	Primary host fish use
Pleurobema (5 spp.)	Riverine minnows
Ptychobranchus subtentum	Darters
Quadrula cylindrica	Riverine minnows
Quadrula fragosa	Catfishes
Quadrula intermedia	Riverine minnows
Strophitus undulatus	Generalist
Villosa iris	Black basses, rock bass
Villosa taeniata	Rock bass
Villosa trabalis	Darters

recruited prior to impoundment and continue to exist in some streams today. The widespread persistence of these relict faunas shows that adults of many species could survive even in highly modified riverine habitats but the lack of younger individuals shows that recruitment effectively ceased soon after impoundment.

Many species that were eliminated from large river habitats also occurred in medium-sized tributary streams and therefore survived the destruction of large rivers. However, impoundment of tributaries also resulted directly in extinction of some species with limited ranges. By the late 1960s, *Epioblasma lenior* and *Epioblasma turgidula* had both been reduced to single populations, in the Stones River, Tennessee, and the Duck River, Tennessee, respectively (Stansbery 1970, 1976), until construction of J. Percy Priest Reservoir (Stones River) in 1967 and Normandy Reservoir (Duck River) in 1976 eliminated the last habitat for these species. However, impoundment of tributary streams, although widespread, was in general less complete than the impoundment of large rivers, and many tributaries remained free-flowing and continued to support diverse mussel faunas.

Although most extinctions during this period were due directly to elimination of all suitable habitat by dams, other factors were responsible for extinctions of species with very small natural ranges (Table 5.3). *Alasmidonta wrightiana* is known only from 15 specimens collected at two sites in the Ochlockonee River, Florida, prior to 1932 (Clarke 1981; Williams and Butler 1994). The construction of Talquin Dam in 1927 inundated one of these sites but left the other intact. Reduction of the already small range of this rare species may have reduced the population size below a viable level; alternatively the dam may have eliminated an anadromous host fish required for reproduction by this species (R. Butler, personal communication). *Lampsilis binominata* occurred historically only in the upper Flint and Chattahoochee Rivers, Georgia, mostly above the Fall Line demarcating uplands from the Coastal Plain (Brim Box and Williams 2000). This species was extirpated from the Chattahoochee River by the 1940s but persisted in the upper Flint River until expansion of the Atlanta urban area in the 1970s degraded the remaining habitat (Gillies *et al.* 2003). Other species with very small historical ranges may have been relicts that went extinct naturally before significant human impacts to their habitat. *Quadrula couchiana* was endemic to the Rio Grande system in Texas and Chihuahua, Mexico. This species is frequent in the recent fossil record but only a few living individuals were ever found, the last in 1898 (Howells *et al.* 1996), suggesting that this species became extinct naturally before major human impacts to the Rio Grande in the 1900s.

Most species that went extinct during the period of systematic habitat destruction were morphologically distinctive (see Fig. 5.1), were well known to early naturalists, and are well represented in historical museum collections. However, a precise accounting of the number of extinctions during this period will never be possible because of uncertainty about the phylogenetic status of some morphologically similar or poorly known taxa. Of the numerous species of *Pleurobema* described from main-channel habitats in the Mobile Basin, especially the Coosa River, Alabama, as many as 14 of these taxa are now considered extinct (Neves *et al.* 1997; Turgeon *et al.* 1998). However, some of these taxa likely represent expressions of clinal variation (Turgeon *et al.* 1998) within both extant and extinct species, resulting in an overestimate of extinction in this group. Other species previously considered extinct based on the existence of only one or two historical specimens (e.g. *Medionidus macglameriae*; Neves *et al.* 1997; Turgeon *et al.* 1998) were based on misidentifications of extant species (Williams *et al.* 2008). The validity of other poorly known taxa that were restricted to large rivers and now considered extinct (e.g. *Pleurobema bournianum*, *Quadrula tuberosa*) will never be known because tissues from these animals are unavailable. On the other hand, current species concepts prevalent in freshwater mussel taxonomy may underestimate diversity and past extinctions in other mussel groups due to the presence of previously unrecognized cryptic species (e.g. Jones *et al.* 2006a; Serb 2006).

By the close of the most intensive era of dam construction, the mussel fauna of North America had been changed radically. At least 12 species that occurred only in free-flowing large rivers had been rendered extinct directly by elimination of all existing habitat (Table 5.3), and a large number of other species were eliminated from these habitats and reduced to smaller populations in tributary streams. Large-stream mussel faunas were now composed of a smaller number of species that could adapt to impounded riverine habitats, and the highest mussel diversity now occurred in tributary streams not directly affected by impoundment. Although the mussel fauna of many tributaries was eliminated or greatly reduced by impoundment as well as other impacts, as a whole the small and medium-sized stream fauna of North America remained largely intact. However, this diverse fauna was now highly fragmented by dams or by long stream reaches that no longer provided suitable mussel habitat. Therefore, even though diverse and abundant mussel faunas remained in many streams, these assemblages were now composed of isolated and highly vulnerable populations.

5.3.3 The post-dam construction era: fragmentation, isolation, and the extinction debt

Widespread recognition of the extinction of mussel species and the endangered status of others occurred by 1970 (Stansbery 1970, 1971) and conservation efforts to protect mussel diversity began in earnest with passage of the US Endangered Species Act in 1973. Although the vulnerability of isolated populations of rare species was recognized, the long-term effects of habitat fragmentation have not become clear until recently. In isolated populations, declines in population size due to stochastic events, whether natural or human-caused, cannot be offset by colonization from other populations; therefore, a single major impact or a series of lesser impacts can cast a population into a slow, downward spiral of abundance from which it may never recover (Gilpin and Soulé 1986). Unlike direct, immediate effects of habitat destruction, extinctions due to habitat fragmentation have a time lag during which the isolated community bears an extinction debt to be repaid in the future (Tilman *et al.* 1994; Hanski and Ovaskainen 2002). Although extinction of an isolated population may ultimately be caused by a single factor unrelated to fragmentation, the true cause is an accumulation of impacts over time, beginning with the initial fragmentation of the species' habitat. Because connectivity can be as important as the size of the population in determining vulnerability to extinction (Paquet *et al.* 2004), the complete isolation of many remnant mussel populations following dam construction predicted a large number of delayed extinctions.

In some cases, payment of the extinction debt created by habitat fragmentation is already underway. Due mostly to widespread impoundment of their habitat, by the 1960s, *Epioblasma haysiana* and *Epioblasma torulosa gubernaculum* both survived only in a single free-flowing section of the Clinch River in Tennessee and Virginia, upstream of Norris Reservoir. In 1967 and 1970, industrial chemical spills occurred within this section of river, eliminating all molluscs for over 18 river kilometres, reducing mussel abundance for at least 124 km (US Fish and Wildlife Service 1983), and killing nearly all fishes for over 100 km (Jenkins and Burkhead

1993). Other mussel species that were distributed more widely within the river or had larger population sizes before the spill survived, but *E. haysiana* was never seen again after the spill, and *E. t. gubernaculum* was evidently reduced to a population size too small for recovery. Only two individuals of *E. t. gubernaculum* were found after the spill, with the last individual seen in 1982 (US Fish and Wildlife Service 1983).

In the Tombigbee River system, elimination of main channel habitat by the Tennessee-Tombigbee Waterway reduced two obligate large-river species, *Pleurobema curtum* and *Quadrula stapes*, to single, small populations, each of which occurred in the lower reaches of two tributaries, the East Fork Tombigbee River (Mississippi) and the Sipsey River (Alabama) respectively (McCullagh *et al.* 2002; Haag 2004b). These populations each persisted for at least 10 years after destruction of the Tombigbee River, but recent intensive surveys in these and other unimpounded tributaries have failed to find the species and they are now considered extinct. The East Fork Tombigbee and Sipsey Rivers both continue to support diverse mussel assemblages composed of species not restricted to main-channel habitats (McCullagh *et al.* 2002). However, tributary populations of *P. curtum* and *Q. stapes* were probably sinks that were sustained solely by immigration from source populations in the main river. After the loss of the source populations, tributary populations were too small to be reproductively viable and disappeared as remnant adults reached the end of their lifespan.

Epioblasma metastriata became extinct when populations were reduced and fragmented by reservoirs, followed by gradual deterioration of habitat for the few remaining populations. Most habitat for *E. metastriata* was destroyed by impoundment of the Black Warrior and Coosa Rivers. By the 1960s, the species survived in three tributary streams in the Cahaba and Upper Coosa River basin, but these populations declined steadily and living animals have not been found since 1973 (Cahaba) and the 1980s (Coosa) (US Fish and Wildlife Service 2000, Williams *et al.* 2008). The decline of these remnant populations is attributable to a combination of sub-acute stressors (e.g. sedimentation and other non-point-source contaminants) within the watersheds that occurred at different times over the last 40 years. In a large, interconnected watershed such as existed before dams, these populations could have been sustained or rescued by immigrants from other populations, allowing them to recover from periodic, localized stress. However, without the potential for immigration, the steady erosion of these populations over time ultimately resulted in the extinction of the species.

With the exception of these and a handful of other recently extinct species (Table 5.3), most of the fauna that remained after the era of dam building survives to the present day. However, the major portion of the principal on the extinction debt held by this fauna remains unpaid. The combination of indirect effects of fragmentation and isolation coupled with an array of acute and chronic stressors, similar to that which led to the extinction of *E. metastriata*, has set the stage for a second wave of mussel extinctions that will likely surpass the first extinction wave caused by direct habitat destruction. Currently, at least 31 species survive only as one or two populations (Table 5.5); in the remainder of this chapter I refer to these species as critically imperiled. For many of these species, only short reaches of habitat remain, and populations are extremely small for all. For example, *Lampsilis streckeri* is known to inhabit only about 10 km of stream in Arkansas (US Fish and Wildlife Service 1991). Only five living individuals of *Medionidus simpsonianus* have been seen in the last 35 years, including two in 1974, one in 1993, and two in 2007 (US Fish and Wildlife Service 2003, J. Williams personal communication). All of these critically imperiled species are in imminent danger of extinction due to stochastic events. For some, it is likely that abundance has already fallen below a threshold necessary for successful reproduction, and those species can be considered functionally extinct.

In addition to these critically imperiled species, a large number of other mussel species persist as a handful of isolated populations, only one or two of which are large and robust. In these cases, natural or human-caused stochastic events can quickly degrade a species' status to critically imperiled. Prior to 1990, the Little South Fork Cumberland River and Horse Lick Creek, both in Kentucky, supported two of the most important remnants of the diverse and highly endangered mussel fauna

Table 5.5 Critically imperiled mussel species in North America. A population is defined here as a collection of occurrences within a contiguous stream system that can conceivably exchange immigrants. For most species on this table, populations do not extend for more than approximately 80 stream kilometres. Species that occur only in one large population but covering much more than 80 km of contiguous habitat (e.g. several endemic mussel species in Altamaha River, GA) are not included in this table. Host use was determined as described for Table 5.4, with the exception that host use of many of these species is not known; for these species host use was inferred based on information for congeners (see Haag and Warren 2003). Species for which no host-use information was available for congeners or closely related species are listed as unknown. Region of occurrence gives the general biogeographic affinity of each species: Gulf of Mexico refers to Gulf drainages not including the Mississippi and Mobile basins; Ohio River encompasses all tributaries of this river including the Cumberland and Tennessee River drainages.

Species	Number of extant populations	Primary hosts	Region of occurrence
Alasmidonta triangulata	2	Generalist	Gulf of Mexico
Amblema neislerii	2	Generalist	Gulf of Mexico
Arkansia wheeleri	2	Generalist	Lower Mississippi Basin
Dromus dromas	2	Darters[1]	Ohio River
Elliptio chipolaensis	1	Generalist	Gulf of Mexico
Elliptio spinosa	1	Generalist	Atlantic coast drainages
Epioblasma obliquata obliquata	1	Darters[1]	Ohio River
Epioblasma obliquata perobliqua	1	Darters[1]	Great Lakes Basin
Epioblasma penita	1	Darters[1]	Mobile Basin
Epioblasma florentina walkeri	2	Darters[1]	Ohio River
Fusconaia rotulata	1	Unknown	Gulf of Mexico
Hemistena lata	2	Unknown	Ohio River
Lampsilis streckeri	1	Black basses[3]	Lower Mississippi Basin
Lampsilis virescens	1	Black basses[3]	Ohio River
Margaritifera marrianae	2	Madtom catfishes[4]	Gulf of Mexico/Mobile Basin
Medionidus parvulus	2	Darters[1]	Mobile Basin
Medionidus simpsonianus	1	Darters[1]	Gulf of Mexico
Medionidus walkeri	1	Darters[1]	Gulf of Mexico
Obovaria retusa	1	Darters[1]	Ohio River
Plethobasus cicatricosus	1	Sauger[5]	Ohio River
Plethobasus cooperianus	2	Sauger[5]	Ohio River
Pleurobema furvum	2	Riverine minnows[2]	Mobile Basin
Pleurobema gibberum	2	Riverine minnows[2]	Ohio River
Pleurobema taitianum	1	Riverine minnows[2]	Mobile Basin
Ptychobranchus jonesi	1	Darters[1]	Gulf of Mexico
Quadrula intermedia	2	Riverine minnows[2]	Ohio River
Quadrula petrina	2	Catfishes[6]	Gulf of Mexico
Quadrula sparsa	2	Riverine minnows[2]	Ohio River
Quincuncina burkei	2	Unknown	Gulf of Mexico
Toxolasma cylindrellus	1	Sunfishes[7]	Ohio River
Truncilla cognatus	1	Freshwater drum[8]	Gulf of Mexico

[1]Percidae: *Ammocrypta, Etheostoma, Percina*; [2]Cyprinidae: e.g. *Cyprinella, Erimystax, Nocomis, Notropis*; [3]Centrachidae: *Micropterus*; [4]Ictaluridae: *Noturus*; [5]Percidae: *Sander canadensis*; [6]Ictaluridae: *Ictalurus, Pylodictis olivaris*; [7]Centrachidae: *Lepomis, Micropterus*; [8]Sciaenidae: *Aplodinotus grunniens*.

endemic to the Cumberland and Tennessee River systems. In less than 15 years, a series of temporally successive impacts to these streams from coal mining and oil extraction resulted in the near total loss of the mussel fauna from both streams, including some of the largest remaining populations of *Pegias fabula, Pleurobema oviforme, Ptychobranchus subtentum,* and *Villosa trabalis* (Warren and Haag

2005). Because these streams are isolated by impoundments downstream, precluding recolonization from other populations, these vital free-flowing habitats are now lost as conservation refugia. Similarly, in the 1990s, free-flowing streams in Bankhead National Forest, Alabama, supported the largest and most intact example of an upland mussel fauna endemic to the Mobile Basin, including the largest remaining populations of *Hamiota perovalis*, *Pleurobema furvum*, and *Ptychobranchus greeni*. In 2000 a record drought resulted in a reduction of mussel abundance in these streams by as much as 80% (Haag and Warren 2008). Although all species survived the drought, these populations are isolated by a large reservoir downstream, and the post-drought abundance of some may now be too low for natural recovery.

Even for species considered relatively common today, their current distribution may create an illusion of future security. *Elliptio arca* remains widely distributed in the Mobile Basin, but only one large population exists, in the Sipsey River, Alabama (Haag 2004c). Loss of the Sipsey River population from either a natural or human-caused event would suddenly leave this species represented by only by a handful of small, widely scattered populations and thus highly vulnerable to extinction. Similarly, after the decline of *Ptychobranchus subtentum* in the Little South Fork of the Cumberland River (see above), this species now remains abundant only in the Clinch River, Tennessee and Virginia, even though a number of other, small populations persist. This type of distribution is now characteristic of a large number of North American species, portending that the list of critically imperiled species will increase in the future.

Unlike the first wave of extinctions, the second extinction wave will not be limited to obligate large-river species or species with specific life-history traits, such as host-fish use, but will encompass a wide range of mussel diversity. Less than half of critically imperiled species are obligate large-river species; critically imperiled mussels include species restricted to headwater streams as well as species formerly widespread in a range of stream sizes and habitats. Collectively, these critically imperiled species use a wide range of host fishes and, unlike species extirpated from large rivers by impoundment, only about half of these species use darters or riverine minnows as hosts (Table 5.5). Extinction risk in isolated tributary streams will mostly be a function of initial post-fragmentation population size and the extent and connectivity of remaining habitat that is occupied by a particular species (e.g. Loehle and Li 1996; Hanski and Ovoskainen 2002). In tributary streams that experienced mussel declines due to either natural or human-caused impacts, all species were affected non-selectively and the probability of persistence was primarily a function of predisturbance population size rather than interspecific differences in tolerance to disturbance (Warren and Haag 2005; Haag and Warren 2008). Isolated tributary streams can therefore be viewed as having a temporary excess of rare species which will be lost as the communities pay their extinction debt and reach a new equilibrium corresponding to the extent of habitat remaining after fragmentation.

With the exception of a small group of species that thrive in disturbed habitats (e.g. *Anodonta suborbiculata*, *Ligumia subrostrata*, *Pyganodon grandis*, *Toxolasma parvus*, *Utterbackia imbecillis*), mussel species that now have the greatest chance of long-term survival are large-river species that have adapted to run-of-the-river conditions present in some impounded streams. Because of the high volume and large catchments of these rivers, large-river habitats are less sensitive to single point-source impacts or natural disturbances than smaller streams (Petts 1994). Further, widespread construction of navigation dams on large rivers in the central and south-eastern USA created thousands of kilometres of contiguous habitat for impoundment-tolerant species. Unlike populations in isolated tributary streams, populations in impounded riverine habitats can recover from periodic local extinctions or declines due to the potential for recolonization from other populations.

In today's landscape, long-term survival of species in isolated streams is probable only for those having populations large enough, both in size and geographic extent, to allow them to weather periodic natural disturbances and an ever-increasing array of human impacts. The establishment of additional populations of a species in parts of its historical range that have recovered from past insults can greatly reduce the risk of extinction. Short-term survival of some species can potentially be prolonged by intensive captive propagation and stocking programmes. However, in addition to the risks of artificial selection and other genetic hazards (e.g. Jones *et al.* 2006b) the major disadvantage of stocking as a method of sustaining small, isolated populations is the necessity of continuing these programmes in perpetuity given the continued isolation of these populations. In the long run, the extinction debt for North American freshwater mussels can be reduced while minimizing additional species losses only by increasing the extent and connectivity of suitable habitat.

Summary

The world's most diverse freshwater mussel fauna is currently experiencing a massive extinction event. In the last 100 years, at least 26 but potentially more than 40 taxa have gone extinct. An exact accounting of extinctions is impossible due to irresolvable taxonomic problems with several described species that are morphologically similar or were known historically by only a small number of specimens; most of these problematic taxa have not been seen in over 50 years. An accounting of extinctions is also hampered by the difficulty of concluding with certainty that rare species have slipped into extinction. Within the last 15 years, at least one taxon previously considered extinct (*Epioblasma obliquata obliquata*; see Neves *et al.* 1997; Turgeon *et al.* 1998) has been rediscovered, albeit in a small, isolated population (Hoggarth *et al.* 1995). Despite these uncertainties, the severity of this extinction event is clear. Most extinct species were morphologically distinctive (see Fig. 5.1) and specimens are well represented in historical collections as well as in prehistoric middens. Furthermore, mussel research efforts have increased nearly 10-fold since the 1970s (Strayer *et al.* 2004) and in the vast majority of cases intensive survey efforts have corroborated the apparent extinction of species. At least 31 surviving species are in imminent danger of extinction, and a large number of others are highly vulnerable to extinction in the long term. In total, it is likely that 25% or more of the North American mussel fauna will be extinct within a human generation.

Most recent extinctions have occurred in the region of highest mussel diversity, especially the Ohio (including the Cumberland and Tennessee systems) and Mobile River basins (Table 5.3). Similarly, critically imperiled species (i.e. species surviving as only one or two populations) are concentrated in these systems but also occur predominantly in other Gulf of Mexico drainages (e.g. Apalachicola, Escambia, Rio Grande; Table 5.5). There are currently no known species extinctions or critically imperiled species (as defined here) in the Pacific Northwest or the upper Mississippi River basin and few in the lower Mississippi River basin, Great Lakes, or Atlantic coast drainages, even though all of these areas have experienced extensive habitat loss and fragmentation, and widespread declines in mussel abundance. The low rate of extinction in these areas can be explained by three factors: (1) lower initial diversity (e.g. Pacific Northwest, northern Atlantic coast drainages), (2) no species restricted to a single habitat type (e.g. large rivers) that has been systematically destroyed, and (3) few species with highly restricted ranges (e.g. upper Mississippi River basin). However, estimates of rates of extinction or imperilment may be artificially low for some regions due to the presence of unrecognized cryptic diversity, especially within the genus *Elliptio* in southern Atlantic coast drainages. In all regions, imperilment can be expected to increase as human impacts continue to accelerate, exacerbating effects of habitat fragmentation and population isolation.

Although a small number of species may have gone extinct naturally, most mussel extinctions are a result of human impacts. Humans have exerted substantial pressures on mussel populations for over 5000 years. However, human impacts from pre-Columbian times until the early twentieth century, including intense harvest and degradation of stream habitat, resulted in no documented species extinctions despite a remarkably complete record of mussel distribution and abundance throughout this lengthy time period. The first wave of mussel extinctions occurred rapidly in the mid-twentieth century in response to large-scale, systematic destruction of large river habitat by the construction of dams and was composed mostly of species that were restricted to this specialized habitat. In the late twentieth century a second, as yet smaller wave of extinctions followed due to a variety of proximal causes but ultimately related to indirect effects of habitat fragmentation caused by habitat destruction earlier in the century. As a result of the time lag associated with the manifestation of these indirect effects, habitat fragmentation has created a large extinction debt, the payment of which has only begun. Over time, this second extinction wave due to indirect effects of fragmentation can be expected to eclipse the first wave due to direct habitat destruction. In contrast to the first extinction wave, the second wave will not be limited to species with specific ecological attributes; rather, probability of extinction will be primarily a function of each species' initial post-fragmentation population size and the extent and connectivity of its remaining habitat, ultimately resulting in extinction of a broad cross-section of freshwater mussel diversity. Although the wide variety of impacts associated with an increasing human population would have probably caused at least some extinctions even if stream systems had retained connectivity, the legacy of fragmentation due to stream impoundment has underpinned and magnified all other insults to streams and has greatly intensified the current freshwater mussel extinction crisis in North America.

Acknowledgements

I am grateful to the following people for their various contributions to this chapter. Bob Butler, Jim Theler, and Melvin Warren reviewed drafts of the manuscript. Melvin Warren, Paul Hartfield, Jim Williams, Art Bogan, and Jeff Garner provided helpful suggestions and information. Mickey Bland and Amy Commens helped in the laboratory and in assembling pertinent literature.

CHAPTER 6

Holocene extinctions in the sea

Nicholas K. Dulvy, John K. Pinnegar, and John D. Reynolds

6.1 Introduction

The Holocene is bounded at either end by pulses of terrestrial extinction. The beginning was marked by megafaunal extinctions associated with colonizing waves of hunter-gathering humans. The current pulse of terrestrial extinction is at least one to two orders of magnitude higher than the background rate, mainly due to large-scale habitat modification. By comparison, few extinctions of marine species have been recorded, and these have been of range-restricted species, mainly mammals and birds, over the past 300 years. Today's extensive overexploitation of global fisheries has a historical and prehistoric precedent in archaeological evidence for the local collapse of many fisheries and shellfish beds, and regional extinction of populations such as the Atlantic gray whale. Marine extinctions may be more widespread than is currently appreciated, largely because it is very difficult to observe the last days of the last individual of a marine species, and because of a fallacious but widespread perception that marine organisms cannot be driven to extinction. The observed human capacity for causing rapid and widespread terrestrial extinctions combined with the rapidly increasing scale of human impact on the sea forewarn of an impending marine extinction event. The scale of this may be the equivalent of concatenating both of the terrestrial Late Quaternary extinction waves into a much shorter time frame. Fortunately the opportunity to forestall major loss of ocean biodiversity has never been greater. While much megafaunal biomass has been depleted and some habitat has been lost, society, scientists, and managers are acutely aware of these problems and legislation and institutions are strengthening in response.

6.2 The origins of human exploitation of the sea

The Holocene is marked by a major transition in human social organization, from hunter-gathering to a settled agrarian lifestyle. This was made possible by domestication of a wide range of crops and animals. Prior to this, the expansion of hunter-gatherer populations across continents and islands led to waves of extinction of megafaunal mammals and birds (see Chapter 2 in this volume). Around 11 000 years ago prehistoric hunters in North America were probably responsible for causing the extinction of 34 genera of large-bodied mammals (including mastodons, ground sloths, artiodactyls, and perissodactyls) within a millennium of hunting (Martin and Steadman 1999). In Australia, human arrival around 46 000 years ago corresponds with the extinction of all mammal, reptile, and bird species weighing more than 100 kg (Roberts *et al.* 2001). Polynesians exterminated as many as 2000 bird species from Pacific Islands (Steadman 1995), and hunted 10 moa species to extinction soon after colonizing New Zealand (Holdaway and Jacomb 2000; see Chapter 2 in this volume). One-fifth of all the bird species extant at the beginning of the Holocene are now extinct (Milberg and Tyrberg 1993).

Environmental impacts associated with settled agrarian lifestyles led to a second wave of terrestrial extinctions towards the end of Holocene. At least 844 extinctions have been documented since AD 1500, and almost all have been of terrestrial plants and animals (Baillie *et al.* 2004). The main drivers of the recent extinction wave are habitat loss caused largely by forest clearance for agriculture, and predation and habitat degradation caused by invasive species such as dogs, cats, rats, mongooses, pigs, and goats (Baillie *et al.* 2004). The terrestrial extinction rate is now one to two orders of magnitude higher than the background rate inferred from the fossil record (Mace *et al.* 2005).

Today the oceans may be in a state more akin to terrestrial ecosystems 10 000 to 50 000 years ago, at the Late Pleistocene or advent of the Holocene. There has been large-scale biomass removal of ocean megafauna, increasing extinctions of populations and species, and a rapid increase in the domestication of marine animals and plants (Dulvy *et al.* 2003; Lotze *et al.* 2006; Duarte *et al.* 2007). A hunting-to-cultivation transition is now gaining pace in the oceans, similar to the transition that began on land thousands of years ago (Fig. 6.1). One hundred million tonnes of food fish are extracted from oceans and coastal waters each year (2000–2006); most (85%) is still hunted from the wild, with the remainder provided by aquaculture of domesticated fish and invertebrates. Currently, the yield of aquaculture is small, but the rate of species domestication is rapid and the relative yield contribution has increased year on year since the 1980s (FAO 2007). Most (90%) cultivated marine species were domesticated in the last decade, whereas the majority of terrestrial species in cultivation were domesticated over 2000 years ago (Fig. 6.1). Aquaculture is predicted to provide as much as 41% of global fish production by 2020, up from 31% in 1997 (Delgado *et al.* 2003).

This chapter documents the changing state of coastal and high-seas ocean ecosystem biodiversity, particularly over the last 1000 years, with occasional insights provided from the deeper Holocene. We summarize the main types of methods and data sets used to measure historical changes in marine biodiversity, and use this summary to justify limiting this chapter to the last millennium, instead of providing detailed treatment of the whole Holocene. We summarize the main causes of marine biodiversity loss, with a focus on overexploitation as the main driver of Holocene extinctions. Next, we outline the evidence for the spread of fishing impacts and habitat loss across the world and deeper into the oceans, and the sequential depletion of marine megafauna. Finally, we sketch out current knowledge of the number and taxonomic scale of marine extinctions.

6.3 Measuring marine biodiversity change over the Holocene

It remains difficult for scientists to identify the causes of the major waves of terrestrial megafaunal

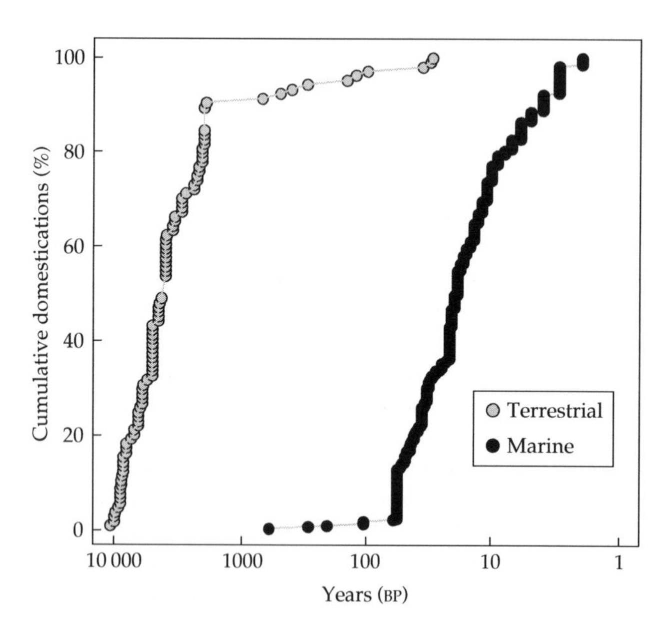

Figure 6.1 The cumulative number of domestications of terrestrial and marine plants and animals over the last 100 000 years. Most terrestrial domestications occurred around the beginning of the Holocene, whereas most domestications of marine species occurred in the last 100 years. Redrawn from Duarte *et al.* (2007).

extinction because these events took place several millennia or tens of millennia ago (Burney and Flannery 2005). The major loss of marine megafauna occurred more recently and largely within the period of increasing scientific knowledge, yet reconstructing the recent historical changes in marine biodiversity remains particularly challenging (Pinnegar and Engelhard 2008).

There is a relative paucity of observational data and knowledge of ecological conditions in the oceans for the prehistoric Holocene and Late Pleistocene (>1000 years before present). Humans inhabit the terrestrial portion of the Earth and frequent ocean travel has been largely restricted to the last millennium. The oldest perspective is provided by comparisons between fossil and modern coral reefs, providing insights into changing community structure from 125 000 to 17 000 years BP (Aronson *et al.* 2002; Pandolfi and Jackson 2006). Subfossil remains, such as fish bone deposits and kitchen middens (human refuse dumps), extend back several thousand years and can be used to demonstrate changing average fish size, changing fish community structure, and human economic and technological advances (Jackson *et al.* 2001; Wing and Wing 2001; Barrett *et al.* 2004b). However, many coastal archaeological and historical sites are now under water due to sea-level rise.

Historical documents and art, such as pottery and sculpture, depict catches and may hint at long-lost species (Pinnegar and Engelhard 2008). Documentary history, for example of trade and tax records, can often provide more complete information and may in exceptional cases be used to reconstruct the fate of populations, such as the 500 year span of Newfoundland cod catches (Rose 2004) or the 300 year span (AD 1650–1950) of Mediterranean tuna catches (Ravier and Fromentin 2001). The spread of exploration and rising interest in the natural world during the fifteenth to eighteenth centuries provided detailed taxonomic inventories and historical species distributions. Such information, when compared to modern surveys, is a major source of our knowledge on recent marine extinctions (e.g. Jackson *et al.* 2001; Dulvy *et al.* 2003).

The modern marine scientific era provides detailed fisheries and research surveys spanning large areas of coastal and oceanic seas, especially over the past 25–50 years (FAOSTAT 2004). Research surveys have provided considerable insight into the scale of human impact, particularly in temperate waters of the northern hemisphere, but they are rarely powerful enough or offer sufficient taxonomic resolution to be useful for detecting marine extinctions (Maxwell and Jennings 2005). Research time series are largely absent from the tropical oceans of developing nations. These knowledge gaps are now being filled through interviews with members of coastal communities for their traditional knowledge of species presence, behaviour, and ecology; for example, in Brazil, China, Palau, and Fiji (Johannes 1981; Sadovy and Cheung 2003; Dulvy and Polunin 2004; Silvano *et al.* 2006).

6.4 A millennium and more of fishing

Fishing or hunting is the greatest cause of threat and population extinctions in the sea, followed by habitat loss, pollution, and invasive species (Fig. 6.2). The last thousand years have seen great technological advances in fishing power and demand for fish from burgeoning human populations (Pauly *et al.* 2005). Industrial fishing fleets have expanded out over the oceans and into deeper waters, and the increasing densities of subsistence fishers make sustainability unlikely (Newton *et al.* 2007). Over a third of the human global population inhabits the seaward margins of the terrestrial realm (Cohen *et al.* 1997). The coastal seas provide abundant and easily accessible food that can be gleaned from tide pools, or caught in nets and traps or using lines of baited hooks. Fishing provides many nations with a large proportion of their dietary animal protein intake. For example, fish provide nearly two-thirds of the animal protein to people in the West African countries of Gambia, Ghana, and Sierra Leone, and over a third of the intake of the Asian countries of Vietnam, Malaysia, Thailand, Cambodia, and Bangladesh. In island nations fish are typically the major source of the average daily protein intake, for example for the Maldives (84%), Comoros (64%), Indonesia (57%), and Sri Lanka (52%) (FAO 2004b).

In some parts of the world there is good reason to believe that human dependence on marine fish, molluscs, and crustaceans for food and dietary

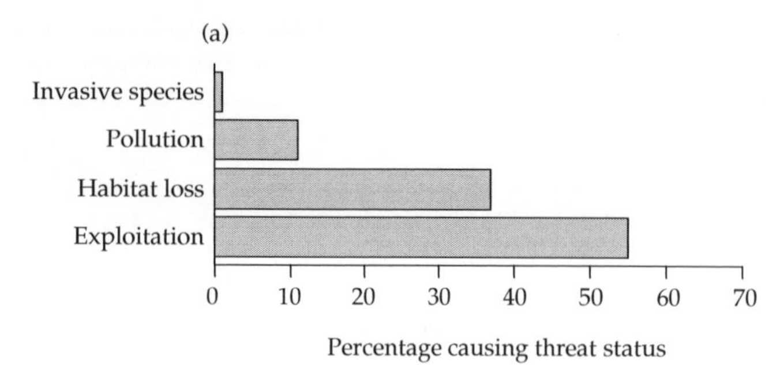

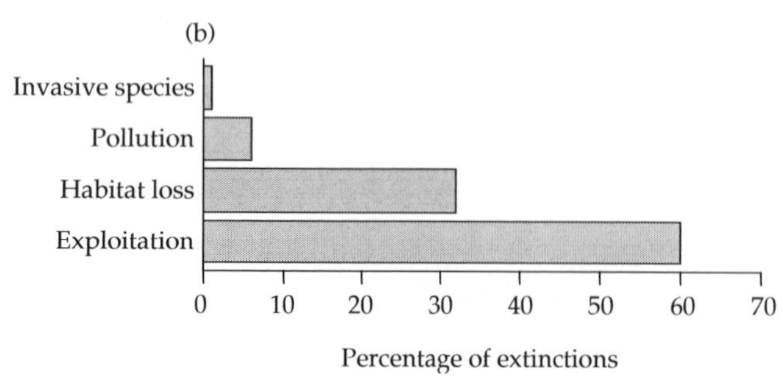

Figure 6.2 The main causes of (a) threat and (b) extinction risk for marine fish populations and species. The figures include cases where more than one cause of threat has been identified for a given population or species. (a) North American species threatened with extinction ($n=82$), including those considered Vulnerable, Endangered, or Critically Endangered (Musick *et al.* 2000). (b) Local, regional, and global marine fish extinctions ($n=65$) (Dulvy *et al.* 2003). In all cases exploitation and habitat loss were the primary causes of threat.

protein was just as great in our recent historical past. Pacific Island reef and lagoon fisheries resources have been continuously exploited for many centuries, and exploitation has been occurring in western Melanesia for 20 000–30 000 years (Dalzell 1998). Molluscs appear to have been extremely important as a food source for early Pacific Island human populations (Dalzell 1998). In some instances, declines in mollusc resources forced early human populations to increase exploitation of other marine resources, and to rely increasingly on agriculture.

The oldest evidence of marine harvesting is the presence of shellfish remains in two middens in Saldanha Bay, South Africa, dating from 60 000–70 000 years ago (Volman 1978). The earliest hunter-gatherers collected shellfish opportunistically, and hunted slow-moving terrestrial reptiles such as tortoises (Klein *et al.* 2004). By the early Holocene (11 500–8500 years BP), fishing technology had advanced considerably, broadly concurrent with the development of agriculture and crop domestication on land. The use of boats, hooks, and lines are known from a number of locations in the prehistoric Holocene, including the Northern Channel Islands, California (Rick *et al.* 2001). In Parita Bay, Panama, a comparison of fish faunas from Cerro Mangote (6000 years BP) and Sitio Sierra (1800 years BP) suggests that regional fishing methods shifted between earlier and later periods from a shore-based, netless technique to a more complex one based on fine-meshed gill-nets and watercraft (Cooke 2001). Similar observations have been made for southern Taiwan (Kuang-Ti 2001) and northern Scotland (Barrett *et al.* 1999). At an early settlement on Cyprus, middens dated to 8000 years BP revealed that large individuals of certain species, notably sea breams (Sparidae) and groupers (Serranidae), were much more common during the Neolithic than they are now (Desse and Desse-Berset 1993). Similarly, fish faunal diversity and fish body size decreased over a 12 000 year period at a site in southern Spain (Morales *et al.* 1994).

Up to the turn of the first millennium AD, marine fisheries were a minor affair in Europe. For example, exploitation of fish resources in Britain during this period focused mainly on freshwater species such as northern pike *Esox lucius*, and migratory species such as European eel *Anguilla anguilla*, Atlantic salmon *Salmo salar*, and European sturgeon *Acipenser sturio*. The advent of the second millennium in Britain is marked by increasing quantities of marine fish remains in coastal and inland middens

(Barrett *et al.* 2004b). This increase in the consumption of marine fishes was repeated at a similar time or soon after in mainland Europe (Barrett *et al.* 2004a; Pauly 2004). The most parsimonious explanation for this transition was that increasing urbanization of European human populations led to increased food demand, concomitant with declining availability of freshwater fish. This led to the development of marine fisheries for Atlantic herring *Clupea harengus*, and Atlantic cod *Gadus morhua* (Barrett *et al.* 2004b). The decline of freshwater fisheries is thought to have stemmed from a combination of pollution from agricultural run-off, overexploitation, and damming (Hoffmann 1996; Barrett *et al.* 2004a, 2004b). Salmon were heavily overexploited, and populations of other freshwater fish species disappeared completely. The burbot *Lota lota* is common in archaeological deposits and was eaten in large numbers throughout Britain, but is now regionally extinct. Sturgeon were virtually extinct across much of northern Europe by the fourteenth century due to overexploitation, damming, and diking of key habitats (Hoffmann 1996), although they were still commercially exploited into the nineteenth and possibly the twentieth century in The Netherlands, Germany, and other countries (W.J. Wolff, personal communication). Climatic variability is not thought to have contributed significantly to the transition to marine fisheries; the transition occurred when environmental conditions were unlikely to promote such a switch, when local productivity of cod and herring in the southern North Sea was probably reduced, conditions which would be expected to have instead supported agricultural expansion (Barrett *et al.* 2004b).

Across the Atlantic Ocean, some subsistence Caribbean island fisheries had already begun to deplete their marine resources. Comparison of faunal remains between two time periods (early and late, 1850–1280 and 1415–560 years BP, respectively) on Puerto Rico, St Thomas, St Martin, Saba, and Nevis indicates that the average weight of reef fish declined between early and late periods on each island, with a decrease in representation of inshore reef fishes and increase in representation of offshore pelagic fishes (Wing and Wing 2001). These changes in species composition and average size resulted in a decrease in mean trophic level of the fish assemblage at each island between early and late periods. However, this pattern is not widespread throughout the Caribbean, as there is evidence for sustainable fisheries (i.e. no change in average fish size or range in fish sizes) from AD 600/800 to AD 1500 on Anguilla in the northern Lesser Antilles (Carder *et al.* 2007).

Around the same time, Europeans were discovering new countries and new fishing grounds in the north-west Atlantic. In the eleventh century, Basque whalers from Spain and Portugal hunted around the coast of the Bay of Biscay, but from the twelfth century up to 600 Basque whalers caught bowhead *Balaena mysticetus* and North Atlantic right whale *Eubalaena glacialis* off the Labrador coast for their oil (Cumbaa 1986). The remnant bowhead population which survives in Baffin Bay–Davis Strait is thought to number between 450 and 1000 adults (IUCN 2006). The adjacent East Arctic bowhead population was exploited to near extinction by commercial whaling fleets from AD 1611 onwards (Allen and Keay 2006); this population now contains fewer than 100 individuals, and possibly fewer than 50 mature adults, and is listed as Critically Endangered (IUCN 2006). Portuguese and Norwegian fishermen also crossed the Atlantic to Newfoundland and stayed for the short summer season to fish the bountiful cod, drying and salting their catch before returning each winter (Kurlansky 1998). This was one of the world's largest ever fisheries and has been driven to collapse in 500 years or so, culminating in closure in 1992, with many sub-fisheries yet to be reopened (Hutchings 1996; Myers *et al.* 1996). As with the extinction of terrestrial megafauna, there has been debate over whether the ultimate cause of the decline of the fishery was environmental change or human exploitation. Long time series of Atlantic cod catch data from Newfoundland between 1505 and 2004 can be used to explore the relative explanatory power of climate and overfishing hypotheses (Rose 2004). Fishing or climate variability (as represented by interannual variation in tree-ring growth) alone did not mimic the observed catch trends; the model best describing the collapse and non-recovery of Newfoundland cod since 1505 incorporates climate variability, fishing mortality, and depensation (negative population growth at small population sizes) (Fig. 6.3).

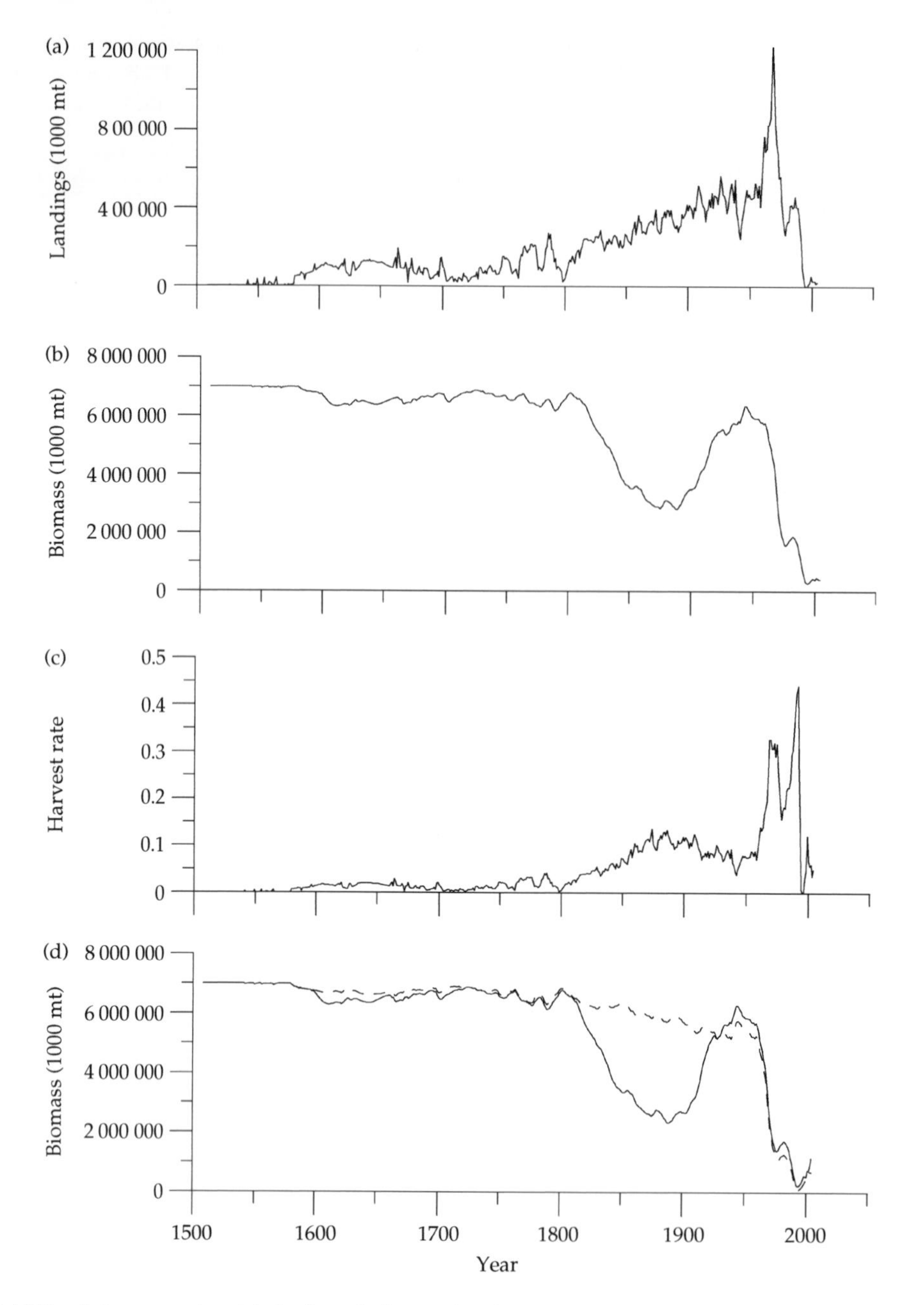

Figure 6.3 (a) Historical reconstruction of the landings of Atlantic cod *Gadus morhua* in Newfoundland and Labrador waters from 1505 to 2004; (b) stock biomass estimates based on a surplus production model incorporating climate forcing and depensation (Allee effects); (c) the annual harvest rate (landings per unit of biomass); and (d) stock biomass estimates derived from a surplus production model assuming constant r and K parameters (dashed line) and the climate forcing alone (solid line). The surplus production model was climate forced using a composite tree-ring data set, representing annual temperatures for northern North America (from Alaska to Quebec) from 1600 to 1974. Redrawn from Rose *et al.* (2004).

6.5 The expansion of fishing into deep water

In recent years, fishing in deep waters (>400 m) has increased as traditional shallow-water stocks have declined (Devine *et al.* 2006). The target deep-water fish (e.g. roundnose grenadier *Coryphaenoides rupestris,* and orange roughy *Hoplostethus atlanticus*) are often long-lived and late-maturing, and hence intrinsically vulnerable (Morato *et al.* 2006a, 2006b). Orange roughy can live to over 125 years of age and may not mature until 20 years. Fishing by factory trawlers and modern long-line fleets started in the late 1960s. Analyses of several of the most important deep-sea fishes, using a widely used index of abundance (catch-per-unit fishing effort, CPUE), have indicated a clear declining trend in abundance. For orange roughy in the north-east Atlantic, the CPUE in 1994 was only 25% of initial catch rates when the fishery commenced in 1991 (ICES 2003). Since 1964, deep-water fisheries have contributed 800000–1000000 tonnes annually to global marine fish landings. The average depth from which catches of both pelagic and bottom-dwelling species are taken has been deepening over time across all oceans (Fig. 6.4). This trend has been accelerating since 2001 (Pauly *et al.* 2003; Morato *et al.* 2006b).

6.6 Declines of marine megafauna

6.6.1 The great whale hunt

As with terrestrial extinctions (Stuart 1991), population declines of large-bodied, long-lived animals are typical of historical and modern fisheries. In some cases, this has led to regional extinctions. The gray whale *Eschrichtius robustus* now occurs only in the North Pacific, but this species first became known to science through the discovery of subfossil remains in England and Sweden (Bryant 1995). Radiocarbon dating of fossil and subfossil remains indicates that a gray whale population existed in the Atlantic until the seventeenth century AD. Coastal whaling has occurred in the North Atlantic since at least AD 1000, and historical accounts suggest that the Atlantic gray whale may have been among the species hunted by the first whalers (Bryant 1995).

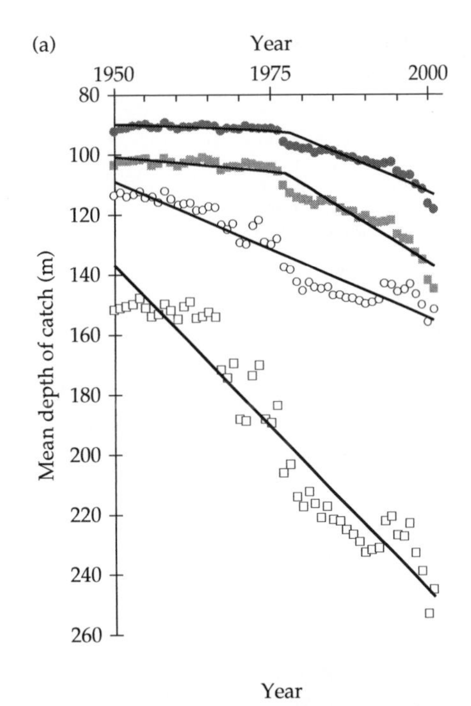

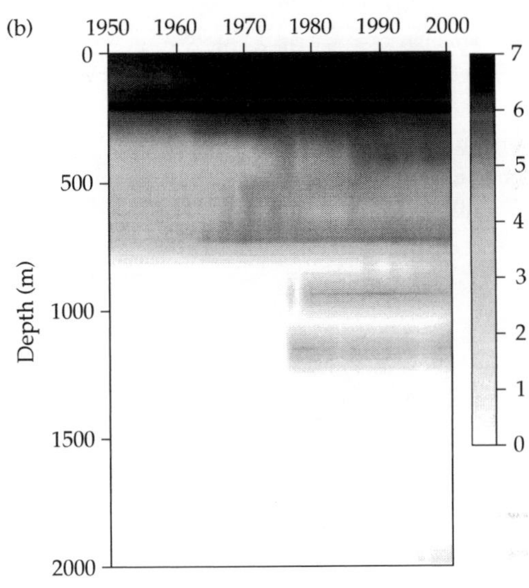

Figure 6.4 (a) Deepening of the global marine fisheries catches from 1950 to 2001 for surface dwelling (pelagic) fishes (dark grey circles) and for bottom marine (demersal) fishes (light grey squares). Open symbols are corresponding estimates for high-seas areas only (beyond Exclusive Economic Zones). (b) Time series of world marine bottom fisheries catches by depth strata. Catch in tonnes are $\log_{10}$ transformed. Redrawn from Morato *et al.* (2006b).

From AD 1059 onwards Basque whalers killed large numbers of whales as they migrated close to shore through the Bay of Biscay. By the fifteenth century, Basque whalers travelled as far as Iceland, Greenland, and Canada in search of whales. It is widely assumed that the primary target species for these whalers was the North Atlantic right whale; however, if the gray whale inhabited nearshore waters in the Atlantic, as surviving populations do in the Pacific, then it is plausible that they may have been an even likelier target for Basque whalers (Bryant 1995).

More recently, industrial whalers fished down and sequentially depleted the great whales in 50 years or less. Depletion of the largest species (the blue whale *Balaenoptera musculus*) occurred first, followed by the North Pacific right whale *Eubalaena japonica*, humpback whale *Megaptera novaeangliae*, fin whale *Balaenoptera physalus*, and eventually moving on to the smaller sei whale *Balaenoptera borealis*, and minke whale *Balaenoptera acutorostrata* (Gulland 1974) (Fig. 6.5). Many populations of these species are now at a fraction of their former abundance, and are listed in one of the three threat categories (Vulnerable, Endangered, or Critically Endangered) on the IUCN Red List. Although cetaceans are the subject of a great deal of attention and controversy, it is interesting to note that out of the 84 known species, only one (freshwater) species is believed to have become globally extinct, and one coastal species is Critically Endangered. While not a marine species, it is worth considering the decline of the Yangtze River dolphin or baiji *Lipotes vexillifer*. This dolphin is endemic to the middle-lower Yangtze River system in eastern China and was long recognized as one of the world's rarest and most threatened mammal species. Chinese scientists reported a steady decline in the baiji population from an estimated 400 individuals in 1979–1981 to as few as 13 individuals in 1997–1999 (Zhang *et al.* 2003), due largely to by-catch in local fisheries, pollution, boat collisions, and dam construction instead of direct persecution. Even though efforts were proposed to conserve the species, an expedition towards the end of 2006 failed to find any baiji in the river. In 2007 the organizers were forced to conclude that the baiji is now very likely to be extinct (Turvey *et al.* 2007b). The Critically Endangered vaquita *Phocoena sinus* is a small porpoise endemic to the northern Gulf of California (Sea of Cortez). This species numbers in the low hundreds, and again suffers from incidental by-catch mortality from fisheries; it is predicted to become extinct within the next few years without intensive conservation efforts (Rojas-Bracho *et al.* 2006; Jaramillo-Legorreta *et al.* 2007).

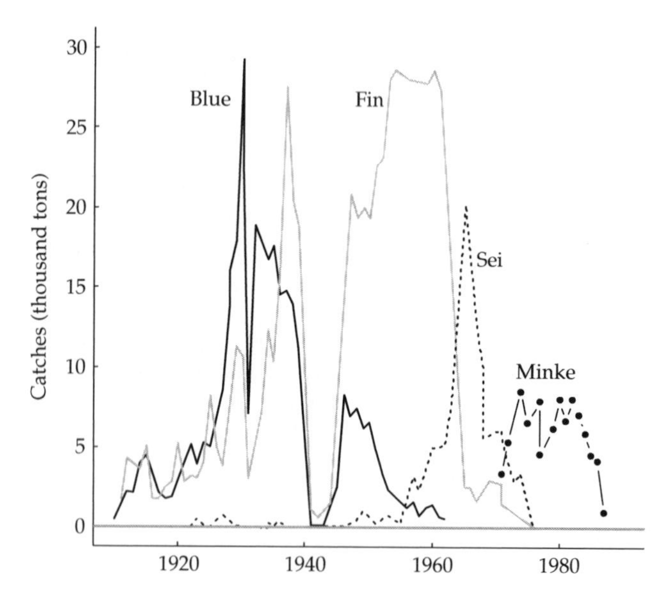

Figure 6.5 Sequential depletion of the great whales in the Antarctic Ocean. Redrawn from Allen (1980).

A striking feature of the Red List status of the world's cetaceans is that the IUCN has been unable to determine the status of 48% of the world's 84 species. One reason for this is that it has proven difficult to obtain good estimates for current or historical population sizes of many of the world's beaked whales, dolphins, and porpoises. One approach that has been used with some success for baleen whales has involved estimates of long-term effective population sizes based on genetic diversity and rates of gene substitution. In the North Atlantic, the historical population estimates of humpback, fin, and minke whales are 240 000, 360 000, and 265 000 respectively. Current population sizes (and overall percentage decline) are 10 000 (96%), 56 000 (84%), and 149 000 (44%) (Roman and Palumbi 2003). Records of historical catches from ship logbooks during the eighteenth and nineteenth centuries are regularly used by the International Whaling Commission (IWC) to reconstruct the population dynamics of whales before, during, and after exploitation (Baker and Clapham 2004). The historical trajectories for southern right whale *Eubalaena australis*, one of the most vulnerable species, show a sharp decline during the mid-1800s, with a slow increase following international protection in 1931 and another decline resulting from illegal Soviet catches during the 1960s. The lowest point of population abundance was in 1920, when as few as 60 adult females were estimated to have survived.

6.6.2 Sea cows, seals, and otters

Only 27 years after the discovery of Steller's sea cow *Hydrodamalis gigas* in 1741, this species was driven to extinction as a result of excessive, and wasteful, hunting to provision Russian fur-hunting expeditions (Anderson 1995; Turvey and Risley 2006). The four extant sirenian species (dugong and manatees) are currently listed as Vulnerable by the IUCN Red List. The dugong *Dugong dugon* was once distributed widely throughout the tropical South Pacific and Indian Oceans. The primary causes for population declines include hunting, habitat degradation, and fishing-related fatalities. Along the coast of Queensland, where the most robust quantitative data on population trends are available, analyses have suggested that the region supported 72 000 dugongs in the early 1960s compared with an estimated 4220 in the mid-1990s (Marsh *et al.* 2005). Similar declines have been experienced by the West Indian manatee *Trichechus manatus*, which is particularly at risk from boat strikes (Marmontel *et al.* 1997).

The Caribbean monk seal *Monachus tropicalis*, the only seal known to be native to the Caribbean Sea and the Gulf of Mexico, is now considered globally extinct. This species was estimated to have originally consisted of more than a quarter of a million individuals divided among 13 populations spread throughout the Caribbean (McClenachan and Cooper 2008). This species was hunted for food and oil by European explorers and plantation settlers. Hunting rapidly eliminated the outer populations, substantially contracting the spatial extent of the seal's range (Fig. 6.6). The last reliable sighting of the Caribbean monk seal was of a small colony at Seranilla Bank, Jamaica, in 1952, but it had been substantially depleted throughout the Caribbean since at least the 1850s (Timm *et al.* 1997). The world's two other monk seal species are also considered highly threatened. The Mediterranean monk seal *Monachus monachus* is believed to be the world's rarest pinniped and one of the most endangered mammals on Earth. It is listed as Critically Endangered (IUCN 2006). A dramatic population decrease over time has been attributed to several distinct causes, in particular commercial hunting (especially during the Roman period and the Middle Ages) and eradication by fishermen during the twentieth century, but also coastal urbanization. As a result of these factors, the entire population is estimated to consist of fewer than 600 individuals scattered throughout a wide geographic range (Forcada *et al.* 1999). The Hawai'ian monk seal *Monachus schauinslandi* has also suffered severe population declines in recent years, due to the spread of human activity to even the most remote and isolated areas in the north-west Hawai'ian Islands. It is estimated that fewer than 1400 Hawai'ian monk seals exist today (Antonelis *et al.* 2006).

Sea otter *Enhydra lutris* populations were hunted for their fur, initially by indigenous Aleut people and later on more extensively by Europeans, and were reduced to local extinction in many parts of

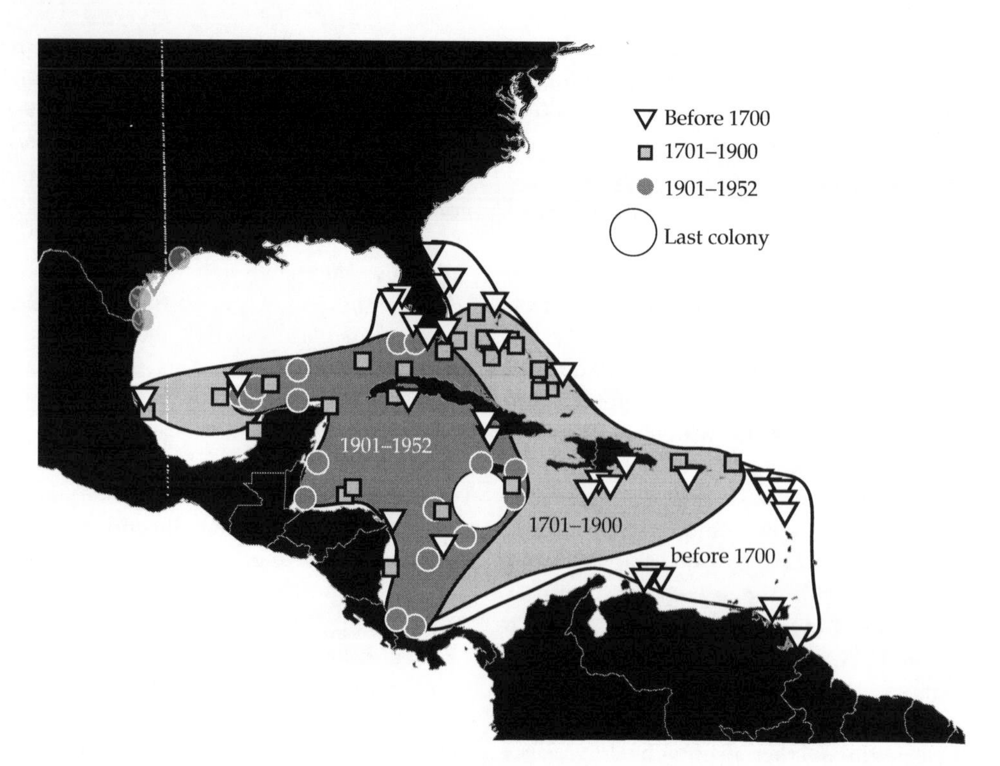

Figure 6.6 Decline in the total geographic extent of the Caribbean monk seal over time. Early observations (triangles, before the eighteenth century) were recorded as far east as the Lesser Antilles and Guyana. Observations from the eighteenth and nineteenth centuries (squares) were recorded in most of the Caribbean basin, but by 1900 observations (small circles) were restricted to the western Caribbean and Gulf of Mexico. The most persistent population (large circle, last colony) was found on the Serranilla Bank. Observations in the western Gulf of Mexico are unconfirmed. Redrawn from McClenachan and Cooper (2008).

their historical range, for example Mexico and British Columbia (Simenstad *et al.* 1978). By 1911 the global population was estimated to be only 1000–2000 individuals (mostly in the Aleutian Islands), compared to as many as 300 000 before the years of the great hunt (Kenyon 1975). So few individuals remained that many authorities assumed they would become extinct. However, in 1938 biologists found a small group of about 50 sea otters along the coast south of Carmel, California. These few animals (together with the last remaining animals in Alaska) formed the nucleus of a breeding population for restoration efforts. Currently the sea otter is listed as Endangered; the current global population estimate for *E. lutris* is approximately 108 000, although the Alaskan and Californian populations are declining due to killer whale *Orcinus orca* predation and disease respectively. The sea otter remains regionally extinct in Mexico and Japan (Springer *et al.* 2003; IUCN 2006).

6.6.3 Seabirds

Many species of seabirds are severely threatened. For example, the long-lived ocean-going albatrosses (Diomedeidae) are threatened from high-seas long-line fisheries, such as those that target southern bluefin tuna *Thunnus maccoyi* and albacore *Thunnus obesus* in the Southern Ocean. The birds are attracted to the baited hooks as they are deployed from the fishing vessels, and are often hooked and drown. The total reported fishing effort was at least 60 million hooks set per year in the 1960s and is presently greater than 180 million hooks and increasing (Tuck *et al.* 2001). The Red List threat status of albatrosses worsened in the decade after 1994 (Butchart *et al.* 2004). All 21 species of albatross are now listed as globally threatened (compared to just three species in 1996 and 16 species in 2000) (IUCN 2006).

Studies of other species of seabirds have shown that impacts of fisheries can be mixed. For example, the Balearic shearwater *Puffinus mauretanicus*, which breeds in the Mediterranean, is listed as Critically Endangered. Population models suggest that by-catch of adults by long-line fisheries is probably the main cause of a declining population trend (Oro *et al.* 2004). Yet the birds benefit from foraging on small fish that are discarded by fisheries, with over 40% of the energetic requirements of chicks being met from this source (Arcos and Oro 2002). Such subsidies from fisheries discards are typical for a range of seabirds (Lewison *et al.* 2004). There is now concern that efforts by the European Union to reduce the amount of fisheries discards may remove a critical food source and push this species more quickly towards extinction.

6.6.4 Turtles

Jackson (1997) highlighted the difference between how we see the seascape today and how early Europeans visiting America witnessed it. Old hunting data from the Cayman Islands together with reports from early explorers indicate that green turtle *Chelonia mydas* populations in the Caribbean may have declined by at least 99% since the arrival of Christopher Columbus in 1492. Turtles suffer many threats worldwide; chief among these is by-catch in offshore long-line fisheries, either on baited hooks or through entanglement, and in inshore shrimp trawls. A recent global estimate of the effect of fisheries on marine turtles suggests that some 260000 loggerheads and 50000 leatherbacks are captured incidentally by long-lines each year, a large proportion of which die as a consequence (Lewison *et al.* 2004). In addition, turtles suffer overexploitation of eggs and habitat loss from human development of their nesting beaches. Analyses of green turtle subpopulation changes at 32 index sites across the world suggest a 48–67% decline in the number of mature females nesting annually over the last three generations (IUCN 2006). Analysis of published estimates for leatherback turtle *Dermochelys coriacea* suggests a reduction of over 70% for the global population of adult females in less than one generation (Pritchard 1982; Spotila *et al.* 1996).

All five species of sea turtles in US waters are listed as threatened or endangered under the US Endangered Species Act. A major source of mortality for these turtles is drowning in shrimp trawls. Most (70–80%) strandings of dead turtles on US beaches are thought to be related to interactions with this fishery (Crowder *et al.* 1995). Efforts are underway to introduce turtle excluder devices (TEDs) in trawl nets in both the USA and northern Australia, although trawl fisheries remain a major problem for turtles elsewhere (FAO 2004a). It has been estimated that the US shrimp fleet alone caught 47000 sea turtles each year prior to the introduction of TEDs in 1989 (FAO 2004a).

6.6.5 Large predatory fishes

Many fishes, particularly the larger-bodied predatory species, have declined massively. This has become particularly apparent in the past half century. The Food and Agriculture Organization (FAO) of the United Nations, the collector of world fishery statistics, has calculated that more than 77% of the world's fisheries are fully or overexploited, 8% have collapsed, and only a quarter remain 'underexploited' (Garcia and Newton 1995; FAO 2007). An independent analysis of the same data suggests that one quarter (366 of 1519) of fish stocks have collapsed in the last 50 years (Mullon *et al.* 2005). These figures may be conservative, as discarded fishes and other animals go unreported in these statistics, as do the catches of artisanal and subsistence fishers (Sadovy 2005; Zeller *et al.* 2006; Andrew *et al.* 2007). While trends in aggregated taxa are widely available, there are few data on the fate of individual species and populations. A more detailed picture of the fate of populations and species comes from the analysis of assessed exploited stocks of the northern temperature fisheries of Europe, Canada, and the USA. Of these 232 stocks (populations) the median decline in adult abundance has been 83% from known historical levels; however, these declines are usually measured from the beginning of the time series, which often started long after exploitation began. Few of these populations have recovered 15 years later (roughly equivalent to three generations; a time scale used in the extinction risk assessments of the IUCN Red List criteria) (Hutchings 2000; Hutchings and Reynolds 2004). A large number of these populations exhibit reduction in age and size of maturity, consistent with an evolutionary response to the effects of overfishing of adults (Law and Grey 1989; Olsen *et al.* 2004; Hutchings 2005; Hutchings and Baum 2005).

Large predatory fishes have undergone the steepest declines due to their lower intrinsic rate of population increase and hence lower resilience to fishing mortality (Reynolds *et al.* 2005). The average trophic level of this global catch has declined as predatory fishes have been sequentially depleted and fishers target more productive species at lower trophic levels (Pauly *et al.* 1998; Essington *et al.* 2006). In the North Atlantic, predatory fishes have declined by two-thirds over the twentieth century (Christensen *et al.* 2003). A compilation of research survey data suggests that severe reductions in populations of the largest fishes span all oceans. More than 70–90% of the biomass of predatory fishes has been removed in the first 15 years after surveys began (Myers and Worm 2003); however, scientific surveys typically begin long after the onset of fishing, and the true extent of decline may again have been underestimated (Pinnegar and Engelhard 2008). It is incredibly difficult to go back much further in time to estimate the true extent of the decline in predatory fishes. Some insight of the overall impact of fishing compared to the ecological baseline comes from a macroecological energetic analysis that does not suffer from a limited time horizon of data availability. Such analysis suggests that fishing has resulted in a 99.9% decline in North Sea fish ranging in size from 16 to 66 kg (Jennings and Blanchard 2004).

6.6.6 Sharks, rays, and chimaeras

Many sharks and rays (elasmobranchs) are large and feed at or near the top of food webs (Cortés 1999; Stevens *et al.* 2000). Many elasmobranchs are taken as incidental by-catch of the high-seas fisheries for tuna and billfishes, and the great mechanized fisheries targeting cod and other bottom-dwelling fishes.

Some of the great pelagic sharks in the north-west Atlantic, such as the great white shark *Carcharodon carcharias*, scalloped hammerhead *Sphyrna* spp., and thresher shark *Alopias* spp., have declined by approximately 75% in 15 years, which is less than the typical pelagic shark generation time (Baum *et al.* 2003; Myers *et al.* 2007). The oceanic white-tip *Carcharhinus longimanus* and silky shark *Carcharhinus falciformis* have declined by 99% and 90% respectively in the Gulf of Mexico (Baum and Myers 2004). Overall, three-quarters (16 of 21) of the species of oceanic pelagic sharks and rays face an elevated risk of extinction. Many of these species are caught regularly as incidental by-catch in widespread long-line, purse seine, and gill-net fisheries targeting more productive tunas, swordfishes, and other billfishes, as well as in midwater trawl fisheries for small pelagic fishes in boundary current systems, and swordfish fisheries on the high seas (Dulvy *et al.* 2008). Some elasmobranch species are also increasingly targeted for their meat, such as the shortfin mako *Isurus oxyrhinchus*, porbeagle *Lamna nasus*, and blue shark *Prionace glauca*. However, shark fins are often worth more than the meat and these are removed (and the body is discarded); the fins are then dried and sold on in the lucrative Asian shark-fin soup trade (Clarke *et al.* 2006b). There is strong concern about directed fishing to support the demand of the shark-fin soup trade in China and Hong Kong (Clarke *et al.* 2006a, 2006b). The weight of fins imported to Hong Kong each year amounts to approximately 5930 tonnes and the amount traded has been growing by approximately 6% per year (1991–2000) (Clarke 2004). It is estimated that 38 million individuals weighing a total of 1.7 million tonnes are killed each year and pass through the Hong Kong shark-fin market (Clarke *et al.* 2006b).

Smaller bottom-dwelling sharks, skates, and rays have declined severely as a result of incidental capture in bottom-trawl fisheries. The 2 m-long barndoor skate *Dipturus laevis*, formerly widespread along the north-west Atlantic coast of the USA and Canada, has largely been eliminated from Canadian shelf seas, but still persists on deeper slopes 1200 m deep beyond the reach of most fisheries and around protected areas on Georges Banks (Simon *et al.* 2002). On the other side of the Atlantic the largest skate in the world, the unfortunately named common skate *Dipturus batis*, has been eliminated from much of its range (Brander 1981; Dulvy *et al.* 2000). The angel shark *Squatina squatina* was the original monkfish, but as their catches declined they were substituted by anglerfishes *Lophius* spp. This shark species disappeared virtually unnoticed from around the north-west Atlantic and the Mediterranean Sea. Research surveys throughout the west, north, and east Mediterranean Sea suggest that fewer than a couple of hundred adults remain (Baino *et al.* 2001). More recent surveys did not find any around the Balearic Islands, where the last known catches were taken (Massuti and Moranta 2003).

Sharks, rays, and chimaeras are one of the first marine groups subject to comprehensive assessment of threat status. The World Conservation Union (IUCN) Global Shark Assessment has documented a large number of local regional and global declines and near extinctions of oceanic and coastal sharks and rays (Cavanagh and Dulvy 2004). To date the IUCN/SSC Shark Specialist Group has assessed almost half (591 species) of the world's 1100 species of sharks, rays, and chimaeras and found that 21% are threatened (Dulvy *et al.* 2008).

6.7 Our emerging understanding of marine habitat loss

Habitat loss is currently the major driver of terrestrial extinctions, and is an increasingly important cause of threat and extinctions in the sea (Fig. 6.2). Our understanding of the scale of marine habitat loss and degradation is hampered by the difficulty of measuring and monitoring marine habitats, especially those beyond the view of satellite and airborne remote-sensing cameras (Green *et al.* 1996). Coral reefs, mangroves, and temperate estuaries provide the best understood examples of marine habitat loss and the consequences for marine biodiversity, and here we highlight some case studies.

6.7.1 Coral reefs

Coral reefs are among the most diverse ecosystems and also one of the most threatened, suffering multiple human impacts that occur at a range of spatial and temporal scales (Côté and Reynolds 2006). The scale of human impact on coral reefs over the last century is unprecedented in recent geological history. Recent human impacts have changed modern coral reef structure in a manner not previously observed in a 220000 year sequence of fossil reefs in the Bahamas (Pandolfi and Jackson 2006). Unprecedented rates of coral reef loss have resulted from climate change-induced coral bleaching, the cascading effects of overfishing, emerging coral diseases, pollution, and hurricane disturbance (Côté and Reynolds 2006). In addition to these more immediate threats, the increase in anthropogenic CO_2 will largely be absorbed by the oceans, resulting in the acidification of surface waters and a reduction in the saturation state of the carbonate mineral aragonite by 30% by 2100 (Kleypas *et al.* 1999). The calcification of coral reefs is highly correlated with the aragonite saturation state, and the predicted 10–20% deficit in calcium carbonate accretion will render reef-building corals increasingly rare (Kleypas *et al.* 1999; Hoegh-Guldberg *et al.* 2007). Fish catches from island coral reefs are currently 64% higher than can be sustained (Newton *et al.* 2007). Overfishing may result in habitat degradation through herbivore removal, which reduces grazing pressure on algae that compete with and overgrow hard corals, and the release of coral predators resulting in coral mortality (McClanahan 1995; McCook *et al.* 2001; Dulvy *et al.* 2004). The rate of coral reef degradation has increased with predictions that up to 60% of reefs may be lost by 2030, not least due to increasing frequency of coral bleaching events (Hoegh-Guldburg 1999; Wilkinson 2000; Pandolfi *et al.* 2003; Sheppard 2003). The loss of hard coral cover, particularly due to global coral bleaching events associated with the 1998–2000 El Niño/La Niña Southern Oscillation, has led to the local extinction of fishes that specialize in feeding on corals or dwell within corals. The harlequin leatherjacket *Oxymonacanthus longirostris* is an obligate corallivore that disappeared from small study sites in southern Japan and elsewhere soon after a coral bleaching event (Kokita and Nakazono 2001; Dulvy *et al.* 2003). Coral-dwelling gobies (Gobiidae) and hawkfishes (Cirrhitidae) dwelling among branching *Acropora* corals declined by 59% between 1996 and 1997

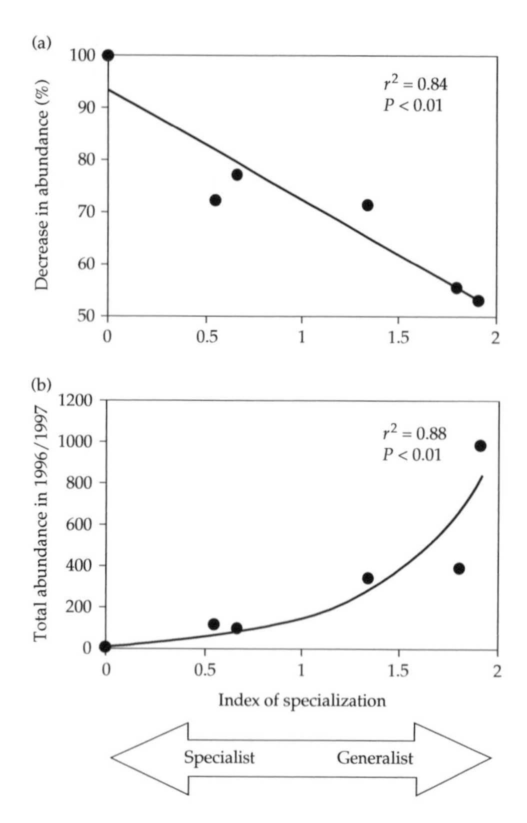

Figure 6.7 Decline (%) in abundance of coral-dwelling fishes compared to their habitat specialization (Shannon index of the diversity of coral species inhabited). (b) The relationship between the habitat specialization index and the total number of gobies of each species recorded in Kimbe Bay in 1996–1997. Redrawn from Munday (2004).

and 2003 in Kimbe Bay, Papua New Guinea due to coral bleaching and sedimentation (Munday 2004). This resulted in a 59% decline in the abundance of coral-dwelling gobies. The most specialist species typically had the smallest initial population sizes and suffered the most from the coral mortality. One undescribed goby species endemic to Kimbe Bay (*Gobiodon* sp. C.) may well have become globally extinct and another more widespread species (*Gobiodon* sp. A.) suffered local extinction (Munday 2004) (Fig. 6.7).

Reef-building corals also exist as patch reefs and mounds in deep coldwater habitats, such as continental shelf edges and seamounts (Roberts *et al.* 2006). There are more than 800 species of shallow-water reef-building corals; in the deep sea there are eight described reef-building species as well as a wide array of soft corals, sea fans and bamboo corals (Octocorallia), black corals (Antipatharia), and hydrocorals (Stylasteridae) that can form or contribute to large structures. Deep-water coral reefs harbour considerable diversity: for example, over 1300 species have been found on *Lophelia pertusa* reefs in the north-east Atlantic (Roberts *et al.* 2006). Deep-water coral reefs are threatened by bottom trawling, oil exploration, and a shallowing aragonite saturation horizon due to ocean acidification (Roberts *et al.* 2006; Turley *et al.* 2007). The corals comprising or associated with deep-water reefs are very long lived and therefore unlikely to recover from impacts within a human time frame (Roberts *et al.* 2006). Hawai'ian coral-like species were radiocarbon-aged and found to be 450–2742 years old (Roark *et al.* 2006), New Zealand deep-sea corals were aged between 45 and 1200 years (Sikes *et al.* 2008), and dead portions of reef matrix taken in by-catch from commercial trawls targeting deep-water fishes off western Ireland were found to be at least 4500 years old (Hall-Spencer *et al.* 2002). However, a large Norwegian deep-water reef on the Sula ridge comprised of *Lophelia pertusa*, 13 km long and 10–35 m in height, was found to have reef-accumulation rates comparable to shallow tropical coral reefs (Freiwald *et al.* 1999). The most widespread and pressing threat is bottom trawling; in trawled areas reef-building corals are broken and dislodged by the heavy otter boards of the trawl gear. Several nations have acted swiftly to close newly discovered reefs to fishing activity. However, much habitat exists beyond the 200 mile limit of Exclusive Economic Zones and outside national jurisdictions (Roberts *et al.* 2006).

6.7.2 Estuaries and coastal seas

Estuaries have suffered multiple impacts over the course of human history. They provide focal points for human activity, providing river access to the interior of countries and sheltered harbours and ports allowing the development of sea-going trade. Consequently, major settlements and ports have expanded rapidly along estuarine coastlines and into salt marshes. A well-documented case study is the Wadden Sea in The Netherlands, Germany, and

Denmark. Large-scale embankments and drainage of coastal marshes began approximately 1000 years ago (Wolff 2000a; Lotze 2005; Reise 2005). We will never know precisely how many species became extinct over any region across such a large time scale, but there is good evidence for extinction or severe depletion of 144 species in the twentieth century, with at least nine species having been lost in earlier times (Wolff 2000b; Lotze 2005). As is typical for most extinctions around the world (but see below), habitat loss has been the most important extinction driver, especially for invertebrates, seaweeds, and birds, followed by exploitation (invertebrates, bird, fishes, and marine mammals). So far, invasive species have not yet been blamed for the loss of any native species. However, the rate of invasion is increasing rapidly to the extent that, in some European port and harbour areas, non-native species may represent as much as 60% of the biomass of all species present (e.g. Eno *et al.* 1997).

An analysis of depletions and extinctions in 12 major coastal seas and estuaries around the world suggests that the Wadden Sea may be typical of heavily industrialized regions (Lotze *et al.* 2006). Key species groups from all regions showed similar patterns of gradual decline until the last 150–300 years, followed by recent rapid depletion of over 90% of formerly important species. For the species in this study, exploitation was responsible for approximately 95% of depletions and extinctions, followed by habitat loss.

6.7.3 Mangroves

Mangroves trap sediment along tropical coastlines, creating natural barriers to sea-level rise and storm surges and saltwater intrusion into coastal soils and estuaries (Spalding *et al.* 1997; Danielson *et al.* 2005). They also function as key nursery habitats for fishes and invertebrates and are likely to contribute to ecosystem resilience and fisheries productivity (Mumby *et al.* 2004). Mangroves are threatened by deforestation for firewood, coastal development, the expansion of shrimp aquaculture, and rising sea levels. The global extent of mangrove forest has declined by a third over the last 50 years (Alongi 2002). The greatest cause of deforestation is shrimp aquaculture: the rich mangrove soils support highly productive growing conditions for shrimp, but productivity declines within a few years and the farming operation moves on, clearing more mangrove forest (EJF 2004). Shrimp farming has caused the loss of 20–50% of mangroves worldwide, particularly in developing countries where mangroves are predicted to decline by another 25% by 2025. The loss of mangroves has affected local populations of plants and animals, but it is not yet known whether deforestation has led to extinctions (Dulvy *et al.* 2003). Mangrove loss is expected to increase due to anthropogenically induced sea-level rise; mangrove habitat will be trapped between rising sea levels and coastal development. In the Pacific Ocean sea level is (conservatively) predicted to rise by 0.5–0.8 m by 2100 (Church *et al.* 2001) and is predicted to reduce mangrove area by 12% by 2100 (Gilman *et al.* 2007). However, there is a possibility that sea-level rise may be an order of magnitude greater, as these estimates do not incorporate emerging evidence of rapid dynamic melting of west Antarctic and Greenland ice sheets (Hansen 2007).

6.7.4 Overfishing-induced habitat transformation

The effects of overfishing and habitat loss may be more difficult to disentangle, as the depletion of predatory fishes has led to habitat degradation and transformed the production base of some marine ecosystems through trophic cascades (Pace *et al.* 1999; Pinnegar *et al.* 2000; Tittensor *et al.* 2008). A top-down view of many ecosystems is emerging whereby top predators control herbivore abundance, biomass, and behaviour with cascading effects on the structure and dynamics of the resource base (Micheli 1999; Shurin and Seabloom 2005; Heithaus *et al.* 2008). Overfishing-induced proliferations of urchins and starfish have transformed coral communities into algal-dominated states (Carreiro-Silva and McClanahan 2001; Dulvy *et al.* 2004). For example, hunting and elevated predation on Pacific sea otters has led to urchin proliferation and shifts from kelp forests to coralline algal barrens (Estes 1998; Steneck *et al.* 2003). While it is difficult to demonstrate causality, compelling evidence suggests the massive-scale population collapses of the great whales by post-World War II industrial whaling caused killer whales to begin feeding more intensively on smaller marine mammals such as sea otters (Springer *et al.* 2003). The collapse of Atlantic cod is associated with increases in northern snow crab *Chionoecetes opilio*, northern shrimp *Pandalus borealis*, urchins, and small pelagic fishes (Worm and Myers 2003; Frank *et al.* 2005). The increase in urchins has denuded coastal kelp forests in the Gulf of Maine (Jackson *et al.* 2001) and increase in small pelagic fishes was associated with lower abundance of large zooplankton and elevated phytoplankton abundance on the eastern Scotian shelf, off Nova Scotia, Canada (Frank *et al.* 2005). Such trophic cascades are relatively commonplace and have been reported for marine ecosystems all around the world (for recent reviews see Pinnegar *et al.* 2000; Lees *et al.* 2006).

6.8 A brief overview of known marine extinctions

There is unequivocal evidence for at least 20 global marine extinctions during the historical era (Table 6.1), an increase on the last estimate of 12 reported in 1999 (Norse 1993; Vermeij 1993; Carlton *et al.* 1999). As far as we are aware there have not been any global marine extinctions in the past two decades. The most recent extinction, that of the Galapagos damselfish *Azurina eupalama*, occurred at some point after 1983. The cause of the increase in the number of documented global marine extinctions is instead due to our more detailed understanding of the taxonomy of extinct species and the discovery of previously unknown historical extinctions. For example, the Japanese sea lion *Zalophus japonicus* was previously thought to be a subspecies of the California sea lion *Zalophus californianus*, but is now recognized as a separate species (Wilson and Reeder 2005). Among the newly discovered extinctions is Bennett's seaweed *Vanvoortsia bennettiana*, which was last recorded in Sydney harbour in 1916. This extinction was only uncovered by the diligent efforts of a taxonomist compiling a regional species list. Not included in this list of marine extinctions are the species on 'death row'; these include European sturgeon and white abalone *Haliotis sorenseni* (Dulvy *et al.* 2003), neither of which has bred successfully in the past two decades. There may be hope for the white abalone, as captive breeding appears increasingly feasible. However, the prognosis for the European sturgeon is poor; it is now only found in one river system in Europe, the Gironde system in France, and is also threatened by the accidental escape of Siberian sturgeon *Acipenser baerii* into this river.

More biodiversity has been permanently lost during the Holocene than might be inferred from this relatively low number of known historical-era marine species extinctions (Knowlton 1993; Reaka-Kudla 1997). The 20 known marine species extinctions documented here include only two procellariiform species: the St. Helena Bulwer's petrel *Bulweria bifax*, and the large St. Helena petrel *Pterodroma rupinarum* (Table 6.1). However, a further 11 species and 79 populations of procellariiforms may have gone extinct in the prehistoric Holocene (see Chapter 4 in this volume). The recent survey of Dulvy *et al.* (2003) has highlighted that numerous population extinctions have occurred at the local and regional level, and there may be more impending global-scale extinctions that have yet to be discovered. This survey focused on local and regional population-scale extinctions for four reasons. First, populations are often morphologically and genetically distinct (Carlton *et al.* 1999; Ruzzante *et al.* 2000; McIntyre and Hutchings 2004). Second, source populations may also rescue other sink populations contributing to the resilience of the species as a whole (Smedbol *et al.* 2002). Third, population extinctions usually precede global extinction (King 1987; Pitcher 1998). Finally, impacts and management typically occur at the population scale. This survey uncovered evidence for 133 local, regional, and global extinctions. Local- and regional-scale extinctions cover the scale of small semi-enclosed seas such as the Irish Sea up to the Mediterranean Sea and ocean quadrants. There was evidence for at least seven new possible global extinctions of fishes, corals, and algae. Four of these are now recognized are global marine extinctions (Table 6.1), leaving three whose status has yet to be confirmed (two eastern Pacific corals, *Millepora boschmai* and *Siderastrea glynni*, and Turkish towel algae *Gigartina australis*) (Dulvy *et al.* 2003).

A key assumption of this analysis is that the populations have truly become extinct. A proposed

Table 6.1 Twenty historical-era global marine extinctions of mammals (4), birds (8), fishes (3), molluscs (4), and algae (1).

Common name (order, family: species name)	Historical range	Last known date of occurrence	Cause of extinction	Source
Mammals				
Steller's sea cow (Sirenia, Dugongidae: *Hydrodamalis gigas*)	Commander Islands (Bering Sea, north-west Pacific Ocean)	1768	Overexploitation	Anderson (1995); Carlton *et al.* (1999); Turvey and Risley (2006)
Sea mink (Carnivora, Mustelidae: *Neovison macrodon*)	Canadian (New Brunswick) and USA (Maine) coasts	1860	Overexploitation	Campbell (1988); Youngman (1989); Carlton *et al.* (1999); IUCN (2006); Sealfon (2007)
Japanese sea lion (Carnivora, Otariidae: *Zalophus japonicus*)	Japan (Sea of Japan), Russia (Kamchatka)	No credible sightings since late 1950s		Rice (1998); Carlton *et al.* (1999); Wilson and Reeder (2005); IUCN (2006)
Caribbean monk seal (Carnivora, Phocidae: *Monachus tropicalis*)	Coastal Caribbean Sea and Yucatan, including Mexico, Bahamas, Guadeloupe, Jamaica, Puerto Rico, USA (Florida)	1952	Overexploitation	Carlton *et al.* (1999); Wilson and Reeder (2005); IUCN (2006); McClenachan and Cooper (2008)
Birds				
Pallas's cormorant (Pelecaniformes, Phalacrocoracidae: *Phalacrocorax perspicillatus*)	North-west Pacific	*c.*1850	Overexploitation	Greenway (1967); Carlton *et al.* (1999)
Tasman booby (Pelecaniformes, Sulidae: *Sula tasmani*)	Lord Howe and Norfolk Islands (Australia)	Nineteenth century?; last seen in 1788	Overexploitation and introduced species	BirdLife International (2004); IUCN (2006)
St. Helena Bulwer's petrel (Procellariiformes, Procellariidae: *Bulweria bifax*)	St. Helena, central Atlantic	Sixteenth century	Overexploitation	BirdLife International (2004); IUCN (2006)
Large St. Helena petrel (Procellariiformes, Procellariidae: *Pterodroma rupinarum*)	St. Helena, central Atlantic	Sixteenth century	Overexploitation and introduced species	BirdLife International (2004); IUCN (2006)
Auckland Island merganser (Anseriformes, Anatidae: *Mergus australis*)	South-west Pacific	1902	Overexploitation	Carlton *et al.* (1999)
Labrador duck (Anseriformes, Anatidae: *Camptorhynchus labradorius*)	Breeding habitat in Gulf of St. Lawrence and coastal Canada, north-west Atlantic	1875	Overexploitation of adults and eggs	Carlton *et al.* (1999); BirdLife International (2004); IUCN (2006)
Great auk (Charadriiformes, Charadriidae: *Pinguinus impennis*)	North Atlantic	1844	Overexploitation	Carlton *et al.* (1999)
Canary Islands oystercatcher (Charadriiformes, Charadriidae: *Haematopus meadewaldoi*)	North-east Atlantic	1913	Invasive species	Carlton *et al.* (1999)
Fishes				
Galapagos damselfish (Perciformes, Pomacentridae: *Azurina eupalama*)	Galapagos Islands	1982	Habitat loss, climate change	Jennings *et al.* (1994); Roberts and Hawkins (1999); G.J. Edgar *et al.* (unpublished work)

Table 6.1 *Continued*

Common name (order, family: species name)	Historical range	Last known date of occurrence	Cause of extinction	Source
Mauritius green wrasse (Perciformes, Labridae: *Anampses viridis*)	Mauritius	1839	Unknown	Hawkins *et al.* (2000)
New Zealand grayling (Salmoniformes, Retropinnidae: *Prototroctes oxyrhinchus*)	New Zealand	1923	Exploitation and invasive species	Balouet and Alibert (1990); McDowell (1996)
Invertebrates				
Atlantic eelgrass limpet (Archaeoastropoda, Lottidae: *Lottia alveus*)	North-west Atlantic	1929	Habitat loss	Carlton *et al.* (1991, 1999); Carlton (1993)
Rocky shore limpet (Archaeoastropoda, Nacellidae: *Collisella edmitchelli*)	North Pacific	1861	Habitat loss	Carlton (1993); Carlton *et al.* (1999)
Horn snail (Gastropoda, Cerithideidae: *Cerithidea fuscata*)	North-east Pacific	1935	Overexploitation	Carlton (1993); Carlton *et al.* (1999)
Periwinkle (Mesogastropoda, Littorinidae: *Littoraria flammea*)	China	1840	Habitat loss	Carlton (1993); Carlton *et al.* (1999)
Algae				
Bennett's seaweed (Ceraniales, Delesseriaceae: *Vanvoortsia bennettiana*)	Sydney Harbour, eastern Australia	1916	Habitat loss	Millar (2001); IUCN (2006)

alternative is that they represent shifts in dynamic geography, whereby reduced abundance is associated with reduced spatial occupancy, and these disappearances merely represent temporary extinction and recolonization events which may be particularly likely at the edge of a species' geographic range (MacCall 1990; Hanski 1998; del Monte-Luna *et al.* 2007; Webb *et al.* 2007). However, permanent range contractions at local scales are the stepping stones toward species extinction (King 1987). The population extinctions reported by Dulvy *et al.* (2003) are unlikely to be temporary patch extinctions. These extinctions are long-standing; the local extinctions have persisted on average for 64 years, and global extinctions for 77 years on average (Fig. 6.8a). Comparatively less time (33 years) has elapsed for regional extinctions, largely due to the inclusion of numerous recent fish population extinctions (Fig. 6.8b).

Since the review by Dulvy *et al.* (2003) was published there have been several recolonizations in the Dutch Wadden Sea, including grey seal *Halichoerus grypus*, eider duck *Somateria mollissima*, common gull *Larus canus*, and lesser black-backed gull *L. fuscus*. All of these cases may be attributed to strongly improved protection (W.J. Wolff, personal communication). While these recolonizations are a important sign of changing management focus and efficacy, they are unlikely to mitigate against the likely loss of genetic, morphological, and behavioural diversity associated with the original population extinctions. Population extinctions are being recognized with greater frequency, as predicted in the original study. The authors originally stressed that this data set was far from definitive, because of the problems of recognizing and defining extinctions; however, they provided the first systematic review of the evidence

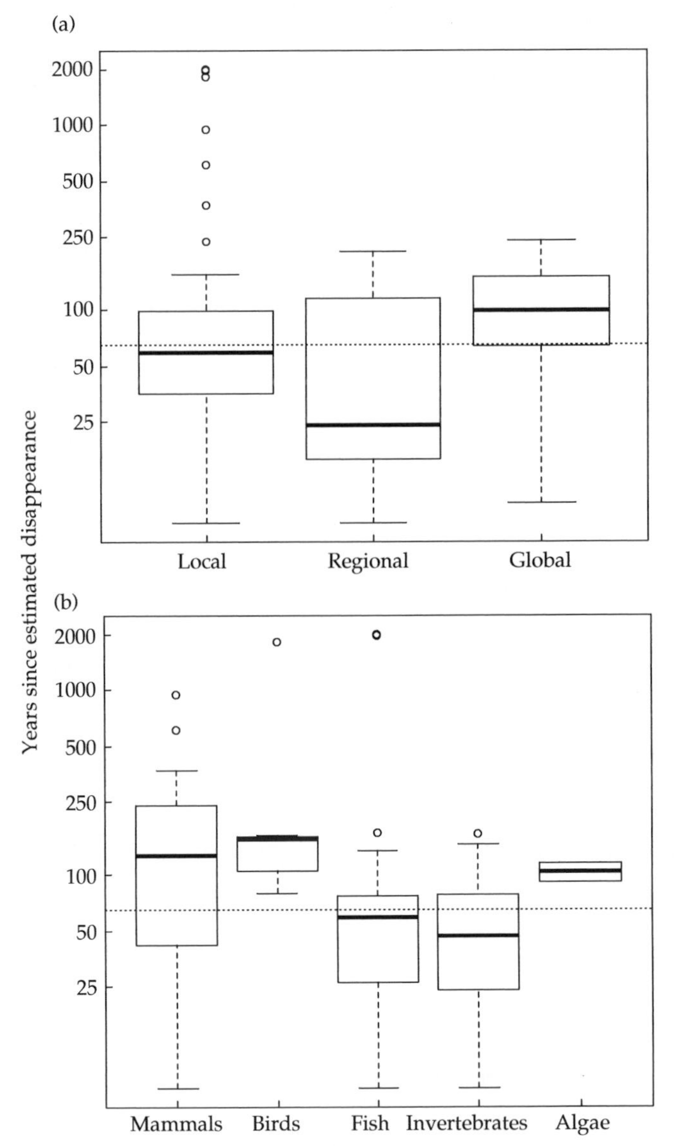

Figure 6.8 The persistence of marine extinctions. Time elapsed since the estimated extinction date, split by (a) geographical scale and (a) taxon. The average time elapsed since the estimated extinction date is 65 years, and is represented by the dotted line. There is no significant difference in the time elapsed since estimated extinction among the differing spatial scales of extinctions ($F_{2,97}=0.7$, $P=0.48$). Bird and mammal disappearances were detected earlier and/or happened longer ago compared with the estimations of fishes and invertebrates ($F_{4,95}=5.3$, $P<0.001$), but only pairwise comparisons between birds versus either fish or invertebrates are significant (Tukey's HSD, $P<0.05$). This test is robust to the exclusion of algae. Estimated extinction dates were available for 13 of the 14 mammals, 11 of the 12 birds, 57 of the 65 fishes, 17 of the 31 invertebrates, and 2 of the 12 algae. Data from table 1 in Dulvy *et al.* (2003).

and concluded that marine extinctions were being overlooked (Dulvy *et al.* 2003). The detection and reporting of marine extinctions lags behind the date of actual extinction by about 50 years. However, this detection lag is becoming shorter over time, suggesting that scientific capacity to detect marine extinctions is steadily improving (Dulvy *et al.* 2003). Indeed, since the original review was published, additional population-level extinctions have been discovered, particularly on coral reefs. These include: the disappearance of the rainbow parrotfish *Scarus guacamaia* from the coastline of Brazil (Ferreira *et al.* 2005); population extinctions in the world's largest parrotfish, the giant bump-head parrotfish *Bolbometopon muricatum*, from Guam and the Marshall Islands (Bellwood *et al.* 2003; Hamilton 2003; Donaldson and Dulvy 2004; Dulvy and Polunin 2004); the local and near-global extinction of two coral-dwelling gobies (Gobiidae) (Munday 2004); and the probable global extinction of the endemic Galapagos stringweed *Bifurcaria galapagensis*, and possible local extinction of another seven species (G.J. Edgar *et al.*, unpublished work). The recent increase in the number of documented global extinctions and these newly discovered extinctions may reflect an elevated marine extinction rate, and they refute recent assertions by del Monte-Luna *et al.* (2007) that the marine extinction rate is overestimated and overstated. The increase in extinctions is a more likely hypothesis for two reasons: evidence for accelerating threats such as exploitation, climate change, and habitat loss (Edgar *et al.* 2005), and increased likelihood of extinction detection due to greater awareness of the potential for marine extinctions (Dulvy *et al.* 2003).

A major barrier to raising awareness of the likelihood that marine populations and species have gone extinct is a highly risk-averse attitude towards evaluating evidence for marine extinctions. Some scientists believe that species extinctions should not be highlighted until sufficient evidence has been accumulated (del Monte-Luna *et al.* 2007). Raising false alarms—incorrectly flagging the extinction of an extant marine population or species—should be avoided at all costs, because a high rate of false alarms would devalue the credibility of threat assessments and be used to question the integrity of conservation and management policies (del Monte-Luna *et al.* 2007). However, while a more stringent approach to defining and documenting marine extinctions appears highly risk-adverse, this strategy runs the risk of allowing extinctions to go unnoticed and undocumented (Peterman and M'Gonigle 1992). Given (1) the rise in the scale of human activity in the seas, (2) the clear link between human-induced habitat transformation and terrestrial extinctions over the last 10 000 years, and (3) the documented evidence that marine extinctions are underestimated, it seems less precautionary to wait until sufficient evidence is available to ensure the accurate documentation of a species extinction. Instead it may be more appropriate to focus on identifying local marine extinction and biodiversity loss with a view to conserving and managing remaining populations (Peterman and M'Gonigle 1992). The choice lies somewhere between providing sound defensible assessments based on the currently available evidence that can be used to prioritize conservation action, or waiting until *all* evidence is available and taking the risk of presiding over a post-mortem of marine biodiversity loss.

While scientists can provide and describe the consequences of these options, to a large degree the choice lies with society rather than scientists. This choice depends on the degree to which human societies are able to move towards bearing the social and economic cost of the lost opportunity to exploit provisioning ecological services, such as fisheries, to ensure that all biodiversity is preserved (Jennings 2007). Historically, (European) society has been more concerned with maintaining food supply and minimizing conflict (e.g. Icelandic cod wars) in fisheries (Holden 1992), although now there is an increasing and concerted effort to ensure biodiversity protection and maintenance of ecosystem function and processes through the Ecosystem Approach to Fisheries (Sainsbury and Sumaila 2003; Pikitch *et al.* 2004; ICES 2005). However, we recognize that other less developed nations, which are yet to experience the human demographic transition and reduced population growth rates, may face a more difficult or delayed transition towards sustainability.

6.9 Can we avert a double extinction wave in the sea?

The evidence for loss of marine megafaunal biomass and alterations to marine ecosystems over the last thousand years is incontrovertible. So far most marine extinctions have been of local and regional populations: relatively few species extinctions have been documented. The depletion of marine biomass is a price paid as a consequence of supplying considerable provisioning ecosystem services for the development and benefit of human societies (Pauly and Alder 2006). There is increasing recognition that even greater benefits can be derived while sustaining rather than depleting biodiversity (Worm *et al.* 2006). For example, a large stock size and a multistock portfolio policy allows sustainable yields to be taken in perpetuity while minimizing the risks and insecurity of environmental variability and long-term change (Hilborn *et al.* 2003; DEFRA 2004). Not only will reduction of fishing capacity and effort on target stocks help build resilience; the integration of fisheries considerations within a wider management framework is well advanced, for example through development of the ecosystem approach to fisheries management, thereby addressing the World Summit on Sustainable Development 2010 target to halt the loss of global biodiversity by 2010 (Sainsbury and Sumaila 2003; Pikitch *et al.* 2004; ICES 2005). The scientific knowledge, institutional capacity, and political commitment available to avert a marine megafaunal extinction wave is greater now than in the past, and our capacity to recognize and avert serious environmental damage has never been greater in the Holocene. The first fishes have been listed on international conventions typically used to restrict the trade in endangered plants, mammals, and birds (including sea horses, humphead wrasse *Cheilinus undulatus*, basking shark *Cetorhinus maximus*, whale shark *Rhincodon typus*, and great white shark). Regionally and nationally there is increasing use of wildlife conservation legislation to assess and protect exploited fishes, such as the Endangered Species Act (USA), the Species at Risk Act (Canada), the Wildlife and Countryside Act (UK), and the Bern Convention (Europe). This is clearly an optimistic view and one more typical of European and North American fisheries science; research capacity and the strength of institutions may not be as favourable elsewhere in the world, but there is still scope for the development of novel fisheries diagnostics and management (Andrew *et al.* 2007).

Some populations of marine species have been brought 'back from the brink', notably sea otters in the western Pacific and some whale species. This has required draconian measures, including a complete moratorium on hunting or whaling. Several species of marine mammals and birds in the countries around the North Sea have recovered spectacularly thanks to a range of concerted conservation measures in these countries (W.J. Wolff, personal communication). Similarly, a few fish and invertebrate stocks have been allowed to recover from overexploitation (Caddy and Agnew 2004; Mace 2004). More successes have been documented for invertebrate pelagic fisheries than demersal species and depletions aggravated by multispecies fisheries, and unfavourable climatic regimes will always be difficult to reverse (Piet and Rice 2004; Brander 2007).

We are now catching up with the terrestrial preoccupation with captive rearing, as numerous forms of aquaculture are growing rapidly while global yields from wild-capture fisheries have been stalled for the past 15 years (FAO 2007). The current transition to aquaculture co-occurs with a period of uncertainty for the future of marine biodiversity. Whereas the transition to farming on land eventually led to the elimination of significant hunting of wild animals for food in most developed countries, there is little sign yet that the growth of aquaculture is relieving fishing pressure on wild stocks. Furthermore, fish farming often involves high-trophic-level carnivores, such as salmon, sea bass, and tuna. These farming operations create their own markets for wild-caught fishes to be converted to fish meal (Naylor *et al.* 2000). Thus, the challenges facing biodiversity in the sea show no signs of abating in the near future, and as with all environmental problems, our best hopes lie with a concerted global focus on the core underlying drivers of change: human population growth and increasing per-capita demands for resources.

While a major extinction wave driven by exploitation and habitat loss occurred on land during the Late Pleistocene and prehistoric Holocene, a similar process may now be unfolding in the oceans. The challenge will be to limit the scale of any impending marine extinction wave. The opportunity to forestall a major loss of ocean biodiversity has never been greater. Society, scientists, and managers are acutely aware of these problems and legislation and institutions are strengthening in response. However, several 'ratchet-like' processes, including a growing global population and international markets for marine products, make it very difficult to return to a more 'natural' state (see discussion in Pinnegar and Engelhard 2008).

Acknowledgements

The authors are grateful to Andy Cooper, Graham Edgar, Rodrigo Bustamante, Heike Lotze, Loren McClenachan, Telmo Morato, Heike Lotze, and Phil Munday for providing access to documents and figures. We are grateful to Wim Wolff for providing insightful and detailed comments which considerably improved the manuscript. The authors are grateful to the Department for Environment, Food and Rural Affairs and the Natural Environment Research Council UK and the European Union (MarBEF), work-package 2 (shifting baselines) of the Framework VI project INCOFISH (INCO project number 003739), and the Natural Sciences and Engineering Research Council of Canada.

CHAPTER 7

Procellariiform extinctions in the Holocene: threat processes and wider ecosystem-scale implications

R. Paul Scofield

7.1 Introduction

Procellariiformes are the order of tube-nosed seabirds that includes the albatrosses, petrels, and shearwaters. Members of the order breed mainly on islands; individual species are often extremely widely distributed, with populations on many islands within many island groups, and frequently in different oceans. The order has a global distribution, and contains four extant families: Pelecanoididae (diving-petrels), Procellariidae (shearwaters, petrels, and fulmars), Diomedeidae (albatrosses), and Hydrobatidae (storm-petrels), together with either one or two extinct families. Species are generally either pelagic scavengers primarily taking squid and fish, or are planktivorous. As basal members of the Neoaves, the group has a physiologically restrained breeding system in which birds only produce a single egg, usually annually, with no relaying, do not have complex nest structures, breed either at or below ground level, and generally have a tightly constrained breeding season. These traits leave them open to predation by mammalian predators, and it is hypothesized that for this reason their breeding biology is characterized by a number of behaviours that have evolved to reduce the effect of such predation. On islands where petrels are without predators they occur in huge numbers; indeed, one of the largest concentrations of breeding animals in the world is a petrel colony (Reyes-Arriagada *et al.* 2007). Although these vast colonies are susceptible to the invasion of exotic mammals, and can disappear rapidly (see Atkinson 1985 for examples), procellariiform populations also typically spawn offshoots of a few hundred individuals on tiny offshore islands or stacks, and when invasions occur it is rare for populations to disappear entirely. Furthermore, a majority of petrel species arrive nocturnally on their breeding grounds, and burrow-nesting and extensive habitat gardening are the norm.

Despite being susceptible to mammalian predation, the procellariiforms are an ancient group that has survived comparatively unchanged since the earliest Cenozoic. Although procellariiform communities are dynamic over time (Warheit 2002), responding to environmental changes, mammalian invasions and the emergence of new islands, the taxa themselves have shown remarkable resilience. For example, the short-tailed albatross *Phoebastria albatrus*, which is rare but still extant today, has managed to survive since at least the Pliocene (Olson and Rasmussen 2001) and move from the Atlantic, where it is now extinct (Olson and Hearty 2003), to the Pacific. Here I will attempt to determine reasons for the group's longevity, and discuss their ecosystem impacts and extinction drivers in the Holocene.

7.2 The earliest fossil record of procellariiforms

7.2.1 The Tertiary record (and earlier)

Fossil evidence suggests that procellariiforms have survived comparatively unchanged since the earliest Cenozoic, and there is limited (and somewhat contentious) evidence of procellariiform-like birds at the end of the Mesozoic. The late Maastrichtian Lance Creek Formation of Wyoming contains two species of the genus *Lonchodytes* described by Brodkorb (1963a). Hope (2002) argued that *Lonchodytes* is likely to be a procellariiform (most closely resembling procellariids) although this has been questioned (Scofield *et al.* 2006). Of a similar age, and equally enigmatic, is a fragmentary clavicle from the Nemegt Basin in Mongolia, which Kurochkin (1995, 2000) assigned to the Diomedeidae. Olson and Parris (1987) described *Tytthostonyx glauconiticus* from the Late Cretaceous or early Paleocene of New Jersey, placing it in its own family (Tytthostonichidae) and suggesting that it appeared to be either a basal procellariiform or close to the origin of the Fregatidae and the Pelecaniformes, the later opinion with which Feduccia (1996) concurred.

Eopuffinus kazachstanensis, a species assigned to the Procellaridae, is known from the Paleocene of Kazakhstan, although it is currently considered of uncertain status. The earliest unequivocal procellariiform appears to come from the Late Eocene of Louisiana. A distal tibiotarsus has been accepted as morphologically close to the extant genus *Pterodroma* in the Procellaridae (Feduccia and McPherson 1993). Procellariiforms and petrel-like taxa are known from Eocene deposits from Uzbekistan, Louisiana, and perhaps the London Clays (but see Mayr *et al.* 2002). *Murunkus subitus*, based on a carpometacarpus from the mid Eocene of Uzbekistan, has been placed in the Diomedeidae (Panteleyev and Nessov 1987), as has *Manu antiquus* from the mid to Late Oligocene of Otago, New Zealand (Marples 1946), although both assignments are tentative. Another extinct procellariiform family, the Diomedeoididae Fischer, 1985, has recently been described from the Oligocene of Germany, and subsequent work by Mayr *et al.* (2002) has shown this group to indeed warrant family status and include a number of species originally described as albatrosses.

Procellariidae are reliably reported in the Oligocene. *'Larus' raemdonckii* from the early Oligocene (Rupelian) of Belgium was placed in the extant genus *Puffinus* (Procellariidae) by Brodkorb (1963b). Three extinct *Diomedea* species (Diomedeidae) are known from the North Pacific Neogene, and several more indeterminate Tertiary records exist for the genus (Chandler 1990). The Upper Miocene of Australia has produced a *Diomedea* species described from an unguis (Wilkinson 1969).

The fossil record of Hydrobatidae and Pelecanoididae is much younger. The first storm-petrel is found in the upper Miocene of California (Olson 1985c), while Scofield *et al.* (2006) and Worthy *et al.* (2007) have recently described diving-petrels differing little from modern species from the Miocene of southern New Zealand. A diving-petrel has also been described from the early Pliocene of South Africa (Olson 1985c).

7.2.2 Pre-human petrel extinctions

In Table 7.1, I summarize a list of published procellariiform taxa that became extinct in the Pliocene and Pleistocene. While undoubtedly incomplete, it is indicative of comparatively low levels of extinction in the group. Our understanding of extirpations of extant species populations is more incomplete for this interval; however, the Pleistocene fossil record for the Palaearctic has been comprehensively surveyed by Tyrberg (1998), and the Quaternary in the Mediterranean Region was examined by Sánchez Marco (2004), and similarly low levels of population extinction have been recorded (Table 7.2). Thus, from what is currently known about the avian fossil record, it would seem that comparatively few species of procellariiforms have become extinct in the last 4 million years, before humans began to impact global ecosystems. However, it is worth noting that, in being primarily pelagic, the bones of procellariiforms will usually only be preserved when birds return to the land to nest. Even then conditions have to be suitable for preservation to occur, as is evident from the fact that the majority of Pleistocene procellariiform fossils are known from limestone or karstic environments. Where these conditions are absent, fossil seabirds are rare. Many modern-day petrel colonies occur on peat soils, which are often extremely acidic because petrels acidify the soil though the nitrification of their guano. Petrel biology therefore typically ensures the destruction of their bones. The outcome of this

Table 7.1 Procellariiform species that became extinct in the Pliocene and Pleistocene.

Species	Locality	Reference
Calonectris krantzi	Lee Creek (USA)	Olson and Rasmussen (2001)
Phoebastria anglica	UK; USA	Olson and Rasmussen (2001)
Phoebastria rexsularum	Lee Creek (USA)	Olson and Rasmussen (2001)
Pterodroma kurodai	Aldabra (Indian Ocean)	Harrison and Walker (1978)
Pterodromoides minoricensis	Menorca; Lee Creek (USA)	Olson and Rasmussen (2001); Seguí *et al.* (2001)
Puffinus nestori	Ibiza (Mediterranean)	Alcover (1989)
Puffinus pacificoides	St. Helena (South Atlantic)	Olson (1975)
Puffinus tedfordi	Western North America	Howard (1971)

Table 7.2 Procellariiform populations that have been documented as becoming extinct in the Pliocene and Pleistocene.

Species	Locality	Reference
Calonectris diomedea diomedea	Bermuda	Olson *et al.* (2005b)
Phoebastria cf. *albatrus*	Bermuda; Lee Creek (USA)	Olson and Rasmussen (2001); Olson and Hearty (2003)
Phoebastria aff. *immutabilis*	Lee Creek (USA)	Olson and Rasmussen (2001)
Phoebastria aff. *nigripes*	Lee Creek (USA)	Olson and Rasmussen (2001)
Puffinus mauretanicus	Pityusic Islands (Mediterranean)	Alcover (1989)

is that our understanding of the extinction of many Neogene procellariiform species (e.g. *Puffinus felthami* and *Puffinus kanakoffi*, described from the Pliocene of California; Chandler 1990) remains incomplete, and these species could well have survived considerably later into the human-impacted Late Pleistocene. For these reasons, it is possible that putative Pleistocene (and even Pliocene) extinctions were in fact human-induced.

7.3 An unparalleled series of extinctions?

The Holocene fossil record of the Procellariiformes is considerably more complete than that of the Pliocene and Pleistocene. However, many seemingly geologically young sites are not reliably dated, and may actually be pre-Holocene in age. Some subfossil dune deposits dated as Holocene may include birds that have been forced ashore by storms or by vagrancy, and so these records may not actually indicate breeding populations. Furthermore, archaeologists frequently interpret the presence of species in middens to indicate nearby breeding sites, but the occurrence of species known to be high-latitude specialists in tropical cultural sites (Steadman 2006b) indicates that that some indigenous populations may have exploited vagrant or migratory populations. Nevertheless, there is strong evidence that species-level extinctions have occurred more often in the Holocene than is documented in the Pliocene and Pleistocene (see Chapter 4 in this volume), and 56% (76 of 136; Onley and Scofield 2007) of Holocene procellariiform species have lost populations (Table 7.3).

'In terms of its rate and geographical extent, its potential for synergistic disruption and the scope of its evolutionary consequences, the current mass invasion event is without precedent and should be regarded as a unique form of global change'

Table 7.3 Procellariiform populations that have become extinct in the Holocene (some of these populations have since reintroduced themselves).

Species	Locality	Likely cause of extinction[1]	Pre- or post-European?	Reference
Diomedea exulans	Macquarie Island	1	Post	de la Mare and Kerry (1994)
Diomedea sanfordi	Pitt Island (Chatham Group)	1	Pre	Millener (1999)
Phoebastria albatrus	Agincourt Island and Pescadore Islands (Taiwan); Izu, Bonin, Daito, Senkaku and western volcanic groups of Japan	1	Post	Hasegawa and DeGange (1982)
Phoebastria immutabilis	Johnston, Marcus, and Wake Islands, Izu Islands	1	Post	Rice and Kenyon (1962)
Phoebastria nigripes	Johnston, Marcus, Volcano, Wake, and Marshall Islands, and Northern Marianas	1	Post	Rice and Kenyon (1962)
Thalassarche bulleri	North Island (New Zealand)	1	Pre	Worthy and Holdaway (2002)
Macronectes halli	Southern New Zealand	1	Post	Worthy and Holdaway (2002)
Halobaena caerulea	Macquarie Island	1	Post	Clarke and Schulz (2005)
Pachyptila turtur	North Island (New Zealand), Amsterdam Island	1	Pre/post	Worthy and Jouventin (1999); Worthy and Holdaway (2002)
Pachyptila vittata	Southern New Zealand, Main Chatham Island	1	Post	Imber (1994); Worthy and Holdaway (2002)
Bulweria bulwerii	Main Hawai'ian Islands, Midway Atoll, and many southeast Pacific Islands; islands off China; Tenerife and islands off Lanzarote (Canary Islands)	1	Pre/post	Megyesi and O'Daniel (1997); Steadman (2006b)
Pseudobulweria aterrima	Amsterdam Island	1	Post	Worthy and Jouventin (1999)
Pseudobulweria becki	?Solomon Islands	1	Post	BirdLife International (2007a)
Pseudobulweria rostrata	Ofu (Samoa), Aitutaki, Huahine	1	Pre	Steadman (2006b)
Pterodroma alba	Huahine; Ua Huka	1	Pre	Steadman (2006b)
Pterodroma arminjoniana	Amsterdam Island	1	Post	Worthy and Jouventin (1999)
Pterodroma axillaris	Pitt Mangere Island (Chatham Islands)	1	Pre	Tennyson and Millener (1994)
Pterodroma baraui	Amsterdam Island	1	Post	Worthy and Jouventin (1999)
Pterodroma brevipes	Many central Pacific Islands	1	Pre/post	BirdLife International (2007b)
Pterodroma cahow	Bermuda	1	Post	Olson *et al.* (2005b)
Pterodroma cervicalis	Raoul Island (Kermadec Islands)	1	Post	Veitch *et al.* (2004)
Pterodroma cookii	North Island (New Zealand)	1	Pre	Worthy and Holdaway (2002)
Pterodroma defilippiana	Isla Robinson Crusoe (Juan Fernández Group)	1	Post	Brooke (1987)
Pterodroma cf. *feae*	Scotland, The Netherlands, Denmark, Sweden	1	Post	Lepiksaar (1958); Ericson and Tyrberg (2004); Leopold (2005); Serjeantson (2005)
Pterodroma gouldi	North Island (New Zealand)	1	Post	Worthy and Jouventin (1999)
Pterodroma hasitata	Guadeloupe, Martinique	1	Post	Bent (1922)
Pterodroma hypoleuca	Main Hawai'ian Island and islands of north-west chain (including Kure)	1	Pre/post	Kepler (1967); Olson and James (1982)

Pterodroma inexpectata	New Zealand mainland	1	Pre	Worthy and Holdaway (2002)
Pterodroma lessonii	Campbell Island	1	Post	Taylor (2000)
Pterodroma macroptera	Amsterdam Island	1	Post	Worthy and Jouventin (1999)
Pterodroma cf. *madeira*	El Hierro (Canary Islands)	1	Post	Rando (2002)
Pterodroma mollis	Amsterdam Island, Macquarie Island	1	Post	Worthy and Jouventin (1999); Clarke and Schulz (2005)
Pterodroma neglecta	Raoul Island (Kermadec Islands); Isla Robinson Crusoe (Juan Fernandez Group)	1	Post	Veitch *et al.* (2004)
Pterodroma nigripennis	Raoul Island (Kermadec Islands), Henderson Island, many central Pacific Islands	1	Post	Wragg and Weisler (1994); Steadman (2006b)
Pterodroma phaeopygia	Easter Island	1	Pre	Steadman (2006b)
Pterodroma ?pycrofti	Lord Howe Island, Norfolk Island	1, 2?	Post	Meredith (1991); Holdaway and Anderson (2001)
Pterodroma sandwichensis	Oahu and other Hawai'ian Islands	1		Olson and James (1982)
Pterodroma solandri	Norfolk Island	1	Pre	Medway (2002)
Pterodroma ultima	Tahuata, Easter Island	1	Pre	Steadman (2006b)
Procellaria cinerea	Amsterdam Island	1	Post	Worthy and Jouventin (1999)
Procellaria parkinsoni	New Zealand mainland	1	Pre	Worthy and Holdaway (2002)
Procellaria westlandica	Northern South Island (New Zealand)	1	Pre	Worthy and Holdaway (2002)
Calonectris diomedea	Giraglia Island (Mediterranean)	2	Post	Thibault and Bretagnolle (1998)
Calonectris leucomelas	Islands off southern Japan and south-east Russia	2	Pre/post	Oka (2004)
Puffinus assimilis	Amsterdam Island	1	Post	Worthy and Jouventin (1999)
Puffinus auricularis	San Benedicto Island (Mexico)	3	Post	Jehl and Parkes (1982)
Puffinus bailloni	Lifuka and Hu'ano (Tonga), Ofu (Samoa), Henderson, and many southeast Pacific Islands	1	Pre	Wragg and Weisler (1994); Steadman (2006b)
Puffinus bannermani	Main islands of Bonin Group	1	Post	Hasegawa (1991)
Puffinus bulleri	North Island and some offshore islands (New Zealand)	1	Pre	Worthy and Holdaway (2002)
Puffinus carneipes	Amsterdam Island	1	Post	Worthy and Jouventin (1999)
Puffinus elegans	Chatham Island	1	Pre	Millener (1999)
Puffinus gavia	North Island and northern South Island (New Zealand)	1	Pre	Worthy and Holdaway (2002)
Puffinus gravis	New Zealand mainland	1	Pre	Worthy and Holdaway (2002)
Puffinus griseus	New Zealand mainland, Main Chatham Island	1	Pre	Imber (1994); Worthy and Holdaway (2002)
Puffinus huttoni	Northern South Island (New Zealand)	1	Pre	Worthy and Holdaway (2002)
Puffinus lherminieri	Bermuda	1	Pre	Olson *et al.* (2005b)
Puffinus mauretanicus	Cabrera, Formentera (Balearics)	1	Post	Oro *et al.* (2004)
Puffinus nativitatis	Ogasawara Island, Marcus Island, Wake Island, Henderson Island, Pitcairn Island, and some Marquesas Islands	1	Pre/post	Wragg and Weisler (1994); Seto (2001); Steadman (2006b)
Puffinus newelli	O'ahu, Maui, and Laana'i (Hawai'i)	1	Post	Ainley *et al.* (1997)
Puffinus opisthomelas	Guadalupe Island (Mexico)	1	Post	Everett and Anderson (1991)

Table 7.3 *Continued*

Species	Locality	Likely cause of extinction[1]	Pre- or post-European?	Reference
Puffinus pacificus	Main Hawai'ian Islands, Tonga and many southeast Pacific Islands	1	Post	Olson and James (1982); Steadman (2006b)
Puffinus puffinus	Bermuda; many European islands	1	Post	Brooke (1990)
Puffinus yelkouan	Mediterranean Islands	1	Post	Martin *et al.* (2000)
Pelecanoides georgicus	Macquarie Island	1	Post	Clarke and Schulz (2005)
Pelecanoides urinatrix	Macquarie Island, Chatham Island, New Zealand mainland, Amsterdam Island	1	Pre/post	Worthy and Jouventin (1999); Worthy and Holdaway (2002); Clarke and Schulz (2005)
Garrodia nereis	Main Chatham Islands	1	Post	Imber (1994)
Pelagodroma marina	Amsterdam Island	1	Post	Worthy and Jouventin (1999)
Fregetta grallaria	Henderson, Ua Huka, Tahuata	1	Pre	Steadman (2006b)
Fregetta tropica	Rapa Island	1	Post	Murphy and Snyder (1952)
Nesofregetta fuliginosa	'Eua (Tonga), Samoa, Mangaia, Tahuata, Easter Island, Henderson Island	1	Pre	Wragg and Weisler (1994); Steadman (2006b)
Hydrobates pelagicus	Many islands and one mainland site around coast of Europe	1	Post	Martin *et al.* (2000)
Oceanodroma castro	Main islands of Azores, Canary Islands and Hawai'i	1	Pre	Olson and James (1982); Rando (2002)
Oceanodroma furcata	Central Kuril Island (Russia); some islands in north-western California	1	Post	Boersma and Silva (2001)
Oceanodroma leucorhoa	Many islands and one mainland site around coast of North America and Ireland	1	Post	Podolsky and Kress (1989); Huntington *et al.* (1996)
Oceanodroma melania	Many Californian Islands	1	Pre and post	Everett and Anderson (1991)
Oceanodroma microsoma	Numerous islands in Gulf of California	1	Post	Donlan *et al.* (2000)
Oceanodroma tristrami	Midway Island, Izu Islands	1	Post	Hasegawa (1984); Baker *et al.* (1997)

[1]Key to likely causes of extinction: 1, introduction of predators; 2, hunting; 3, volcanic eruption.

(Ricciardi 2007). Being burrowing species with low fecundity, procellariiform populations have of necessity evolved in the absence of mammalian predators. The arrival of humans on oceanic islands during the Late Pleistocene and Holocene precipitated a wave of extinctions among birds, especially seabirds, caused largely by the introduction of exotic mammals (see also Chapter 2 in this volume). The magnitude of this extinction event varies markedly between islands, and correlates of recent bird extinctions are becoming increasingly understood (e.g. see Blackburn *et al.* 2004; Duncan and Forsyth 2006; Chapter 12 in this volume). But what are the mechanisms driving procellariiform extinctions?

7.4 Human-induced petrel extinction

7.4.1 Hunting versus introduced predators

Evidence for the effect of hunting by prehistoric settlers being responsible for local or global extinction of petrel species is limited. Mourer-Chauviré and Antunes (2000) found the bones of the extinct shearwater *Puffinus holeae* associated with Upper Pleistocene Neanderthal middens in Portugal, and Rando and Alcover (2007) found cut marks and burning on bones of the extinct lava shearwater *Puffinus olsoni* on the Canary Islands (Fig. 7.1). However, while these observations indicate that at least some now-extinct procellariiforms were actively persecuted by humans before the historical period, they provide little information on the intensity of anthropogenic persecution or whether it represented a significant factor in the disappearance of these species.

It is also rarely possible to tell from the fossil record how quickly petrel populations disappeared, although some evidence can be gained from St. Helena (discovered in 1502 and first described in written accounts in 1588), where the St. Helena petrel *Pseudobulweria rupinarum* became extinct shortly after human arrival. Here the species was extinct or very rare before any documented record could be made of it (Olson 1975). In another more recent case feral cats *Felis catus* probably reached

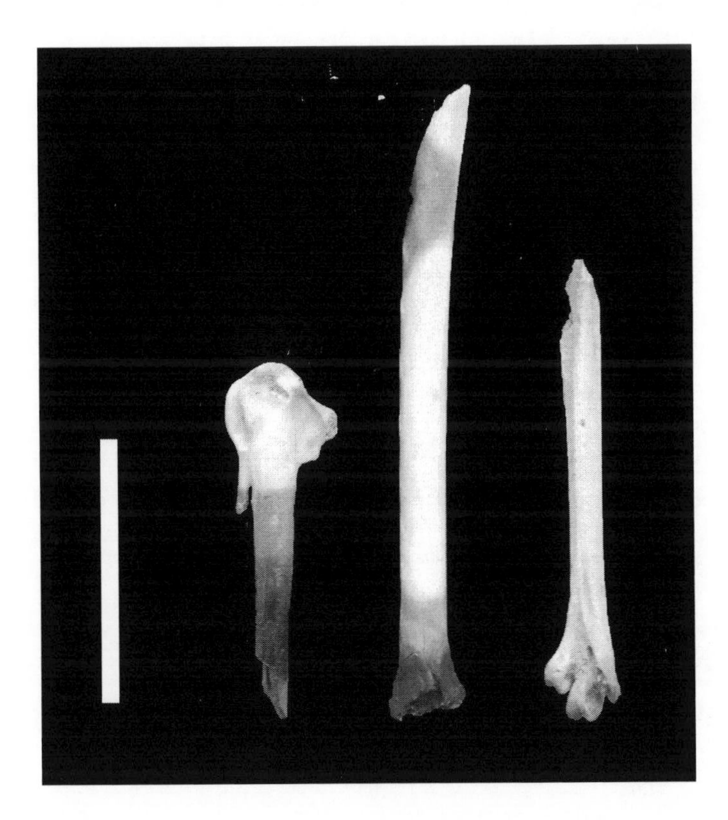

Figure 7.1 Burnt and cut bones of extinct *Puffinus olsoni* from the Canary Islands, indicating that the species experienced prehistoric exploitation by humans. Scale bar = 2 cm. Courtesy of J.C. Rando and J.A. Alcover.

Little Barrier Island, off New Zealand's North Island, in about 1870. By the 1980s the population of Parkinson's petrel *Pterodroma parkinsoni* was functionally extinct (Veitch 1999).

7.4.2 Introduction of predators

7.4.2.1 Rats

Invasive rats are some of the largest contributors to seabird extinction and endangerment worldwide (Jones *et al.* 2008a). The most problematic species of the genus *Rattus* are native to Asia. Rats spread out of Asia at times of major human diaspora, with the Pacific rat or kiore *Rattus exulans* being spread through the Pacific by Melanesian and Polynesian peoples in the Late Pleistocene and Holocene (Spennemann 1997; Matisoo-Smith *et al.* 1998). The black rat *Rattus rattus* reached Europe in Roman times while the brown or Norway rat *Rattus norvegicus* arrived in the Middle Ages (Kurtén 1968). Thus, although rats are generally thought of as ubiquitous, their presence is a relatively new phenomenon worldwide. Being commensal with humans, rats generally accompany humans accidentally wherever they settle, including islands representing important sanctuaries for procellariiforms. The introduction of rats to island communities is frequently devastating. For example, the accidental introduction of black rats to Big South Cape Island (Taukihepa) (Atkinson and Bell 1973) led to a plague of rats, the rapid extinction of three species of landbird, and the decline to virtual elimination of two seabird species within years of their introduction (Bell 1978). Among predators, *R. rattus* and to a lesser extent *R. norvegicus* are well known as the leading agents of bird extinction on islands, but the effects of *R. exulans* introduced by Polynesians throughout Remote Oceania is now thought to be significant, especially on seabirds (Steadman 2006b). Whereas the larger rat species kill the adults of smaller species of petrel, all species kill chicks and eggs. Modern techniques such as stable isotope analyses (e.g. Hobson *et al.* 1999; Stapp 2002) have not only confirmed that rats do actually feed on seabirds, but have also suggested that the traditional studies of rat stomach contents were underestimating the importance of seabirds in the diet of rats. One study looking at the relative effects of predation by rats on species of differing size showed that small species of petrel are most susceptible to egg and adult predation whereas larger species such as Cory's shearwater *Calonectris diomedea* are more likely to be affected by chick predation (Igual *et al.* 2006). These studies indicate that our knowledge of the way rats impact island populations is limited and that more work is needed. There can be no doubt, however, that rats severely impact seabirds, reducing their populations and in many cases triggering their local extinction (Atkinson 1985). Wherever archaeological evidence is examined in the Pacific, extinctions begin with the introduction of kiore, and, whether prehistoric or recent, seabird extinctions on islands normally occur very shortly after the introduction of rats.

7.4.2.2 Mice

The ubiquitous, commensal house mouse *Mus musculus* is the most widely introduced of all mammals, but its effect on native biota is poorly known. On Gough Island in the South Atlantic, unlike most other sub-Antarctic islands, the house mouse is the only introduced mammal. Mice were known to affect populations of the small storm-petrels and diving-petrels but were thought to pose little population level risk to large seabirds. However, recent video evidence has shown house mice killing chicks of the Tristan albatross *Diomedea dabbenena* and Atlantic petrel *Pterodroma incerta*, and has indicated that mouse-induced mortality is another significant cause of poor breeding success in these species. Population models show that the levels of predation recorded there are sufficient to cause population decreases (Wanless *et al.* 2007). It is suggested that mice may not be a significant issue to larger seabirds on islands when constrained by other introduced predators, but when these mouse populations are released from the ecological effects of predators and competitors by eradication and restoration programmes, they too may become predatory on seabird chicks (see section 7.4.2.9, below).

7.4.2.3 Cats

Feral cats have a devastating effect on seabird colonies. Researchers have estimated that cat mortality on Marion Island prior to rat eradication was 450000 seabirds per annum (van Aarde 1980), whereas on Macquarie Island 47000 broad-billed prions *Pachyptila vittata* and 110000 white-headed petrels *Pterodroma lessonii* were estimated to be killed annually (Jones 1977), and on Kerguelen Island 1.2 million seabirds were estimated to be killed annually (Pascal 1980). This level of mortality has dramatic effects on the population viability of seabird breeding colonies. For example, cats have been responsible for the local extinction of 10 petrel species on the Crozet Islands in the last century (Derenne and Mougin 1976), whereas in New Zealand the extinction of most seabirds from main Chatham Island has been attributed to cats (Imber 1994). However, the effects of cats are not always predictable, and their removal may upset delicate (though artificial) ecosystems that have developed on some islands (see section 7.4.2.9, below).

7.4.2.4 Foxes

More than 10 million seabirds belonging to 29 species formerly bred on the Aleutian Islands off Alaska (Croll *et al.* 2005). The region lacked large terrestrial predators prior to the arrival of humans, but in the past 5000 years Arctic foxes *Alopex lagopus* have been introduced to at least 400 of these islands. Being primarily carnivorous, foxes preyed on the local seabirds that had evolved in the absence of predators, and only those species that nested on unreachable cliff faces were able to survive. Burrow-nesting species and surface-nesters (e.g. gulls) were predated easily, and their populations became locally or regionally extirpated. Fox removal from 40 islands has resulted in an increase by two orders of magnitude in whiskered auklet *Aethia pygmaea* and other seabird populations. Similarly, nine fox-free islands have nearly 100 times more seabirds than nine geographically and ecologically similar but fox-infested islands (Croll *et al.* 2005).

7.4.2.5 Pigs

Pigs *Sus scrofa* are a frequently ignored but extremely important predator of burrowing petrels. They were responsible for the eradication of Buller's petrel *Puffinus bulleri* on one of the two main islands in The Poor Knights group off northern New Zealand (Medway 2001), they have been implicated in the decline of Galapagos petrel *Pterodroma phaeopygia* in the Galapagos Islands (Cruz and Cruz 1987), and they have been responsible for the degradation of habitat on the main Auckland Islands and (with cats) caused the extinction of most breeding seabirds on the large main island in this island group (Challies 1975).

7.4.2.6 Ungulates

The indirect effects on ungulates on habitat are discussed below. Livestock on some isolated islands may not only trample burrows, but also exhibit unusual behaviours caused by an absence of nutrients that can directly affect seabird survival. In particular, they may eat chicks and/or fledglings in order to ingest nutrient-rich bone. On Foula in the Shetland Islands, sheep *Ovis aries* have been observed biting off the legs, wings, or heads of unfledged young Arctic tern *Sterna paradisaea* and Arctic skua *Stercorarius parasiticus* chicks, and on Rhum in the Inner Hebrides, red deer *Cervus elaphus* have been seen biting the heads off chicks of Manx shearwater *Puffinus puffinus* and occasionally also chewing the chicks' legs and wings (Furness 1988).

7.4.2.7 Mongooses

The mongoose *Herpestes auropunctatus* was introduced to many tropical islands to control rats and other species in sugar cane plantations. This has severely affected many ground-dwelling species and especially seabirds. For example, in some years more than 60% of all egg and chick mortality in the Hawai'ian petrel *Pterodroma sandwichensis* on Oahu is caused by cats and mongooses. Although rats also prey on *P. sandwichensis* eggs, the major threat that they pose is providing a prey base for these larger exotic predators. The lower limit of the Hawai'ian petrel's breeding range on Maui now coincides with the upper limit of permanent mongoose infestation. In contrast, the species has a high breeding success rate on Kaua'i, where mongoose are not established (Simons 1983).

7.4.2.8 Snakes

The accidental introduction of the brown tree snake *Boiga irregularis* to Guam around 1950 induced a cascade of extirpations that may be unprecedented among historical extinction events in taxonomic scope and severity. Birds (including wedge-tailed shearwater *Puffinus pacificus*), bats, and reptiles were affected, and by 1990 most forested areas on Guam retained only three native vertebrates, all of which were small lizards (Fritts and Rodda 1998). It is clear that some petrels and shearwaters can coexist with snake species (e.g. tiger snakes *Notechis* spp. with wedge-tailed shearwater and short-tailed shearwater *Puffinus tenuirostris* in Australia), so it would appear that it is the invasive nature of the *Boiga* population and the Guam fauna's lack of adaptation to snake predation that has caused extirpation.

7.4.2.9 Mesopredator release

The example of Little Barrier Island, in New Zealand's northern Hauraki Gulf, demonstrates the complexity of the delicate (though artificial) ecosystems that have developed on some islands. Cats were introduced in the 1870s to an island that previously only hosted an introduced population of kiore. Contrary to previous expectations that cats would preferentially forage on smaller prey, cats virtually exterminated not the much smaller Cook's petrel *Pterodroma cooki*, but the large Parkinson's petrel from the island. It has been suggested that the timing of Parkinson's petrel chick emergence from their burrows, during a period of low food availability for cats, led to their disproportionate decline (Veitch 1999). Cook's petrel survived in large numbers on the island until cats were eradicated in 1990. However, the initial eradication of cats on Little Barrier Island led not to an increase but a decrease in Cook's petrel, due to the reduced breeding success of these small petrels which are vulnerable to predation by kiore (Rayner *et al.* 2007). What led to this apparently counter-intuitive situation? Elimination of the top introduced predators from islands can, in fact, lead to the decline of smaller prey populations through the ecological release of smaller introduced predators, in a process termed mesopredator release (Crooks and Soulé 1999).

7.4.3 Habitat destruction

The indirect effects of ungulates on burrowing petrels are poorly documented. Cattle *Bos taurus* and goats *Capra aegagrus hircus* are very destructive to vegetation; by browsing on seedlings they slow or even halt the regeneration of the forest canopy and reduce native plant diversity. Cattle in particular can damage and destroy burrows and consolidate soils, and have been implicated in the decline of the Galapagos petrel (Cruz and Cruz 1987). Goats are present on at least nine island groups in the Pacific and four in the sub-Antarctic. Wherever they are present in dense populations they cause great destruction to vegetation and landscapes, typically compounded by subsequent soil erosion, which often results in total habitat loss. Forested environments can therefore be converted into degraded grasslands (e.g. on Isabela Island, Galapagos) or become more vulnerable to further invasion by weeds or to cyclone damage, all of which can adversely affect burrowing seabirds as well as wider island ecology.

7.4.4 Fisheries interactions

Fishery waste is unquestionably an important food source for seabirds, with about 6 million birds including procellariiforms being supported by this resource in the Baltic Sea alone (Garthe and Scherp 2003). An interesting and unexpected outcome of such high levels of food availability is that numbers of generalist gulls (*Larus*) have reached unnaturally high population levels, and exclude more specialized seabird species (e.g. Manx shearwater) from resources such as limited nesting grounds in these highly populated areas (Garthe and Scherp 2003). However, fisheries interactions are not always negative; northern fulmar *Fulmarus glacialis* was historically restricted to Arctic regions, but a gradual southward expansion began during the mid-1700s from Iceland through the Faeroes, Shetland and Orkney, and down the British and Irish coasts to the Channel Islands and France (Fisher 1952). This expansion has been postulated to have resulted from increased availability of offal from whaling, and more recently due to fisheries discards (Fisher and Lockley 1954), although other explanations have also been proposed; some researchers have suggested that a behavioural or genetic transformation may have occurred and a colonizing phenotype emerged among the boreal fulmar population which was able to spread into lower latitudes (Wynne-Edwards 1962), or that a gradual change in sea temperature or oceanographic conditions may have taken place (Salomonsen 1965). Evidence now suggests that the spatial overlap between fulmars and commercial fisheries is far from complete, and while it is indisputable that northern fulmars are major consumers of fishery waste in the southern part of their range, the extent to which their distribution is or was constrained by the availability of this resource is debatable (Phillips *et al.* 1999). The real question arises about what will happen when supplementary fisheries offal is inevitably removed by overfishing. Will fulmar populations stay high?

7.4.5 Direct fisheries mortality

The recent decline in procellariiforms in the Southern Ocean is thought to be largely driven by birds getting accidentally caught on fishing long lines or tangled in trawling gear. It is believed that the high levels of mortality offset any population increases accrued by the ready availability of fishery waste. The magnitude of by-catch mortality is exemplified by black-browed albatross *Thalassarche melanophris*, the IUCN Red List status of which rose from Least Concern in 1998 to Near Threatened in 2000, Vulnerable in 2002, and Endangered in 2003, an increase greater than virtually any other bird species. Many different methods have been developed to mitigate fisheries by-catch, but these can only be enforced rigorously within Exclusive Economic Zones of individual countries, or within areas designated by international agreements such as the Commission for the Conservation of Antarctic Marine Living Resources. Implementation in international waters, where many of these pelagic species feed primarily, is currently politically difficult.

7.4.6 Pollution

Many studies have shown that procellariiforms are sensitive to marine pollution. For this reason, procellariiform species have frequently been used as monitors of pollution, especially oil pollution. Beached bird surveys provide important evidence of geographical and temporal patterns, and, for example, show that there have been consistent declines in oil release into the southern North Sea in recent years (Furness and Camphuysen 1997). As predators high in marine food webs, procellariiforms can also be affected by pollutants that accumulate at higher trophic levels. Recent work on mercury in seabirds has permitted an analysis of spatial patterns and of the rates of increase in mercury contamination of ecosystems over the last 150 years, since mercury concentrations in feathers of museum specimens can be used to assess contamination in the birds when they were alive. This work has led to hypotheses about the cause and effect of these levels of pollutants, but specific mechanisms and pathways remain unclear. Surprisingly, pelagic procellariiforms show higher pollutant loads than coastal seabirds, and increases have been greatest in seabirds feeding on mesopelagic prey (Furness and Camphuysen 1997). Floating plastic is a pollutant of considerable concern, especially in the Pacific, where voluntary ingestion in surface-feeding birds may occur due to floating particles of plastic being confused with prey items, or through plastic already being incorporated within the bodies of prey species. Plastic is often passed from parents to chicks in regurgitated food (Blight and Burger 1997). The effects of this pollution on procellariiforms is poorly understood, but assimilation of polychlorinated biphenyls (PCBs) is recognized as inhibitory to fecundity (Powell *et al.* 1996), and chicks that ingest plastic fledge at lower weights and are liable to death from dehydration (Sievert and Sileo 1993).

7.5 Financial impacts of a decline in procellariiforms

As well as ecosystem impacts, economic impacts may result from alterations in seabird abundance.

7.5.1 Guano production

Guano is not primarily produced by procellariiforms, but they are a contributor to this valuable economic resource. The growth of seabird populations from 1925 to 1955 in the Peruvian Current was probably a response to increased productivity of the Peruvian upwelling system, but a subsequent drastic decline in seabird abundance was likely to be due to competition for food with the large-scale regional fishery, which caught approximately 85% of the region's anchovies that would otherwise have been available for seabirds. This crash led to a financially devastating decline in guano production on the islands of the Peruvian coast (Jahncke *et al.* 2004). The decline in guano led to economic depression and political turmoil in western South America that still has repercussions today.

7.5.2 Impact of birds on fisheries

Whereas the impacts of fisheries on procellariiforms are widely published, the impacts of procellariiforms on fisheries are less well known. Prior to effective mitigation it was estimated that albatrosses and petrels removed over 20% of all baits from tuna long-line hooks prior to their deployment (Tasker *et al.* 2000). Brooke (2004) estimated that petrels take approximately 16 million tonnes of prey per annum (approximately twice the annual take of the Japanese fishing fleet), whereas annual pelagic marine fisheries take 70 million tonnes. Brooke (2004) further estimated the trophic levels at which the majority of fisheries and petrels operate, and concluded that seabirds and fisheries overlap considerably. Whether seabirds and fisheries are actually in competition is arguable, though, as they could be taking fish of different ages and size ranges.

7.5.3 Harvest

Three significant human harvests of procellariiforms still occur—northern fulmar on the Faeroe Islands and Iceland; short-tailed shearwater in southern Australia; sooty shearwater *Puffinus griseus* in southern New Zealand (Mallory 2006)—and at least 10 other populations are quasi-legally or illegally harvested. These harvests require mechanisms to be put in place to ensure that they remain sustainable and economically viable. However, much ongoing seabird harvesting is sporadic and unobserved.

In particular, poaching and difficulties of enforcing harvest prohibitions make it clear that existing legislative restrictions and reservations are not sufficient safeguards to halt ongoing risk to seabirds and the economies they sustain.

7.6 The impact of procellariiforms on ecosystems

Little is known about the potential consequences of widespread disappearance of fish-eating and scavenging bird species. There is an urgent need to investigate whether ongoing declines in seabird populations may have unanticipated top-down or bottom-up consequences as a result of trophic cascades or significant reductions in nutrient deposition (Sekercioglu *et al.* 2004).

7.6.1 Trophic cascades

This term has been coined to describe a situation where predators in a food chain suppress the abundance of their prey, thereby releasing the next lower trophic level from predation or herbivory (Paine 1980). When human or other agencies alter this balance, then large-scale ecological shifts may occur. In a classic example, when the abundance of the northern fulmar, a large piscivorous seabird, increased in the North Atlantic, the abundance of its prey (capelin *Mallotus villosus*, a zooplanktivorous fish) decreased, large zooplankton abundance increased and phytoplankton biomass decreased. The combined effects of environmental conditions and overfishing have led to dramatic fluctuations in pelagic fish stocks. These changes have, in turn, induced large decreases in some seabird populations (although no information is available for northern fulmars) (Cherel *et al.* 2001). It would appear that birds reliant on single prey species are more likely to represent such keystone species (Bruno and O'Connor 2005).

7.6.2 Allochthonous nutrients

There is increasing recognition that a diverse range of terrestrial ecosystems are in fact supported by nutrients that originate from marine systems (Polis *et al.* 1997; Erskine *et al.* 1998; Vitousek 2004; Ellis 2005; Ellis *et al.* 2006; Harrow *et al.* 2006). Seabird droppings are enriched in important plant nutrients such as calcium, magnesium, nitrogen, phosphorus, and potassium. By feeding in productive oceanic waters and defecating inland, seabirds act as a primary vector for transferring nutrients to terrestrial systems. The size of these nutrient inputs cannot be underestimated. It is estimated that seabirds around the world transfer more than 10^4–10^5 tonnes of phosphorus from sea to land every year (Murphy 1981). Ironically, many marine currents that facilitate spectacular marine productivity (e.g. Benguela, California, Humboldt) also create temperature inversions that result in low-productivity deserts on nearby landmasses, and seabirds help offset this inbalance by providing allochthonous inputs in the form of guano and carcasses.

The effects of seabird nutrient inputs on terrestrial systems can be ably illustrated by several examples. The two species of seabird on 17 ha Heron Island in Australia's Great Barrier Reef contribute 129 tonnes of guano per annum, including 9.4 tonnes of nitrogen and 1.4 tonnes of phosphorus (Staunton Smith and Johnson 1995). These high nutrient levels (over three times the recommended levels of nitrogen fertilization for nearby agricultural areas) have created unique soil environments that cannot be replicated by terrestrial processes. On the Aleutian Islands, Croll *et al.* (2005) found that guano was the main source of fertilizer for terrestrial vegetation. When introduced arctic foxes nearly eliminated breeding seabirds in the region, the annual input of guano was reduced from 362 to 5.7 g/m^2, resulting in substantial declines in soil phosphorus, marine-derived nitrogen, and plant nitrogen content, and triggering an ecosystem switch from grassland to maritime tundra on fox-infested islands. Similarly, increased numbers of snow geese *Chen caerulescens* due to intensification in agricultural practices along migration pathways in North America has altered plant productivity and community structure on their Arctic breeding grounds thousands of kilometres away (Jefferies *et al.* 2004). 28–38% of the nitrogen in the biota of streams near Westland petrel *Procellaria westlandica* breeding colonies in the South Island of New Zealand is marine-derived (Harding *et al.* 2004), and possible correlations have been shown between Westland petrel numbers and growth rates of long-lived tree species (Holdaway *et al.* 2007). These effects are also of critical importance today in islands in the Southern Ocean. On high-latitude Marion Island, plants influenced by seabirds were 55% more enriched in nitrogen (1.59 compared with 2.46% N) and 88% in phosphorus (0.17 compared with 0.32% P) than plants away from guano areas (Smith 1978). Plants collected from a range of sites on sub-Antarctic Macquarie Island varied by up to 30% in their leaf δ^{15}N values, with the majority of nitrogen utilized by plants growing in the vicinity of animal colonies or burrows being animal-derived (Erskine *et al.* 1998). Other studies have demonstrated that the presence of breeding seabirds increases plant productivity (Bancroft *et al.* 2005; Wait *et al.* 2005), leaf nutrient status (Anderson and Polis 1999; García *et al.* 2002), and, probably most telling of all in terms of understanding biodiversity, insect abundance (Sanchez-Pinero and Polis 2000).

Any disruption to these nutrient imports may drastically affect ecosystems at both small and large scales (Vanni *et al.* 2004), and this is especially true in coastal areas and unproductive island systems. In particular, vast colonies of burrowing petrels and shearwaters existed on the New Zealand mainland before human contact (Holdaway 1989), but today these have been virtually eradicated (Worthy and Holdaway 2002). Although New Zealand and many other islands are generally interpreted by conservation planners and restoration ecologists as being naturally nutrient-poor systems, the removal of huge numbers of inland seabird colonies by introduced mammalian predators during the Holocene has undoubtedly had a massive impact on island nutrient cycling and productivity, which needs to be recognized in future conservation management (Harding *et al.* 2004). However, it remains extremely difficult to make meaningful inferences about ecosystem-level changes that lack direct scientific observation data, even from the recent past (see also Chapter 10 in this volume). For example, several studies have attempted to address the impact of the now-extinct passenger pigeon *Ectopistes migratorius* on North American ecosystems (Webb 1986; Ellsworth and McComb 2003). Between the seventeenth and nineteenth centuries, 'countless numbers' and 'infinite multitudes' of these birds were described on their annual migration from eastern and central Canada and the north-east USA to the southern USA, with flocks sometimes a mile or more in width and taking several hours to pass overhead (Lewis 1944), but the species was extinct in the wild by 1900. The role that such huge numbers of birds played in shaping their environment cannot be underestimated, but calculating factors such as the effect of nutrients from passenger pigeon droppings on plant growth is confounded by the actual physical impact of such huge numbers of birds on vegetation, and the effects of other large-scale vegetational changes brought about by agricultural changes over the same time period. Similarly, the consequences of the loss of marine nutrient inputs to the main islands of New Zealand are unknown; given the unique character of the archipelago's now-compromised combination of colonial seabirds breeding in mammal-free temperate forests on a relatively large landmass, predicting the consequences for New Zealand terrestrial ecosystems of the extinction of most burrow-nesting seabirds from such studies is difficult. Indeed, the effects of seabird removal on the soil chemistry and ecology of larger areas are often poorly understood. For example, on the New Zealand mainland, recent work has shown that topography and underlying soil chemistry influence the effects of nitrification (Harrow *et al.* 2006). Furthermore, the effects of nitrification are difficult to predict because the responses of different tree taxa to nutrient enhancement vary considerably (Islam and Macdonald 2005), and excessive nutrient inputs may inhibit growth, change species composition, or even kill certain species of plants (Hogg and Morton 1983).

7.6.3 Environmental modification

The term 'ecosystem engineer' describes a species that modulates resource flows and species composition within an ecosystem through the physical modification of habitat (Jones *et al.* 1994). Procellariiforms are classic ecosystem engineers, and burrow building is a common example of ecosystem engineering, as burrows may provide habitats for a range of other burrowing and non-burrowing species. High burrow densities can also reduce plant growth rates and seedling establishment (Mulder and Keall 2001). Biopedturbation has been shown to drive decreased diversity and structural complexity in island ecosystems (Bancroft *et al.* 2005), and influences entire island ecosystems through the effects of burrowing and underground deposition of vegetation on biotic and abiotic island processes (McKechnie 2006). Comparison of rat-free and rat-invaded offshore islands in New Zealand has shown that predation of seabirds by introduced rats has led to altered soil properties, thereby structuring plant and animal communities (Bancroft *et al.* 2004, 2005; Fukami *et al.* 2006).

7.6.4 Seabirds as scavengers and their role in reprocessing nutrients

Although not generally obligate scavengers, most seabird species will scavenge opportunistically. In particular, giant petrels *Macronectes* spp. together with skuas *Catharacta* spp. are considered the marine equivalents of vultures on sub-Antarctic islands, although giant petrels also feed on marine invertebrates and chicks of various marine birds (Hunter and Brooke 1992). Albatross and petrel diets contain large amounts of deep-water species of squid and crustaceans that have not been recorded by oceanographers at the surface even at night, and it is hypothesized that other than scavenging at fishing boats the only way such food items could be obtained is through scavenging at the surface following natural death (Croxall and Prince 1994). We know little about the potential ecological consequences of the changes in the numbers of these scavenging seabirds, but it has been theorized that a lack of alternate pathways for the remineralization of nutrients may decrease ecosystem biodiversity and resilience (Nixon 1981).

7.7 The original terrorists?

The structure of procellariiform populations is very like the structure suggested by Louis Auguste Blanqui (1885) for revolutionary human organizations to avoid persecution and survive. By their very nature, the secretive yet widespread

Table 7.4 Procellariiform populations that have become established in the Holocene.

Species	Locality	Reference
Natural		
Diomedea sp.	Macquarie Island	de la Mare and Kerry (1994)
Diomedea antipodensis	Pitt and Main Islands (Chatham Group)	Miskelly *et al.* (2006)
Phoebastria immutabilis	Hawai'ian Islands	Gummer (2003)
Fulmarus glacialis	Iceland to the Faeroes, Shetland and Orkney Islands, and down the British and Irish coasts to the Channel Islands and France	Fisher (1952)
Four species of Antarctic petrel	Antarctic coastline	Hiller *et al.* (1988); Ainley *et al.* (2006)
Human-encouraged		
Phoebastria immutabilis	Hawai'ian Islands	Gummer (2003)
Pachyptila turtur	Mana Island (New Zealand)	Gummer (2003)
Puffinus newelli	Kilauea Point, Kauai (Hawai'i)	Byrd *et al.* (1984)
Pterodroma axillaris	Pitt Island	Gummer (2003)
Pterodroma leucoptera	Boondelbah Island (NSW, Australia)	Priddel *et al.* (2006)
Pterodroma phaeopygia	Santa Cruz Island (Galapagos)	Podolsky and Kress (1992)
Pterodroma pycrofti	Cuvier Island (New Zealand)	Gummer (2003)
Puffinus gavia	Maud and Mana Islands (New Zealand)	Gummer (2003); Bell *et al.* (2005)
Pelacanoides urinatrix	Mana Island (New Zealand)	Miskelly *et al.* (2004)
Oceanodroma leucorhoa	Old Hump Ledge and Ross Island (Gulf of Maine, USA)	Podolsky and Kress (1989)

species of procellariiforms have formed comparable isolated, independent 'cells' which experience limited genetic interchange over time. By existing in these isolated pockets, populations of many species have survived virtually unchanged over long intervals of geological time, and although populations of more than 50% of extant procellariiform species have been documented to have lost populations during the Holocene (Table 7.3), most species still retain isolated populations that are protected from sources of predation due to their Blanquistic behaviour. However, these refuges are now being progressively invaded by the recent demand for coastal property and the increasing use by humans of even very small islands.

7.8 What is being done to protect this ancient group from extinction?

Given the antiquity of the procellariiform lineage, the extreme effects that large seabird breeding colonies have at the landscape and regional levels, and the importance of their ongoing population declines, extirpations and extinctions cannot be overestimated. Seabird fisheries by-catch can and is being addressed within the territorial waters of most countries, and is also being tackled by international initiatives and agreements outside national Exclusive Economic Zones. Recent work has demonstrated that compensatory mitigation can facilitate high-value uses of biological resources and cost-effective conservation gains for species of concern (Wilcox and Donlan 2007). For example, by levying fishers for their by-catch, money can be made available to remove invasive mammals from breeding islands. However, pollution can only be prevented from affecting populations through effective international agreements and monitoring.

It has been shown that removal of invasive predators is more than 20 times more effective in giving percentage increases in population growth per dollar invested than fisheries closures, and is also more socio-politically feasible (Wilcox and Donlan 2007). Rat populations can be reduced by poisoning (Igual *et al.* 2006), but ultimately the only long-term solution on islands is eradication. Work in New Zealand has shown that large-scale rat eradication can be achieved even on comparatively large islands such as Campbell Island (115 km^2) in New Zealand's sub-Antarctic (Towns and Broome 2003). Techniques have also been developed to remove feral cat populations from islands. Over the past two decades, these conservation techniques have prevented the extinction of many island species and restored many island ecosystems (Nogales 2004). Habitat restoration and active transfer of petrel species to islands is further helping to redress the balance (Table 7.4).

Ultimately, the procellariiforms are a largely secretive group of birds that remain difficult to study, and their precarious status is also relatively poorly known. Ironically, their survival may ultimately depend on publicity and public awareness, and not the secretiveness that has enabled them to survive for so long.

CHAPTER 8

Coextinction: anecdotes, models, and speculation

Robert R. Dunn

8.1 Introduction

Nearly all work in conservation and the study of extinction focuses on vertebrates and plants. In contrast, most species and organisms on Earth are invertebrates, the majority of which are parasites or commensals (e.g. Price 1980; Windsor 1995; Dunn 2005). From the perspective of a biologist who studies invertebrates, it seems as if we have built conservation biology around the study of hosts. As elaborated in other chapters of this book, we know that many of these host species have recently died out. What happens to those species that depended on extinct or simply rare hosts? What else did we lose with the dodo, moa, or elephant bird? When the passenger pigeon went extinct, what happened to its lice, mites, gut parasites, and mutualists?

The answers to these questions do not come easily. In an ideal world we would know enough about the species that lived on the dodo, for example, to look for them once the dodo had died out, and to document whether or not they had become extinct as well. In the real world, most invertebrate species on Earth are undescribed (e.g. Stork 1988; Dunn 2005). Most of those invertebrate species that have been named have not been well studied, and those that have been well studied tend to be common rather than rare species (Stork 1988). Under such circumstances, the study of the chains of extinction long thought to follow such initial extinctions has advanced through anecdotal studies of particular species losses and theoretical considerations of the number of species that might have gone extinct. To date, anecdotal evidence for coextinction has been limited. In contrast, models of coextinction tend to predict large numbers of coextinctions. Here I consider examples of each of several categories of coextinction, examine attempts to model the numbers and kinds of coextinctions, and then offer some thoughts on future directions for reconciling the results of these two approaches to the study of coextinction.

Most of what we know about coextinction focuses on the Anthropocene: the recent historical period, approximately the last 500 years. To understand the importance or even just the frequency of coextinctions across the Holocene, we must stand on the already shaky ground of our understanding of recent coextinctions and speculate even more wildly. No well-documented examples of prehistoric Holocene coextinctions exist, but in discussing coextinctions more generally I will consider how we might approach those coextinctions that may have occurred across this geological epoch.

8.2 Anecdotes

In this chapter, I will use coextinction to mean the process by which one species goes extinct because of the extinction of another species on which it depends. The term coextinction was originated by Stork and Lyal (1993). These authors considered only events in which parasite species went extinct with their hosts to be examples of coextinction. Here, as elsewhere (Koh *et al.* 2004; Dunn 2005), I use a more inclusive definition of coextinction to include not only the loss of parasites but also the loss of commensals and mutualists with the loss of their partners, such that the terms secondary extinction and coextinction are synonymous.

Coendangered species are, in turn, species endangered due to the endangerment of their hosts. In order to use a consistent terminology across parasitic, commensal, and mutualistic relationships, I follow Koh *et al.* (2004) in sometimes calling the species at risk of coextinction 'affiliates', and the species whose extinction causes the affiliate to undergo coextinction 'hosts.' In mutualisms such as fig wasp mutualisms, the wasp is thus (for my purposes here) an affiliate when we are considering the effects of the extinction of the fig on the wasp, and a host when we are considering the effects of the extinction of the wasp itself on the fig.

8.2.1 Vertebrates and their dependants

Stork and Lyal (1993) suggested that two louse species, *Columbicola extinctus* and *Campanulotes defectus*, went extinct with the passenger pigeon. Just a few hundred years ago, the passenger pigeon *Ectopistes migratorius* was the most abundant bird in North America, with a population size as large as 3–5 billion (Webb 1986). Americans shot, burned, and ate passenger pigeons for food and as a leisure activity. The last wild passenger pigeon was killed in 1900 by a 14 year-old boy in Ohio (Schorger 1955).

The passenger pigeon has long been central to discussions of the cascading consequences of extinctions. When the passenger pigeon was still extant, many other species must have interacted with it, if for no other reason than as a function of its tremendous collective biomass and large geographic range. A variety of authors have suggested that the removal of the foraging pressure of passenger pigeons might have influenced forest dynamics. As historical observations of passenger pigeons tend to indicate, the species was very generalized in its foraging on fruits and seeds, and 'in their travels... [made] vast havoc among the acorns and berries of all sorts, that they waste whole forests in a short time, and leave a famine behind them for most other creatures' (Byrd, cited in Webb 1986). Bird species that consume large numbers of fruits and seeds almost invariably consume seeds of many plant species simply as a function of the number of seeds consumed (e.g. Calviño-Cancela *et al.* 2006). Passenger pigeons ate, and—as Webb (1986) has stressed—may have dispersed the seeds of many tens of plant and particularly tree species (Webb 1986), including American chestnuts and many species of oak (Webb 1986; Ellsworth and McComb 2003). The seasonal migrations of passenger pigeons may have further helped to disperse many plant species (Webb 1986), although the birds predated most of the seeds they consumed. Ellsworth and McComb (2003) suggested that twentieth-century shifts in the relative abundance of different oak species in hardwood forests in eastern North America (primarily an increase and expansion in red oak *Quercus rubra*) may be a consequence of the extinction of the passenger pigeon and its effect as a seed predator, arguing that the birds foraged for seeds and fruits primarily in the early spring when acorns of white oak *Quercus alba* would have been unavailable for predation, leading to the dominance of this species until the last century. In death, the pigeons may also have been important food for the now endangered American burying beetle *Nicrophorus americanus*. Whether the extinction of the passenger pigeon precipitated the decline of the American burying beetle is unknown (Sikes and Raithel 2002). It has been speculated, perhaps wildly, that the Lyme disease outbreak of the early twentieth century might also be linked to the extinction of the passenger pigeon (Blockstein 1998). Lyme disease prevalence is worst in years in which oaks mast, because of booms in mouse populations associated with the glut of acorns. Blockstein (1998) argued that oak masts and hence mouse population booms would have been less pronounced when passenger pigeons consumed acorns.

In short, although there has long been a sense that the loss of the passenger pigeon must have produced a series of ecological consequences, these consequences remain incredibly hard to document well. This will continue to be the case, not only for the passenger pigeon but also for other cases where an apparently ecologically dominant species goes extinct (see also Chapter 10 in this volume). Consequently, when initially proposed, Stork and Lyal's (1993) example of coextinction of two louse species with the loss of the passenger pigeon seemed the first clear-cut example not only of the consequence of the passenger pigeon's extinction, but also more generally for the wider consequences of an extinction event. For parasites, the passenger pigeon was an ideal host. If each passenger pigeon carried a few lice (probably an underestimate), there might have been as many as 10 billion passenger pigeon lice in North America. At 2 mg a louse, this is equivalent to approximately 20 tonnes of pigeon lice. There were probably as many, if not more, passenger pigeon fleas, mites, tapeworms, protists, and bacteria (Dunn 2002).

When the passenger pigeon was extant, its lice were not studied. However, a few specimens had been collected and curated; these were rediscovered and studied by Malcomson (1937), who suggested that one of those species, a feather louse he described as the new species *Columbicola extinctus*, did not appear to be conspecific with any living lice and had apparently become extinct with the passenger pigeon. Thirty years later, Tendeiro (1969) described another passenger pigeon feather louse from the same collection, *Campanulotes defectus*, as a distinct, extinct species. Based on their morphology, *Columbicola extinctus* probably lived between the barbs of flight feathers, and *Campanulotes defectus* probably lived in fine down feathers (Clayton and Price 1999; Dunn 2002).

However, careful analysis of the original passenger pigeon lice specimens together with more lice collected from other museum skins led Clayton and Price (1999) to conclude that *Columbicola extinctus* is actually still extant. This species was not restricted to a single species of bird, but in fact also occurs on the band-tailed pigeon *Patagioenas fasciata*; it remains, at least at the time of the last survey, abundant (Clayton and Price 1999; Dunn 2002). Because parasites are often poorly known, the first records of species are almost inevitably from a single host. As a consequence, it is easy to overestimate the host specificity of parasites and commensals, as was the case with *Columbicola extinctus*.

The story of the other passenger pigeon louse, *Campanulotes defectus*, is less straightforward. *Campanulotes defectus* probably never even lived on the passenger pigeon. Price *et al.* (2000) have concluded that somehow Tendeiro incorrectly identified *Campanulotes defectus* as a passenger pigeon louse, and despite searching many other museum skin collections, no further individuals of *Campanulotes defectus* have been found on passenger pigeons (Price *et al.* 2000). All of the relatives of *Campanulotes defectus* are Australian, and so for obvious biogeographical reasons a North American species of *Campanulotes* seems unlikely. Lice identical to *Campanulotes 'defectus'* have also been found living on a species of Australian pigeon, the common bronzewing *Phaps chalcoptera*. The louse is now correctly identified as *Campanulotes flavens* (Price *et al.* 2000).

The stories of the passenger pigeon lice, initially the best examples of coextinction, together highlight two recurrent themes in the study of coextinction. First, coextinction events are very difficult to document with certainty unless all potential hosts have been checked (they rarely, if ever, are). Furthermore, one ideally needs to not only sample all potential hosts, but to sample them across their ranges. A given bird louse, for example, might occur in some but not all populations of one of its hosts. Second, as the story of *Campanulotes defectus* makes clear, evidence of 'extinct' parasites found on museum specimens of extinct hosts or in museum collections of that parasite taxon must be treated with caution. A great deal might be learned from a thorough search of museum specimens of extinct hosts for parasites, but the burden of proof that those species are uniquely found on extinct hosts is great. Examples of Holocene coextinction events will be particularly hard to document clearly because the number of available soft-tissue specimens of potential hosts, much less parasites or other affiliates, will typically be very low at best in museum collections.

Since the case of the passenger pigeon and its lice, subsequent studies have been reluctant to 'declare' species as extinct in light of the possibility that they might persist on alternative hosts. There are a few plausible examples of coextinction, but these must be considered carefully as they often suffer the same problems illustrated by the passenger pigeon lice; that is, poor sampling of other potential hosts, and few and poorly documented specimens.

8.2.2 Other coextinctions with the loss of animal hosts

A few recent examples of coextinction seem to bear further investigation, and the next steps in their study seem obvious (if labour-intensive).

A species of ferret louse (*Neotrichodectes* sp.) as well as a species of protozoan (reported as 'probably *Eimeria* sp.') may have gone extinct with the black-footed ferret *Mustela nigripes*, either when ferret populations were reduced or when the last surviving ferrets were deloused during captive breeding (Gompper and Williams 1998). To date, however, the black-footed ferret example, though well known, is at best preliminary. No host-specific louse has been documented on the ferret, and the possibility of a host-specific louse (and its extinction) rests on a checklist of North American biting lice where Emerson (1964) suggested that the individuals of the weasel louse *Neotrichodectes minutus* found on the black-footed ferret 'may be a distinct species'. A next step in understanding this story will need to be a systematic treatment of the louse genus *Neotrichodectes* that considers whether the individuals on the black-footed ferret were really unique together with additional sampling of related hosts.

Similar to the example of the ferret lice, the lice of the California condor *Gymnogyps californianus* were intentionally removed through delousing (Fig. 8.1). In the mid-1980s, conservationists brought all remaining California condors in for captive breeding at Los Angeles Zoo and at San Diego Zoo's Wild Animal Park (Snyder and Snyder 2000). One of the first things the researchers did was to treat them with a pesticide dust to remove lice. Mike Wallace, a researcher associated with the effort, recalled that only one or two of the birds had any lice at all and that the lice seemed 'typical' (Patel 2006). The *Chewing Lice: World Checklist and Biological Overview* lists three species of louse as parasites of the California condor (Price *et al.* 2003). Two of these species have been found nowhere else and are presumed extinct, though again related hosts do not appear to have been examined for these parasite species. Similar to the example of the California condor, at least nine bird lice are known from extinct seabird species but are unknown from extant bird species, and so are regarded as extinct (V. Smith, personal communication). Again, the possibility (although in this case more remote because of the relatively more intensive study of seabird lice) remains that some of these species remain to be found on extant hosts.

Regardless of whether the lice associated with the California condor or black-footed ferret are extinct, in both cases they deserved further study before they were extirpated. It is unlikely that either species had much of a negative effect on their respective hosts. Specialist parasites on rare hosts are typically rare (e.g. Anderson and May 1991; de Castro and Bolker 2005; Altizer *et al.* 2007) and so likely to impose relatively little fitness cost. In contrast, generalist parasites are at little risk of coextinction, but have been associated with declines of threatened species (Dobson and Foufopoulous

Figure 8.1 The Critically Endangered California condor *Gymnogyps californianus* and its now apparently extinct louse *Colpocephalum californici*. Condor image from Del Hoyo *et al.* (1994); louse image from Vince Smith.

2000; Gog *et al.* 2002; de Castro and Bolker 2005). A first step in both cases should have been to identify whether the parasites being extirpated were rare host-specific taxa or common generalists.

The only well-documented case of coextinction of an insect with its animal host is of a local rather than global extinction (and a rather different kind of relationship), that of the large blue butterfly *Maculinea arion*. As larvae, large blue butterflies are parasites on a single species of host ant *Myrmica sabuleti*. Biological control of introduced rabbits *Oryctolagus cunicularius* in the UK with myxoma virus during the second half of the twentieth century appears to have reduced the occurrence of open habitats which had formerly been grazed by rabbits. *Myrmica sabuleti* decreased in abundance as the amount of open habitat decreased, which in turn appears to have led to the extinction of the large blue in the UK (Hochberg *et al.* 1992). Whether the local extinction of the large blue led in turn to the extinction of its parasites remains undocumented. The example of the large blue serves to demonstrate both the complexity of coextinctions, and that parasites can become extinct (at least locally) even if their hosts simply decline in abundance. This latter process is equivalent to 'stochastic fade-out' or the critical host population threshold for the invasion or persistence of infectious diseases, which is central to the theory and practice of disease ecology (e.g. Hudson *et al.* 2002; Deredec and Courchamp 2003; see also below).

Whereas it is difficult to document coextinctions, it may be easier in theory (although seldom carried out in practice) to document cases of coendangerment, where an affiliate is endangered because its host or hosts are rare. At least with coendangered species there are living parasite individuals that can be studied and used to understand the potential specificity of the relationship. For example, the most endangered feline in the world, the Iberian lynx *Lynx pardinus*, appears to be the sole host of what is probably the most endangered feline louse in the world (Perez and Palma 2001). Mite biologists have recently described a new species of feather mite, *Titanolichus seemani*, from the endangered orange-billed parrot *Neophema chrysogaster* (Dabert *et al.* 2006). However, the mite was discovered on a museum skin, so it is not clear whether it is still extant. Two of the UK's few endemic taxa, the Lundy cabbage flea beetle *Psylliodes luridipennis* and the Lundy cabbage weevil *Ceutorhynchus contractus pallipes*, are both coendangered because of the endangerment of their host plant, the similarly endemic Lundy cabbage *Coincya wrightii*. A survey of the literature could very easily yield many more examples of such coendangered species. As with examples of coextinction, however, the burden with coendangered species is to show that they do not occur, or more importantly do not reproduce, on alternative hosts.

8.2.3 Plants and their dependants

Defined broadly (e.g. Price 1980), parasites also include herbivores, and a few cases of parasite extinction have been preliminarily documented for host-specific plant feeders such as butterflies and moths. For example, when the chestnut blight attacked American chestnuts *Castanea dentata*, Opler (1978) speculated that seven lepidopteran species might have been lost. Four of those species have subsequently been rediscovered (P. Opler and D. Wagner, personal communication), but three species remain missing and are presumed extinct (IUCN 2006). Host-specific parasitoids and other groups could also have been lost with the chestnut decline, but this remains to be investigated (D. Wagner, personal communication).

Mirroring the story of the large blue and myxomatosis is an equally complex case of a marine extinction linked to the spread of a disease. An outbreak of chronic wasting disease (due to the slime mould *Labyrinthula zosterae*) in the marine eelgrass *Zostera marina* led to the reduction by over 90% of eelgrass cover in the North Atlantic Ocean between 1930 and 1933. The eelgrass survived only in brackish rather than fully salty water. The population reduction of the eelgrass in turn appears to have led to the extinction of a host-specific eelgrass limpet, *Lottia alveus* (Carlton *et al.* 1991; Dulvy *et al.*, this volume). Morphological studies indicate that *L. alveus* possessed a feeding morphology that was apparently specialized for feeding on *Z. marina*, and this limpet does not appear to have been able to survive in brackish water. Therefore, while the host species did not become entirely extinct, all populations of the host living in conditions that were tolerable for *L. alveus* did disappear. More clearly than for other examples, evidence seems to point to the loss of *L. alveus* as a case of coextinction.

8.2.4 Mutualisms

If parasites can go extinct when their hosts go extinct, the same should also be true for mutualists with the extinction of their partner(s), depending on the specificity of the interaction. The case of the dodo *Raphus cucullatus* and the tambalacoque tree *Sideroxylon grandiflorum* was offered as an example of a plant that went extinct with its putative disperser (Temple 1979). We would now consider this a case of coextinction (were it true). However, over time the relationship between the dodo and the tambalacoque tree has begun to look less and less host-specific (see Chapter 10 in this volume for further details). Over the past 300 years, new individuals of the tambalocoque tree have recruited without the dodo. Furthermore, the dodo's beak morphology suggests that it is far more likely to have been a seed predator than a seed disperser (Witmer and Cheke 1991).

However, just because the example of the dodo proved flawed does not mean that the loss of seed dispersers cannot result in extinctions. The best analogy for the scenario that was initially suspected to have occurred with the dodo and the tambalacoque is one involving seed-dispersing ants in the fynbos of South Africa, where the loss of native species due to the invasion of the Argentine ant *Linipithema humile* has led to reduced recruitment of a subset of the plant species that depended on the native ants (Bond and Slingsby 1984; Christian 2001). However, even in this case there was no extinction, whether local or global, and the landscape-scale effect of the invasion remains limited. There exist a handful of other studies on the consequences of changes in the abundance of one mutualist partner on another, but rarely do those consequences seem to include even local extinction (Bond 1994, 1995; Kearns *et al.* 1998).

8.2.5 Chains of extinction

So far I have focused on no more than two kinds of interacting species at a time, hosts and parasites (or commensals, given that bird lice often have minimal fitness consequences for their hosts), or mutualists and their affiliates. Many interspecific interactions, however, include more (sometimes many more) than two species. Consequently, extinctions of species have the potential to lead to far more complex consequences than simply the loss of individual dependent species. For the most part these more complex scenarios have been neglected empirically, although they are often discussed (as with the case of the species potentially dependent on the passenger pigeon). In one scenario, the loss of one host of a parasite species with multiple hosts might lead to changes in the parasite's abundance on its remaining hosts and competition with other parasites. In another scenario, loss of a particularly important keystone host (e.g. a fig tree species) with many affiliates might lead to a disproportionately large loss of mutualists or parasites. In a related case, loss of a pair of mutualists (e.g. fig and fig wasp) might lead to cascading effects (so-called chains of extinction) due to the more general ecological role of the pair of mutualists. Examples of such complex chains of consequences are, like other types of coextinction, scarce and typically poorly documented, particularly at anything larger than local scales.

An additional kind of 'dependent' extinction, not strictly a coextinction as I have defined it here, could occur when the extinction of a predator leads to cascading effects at lower trophic levels. The extirpation of sea otters *Enhydra lutris* from the Pacific coast of North America appears to have led to the collapse of kelp forest communities. This collapse led, at least in some regions, to the replacement of formerly diverse communities by barren habitats (Estes *et al.* 2004; Carter *et al.* 2007). It appears that in the absence of sea otters, sea urchins (the preferred prey species) increased in abundance and overgrazed their primary resource, kelp. Local extinction of kelp led in turn to the local loss of fish and invertebrate species that depended on the kelp forests. It is not clear whether any species were globally lost due to these changes; it has been suggested that the cascading consequences of sea otter extirpations included the extinction of Steller's sea cow *Hydrodamalis gigas* but models of sea cow population dynamics suggest that this extinction event is more parsimoniously explained as a consequence of overexploitation by humans (Turvey and Risley 2006). Similar changes have resulted from the loss of grey wolves *Canis lupus* and grizzly bears *Ursus arctos* in many regions of North America, which led to increased densities of moose *Alces alces* and elk *Cervus elaphus* and associated changes (although apparently no extinctions) in plant communities (Berger *et al.* 2001). Most recently, it has also been argued that the historical-era extinction of several small mammal species in Australia can be linked to the decline (due to persecution) in abundance of the top predator, the dingo *Canis lupus dingo*, and consequent growth in population sizes of mesopredators (Johnson *et al.* 2007).

8.3 How many coextinctions?

To date, anecdotal examples of coextinction seem to offer some measure of how coextinction might occur, but give little insight into the frequency or broader significance of coextinction. The question remains as to whether these coextinctions matter, and how we might develop a more thorough understanding of coextinctions on the basis of so few examples. Are they common enough to be ecologically or numerically important now? Have they ever been common enough to be important to the dynamics of extinctions? Were there cascading coextinctions during the Late Quaternary in response to, for example, the end-Pleistocene continental megafaunal extinctions or the series of Holocene mammal and bird extinctions? These are not easy questions to answer. One way to better understand the process of coextinction and how many coextinctions we should have seen historically (or might see in the future) is to develop models. Several modelling approaches have been employed to date. I treat these approaches in turn and then attempt to reconcile the conclusions of these modelling efforts with our empirical evidence (or lack thereof) for coextinction.

8.3.1 Estimates of the dynamics, rate, or number of coextinctions

One of the first attempts to model coextinctions was that of Nunes Amaral and Meyer (1998), who developed a simulation model with herbivores, consumers, and predators. Each predator (or parasite; the model does not specify) preyed upon some number of prey species, and each prey or host on some number of herbivore species. Each species, regardless of trophic level, then had a probability of giving rise to a new species. Finally, a fraction of species of plants were extinguished at some rate. If a plant became extinct its herbivores also became extinct, and if a herbivore became extinct its parasites/predators also became extinct. This research set the stage for future analyses.

Koh *et al.* (2004) were the first authors to try to estimate the frequency with which coextinctions occur at a global scale. Their model borrowed from the work of Nunes Amaral and Meyer (1998), but focused on the consequences of extinctions of host from host/affiliate pairs (parasite/host, mutualist/partner). Koh *et al.* (2004) applied a probabilistic model to empirical 'affiliation matrices' (host*affiliate presence/absence matrices) to examine the relationship between affiliate and host extinctions across a wide range of coevolved interspecific systems. The 'systems' considered were pollinating fig wasps and figs, primate parasites (*Pneumocystis* fungi, nematodes, and lice) and their hosts, parasitic mites and lice and their avian hosts, butterflies and their larval host plants, and ant-butterflies and their host ants. Table 8.1 shows an example of one of these matrices for figs (*Ficus* spp.) and their pollinating fig wasp (*Ceratosolen* spp.) for one small fig clade and then also for non-pollinating (parasitic) fig wasps for the same figs (data from Cook and Rasplus 2003).

In the first step of their analyses, Koh *et al.* (2004) randomly deleted hosts from their host/parasite matrices (host extinctions). In Table 8.1, this would correspond to the random deletion of rows, if we take the figs as hosts. They then examined the number and proportion of affiliates that went extinct with different numbers of host extinctions for various groups. In the example from Table 8.1, any host extinction would lead to a pollinator extinction (pollinating fig wasps), and some host extinctions would also lead to parasite extinctions (non-pollinating fig wasps). This first model estimates the number of affiliate extinctions as a function of the number of host extinctions, assuming a

Table 8.1 An empirical matrix of host/pollinator (left) and host/parasite (right) relationships for one clade of figs and its associated wasps. On the left, each fig has just one species of pollinating fig wasp. On the right, some figs have multiple parasites, but no parasite has more than one host fig. Numbers in cells simply indicate that a given affiliate is found on a given host, not the abundance of the affiliate (from Cook and Rasplus 2003).

		Pollinating fig wasps (*Ceratosolen*)					**Non-pollinating fig wasps (*Apocryptophagus*)**				
		C. nanus	*C. corneri*	*C. bisulcatus*	*C. dentifer*	*C. hooglandi*	*A.* sp. 1	*A.* sp. 2	*A.* sp. 3	*A.* sp. 4	*A.* sp. 5
Figs (*Ficus*)	*F. pungens*	1									
	F. botryocarpa		1				1				
	F. septica			1							
	F. hispidioides				1			1	1		
	F. bernaysii					1				1	1

random order of host extinction. The model shows that the number of coextinctions depends exclusively on the diversity and host specificity of parasites, and that coextinction levels can be high even for affiliates that are not completely host-specific (e.g. those that have two or three hosts). In specifying the number of coextinctions for a given number of host extinctions, the Koh *et al.* (2004) model essentially answers the question, if parasites are as specialized as data suggest, what number of coextinctions should we have seen? An obvious follow-up is whether and how the predictions of Koh *et al.* (2004) can be reconciled with empirical observations.

Koh *et al.* (2004) went on to develop a separate model (in their terminology, the nomographic model) that could be used to estimate coextinction rates in cases where host specificity was not known but could be approximated. This second model allowed estimation of coextinction rates for a far larger number of groups, including, for example, tropical beetles, for which data on all pairings of hosts and parasites are not available, but for which mean host specificity has been estimated. This second model makes the additional assumption that host-specificity distributions tend to have relatively consistent shapes (lognormal), which was true for the host-specificity distributions they examined. The conclusion from both models was that coextinctions should be at least as common as host extinctions (Table 8.2).

8.3.2 Mutualisms versus parasitism (+commensalisms)

In the same year as the study by Koh *et al.* (2004), Memmott *et al.* (2004) published a related paper focusing on coextinctions (which they termed 'secondary extinctions') in local pollinator/plant networks. These authors focused on the consequences of losses of pollinator species for their plant affiliates in two local pollinator networks: the first in Pikes Peak in the Rocky Mountains of Colorado, USA (data from Clements and Long 1923), and the second from a prairie–forest transition in western Illinois, USA (data from Robertson 1929). Like Koh *et al.* (2004), Memmott *et al.* (2004) simulated extinction, but here focusing only on the extinctions of pollinators from the empirical data matrices representing local communities. In addition to random extinctions, Memmott *et al.* (2004) also employed two other extinction algorithms: in one algorithm, specialized pollinators were removed first; in the other algorithm, generalized pollinators were removed first.

The results of Memmott *et al.* (2004) were similar, at least in general terms, to those of Koh *et al.* (2004). The relationship between pollinator (host) extinction and plant (affiliate) extinction was nonlinear for random extinctions (fig. 2 in Memmott *et al.* 2004). For increasing numbers of extinctions, the numbers of plant extinctions were initially low, but then rose sharply when more than 80%

Table 8.2 Data sets used for calculating the number of historical and projected affiliate extinctions (taken from table S1, Koh *et al.* 2004). Data on relationships between bird mites and their host birds were collected from published literature from the 1800s to 2004.

Affiliate	Host	Mean host specificity	Affiliate species richness	Host species richness	Number of endangered hosts	Number of extinct hosts
Mammal lice	Mammals	2.53	554	4627	1130	78
Butterflies (all families)	Host plants	4.30	17 500	310 000	6279	99
Bird mites (all families)	Birds	2.93	2734	9881	1194	132
Bird lice	Birds	2.04	3910	9881	1194	132
Fish monogenean parasites	Fish	1.25	25 000	28 500	746	90
Beetles	Host plants	6.48	1 100 000	310 000	6279	99

of pollinators were extinguished. Commensurate with predictions of relatively rapid declines of plants with pollinators or pollinators with plants, a recent study (Biesmeijer *et al.* 2006) has shown that the most endangered pollinators (bees and syrphid flies in Britain and The Netherlands) tend to be relatively specialized with respect to the flowers they pollinate. In addition, patterns of spatial variation in pollinator decline are correlated with patterns of plant decline (Biesmeijer *et al.* 2006).

The results of Koh *et al.* (2004) and Memmott *et al.* (2004) bear directly not only on the expected number of coextinctions, but also on where they might be expected to occur. In the models of Koh *et al.* (2004), for example, rates of coextinction are a function of two features of the study systems: host-specificity distributions and the number of parasites or affiliates per host. Of the host-specificity distributions Koh *et al.* (2004) considered, the least host-specific tend to be mutualisms, with the exception of figs and fig wasps (which, it is worth noting, appear to have evolved from parasitic relationships). Similarly, as already mentioned, the plant/pollinator mutualisms studied by Memmott *et al.* (2004) were relatively robust to 'host' extinctions because of their lack of specificity, as has been suggested elsewhere (e.g. Bronstein *et al.* 2004). Although it would be interesting to investigate more examples, it appears that host-specificity distributions for host–parasite relationships tend to be more specific (i.e. less left-skewed) than those of mutualisms. The relative lack of specificity in mutualistic relationships has been noted in the context of species invasions (Richardson *et al.* 2000) and pollination networks (Waser *et al.* 1996), but there are apparently no quantitative comparisons of more general host-specificity levels between parasitic and mutualistic relationships. If mutualisms do tend to be less specialized on average, then they may be much more robust to coextinction than are parasitic relationships.

8.3.3 Community viability models

A second kind of model has been employed to understand the kinds of ecological communities that might be particularly susceptible to coextinction or chains of extinction. In contrast to the work of Koh *et al.* (2004), which is static (populations of individual species are not tracked and are simply coded as 1s or 0s), dynamic community viability models allow species population changes to be tracked, allowing a more diverse range of coextinction events to be studied. Community viability models have been used to explore the consequences of different levels of connectedness within communities, different levels of diversity, the distribution of strengths of interactions among species, trophic position, and other factors. These models essentially examine the consequences of the deletion of particular kinds of species on webs of other species (e.g. Rezende *et al.* 2007; Srinivasan *et al.* 2007). However, many community viability models suffer from even greater disconnection from empirical datasets than do the models of Koh *et al.* (2004) and Memmott *et al.* (2004). While the community viability approach can be used to address interesting questions, results to date remain detached from real communities of interacting hosts and parasites or mutualists.

Developing better links between empirical evidence of coextinctions and such viability models seems to be an important potential step in our understanding of coextinction. As an example of the relative disconnection between these models and empirical reality, in a recent review of community viability models and coextinction Ebenman and Jonsson (2005) mentioned just one empirical example of coextinction (that of sea otters and kelp forests, mentioned above) and did not integrate the example in any way with the models considered. Without stronger links between community viability models and real parameter values from communities (for host specificity, for example), it is difficult to know how much has or can be learned from this body of theory.

8.4 Models of coextinction and the Holocene

To date, no models have explicitly considered levels of coextinction across the Holocene. Koh *et al.* (2004) focused mostly on future extinctions (by assuming that all endangered hosts become extinct), as do most models of coextinction or community networks. Koh *et al.*'s model is simple enough, however, that it lends itself to examination of additional taxa or time intervals and so I extend their analyses here by estimating the number of coextinctions that may have occurred during the Holocene. I begin by assuming that there were approximately 250 mammal extinctions in the Holocene and no fewer than 500 bird extinctions (see Chapters 3 and 4 in this volume). I then assume, as in Koh *et al.* (2004), that there are 9900 extant bird species and roughly 4600 extant mammal species, and so 10400 Holocene bird species and 4850 Holocene mammal species. This yields extinction rates of 500/10400=4.8% for birds and 250/4850=4.9% for mammals for the Holocene. The next step is to combine these 'rates' of extinction with knowledge of host specificities of mammal and bird parasites and commensals to estimate the number of coextinctions with those host extinctions.

I used the equation given by Koh *et al.* (2004) for their nomographic model (where coextinction rates are predicted for a given number of host extinctions and host specificity). Incorporating the above estimates of bird and mammal extinction rates and Koh *et al.*'s estimates of host specificity of different parasite taxa into their equation linking host specificity and coextinction yields estimates of coextinction probabilities as shown in Table 8.3. Using current estimates of the diversity of three different parasite taxa (mammal lice, bird lice, bird mites) in turn leads to estimates of the number of parasite extinctions that might have occurred during the Holocene. Given that these estimates are for only a narrow subset of parasite taxa, it can be predicted that total numbers of Holocene coextinctions could have easily been as numerous as Holocene bird and mammal extinctions. However, it is worth noting that these coextinctions are driven to a large extent by bird parasites and commensals rather than those of mammals, which are relatively less host-specific. Conversely, Table 8.3 ignores many kinds of parasites and commensals, including not only microbial lineages, but also, for example, helminths, which are both diverse and often very host-specific. The average vertebrate species is host to two or more helminth species, such that if helminth host specificity is similar to or greater than that of mites (Poulin and Morand 2000; Koh *et al.* 2004) then we should expect to have seen a similar number of helminth coextinctions.

An additional aspect of the Holocene extinctions in particular that bears on estimates of coextinctions is the phylogenetic clustering of extinctions. In a recent analysis like that of Memmott *et al.* (2004), Rezende *et al.* (2007) conducted simulations of tens of local communities containing either plants and their pollinators or plants and their seed dispersers for which phylogenetic hypotheses were available. Among other findings, their results indicated that if phylogenetically related hosts become extinct first (i.e. extinction is non-randomly distributed on the phylogenetic tree), extinction rates of mutualists will be higher than would be the case with random host extinction. Holocene extinctions of vertebrates were phylogenetically non-random: in particular, island extinctions affected only those clades that tended to radiate on islands, for example rails (see

Table 8.3 Coextinctions of lice and mites with Holocene extinctions of mammals and birds based on the nomographic model of Koh *et al.* (2004).

Host	Parasite	Parasite specificity	Host extinctions	Parasite extinction probability	Parasite extinctions
Mammals	Lice	2.53	252	0.030	17
Birds	Lice	2.04	500	0.029	115
	Mites	2.93	500	0.048	129

also Chapter 14 in this volume). Holocene coextinction rates estimated in Table 8.2 would thus be higher if the phylogenetic non-randomness of host extinctions were taken into account.

Because mutualisms seem less host-specific on average than parasitic and commensal relationships, coextinctions leading to loss of mutualists during the Holocene can be predicted to have been more rare than those for parasites and commensals. Although it is often emphasized that the extinction of the dodo or other species in the Holocene might have led to coextinctions of the plant species they dispersed (e.g. Temple 1979), such extinctions are unlikely simply on the grounds of what we know about current seed-dispersal mutualisms. There is now evidence, for example, that dispersal of seeds by ants is one of the more specialized seed-dispersal mutualisms, with a single genus of ants (*Rhytidoponera*) responsible for most seed-dispersal events of hundreds if not thousands of plant species in Australia (Gove *et al.* 2007). However, even in this mutualism it is unclear whether loss of the keystone ant mutualist would lead to extinction of plant species, or instead just to reductions in local fitness with seeds removed instead by inferior but still functional replacements. Perhaps the best-explored example of extinction effects on seed dispersal is that of the suggested gomphothere–fruit mutualism. Janzen and Martin (1982) suggested that the loss of gomphotheres and other large Late Quaternary megafauna led to the loss of dispersal for large-fruited tree species in Central America. However, many if not all of these megafauna-dispersed trees continue to reproduce effectively, through both dispersal by livestock and other means, even though what has apparently been lost is an entire guild of seed dispersers rather than just a single favoured species (see also Howe 1985).

8.5 Reconciling anecdotes and models

In summary, empirical examples of coextinction remain rare, but models predict that coextinction should be very common. We can reconcile these two observations if we assume that the vast majority of coextinctions are unobserved, but such an assumption seems premature. There have been, no doubt, more coextinctions than have been empirically observed, but whether there have been tens more, thousands more, or hundreds of thousands more seems to be an open question. Alternatively (although these options are not mutually exclusive), the assumptions of existing models of coextinction may bias estimates of coextinction rates upwards (or downwards). I conclude this chapter by considering these assumptions, how we might better inform them, and how they do or could bias estimates of coextinction rates.

8.5.1 Sampling of affiliates is complete and accurate

The work of Koh *et al.* (2004), Memmott *et al.* (2004), and other researchers interested in modelling coextinction is almost exclusively based on field measures of host specificity (the number of hosts on which an affiliate has been found). For example, plant/herbivore data tend to come from fogging studies (Stork 1988) or from studies of leaf collection, where herbivores 'occur on' the plants they are found on (Novotny *et al.* 2006). Such occurrence-based estimates of host specificity have two biases, acting in opposite directions with unknown strength.

The first bias is that such sampling may accidentally collect species of invertebrate herbivores (e.g. butterfly larvae) from plants on which they do not actually feed, but with which they are transiently or accidentally associated (e.g. for climbing). The inclusion of such 'tourists' will tend to artificially reduce the measured host specificity and hence the estimates of coextinction. On the other hand, if communities are sampled incompletely (which all real communities undoubtedly are) then parasites will be missed from hosts on which they can potentially survive. This problem will be particularly acute where there are many rare host species. What one ideally wants to know is the survivorship and reproduction of the species in question on each possible host. Such data are rare for local communities and non-existent for large spatial scales (e.g. continents). Having a better estimate, even for a single taxon of how measured host specificity (according to the standard metric) compares with true host specificity, would be very useful in understanding the net direction of the bias due to measurement error.

Even where host specificity and fitness on multiple hosts are measured well, however, such estimates may not be indicative of the response of species to host loss. For example, some bees can shift 'host' flowers when there is a failure of their hosts to flower (Bronstein *et al.* 2004; Wcislo and Cane 1996). In such cases behavioural decisions on the behalf of parasites or mutualists may allow them to be more flexible than would be apparent in a given season or region. Along similar lines, introduced species have been shown in several cases to be able to substitute for native hosts. Even in the highly specific example of the figs and their specialized fig wasps, for example, native fig wasps will occasionally visit introduced 'non-host' fig species (McKey 1989; Nadel *et al.* 1992; Bronstein *et al.* 2004). If their preferred host declined in abundance, it is conceivable that such visits would increase in frequency. In essence, measures of host specificity are difficult not only because of the difficulty of measuring host specificity, but also because host specificity differs depending on the species that are available as potential hosts. Field measurements of host specificity are analogous to the realized niche in that they reflect both the actual tolerance and preferences of a species (the fundamental niche) and the net results of other processes that prevent a parasite or mutualist from finding and using its possible hosts.

8.5.2 No evolution

A second assumption inherent in models of coextinction (or secondary extinction) is that parasite and mutualist species are unable to evolve new host preferences on the timescale over which extinction is occurring. This assumption seems likely to be wrong. When a host population declines in abundance, there is strong selection on a host-specific parasite to use other hosts, which may result in a host-switching event. This possibility is again supported by some empirical observations. For example, the i'iwi *Vestiaria coccinea* (a Hawai'ian honeycreeper) was historically dependent upon the flowers of Hawai'ian lobelioids, which were formerly a prominent understorey component of the original Hawai'ian forests; however, within the last 100 years, 25% of the archipelago's lobelioids have become extinct and most remaining species are rare and endangered as a result of habitat degradation and herbivory by introduced ungulates. The honeycreeper had a unique, curved bill that was an apparent adaptation for feeding on the long decurved corollas of lobelioids, but within the last century the bird's bill shape has evolved in response to unidirectional selection for shorter bills to allow it to feed from the ohia tree *Metrosideros polymorpha*, a surviving native species with a less curved flower (Smith *et al.* 1995). The frequency of evolutionary changes that lead to such switching events during declines in host populations will remain difficult to know; in part it may depend on the timescale of host decline relative to the generation time of the parasite or affiliate.

Phylogenetic approaches provide one means to study the frequency of host switches during population declines in the primary host. For those parasite taxa where co-cladogenesis is common (e.g. bird lice), such switches might represent instances where a host gains an additional parasite in some way other than through a speciation event on that host (leading to a lack of perfect co-cladogenesis). Such events might be seen in heavily sampled phylogenies even where the old hosts persist, if for example the parasites on rare host species tend to have recently derived lineages on the primary host's congeners. Most notably, recent work on primate lice has provided evidence for a now-extinct hominid that could not be inferred based on evidence from the hosts themselves; the lice appear to have switched to a human host population with the decline of their original hosts (Reed *et al.* 2004). Time and more well-sampled phylogenies will tell whether such events are common.

8.5.3 No population dynamics

Finally, models of coextinction rates for empirical communities tend to ignore population dynamics of hosts and parasites, and in doing so assume parasites become extinct only when hosts become extinct. Theoretical work in disease dynamics suggests that for directly transmitted parasites extinction is likely if host populations decline sharply, even if those host declines do not ultimately lead to host extinction (de Castro and Bolker 2005). In the one empirical test of this prediction, researchers found that endangered primate hosts tended to have fewer parasite species, as would be predicted if their endangerment were due to a population declines (Altizer *et al.* 2007). To the extent that parasite/affiliate extinctions occur with host population declines, estimates of coextinction rates may actually be underestimates.

What is the overall direction of bias of assumptions in coextinction models? If estimates of host specificity are biased through the inclusion of 'affiliates' found on host plants they do not actually use, estimates of coextinction rates will be too low. If, on the other hand, we tend to miss hosts of what are actually relatively generalist species, estimates of coextinction rates will be too high. As for the evolutionary lability of affiliates, we may be assuming less lability of affiliates than actually occurs, which would again tend to overestimate rates of coextinction.

8.6 The way forward in understanding coextinctions in general and Holocene coextinctions in particular

For the moment, in considering coextinctions, Holocene or otherwise, we are left with the striking discrepancy that we have few good examples of coextinctions and yet models predict that coextinctions should be very common. As discussed above, this discrepancy could be due to only a few different things. It may be that coextinctions are very common, but we have missed most of them. It may be that parasites and mutualists are actually much less host-specific than they appear (mutualisms certainly seem less specific, on average, than do parasitic relationships). Finally, it may be that parasites and mutualists are much more able to switch hosts than they appear. Three steps are necessary to reconcile the discrepancies between anecdotes of coextinction and models of coextinction.

First, we need more well-documented empirical examples of coextinction. There are few recorded coextinction events associated with known prehistoric Holocene mammal or bird extinctions (see Chapter 10 in this volume), and there are only a handful of reasonably well-documented coextinction events in the recent historical period. Having more such examples would be invaluable in informing us of how common such extinction events actually are, and would allow a more nuanced understanding of the coextinction process.

Second, we need better estimates of the host specificity of parasite taxa, and explicit comparisons of occurrence-based estimates of host specificity and measurements of host specificity where the fitness of parasites on different hosts is documented. How host specific are, for example, bird mites? How quickly can parasites respond to declines in the abundance of a preferred host? How quickly does host preference evolve or shift?

One final difficulty in studying coextinctions that will ultimately bear on both our understanding of prehistoric coextinctions and on present and future coextinctions is that coextinctions (and, more generally, chains of extinctions) do not occur in a vacuum. Whatever threats put hosts at risk may also directly affect parasites and mutualists. With Pleistocene glaciation events, for example, parasites needed to track both their hosts and their favoured climates. As the example of the limpet *Lottia alveus* makes clear, a parasite or other affiliate may not have the same environmental tolerances as its host. *L. alveus* was able to survive in only a subset of those conditions where its eelgrass host was found. The opposite scenario might also be true, where a generalist (relatively non-host-specific) parasite has a broader environmental tolerance than any one of its hosts. The consequence of these mismatches is that coextinction dynamics may layer on top of climate change dynamics in complex ways that either reduce or increase the likelihood of coextinction events. Throughout the Late Quaternary, glacial–interglacial cycles may have increased the risk to parasites of megafauna by adding an additional risk factor for their survival. Similarly, the global climate change now underway may exacerbate the effects of losses of hosts, select for especially environmentally tolerant parasites (at the expense of less tolerant species), or interact with host–parasite dynamics in other complex ways. Much remains to be learned.

Acknowledgements

The author was supported during this work by an ARC grant, a DOE-PER grant, and a DOE-NICCR grant. Special thanks to Lian Pin Koh, Judie Bronstein, Kevin Gross, and two anonymous reviewers for reading this chapter.

CHAPTER 9

Probabilistic methods for determining extinction chronologies

Ben Collen and Samuel T. Turvey

9.1 Introduction

The global extinction of a species typically represents the end point in a series of population extinctions, during which unique evolutionary history is lost at every stage (Pimm *et al.* 1988; Ceballos and Ehrlich 2002). The disappearance of the final individual of a species represents the irretrievable loss of a unique form of biodiversity (Diamond 1987). In a historical context, investigating the tempo, process, and pattern of extinction can provide us with the means to identify species at high risk, since risk is non-randomly distributed across taxa, and can highlight groups or regions that may be particularly extinction-prone (Pimm *et al.* 1988; Baillie *et al.* 2004; Millennium Ecosystem Assessment 2005). At the palaeontological level, the fossil record has been used to provide insight into subjects as broad as the tempo and mode of evolution, the occurrence of mass extinction events, and changes in Earth systems through time (Kemp 1999; Briggs and Crowther 2001). In combination, modern-day, historical, and palaeontological data can provide an enhanced level of understanding of extinction and its consequences.

However, even modern extinction events are witnessed rarely, if ever, and therefore must be inferred (Diamond 1987). For historic and prehistoric extinctions, there is no option other than to attempt to deduce the timing of extinction of a species from the available evidence. If it is impossible to do this without bias, then correspondingly our understanding of biodiversity patterns and processes will be distorted. Predictions of extinction chronology have been elicited from a range of data, including sightings records, oral history, historical museum records, and the fossil record. All are extremely unlikely to include the last individuals of a species, because species typically become extremely rare before becoming extinct.

Other techniques have been developed for determining whether a species is extinct, but have notable limitations. Extensive survey work is typically used to further investigate recent suspected extinctions. On a practical level, this approach may prove too costly to implement, and the expert opinion involved may have considerable subjective bias (van der Ree and McCarthy 2005), so finding more cost-effective and robust alternatives is appropriate. More fundamentally, absence of evidence for the continued survival of an extremely rare or even merely cryptic species clearly does not automatically constitute evidence of its absence (i.e. extinction). The technical difficulty in discriminating contemporary species extinction therefore lies in answering the question, how long should a species go unseen before we can safely call it extinct (Solow 2005)? Unfortunately the answer is not simple, and relies on the expected rate at which a species would be seen if it were extant. International Union for the Conservation of Nature (IUCN) guidelines state that a species should be considered extinct when 'there is no reasonable doubt that the last individual has died', and that this is reliant on 'exhaustive surveys in known and/or expected habitat, at appropriate times... throughout its historic range' (IUCN 2001). These surveys must take place over a timescale appropriate to the taxon's life cycle, which in practice may rarely occur. Although this is no doubt an improvement on the previously used cut-off which was arbitrarily set at 50 years by the Convention on International Trade in Endangered Species of Wild Fauna and Flora (CITES) (Reed 1996), it remains necessarily ambiguous and open to interpretation.

The problem of identifying contemporary extinction events is analogous to that of estimating time of extinction of a species from its last occurrence in the fossil record. Understanding the true stratigraphic distributions of different fossil taxa constitutes the basis of biostratigraphy, and is integral to the development and investigation of evolutionary and ecological hypotheses in palaeobiology. However, palaeontologists have long recognized that times of first and last occurrence in the fossil record rarely coincide with times of origination and extinction; instead, observed stratigraphic ranges almost certainly represent an underestimate of true range. This palaeontological problem is known as the Jaanusson effect (first occurrence is delayed past the true time of origination) and the Signor–Lipps effect (last occurrence precedes the true time of extinction), and is caused by both temporal and geographical variation in the likelihood of preservation due to the vagaries of the fossil record. These include sampling bias (initial abundance or rarity of target species, as well as subsequent intensity of collection effort), facies bias (target species is ecologically or geographically restricted to particular environments/facies with differing preservational potential), and unconformity bias (sections of stratigraphic record may be absent) (Benton 1994; Foote and Raup 1996; Briggs and Crowther 2001). The most recent palaeontological records for a species of interest can therefore only provide a minimum estimate or *terminus post quem* for an extinction chronology. Although it was long assumed that mass extinctions could be read directly from the fossil record, even such catastrophic events may artefactually appear as gradual and staggered (Benton 1994). Attempting to reconstruct extinction chronologies based on stratigraphically incomplete occurrence data therefore represents part of the wider problem that is faced by all palaeontologists of trying to estimate the incompleteness of the fossil record (see also Chapter 10 in this volume).

Although determining extinction dates from the fossil record has in the past often relied on searching for the youngest reliably dated sample, a range of techniques has been developed to correct for the incompleteness of the fossil record and infer the position of the end point of a stratigraphic range, using approaches such as gap analysis under different models (e.g. random versus non-random fossilization: Strauss and Sadler 1989; Marshall 1994; Marshall and Edwards-Jones 1998; Solow 2003). However, despite some attempts to introduce these approaches into Quaternary palaeobiology (e.g. McFarlane 1999a), most investigations of prehistoric Late Quaternary extinctions still aim only to generate new directly measured 'last occurrence' dates for extinct species through radiometric or other dating techniques, and to match such dates to the time of putative extinction drivers (e.g. prehistoric human impacts). Although multiple dates associated with now-extinct Quaternary species may be available, most researchers still only consider the most recent one to be significant when determining the extinction chronology of the species. Typically, little statistical analysis is provided beyond the measurement error associated with laboratory date estimates. For example, Guthrie (2003) used direct interpretation of radiocarbon ages of Pleistocene fossil remains of woolly mammoth *Mammuthus primigenius* and caballoid horses *Equus caballus* to suggest that these two large mammal species became extinct in Alaska at different times, and that horses disappeared before the arrival of prehistoric human hunters in the region. This hypothesis clearly has important implications for understanding Late Quaternary extinctions. However, Solow *et al.* (2006) and Buck and Bard (2007) independently demonstrated that analysis of the temporal distribution of Alaskan mammoth and horse radiocarbon ages used by Guthrie (2003) did not support this hypothesis, and that although these two extinction events may not have been simultaneous, the possibility that both species survived beyond the arrival of humans cannot be ruled out with confidence.

In attempting to infer recent extinction, conservation biologists are facing similar problems. Methods are required that can reliably infer whether or not a taxon is extinct from very limited data sets, among data that are inconsistent in effort and in taxon population size. A number of techniques have been proposed, principally by Solow (see Solow 2005), in an attempt to make the current qualitative process of classifying species as extinct more quantitative. In estimates of both contemporary and prehistoric extinction, the development of probabilistic methods provides a range of tools with which to assess extinction chronologies. Given the range of data sources of species occurrences, what do they tell us about the timing of extinction and, indeed, about ancient anthropogenic species introductions? Here, we review the probabilistic methods that have been developed, to provide an overview of the possible alternatives, and their strengths and weaknesses.

9.2 Available methods

A number of different statistical methods are available to infer true stratigraphic ranges from observed fossil distributions, and to infer extinction from sightings records. Some of the possible approaches are discussed and compared below. The methods presented are not intended to be a complete review of the possible techniques; rather they represent the key types of analytical tools available.

9.2.1 Fossil data

9.2.1.1 Uncertainty periods

Bover and Alcover (2003, 2008) have suggested that simple 'uncertainty periods' can delimit the temporal range within which an extinction event occurred (i.e. a *terminus post quem* pre-dating the extinction, and a *terminus ante quem* or 'capping layer' post-dating the extinction), based on the extreme values of the 2σ intervals of ^{14}C ages. A narrow uncertainty period for a given extinction event therefore implies a high precision for the extinction date estimate.

However, a major methodological constraint with this technique is that it combines positive evidence (known 'presence' data) with negative evidence (presumed 'absence' data), and requires *a priori* knowledge that the target species is definitely absent from the horizon used to establish the *terminus ante quem* for the uncertainty period. Unfortunately, whereas presence data may be reliably interpreted as reflecting genuine occurrence at a particular time interval (if dating methods are robust), we have far less confidence for absence data, because these may represent either genuine absence or alternatively an artefact caused by the incompleteness of the fossil record during an interval when the target species was actually still present. Site-based studies of conformable sequences containing extinct target species only in older horizons below a well-dated capping layer have been used to estimate several Late Quaternary extinction chronologies (e.g. estimating moa extinction in New Zealand based on chronology of human occupation layers in Monck's Cave, South Island, which lack evidence of moa consumption; Holdaway and Jacomb 2000), but the data may in actual fact only reflect local extirpation rather than global extinction (e.g. in the New Zealand example, moa may have survived for longer in remote refugia such as Fiordland, where other endemic species are known to have persisted; see Anderson *et al.* 2000). It may therefore only be possible to suggest, at most, that a target species is probably (rather than definitely) absent from a stratigraphic section at a particular site, rather than making wider-scale conclusions about extinction chronologies. It is therefore preferable to restrict analysis of extinction dates to positive evidence (known presence data) only. These theoretical problems are comparable to those levelled against stratocladistics (e.g. Heyning and Thacker 1999).

9.2.1.2 Springer–Lilje method

The most straightforward methods for attempting to determine the end point of a stratigraphic range using positive presence data assume that fossilization is random, and base their estimates on properties of the size of the internal gaps between records in the observed range. For a series of n records there will be $n-1$ measurable gaps. Various methods also assume that the distribution of these internal records is exponential (ln (0.05)). Springer and Lilje (1988) observed that, if these assumptions are followed, the stratigraphic interval outside the youngest and oldest observed records that is expected to include the true start and end points is equivalent to the confidence interval, x, of the known stratigraphic range:

$$x = \ln P(x)/(-m) \tag{9.1}$$

where P is the specified confidence level for x, and m is the mean number of gaps per year. Assuming a standard confidence interval of 95%, the lower limit for the extinction date D_{ext} is given as:

$$D_{ext} = D_{min} - \{[\ln(0.05)]/-[(N-1)/D_{max} - D_{min}]\} \quad (9.2)$$

9.2.1.3 Median gap method

For an exponential distribution of random stratigraphic records, McFarlane (1999a) observed that the median gap length, i, is equivalent to the 'half-life' of the exponential curve. Ninety five per cent of all of the gaps are therefore less than $4.32i$, and so the 95% confidence interval expected to include the gap between the youngest observed date and the true end point of a stratigraphic range must have its lower limit at:

$$D_{min} - 4.32i \quad (9.3)$$

9.2.1.4 Strauss–Sadler method

An alternative method for calculating confidence limits on the length of stratigraphic ranges was developed by Strauss and Sadler (1989), again assuming that fossilization is random, but this time assuming a non-exponential distribution for gap lengths. The length of the confidence interval for the estimated end point is calculated as a fraction of observed range. This is given as:

$$r_c = \alpha R \quad (9.4)$$

where r_c is the length of the confidence limit, and R is the observed range. α, assuming a random distribution of fossils, is dictated by the number of fossil horizons in the range H and the confidence level C. Therefore it can be assumed that the estimated end point of the taxon is within distance r_c of the observed end point, with a confidence of C:

$$\alpha = \left[(1-C)^{1/(H-1)} - 1\right] \quad (9.5)$$

9.2.1.5 Solow method

Solow (2003) developed a different method as an extension of the approach first proposed by Strauss and Sadler (1989). Whereas the previous three statistical methods have assumed that fossilization is random across the temporal range of a species, there are many reasons to believe that that the density of fossil finds may not remain constant in the majority of cases. Solow's method recognizes that the confidence limits of the point estimate can be sensitive to this assumption, and instead it allows for dependence between gap length and stratigraphic position.

For the purposes of explanation, $X_1, X_2 \ldots X_n$ are the stratigraphic series of fossil finds, ordered from oldest to youngest between the period 0 to X. To detect the extinction date for any taxon record series, we want to estimate the upper (U) boundary from the lower (L). Ideally, we would estimate U from L, but in practice, because the lowest stratigraphic record is not really the first occurrence of the species, U is estimated from X_1; that is, from the oldest known record.

Solow (2003) demonstrated that a decline in the mean density of finds towards the upper boundary causes an underestimation of the true value, and therefore the calculated confidence limit is smaller than it should be. At greater rates of decline, this observation becomes more pronounced. One solution is to use a jackknife approach, with the point estimate of the upper bound suggested by Robson and Whitlock (1964) as:

$$U = X_n + (X_n - X_{n-1}) \quad (9.6)$$

with the confidence interval for U calculated as:

$$\left[X_n + \frac{1-\alpha}{\alpha}(X_n - X_{n-1})\right] \quad (9.7)$$

Other methods for calculating confidence limits on the end points of stratigraphic ranges are also available based on any continuous distribution of gap sizes rather than just those generated by random processes (e.g. Marshall 1994, 1997). However, these methods require richer fossil records than if random fossilization is assumed.

9.2.1.6 Bayesian approaches to radiometric data

The previous statistical methods all rely on the availability of specific time points on which to base estimates of true stratigraphic range, but this poses a serious problem for studying prehistoric Quaternary extinctions based on radiometric data. Calibration is essential in radiocarbon dating, because the amount of radioactive ^{14}C in the

Earth's atmosphere is known to have varied over time. However, calibrated radiometric dates are not simple single central-point estimates, but instead typically represent complex, highly non-normal, multimodal, and non-continuous high-probability distributions (Fig. 9.1). Although intercept-based methods of generating a point estimate are very popular, other methods are statistically preferable; however, the complex shape of a calibrated radiometric probability density function cannot be adequately described by a single parameter, and this full distribution should be used wherever possible (Telford *et al.* 2004).

Different theoretical approaches are therefore required to infer extinction dates on the basis of radiometric data, but these once again rely on assumptions about underlying processes such as fossilization rate. Methods such as those implemented by Holdaway *et al.* (2002a, 2002b) and Buck and Bard (2007) assume that the sample events can be described as a constant-rate Poisson process; that is, discrete random temporal events of relatively low temporal frequency. Sample-based Bayesian inference is then used to compute the distribution of parameters. An in-depth explanation of the phase model implemented by Holdaway *et al.* (2002a, 2002b) was provided by Nicholls and Jones (2001). Such models have been used to estimate not only the date of final extinction but also the duration and timing of the preceding 'extinction phase', but great care must be taken to interpret model data accurately, and it may not be possible to infer more complex information about the dynamics of extinction events other than the last occurrence date of the target species.

9.2.2 Historical data

Other statistical methods are available to test the likelihood of extinction of a taxon using more contemporary data such as collection specimens or sightings data, although the processes and problems involved share many similarities. Determining the fate of rare and cryptic species can be problematic, and relies on the expected rate at which a species would be seen if it were extant. Two general frameworks are possible:

1. testing whether or not we can reject the null hypothesis that a species is extinct;

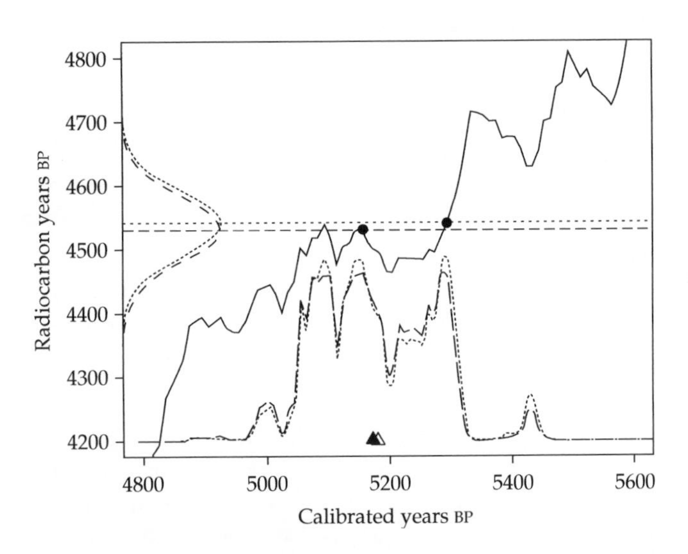

Figure 9.1 Comparison of intercept and weighted average methods for 4530±50 (dashed) and 4540±50 (dotted) ^{14}C years BP. The horizontal lines show the radiocarbon date, with normally distributed errors shown against the *y* axis. The average intercept of each date is marked by a circle and they differ by 138 years. The probability density function for each date is shown along the *x* axis. The weighted averages are marked by a filled and an open triangle for 4530 and 4540 ^{14}C years BP respectively and differ by 8 years. From Telford *et al.* (2004); reproduced with permission from the authors and SAGE Publications.

2. providing a point or interval estimate for a date of extinction.

There are two main possibilities for choosing a model to investigate historical data: parametric and non-parametric methods. Most parametric methods assume the pre-extinction sighting rate is constant over time. If the sighting rate declines prior to extinction (e.g. due to a decline in the abundance of the species in question), then these methods will be incorrect, and will result in a smaller confidence interval than expected. It is possible to model alternative rates of decline (e.g. exponential; see Solow 1993), however, justifying any given parametric model is problematic. We therefore restrict our discussion to non-parametric estimators (see also Solow 2005).

9.2.2.1 Non-parametric estimators

The simplest method which can be applied in this case is given in equation 9.6, with the same confidence interval (equation 9.7). Solow and Roberts (2003) have detailed a P value as a test for extinction:

$$p = \frac{X_n - X_{n-1}}{X - X_{n-1}} \tag{9.8}$$

With this method, the confidence limit is generally much wider than the corresponding parametric version, which makes the technique less useful in inferring extinction chronology for historical species data. Other possible techniques which utilize the same statistical properties of the upper tail of the distribution of sightings include likelihood (e.g. Smith and Weissman 1985) and minimum distance approaches (e.g. Hall and Wang 1999), but utility is restricted due to the non-existence of the upper bound. Therefore, a better option might be optimal linear estimation (Cooke 1980).

The optimal linear estimate of time of extinction, T_E, based on the k most recent sightings, has the form:

$$\hat{T}_E = \sum_{i=1}^{k} w_i t_{n-i+1} \tag{9.9}$$

The weight vector $w = (w_1, w_2, \ldots, w_k)$ is given by:

$$w = \left(e'\Lambda^{-1}e\right)^{-1} \Lambda^{-1}e \tag{9.10}$$

where e is a vector of k 1s and Λ is the symmetric k-by-k matrix with typical element:

$$\Lambda_{ij} = \frac{\Gamma(2\hat{v}+i)\Gamma(\hat{v}+j)}{\Gamma(\hat{v}+i)\Gamma(j)} j \le i \tag{9.11}$$

where Γ is the gamma function and

$$\hat{v} = \frac{1}{k-1}\sum_{i=1}^{k-2} \log \frac{t_n - t_{n-k+1}}{t_n - t_{i+1}} \tag{9.12}$$

is an estimate of the shape parameter of the Weibull extreme value distribution. An approximate P value for testing extinction is given by:

$$P = \exp\left(-k\left(\frac{T - t_n}{T - t_{n-k+1}}\right)^{1/\hat{v}}\right) \tag{9.13}$$

Finally, under the assumption that the species is extinct, the upper bound of an approximate $1-\alpha$ confidence interval for T_E is:

$$T_E^u = \frac{t_n - c(\alpha)t_{n-k+1}}{1 - c(\alpha)} \tag{9.14}$$

where:

$$c(\alpha) = \left(\frac{-\log \alpha}{k}\right)^{-\hat{v}} \tag{9.15}$$

Variations in the sighting rate reflect variations in both how easy the species is to see, and monitoring effort. The main assumption of the technique is that while sighting effort may vary, it never falls to zero, particularly around the time of extinction (Roberts and Solow 2003; Solow 2005).

9.3 Strengths and limitations of probabilistic analysis

For a small number of closely observed species, the last recorded sighting may closely reflect the inferred true date of extinction (Table 9.1). However, in determining extinction chronologies, we must face the limitation that neither the first nor last individual in the recorded time series under consideration represents the true start or end point of that taxon, and that these discrepancies between observed and underlying data may be considerable. Because inference of patterns of taxonomic persistence, extinction timing, and extinction rate is integral both for understanding biodiversity change over time and for constructing predictive frameworks for conservation science (e.g. see Purvis *et al.* 2000b), it is therefore necessary to be able to make robust assumptions from the data available about hypothesized end points or extinction dates. Clearly, different methods for making such conclusions are appropriate for different types of data and different time series. Fig. 9.2 details a conceptual time frame through which both the analytical framework and level at which the analysis is carried out might change. The time frame of different types of data therefore dictates the types of questions that it is possible to ask and the types of analytical tools that might be used.

Of the estimated 10–15 million species on the planet (May 1988), most are not vertebrates, yet vertebrates (and indeed the further restricted subset of birds and mammals) provide most of the tests of determining extinction chronologies, at least from the recent past. Given that the data requirements for newly developed techniques such as optimal linear estimation are not large, this need not be the case. Increasing the breadth of test species and population demographies is important; those which have

Table 9.1 Estimated extinction dates calculated using optimal linear estimation (Solow 2005) for Steller's sea cow *Hydrodamalis gigas* and the passenger pigeon *Ectopistes migratorius*, two extinct species that were relatively closely observed by hunters and/or collectors during the period of their final decline.

Species	Last confirmed record	Sightings/collection data	Extinction date estimate	95% Confidence interval	*P*
Hydrodamalis gigas	1768	Stejneger (1887), Gibson (1999)	1771	1768–1781	<0.001
Ectopistes migratorius	1900	Schorger (1955; all accepted sightings from 1887 or later listed as 'late records')	1902	1900–1906	<0.001

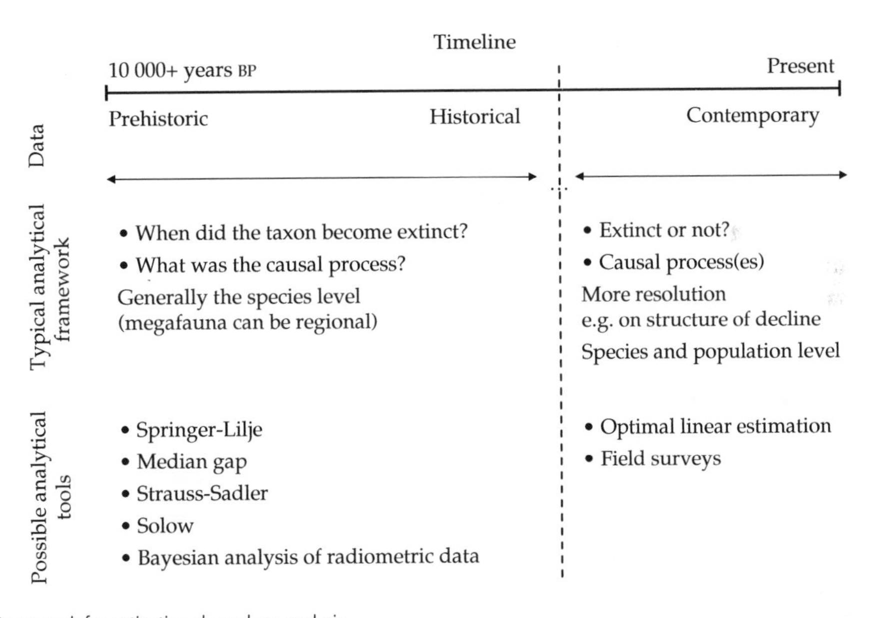

Figure 9.2 Framework for extinction chronology analysis.

been assessed to date may not be typical, so a gain in precision of estimates may be at the loss of generality of findings.

However, all of the probabilistic methods described above possess several important limitations which prevent truly meaningful inferences to be made about real extinction dates. For example, they are inevitably affected by variation in taxonomic coverage over both time and space, which is a problem for analyses that use either historical or palaeontological datasets. All of the described methods rely on knowing (or at least make assumptions regarding) how sighting/occurrence rates vary over time. These parameters are extremely difficult in practice to estimate, as the processes by which data are generated are likely to vary stratigraphically, geographically, temporally, between taxa, and often in a complex manner. Further progress requires a greater understanding of these variations, and whether they can be quantified and controlled for.

Limitations of the fossil record have been widely commented on in the scientific literature (for further discussion see Chapter 10 in this volume). Of particular concern for attempting to identify true extinction dates from stratigraphic occurrence data are underlying biases depending on whether palaeontological sampling methods (both fossil preservation and subsequent collection) are continuous or discrete. Discrete sampling can lead to an underestimation of the true richness of the fossil record, and a meaningful understanding of the use of palaeontological data for inferring extinction dates requires quantitative measurement of these biases; however, the distinction between continuous and discrete sampling is not necessarily clear cut, and this remains a substantial problem (Marshall 1994, 1997; Foote and Raup 1996).

The issues with historical sightings data are similar, but less widely discussed. One of the fundamental assumptions that underlie the techniques used to assess sightings records is that sighting effort may never decline to zero (Solow 2005). This is difficult to assess for any given species, but would be violated if, for example, early collection expeditions that provide much of the historical sightings data in fact consisted of periods of effort interspersed with no effort (which is likely to represent the real sampling situation for many geographical regions and taxonomic groups).

In order to generate meaningful extinction probabilities from recent sightings data, it is also necessary to assess not only data quantity (the number of data points needed to make a robust estimate, and timings of historical sightings) but also data quality. Whereas data quality in the fossil record is influenced primarily by specimen preservation and dating error, both of which can be quantified to some extent, determining the validity of historical records is far from straightforward, and there is rarely any objective way to distinguish true sightings data from misidentifications or hoaxes. This problem with using historical data at face value for such analyses may be intractable in many cases. To use a notorious example, probabilistic methods can tell us nothing new about the possible survival of the thylacine or Tasmanian tiger *Thylacinus cynocephalus*, for which many hundreds of unverified and unverifiable eye-witness sightings have been made subsequent to the death of the last known individual in 1936 (e.g. Heberle 2004), or indeed whether the supposed amateur sighting of a Yangtze River dolphin *Lipotes vexillifer* that was reported after the international announcement of the probable extinction of the species (Turvey *et al.* 2007b) is genuine or otherwise. This problem has been termed the 'thylacine effect' by MacPhee and Flemming (1999).

Similarly, optimal linear estimation has been used to suggest that the timing of mid-twentieth century sightings of the ivory-billed woodpecker *Campephilus principalis* provides independent confirmation of its continued survival into the twenty-first century (Roberts 2006). However, although many mid–late twentieth century sightings were reported by ornithologists and other competent observers prior to the apparent recent rediscovery of the species, the last generally acknowledged North American sighting was made in 1944 (Jackson 2004; Fitzpatrick *et al.* 2005; Gallagher 2005). In the optimal linear estimation analysis, one of the post-1944 sightings was also included. Clearly discerning between data has major repercussions on the conclusions of extinction studies, and indeed a previous analysis (Roberts and Saltmarsh 2006) using a different sighting series suggested that the ivory-billed woodpecker was probably already extinct by 1969. Such sightings can never be categorically verified or discredited, but they can considerably confuse any understanding of true extinction chronologies. This problem becomes more tractable for specimen data from museums and herbaria as identification can be checked, although of course these are not without problems either (e.g. the Meinertzhagen catalogue in the Natural History Museum at Tring, UK; Rasmussen and Prŷs-Jones 2003; Dalton 2005). This greatly limits the utility of such techniques for determining the statistical likelihood of an extinction event.

It is plausible that as current lists of extinct species are based on length of time since last sighting, frequent rediscoveries of reportedly extinct species can undermine conservation efforts (e.g. Roberts 2006). In reality, such lists take much greater information into account than purely time since last sighting (see IUCN 2001; Butchart *et al.* 2006), and require, among other things, exhaustive surveys that are appropriate to the taxon. In some cases, identifying extinctions based solely on the frequency and timing of sighting records can yield misleading results. Numerous species have been rediscovered after an absence of records spanning many decades, simply because they were inadequately searched for, and probabilistic analysis would conclude that these species were already extinct (Table 9.2). The discovery of new species (i.e. discovery of new populations rather than taxonomic splitting or shifting species concepts), even from well-known taxa such as primates (e.g. *Lophocebus kipunji*: Jones *et al.* 2005; *Macaca munzala*: Sinha *et al.* 2005), illustrates the difficulty of complete survey work. None of the above methods can be applied in cases where very few records exist; for example, 62% of the 34 extinct amphibian species listed by IUCN are only known from the type specimen or series (IUCN 2006). Here, the problem is intractable. However, one further component which could also be incorporated into comparisons of extinction chronologies is one of geography. For example, the rediscovery of the New Caledonian owlet-nightjar *Aegotheles savesi* in the remote Rivière Ni valley of south-east New Caledonia 118 years after the only other known record of the species (Tobias and Ekstrom 2002) should come as less of a surprise than if the only global records for this species over this time period were from Regent's Park, London. None of the sighting methods explicitly allow for this, although it would make interspecific comparison considerably more meaningful.

The problem of identifying potentially extinct species is compounded further if a species is cryptic. Scientists are sensibly reluctant to state with certainty if a species is extinct, so as not to facilitate the Romeo effect (giving up on a species too early; Collar 1998) or the Lazarus effect (bringing species back from the dead; Wignall and Benton 1999; Keith and Burgman 2004) (Fig. 9.3). In essence, scientists must decide how long that a species should be expected to remain undetected before it can be confidently interpreted as extinct. The ramifications of making the wrong decision can be far-reaching. For example, the recent apparent rediscovery of the ivory-billed woodpecker in the Big Woods region of Arkansas, USA (Fitzpatrick *et al.* 2005) led to

Table 9.2 Estimated extinction dates calculated using optimal linear estimation (Solow 2005) for three poorly known vertebrate species that have been rediscovered after long time periods when they were unrecorded by scientists.

Species	Last confirmed record before rediscovery	Sightings/collection data	Extinction date estimate	95% Confidence interval	*P*	Date of rediscovery
Forest owlet (*Heteroglaux blewitti*)	1884	Rasmussen and King (1998)	1886	1884–1901	0.004	1997
Bornean bay cat (*Catopuma badia*)	1928	Mohd-Azlan and Sanderson (2007)	1960	1931–2294	0.195	1972
Barbara Brown's titi monkey (*Callicebus barbarabrownae*)	1929	Marinho-Filho and Verissimo (1997)	1933	1929–1956	0.004	1990

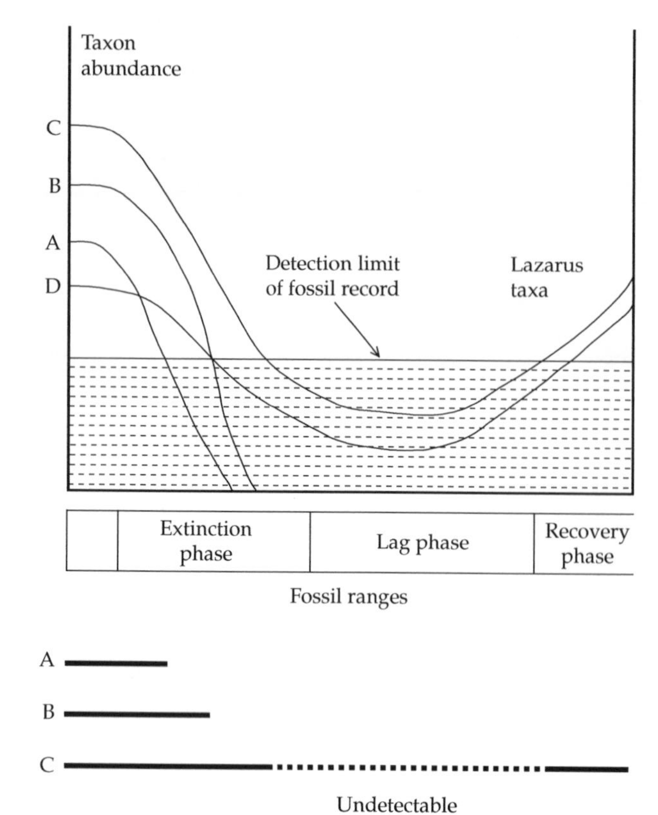

Figure 9.3 Changes in population abundance of four hypothetical taxa (A–D), demonstrating the Lazarus effect. Taxa A and B decline rapidly in abundance and ultimately become extinct, whereas taxa C and D become very rare (and are therefore unlikely to be collected as fossils), but they remain extant. Taxa C and D only 'reappear' in the fossil record when they have once again become sufficiently abundant to cross the detection limit threshold. Redrawn from Wignall and Benton (1999); reproduced with permission from the authors and the Geological Society, London.

the immediate allocation of US$10 million by the US Department of the Interior and Department of Agriculture, for projects to conserve the bird and its habitat (Wilcove 2005). Conservation funds are, understandably, not targeted at species thought to be extinct (Butchart *et al.* 2006), and there are only limited funds available to be divided. Whereas the current precautionary approach towards the classification of extinction is reasonable given the likely repercussions of getting it wrong, underestimating the number of extinct species will nevertheless bias analyses of extinction. It is therefore appropriate that techniques which offer a more objective and quantitative assessment of extinction are tested.

It is also important to recognize that as species become rare during population decline, abundance does not neatly decrease across the entire species range while retaining the original extent of their historical distribution; rather, ranges contract and fragment, according to patterns which remain complex (Channell and Lomolino 2000a, 2000b). The expected form of contraction is likely to vary under different threat processes (Rodriguez 2002). In general, palaeontological data do not afford the resolution or biogeographic coverage needed to reconstruct decline patterns, although there are some rare exceptions where more complete reconstructions might be possible regionally, such as for continental megafauna (e.g. mammoth or the giant deer *Megaloceros giganteus*) in northern Eurasia (MacPhee *et al.* 2002; Stuart *et al.* 2004). Only certain glacial refugia can be reconstructed, and insufficient knowledge can lead to overly simplistic conclusions to be drawn about post-glacial recolonization events (Sommer and Benecke 2004). However, data requirements for such reconstructions are high: for example, Sommer and Benecke (2004) estimated that around 300 subfossil records per species are needed for European mustelids. Whereas historical occurrence/sighting data are typically available at a much greater spatial resolution, they are usually not spatially segregated for probabilistic analysis but treated as though they originate from a single panmictic population. This may have important implications for the interpretation of inferred extinction dates if rare species actually consist of fragmented subpopulations that decline and disappear independently of one another at different rates and times.

9.4 Synthesis

All of the proposed statistical approaches for inferring true extinction dates from observed stratigraphic or historical data represent a major conceptual step forward from the simplistic and inaccurate, but still largely ubiquitous, approach of assuming that the last known record of a target species is equivalent to its last occurrence. Although such methods are becoming increasingly available, considerable opportunity remains for new research to test the assumptions and properties of the different theoretical approaches, and which of these methods are more appropriate to use than others in different situations. Broader taxonomic and geographical coverage, and investigation of different inferences such as sampling potential over time in different natural and non-natural (e.g. zooarchaeological) sites and more recent historical-era contexts, is therefore essential.

However, it is also essential to recognize that even following the robust inference of an extinction date, it is then necessary to interpret this event correctly and incorporate it into a meaningful and accurate wider model of the extinction process. This requires correlation of the extinction event with one or more potential extinction drivers, but identifying cause and effect in prehistoric or historical interactions is also rarely a straightforward process (see also Chapter 10 in this volume). Indeed, the relationship between inferring an extinction date and reconstructing the dynamics of a past extinction event is in many ways analogous to the relationship between constructing a cladogram and reconstructing a family tree and an evolutionary scenario (Eldredge 1979; Kemp 1999); both processes require the generation of increasingly complex theories that require increasing numbers of additional assumptions.

It is also necessary to acknowledge that although some extinct taxa (e.g. mammoths, giant deer, moa) have been the subject of considerable radiometric or other investigations aimed at clarifying the timing of their disappearance, most prehistoric Late Quaternary species still have too few definite dates associated with them to allow meaningful statistical analysis, so extinction estimates remain limited to simplistic use of *terminus post quem* 'last occurrence' dates (e.g. Turvey *et al.* 2007a). Indeed, almost a third of the 255 mammal species listed elsewhere in this volume as having probably become extinct during the Holocene have no direct or indirect associated dates at all, and their recent prehistoric survival is inferred on the basis of their co-occurrence with associated modern faunas and/or the late recorded regional first arrival of human colonists. This problem is more acute for smaller vertebrate species, which have traditionally been understudied. Further understanding of extinction chronologies using statistical inference must therefore also be accompanied by further direct palaeontological investigations into Late Quaternary faunas, about which we still have much to learn.

CHAPTER 10

The past is another country: is evidence for prehistoric, historical, and present-day extinction really comparable?

Samuel T. Turvey and Joanne H. Cooper

History has many cunning passages.

Gerontion, T.S. Eliot

10.1 Introduction

Human activity has profoundly altered global patterns of biodiversity and ecosystem structure and function from the Late Pleistocene onwards, and different sources of data must be used to investigate anthropogenically driven species losses and wider-scale ecological changes over this protracted period. Our understanding of pre-human Late Quaternary ecosystems and the timing and causation of prehistoric human-induced extinction events is based almost entirely on studies of the recent fossil record, with additional information also available from archaeological sites and early historical accounts of now-extinct species. These data from the past have the potential to reveal the full extent of human impacts on global systems over time, providing an invaluable, unique long-term perspective for understanding modern extinction processes, making meaningful predictions about the dynamics and rates of extinction in contemporary systems, and generating a new level of emergent information through meaningful comparisons between species losses in past and present communities.

The recent fossil record can in many cases provide a remarkable level of information about the biology and ecology of extinct species. Whereas exceptional fossil preservation is known from throughout the Phanerozoic in Konzentrat-Lagerstätten and Konservat-Lagerstätten (Briggs and Crowther 2001), the Late Quaternary record can frequently provide novel sources of data that are only rarely available in older deposits. For example, information on skeletal ontogeny and maturation is available from annual growth marks in cortical bone not only for many species of giant moa of New Zealand (Turvey *et al.* 2005) but also for a range of Mesozoic avian and non-avian dinosaurs (Chinsamy-Turan 2005). However, a wealth of additional data are also available on moa soft tissue and plumage, diet (including food remains in preserved gizzards), population structure, wider-scale habitat requirements, and possible ecological interactions and coevolutionary systems, primarily from material preserved in swamp sites and alpine caves (Worthy and Holdaway 2002). A further important consideration is that much of the Late Quaternary record consists of subfossil rather than fully fossilized material, which still retains some of its original organic fraction (although some authors contest the use of this term; e.g. Steadman 2006b). Analysis of subfossil samples can therefore permit both radiocarbon dating and other dating techniques such as amino acid racemization, and phylogenetic analysis using ancient DNA or other biomolecules. The oldest known ancient DNA sequences have been extracted from deep ice cores in southern Greenland between 450000 and 800 000 years old, dating from the mid–Late Pleistocene (Willerslev *et al.* 2007).

Analysis of the rich Late Quaternary fossil or subfossil record has been used to generate substantial new information on patterns and processes of human-driven extinctions in the recent past, which has significant implications for understanding ongoing species losses. Wider-scale analysis of the intrinsic and extrinsic correlates of past extinctions can provide novel insights into patterns of ongoing biodiversity loss, permitting a more inclusive assessment of human impacts, and informing the optimal prioritization of finite conservation resources. The development of mathematical models to investigate the dynamics and drivers of Late Pleistocene and Holocene extinctions has been permitted by the availability of robust data on Quaternary ecosystems and the biology of living relatives (e.g. Mithen 1993; Holdaway and Jacomb 2000; Alroy 2001; Johnson 2002). In some cases, considerable further information on past levels of human hunting is also available from archaeological sites and early historical records. For example, Turvey and Risley (2006) developed a VORTEX model incorporating eighteenth century Russian hunting data to show that Steller's sea cow *Hydrodamalis gigas* was hunted at over seven times the sustainable limit, with overexploitation therefore more than sufficient to exterminate the species without having to evoke a further human-driven collapse of shallow-water kelp ecosystems (Fig. 10.1). The Late Quaternary record is also increasingly being incorporated into conservation planning and decision-making, with data on the faunal and ecological composition of pre-human systems framing the problems of maintaining biodiversity and landscape health in a historical context and indicating optimal goals for ecosystem regeneration (e.g. Bowman 2001). Fossil evidence for prehistoric Holocene occurrence of species, such as the Eurasian eagle owl *Bubo bubo* and the pool frog *Rana lessonae* in the UK, and wider pre-human distributions of other endangered species (such as the presence of Holocene remains of kakapo *Strigops habroptilus* and takahe *Porphyrio hochstetteri* in lowland regions of mainland New Zealand as well as the Southern Alps), are now being integrated into conservation management plans (Bunin and Jamieson 1995; Beebee *et al.* 2005; Stewart 2007a; Carter *et al.* 2008; see also Chapter 13 in this volume). More ambitiously, the proposed 'Pleistocene re-wilding' of large vertebrates to North America, Eurasia, and other regions to restore ecosystem services provided by regionally or globally extinct species during the Late Quaternary has been described as 'an alternative vision for twenty-first century conservation biology' (Donlan *et al.* 2005; Zimov 2005; Caro 2007).

The apparent documentary quality of the fossil record is often interpreted at face value by biologists and conservationists attempting to extend the time frame of observations that are available from the modern era. For example, data on prehistoric extinction rates obtained from the fossil record are frequently considered together with present-day and predicted future extinction rates in analyses of changing patterns of species losses over time (e.g. Millennium Ecosystem Assessment 2005). However, palaeontological data (measured over geological or evolutionary time) and neontological data (measured over ecological time) are fundamentally different in many important regards. In particular, whereas neontological research (as with all scientific endeavour) is also forced to rely on incomplete data, the fossil record—even the relatively well-preserved Late Quaternary record—is notoriously incomplete (e.g. Darwin 1859; Kidwell and Flessa 1995; Briggs and Crowther 2001). Models incorporating such palaeontological information therefore run the risk of overshadowing the data on which they are based, and undermining their potential to be useful (Steadman 1995).

Kemp (1999) identified four major categories of incompleteness in the fossil record: organismic incompleteness, the failure of most attributes of most organisms to fossilize; ecological incompleteness, the failure of fossils and fossil assemblages to accurately reflect the wider ecological context in which extinct species lived; stratigraphic incompleteness, the failure of organisms to be preserved as fossils across all of their stratigraphic range; and biogeographical incompleteness, the failure of organisms to be preserved as fossils throughout their former geographical range. Ecological incompleteness can be further subdivided into autecological incompleteness, or the inability to accurately reconstruct the lifestyle and interpret the adaptations

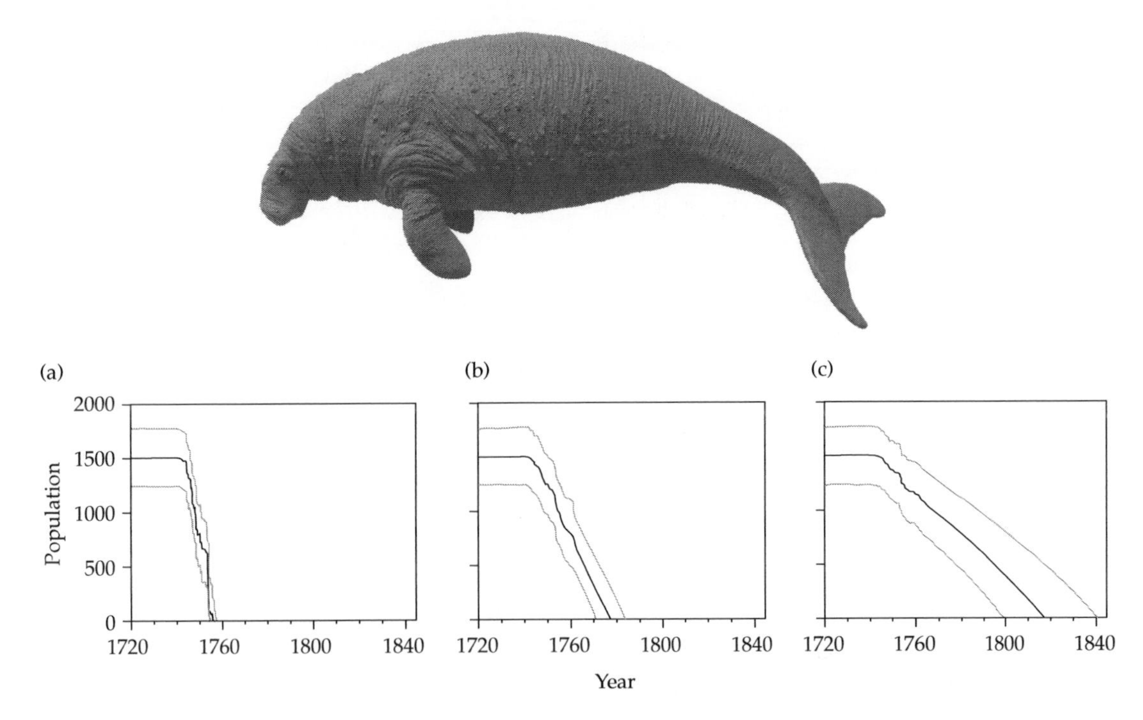

Figure 10.1 Top: Steller's sea cow *Hydrodamalis gigas*. Bottom: projections of Bering Island sea cow population following its initial discovery in 1741 under different hunting regimes. Black line, median number of sea cows; grey lines, 2.5 and 97.5% percentiles. (a) Estimated real level of hunting based on historical records; (b) predicted population survival in the absence of provisioning by sailors for ongoing journeys; (c) predicted population survival in the absence of wasteful hunting (records suggest that five times as many animals were killed as were actually utilized). The real Bering Island sea cow population became extinct by 1768. From Turvey and Risley (2006).

of extinct species which lack living relatives, and synecological incompleteness, the failure of preserved fossil communities to match the living communities they represent. Furthermore, few studies are able to address the full course of an extinction event from first population decline to death of the last individual of a species because an epistemological gap (see Kemp 1999) exists between neontological and palaeontological data. Prehistoric extinctions are typically informative only about temporal, geographical, and taxonomic patterns of extinction, rather than the build-up to these events. Extant threatened species can instead reveal factors responsible for declines and general patterns of range collapse, but are typically uninformative about duration of decline or potential persistence of remnant populations. Historical data can potentially reveal both last occurrence dates for recently extinct species and information about drivers, speed, and magnitude of declines, but in practice are rarely able to do so as the data are usually inadequate.

A particularly illustrative example of the problems faced when trying to interpret the recent fossil record and even the historical record from a neontological perspective is provided by the thylacine or Tasmanian tiger *Thylacinus cynocephalus*, which formerly occurred on mainland Australia and Tasmania and was the only Late Holocene representative of an entire family of large-bodied carnivorous marsupials only distantly related to other dasyuroids. Although this species was last reliably reported from Tasmania as recently as 1936, was only officially declared to be extinct 50 years later in 1986 (IUCN 2006), and has been the subject of considerable scientific and popular interest, remarkably little is understood about its ecology or disappearance: in fact 'pathetically little is known about the biology of thylacines' (Guiler 1985; Paddle 2000). Insights into the hunting behaviour, reproduction, family dynamics, range size, and other key biological characteristics of this unusual mammal are limited to a handful of anecdotal nineteenth century observations made by trappers and farmers rather than trained zoologists. Although the thylacine and the placental wolf *Canis lupus* are widely regarded as a classic example of convergent evolution, these two species instead appear to have had fundamentally different morphological adaptations for locomotion and carnivory, which are still becoming understood (Jones and Stoddart 1998; Wroe *et al.* 2007). The dynamics and timing of the thylacine's extinction are also surrounded by controversy (see Bowman 2001 for further discussion). The introduction of the dingo *Canis lupus dingo* by around 3500 years BP has frequently been blamed for the thylacine's prehistoric disappearance from mainland Australia, but alternative hypotheses based on Late Holocene Aboriginal cultural 'intensification' have also been proposed (Johnson and Wroe 2003; see also Chapter 2 in this volume), and nineteenth century historical records have been interpreted to suggest that the species may have survived on the mainland as late as AD 1830–1840 (Paddle 2000). Recent mathematical models apparently indicate that the species could not have been driven to extinction on Tasmania by recorded levels of human hunting, the standard explanation for its disappearance (Bulte *et al.* 2003; Fig. 10.2), and it remains difficult to evaluate the veracity of the numerous unsubstantiated post-1936 thylacine reports from both Tasmania

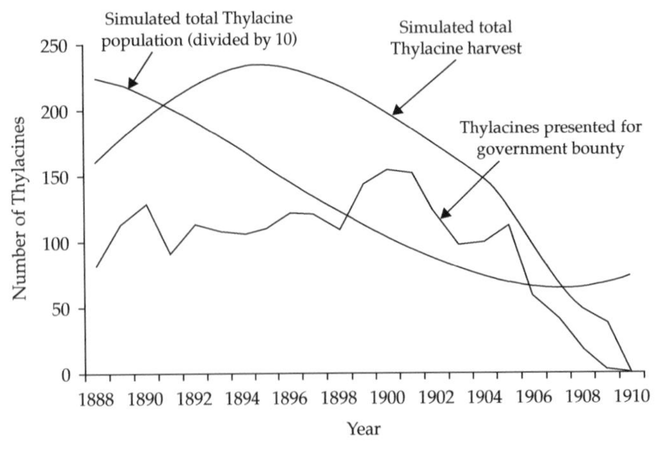

Figure 10.2 Top: the thylacine *Thylacinus cynocephalus*. Bottom: bioeconomic model based on known numbers of thylacines presented for bounty payments, with simulated total thylacine harvest and thylacine population over time. The numerical model estimates an 'all time low' level of abundance of about 632 thylacines, and Monte Carlo simulations suggest that *p*(thylacine extinction) = 0.0004 for a minimum viable population of 100 animals. From Bulte *et al.* (2003).

and the Australian mainland (e.g. Sharland 1940; Rounsevell and Smith 1982; Guiler 1985; Anon 1990; Heberle 2004). Additional Holocene fossil and archaeological records of thylacines are also known from New Guinea, but the taxonomic status of this material remains uncertain, and it may even represent a new, undescribed species (Dawson 1982).

As a result of these concerns over the quality of the fossil and historical records and the degree to which they can usefully be compared with information from modern systems, several researchers have urged caution when combining data from the past and the present. For example, Bennett and Owens (2002) identified species concepts, extinction concepts, methods of inferring the cause of extinction, the survival of evidence, and the biogeographical region of study as five sources of systematic variation between the prehistoric, historic, and contemporary periods that could fundamentally bias the way that extinction estimates were calculated using different sources of data from across the Late Quaternary. It is therefore imperative to give further consideration to the qualitative and quantitative differences in data that may impede our understanding of Quaternary diversity, system-level processes, and extinction dynamics, and the extent to which we can actually characterize changes in Late Quaternary ecosystems over time. This is the necessary first step before we can understand the extent to which pre-human community composition and patterns of prehistoric ecological collapse can be reconstructed, and before we can meaningfully use the past to inform the present.

This review largely restricts itself to a consideration of the quality of the Late Quaternary terrestrial record for mammals and birds, which represent the great majority of extinct species that have been described from the recent fossil and subfossil records. It is important to note that these two vertebrate groups comprise only a very small part of total Quaternary biodiversity, and the disproportionate level of attention they have received reflects not only the preferential preservation of organisms possessing biomineralized skeletons, but also the enduring research interest in these two vertebrate classes among both palaeontologists and neontologists. Much less remains known about patterns of pre-human diversity, community composition, and extinction in non-vertebrate taxa during the Late Quaternary, although it appears that many insular invertebrates (e.g. terrestrial molluscs) experienced similar massively elevated levels of extinction following prehistoric first human arrival due to forest clearance and the introduction of exotic species (Preece 1998; Cowie 2001).

10.2 Taphonomic incompleteness and bias

For modern faunas it is possible, at least in theory, to sample and describe all of the species present in an ecosystem and quantify their relative abundances, niche requirements, and wider ecosystem-level interactions and geographical distributions, although in practice this is typically prevented by logistical difficulties and low detectability of rare and/or cryptic species (see e.g. Thompson 2004). In contrast, fossil assemblages invariably represent only an incomplete and biased subset of the original fauna present in a given region at the time of deposition. Exposed animal remains are usually quickly destroyed by the effects of climate and scavengers unless they are isolated and protected from degradation by surface processes. Even if they become buried, they may still be destroyed and recycled relatively rapidly by unfavourable physical, chemical, and/or biological processes operating in many sediments (e.g. acid soils, calcium- or phosphorus-deficient soils, action of plant roots or other soil organisms). As such, although both mammals and birds would have been widely distributed across most terrestrial ecosystems and geographical regions during the Late Quaternary, only certain environments have been able to preserve high-quality terrestrial vertebrate fossil records. Many rare and/or range-restricted species stand a relatively low chance of having become fossilized at all, and the total number of extant and extinct species identified to date in the recent fossil record is therefore a gross underestimate of the pre-human Quaternary global vertebrate fauna.

Late Quaternary faunas are typically preserved in a restricted range of different natural depositional environments which preserve disproportionately large amounts of fossilized vertebrate remains compared with the relatively small volume of sediment they contain. However, these specialized deposits all have their own distinctive series of taphonomic biases. For example, cave deposits (frequent in karstic landscapes and other regions) considerably enhance the preservation of animal remains, as they possess relatively constant temperature and humidity, and bones are typically buffered from dissolution by carbonate clasts unless acidic guano from bat and cave-bird roosts is present. However, all species present in cave deposits have been transported into the system by one of three types of concentrative mechanism: biotic autochthonous, whereby animals that lived and died in the cave environment, for example foraging, shelter, roosting, and/or hibernation; biotic allochtonous, the accumulation of animal remains from outside the cave environment by predators, for example owls or hyenas; and abiotic allochthonous, where animal remains are transported into the cave by flooding, water-hole traps, pit falls, or other abiotic mechanisms (Andrews 1990; Simms 1994; Schubert *et al.* 2003). Therefore, although a surprisingly wide range of species can become preserved in caves or rockshelters (e.g. Kos 2003), many species present in the local environment will be overrepresented (e.g. preferential owl prey species; cave-dwelling mammals such as bats and the cave bear *Ursus spelaeus*; and flightless or ground-dwelling taxa that are more likely to encounter pit falls) and others will be inadequately sampled or absent. Conversely, swamp sites provide anoxic conditions favourable for vertebrate bone preservation, but are biased not only towards biotic autochthonous species likely to be present in the immediate local environment such as waterfowl, but also towards large flightless taxa (e.g. moa) that had a tendency to become mired while looking for food or water, and predators that became trapped when killing or feeding on other trapped individuals; further biases may develop if swamp sites were only seasonally active, such as at the famous moa swamp site of Pyramid Valley, New Zealand (Holdaway and Worthy 1997).

The relative abundance of different types of natural Late Quaternary fossil deposits varies between different geographical regions. Among island systems, New Zealand and Madagascar possess rich fossil records comprising cave, swamp, and coastal dune sites (Burney 1999; Worthy and Holdaway 2002). In contrast, almost all known Quaternary fossils from the West Indies have been reported from cave deposits, in addition to a small number of anoxic water-filled sinkholes ('blue holes') (Steadman *et al.* 2007) and a single asphalt seep in Cuba (Iturralde-Vinent *et al.* 2000). Continental Late Quaternary terrestrial vertebrate fossils have also been preserved in a wider range of natural depositional settings, notably in permafrost soil in northern hemisphere boreal regions, and in a wider range of freshwater deposits (e.g. fluvial, lacustrine). Anthropogenic midden deposits have also provided an additional rich source of Late Quaternary zooarchaeological material. In many regions, these deposits represent a significant source of information on prehistoric faunas, often because natural fossil deposits remain understudied; for example, the majority of the extinct oryzomyine rice rats of the Lesser Antillean islands and many extinct tropical Pacific birds are known only from prehistoric archaeological horizons (Pregill *et al.* 1994; Steadman 2006b). However, species composition and abundances in middens are again biased with respect to the autochthonous vertebrate assemblages present during the time of deposition, as prehistoric hunters and foragers may have preferentially targeted species of particular nutritional or cultural value (e.g. Duncan *et al.* 2002).

Vertebrate remains undergo considerable physical and biological modification between death, burial, and subsequent long-term preservation within each of these specialized depositional environments. Although soft-tissue preservation is occasionally encountered in prehistoric Late Quaternary environments in both continental and island systems, usually in association with cold-temperature environments (e.g. mammoths and other megafaunal mammals in permafrost soils; upland moa *Megalapteryx didinus* in alpine caves), skeletal elements are typically the only components of the organism that do not decompose quickly after death. Vertebrates have a relatively large skeletal mass consisting of several hundred identifiable bones per individual which typically become dissociated and dispersed unless burial is rapid. These skeletal elements vary greatly in size, shape, and density, and so are differentially affected by most taphonomic processes. Robust elements are more likely to survive early structural or bone density-mediated destructive processes, but smaller elements are more likely to become buried, and can become concentrated in predator faeces and regurgitates; bones also frequently experience hydrodynamic sorting by water currents according to size and density (Behrensmeyer 1991; see also Ericson 1987; Grayson 1989; Livingston 1989; Cruz and Elkin 2003). Different predators may also produce differential representation of skeletal elements in bone assemblages (Andrews 1990; Lyman 1994; Saavedra and Simonetti 1998). However, only a relatively restricted subset of skeletal elements are highly morphologically distinctive and useful as taxonomic identifiers. Further limitations may be presented by the exact taxon under consideration, as the precision of potential identification will vary between orders, families, and genera depending on the degree of osteological conservatism.

Late Quaternary fossil deposits also all generally represent low-energy settings with low sediment-accumulation rates relative to population turnover times, which frequently leads to time-averaging of fossil assemblages. For example, deposits of small vertebrate remains beneath favoured owl roost sites may have accumulated over many thousands of years (Worthy and Holdaway 1994). Further stratigraphic mixing of fossil material originating from different younger and older temporal horizons can also be driven by various biological and physical agents; for example, bioturbation. These processes can lead to extraordinary densities of often fragmentary vertebrate remains in caves, swamps, and other sites that have accumulated over relatively long periods of time (10^2–10^4 years). Although typical scales of time-averaging are short relative to average durations of species in the fossil record (10^6–10^7 years), this process presents substantial further complications for understanding past patterns of ecology and extinction (Graham 1993; Roy *et al.* 1996).

Finally, it is important to recognize that our current understanding of Late Quaternary faunas has been affected not only by preservational biases but also by historical collection and post-collection biases, which vary between different taxa and geographical regions. Not only is a palaeo-community unlikely to be accurately represented by a preserved fossil assemblage, but the fossil assemblage itself is unlikely to be accurately represented by a human-sampled collection of fossil material from a given site. Early collectors were often most interested in obtaining the remains of large charismatic megafaunal species to assemble visually dramatic museum displays. For example, swamp sites in New Zealand were sampled very coarsely during the nineteenth and early twentieth centuries by collectors only interested in the bones of moa and other large flightless birds, using equipment such as 'a large shark hook lashed to the end of a long manuka rod' (Worthy 1989a). Similarly, the Mare aux Songes marsh in Mauritius, first excavated in the nineteenth century, was originally known only for yielding material of large extinct species, notably dodo *Raphus cucullatus* (Cheke and Hume 2008). Few smaller species (e.g. passerines) were recorded from such fossil sites before the advent of more systematic sampling techniques in the twentieth century, notably the use of fine-meshed screens to collect tiny skeletal elements. However, the Mare aux Songes is now recognized as a Konzentrat-Lagerstätten due to the diversity and abundance of species preserved at the site (Rijsdijk *et al.* 2009). Indiscriminate early excavations frequently destroyed details of the stratigraphic and/or geomorphological context of many Late Quaternary fossil sites, preventing subsequent meaningful ecological and temporal reconstruction.

Comparable biases are present in early historical records, which would only be likely to mention the relatively restricted subset of species that were of direct interest to contemporary observers. The most important bias was probably whether or not a given species was good to eat. For example, the author of the only known early historical-era (mid-sixteenth century) record of now-extinct Hispaniolan mammals, the 'hutia' (?*Isolobodon portoricensis*), the 'quemi' (?*Plagiodontia ipnaeum* or *Quemisia gravis*), and the 'mohuy' (*Brotomys voratus*), was primarily concerned with describing what they tasted like (Miller 1929). Whereas large predators such as felids, wolves, and bears would only rarely be present in zooarchaeological deposits relative to typical historical prey species such as ungulates, they were also far more likely to be reported by early chroniclers because of their real or perceived depredations on local communities and livestock. Coverage of Quaternary fossil faunas has also been affected by different research interests in different geographic regions; in particular, reconstructing prehistoric faunas is frequently the main research interest in island systems, whereas in continental regions such as western Europe our understanding of prehistoric Quaternary faunas has frequently developed from zooarchaeological rather than palaeontological research in studies concerned with patterns of early human occupation. It is also important to appreciate that our understanding of past patterns of diversity is dependent upon the formal description and correct identification of adequately preserved fossil collections, but in practice considerable amounts of material may remain undescribed or placed in open nomenclature. There remain relatively few specialists who are able to confidently identify Quaternary palaeontological specimens in comparison with people able to make accurate identifications of mammal and bird species on the basis of living animals or modern material.

10.3 Are living and extinct species comparable units?

Although the evolutionary processes of anagenesis and cladogenesis pose substantial challenges for characterizing discrete species units across intervals of geological time (e.g. Sheldon 1987; Baum 1998; Briggs and Crowther 2001), such problems are relatively limited in Late Quaternary bird and mammal faunas. Phyletic evolution towards smaller body size has been recognized in several Late Pleistocene mammal and bird lineages, notably insular megafaunal mammals (e.g. Lister 1989; Vartanyan *et al.* 1993), Australian marsupials (Murray 1984), African ungulates (Peters *et al.* 1994), and Western Palaearctic birds (Stewart 2007b), in response to environmental shifts between glacial and interglacial periods, preferential human hunting of larger individuals, or resource limitation in isolated island populations. Comparable patterns of anagenetic change in body size across the Late Quaternary have also been documented in some small-bodied terrestrial vertebrates (for example, nesophontid island-shrews on Puerto Rico; McFarlane 1999b), although the ecological drivers remain unclear. These body-size shifts may lead to misinterpretation of Late Quaternary chronospecies and pseudoextinctions (e.g. all apparent African end-Pleistocene large mammal extinctions are probably pseudoextinctions of large-bodied morphs within surviving lineages; Peters *et al.* 1994). However, nearly all vertebrate faunal changes during the Late Pleistocene and Holocene instead represent extinctions, population extirpations, or range shifts in response to human activity and environmental change, and very few new bird and mammal species are believed to have evolved during this interval (e.g. parrot crossbill *Loxia pytyopsittacus*; see Tyrberg 1991a).

It is instead important to consider whether species concepts and our ability to define species differ qualitatively between prehistoric, historical, and present-day systems, and whether extinct Late Quaternary taxa remain incompletely understood relative to modern faunas. Is it possible to reconstruct the same species diversity observed in a present-day ecosystem using only the kinds of data available from the recent fossil record? These questions can be addressed through investigation of the different types of characters available for diagnosing living and long-extinct species. Whereas all fossil bird and mammal species have been formally described on the basis of skeletal characters (typically craniodental or limb element characters), extant species have traditionally been diagnosed largely on the basis of soft-tissue characters (typically pelage or plumage characters), although skeletal characters are often also of substantial importance, particularly for mammals.

The extent to which species can be resolved on the basis of skeletal versus soft-tissue morphology does inevitably differ to some degree, although it is difficult to establish the magnitude of this discrepancy. Many small mammals can still be accurately differentiated on the basis of relatively subtle dental differences (the basis of the 'vole clock' method for defining successive Quaternary strata; Koenigswald and Van Kolfschoten 1995). However, even within the well-studied Western

Palaearctic Late Quaternary avian record, it remains extremely difficult if not impossible to identify many small-bodied passerine genera (e.g. *Anthus, Emberiza, Motacilla, Phylloscopus, Sylvia*; Ericson and Tyrberg 2004) to species level on the basis of skeletal remains. For example, the thorough review of Ericson and Tyrberg (2004) documented only 35% of passerine species compared to 59% of non-passerine species in the recent Swedish avifauna (including vagrants and locally or globally extinct species, e.g. great auk *Pinguinus impennis*, but excluding known exotic introductions) as having been definitely or questionably reported from the Late Pleistocene–Holocene fossil and zooarchaeological records of Sweden. Although many small-bodied passerines are osteologically distinctive, both within the Western Palaearctic avifauna (e.g. *Coccothraustes*) and in extinct island avifaunas (e.g. Worthy and Holdaway 2002; Steadman 2006b), species identification from osteological material in the absence of soft-tissue characters remains a widespread problem even for larger-bodied taxa; for example, *Anas, Larus, Turdus*, and *Bubo* (including *Nyctea*). Problems in osteological identification using fossil material are clearly compounded by the fact that only a restricted subset of the original complement of skeletal elements (which may or may not represent taxonomically diagnostic elements) are typically preserved, recovered, and available for study. Identifying small-bodied species is made more problematic by the diminished preservation potential of minute, fragile bones in many depositional environments, and their rarity in zooarchaeological deposits which may reflect preferential hunting of larger-bodied species. Additionally, associated remains may be rarely recovered, with identifications consequently based on individual elements in isolation. Most Holocene bird extinctions were of island species, and the relatively simple structure of insular avifaunas (in comparison to species-rich continental systems containing many closely related and morphologically similar congeners) may suggest that accurate species identification and community reconstruction on the basis of subfossil material could be a more feasible task in these environments (although see Steadman 2006a). However, insular bird species have often evolved derived morphological characters that not only make specific identification difficult, but generic determination almost impossible.

Other problems in comparing species concepts between modern-day and prehistoric Quaternary systems are generated through different processes of taxonomic inflation (the rapid accumulation of scientific names through processes other than new discoveries of taxa; Isaac *et al.* 2004) being employed by both neontologists and palaeontologists. In neontological studies, behavioural (e.g. birdsong; Irwin *et al.* 2001) and genetic characters, both of which clearly have very limited application in palaeontology, are frequently used in addition to skeletal or soft-tissue characters for identifying cryptic extant species that are otherwise morphologically almost indistinguishable. The use of molecular data is becoming an increasingly widespread and standard taxonomic practice, and molecular studies have been used to support the recognition of often large numbers of cryptic mammal and bird species within existing distributions of known species (e.g. sportive lemurs *Lepilemur* spp.: Andriaholinirina *et al.* 2006; Louis *et al.* 2006; mouse lemurs *Microcebus* spp.: Rasoloarison *et al.* 2000; Yoder *et al.* 2000; giraffes *Giraffa* spp.: Brown *et al.* 2007), with so-called DNA barcoding of reference genes being proposed to further enhance the discovery of new species (Hebert *et al.* 2003; Clare *et al.* 2007; Kerr *et al.* 2007). This 'splitting' of morphologically similar species, typically using phylogenetic species concepts rather than biological species concepts, remains the subject of considerable debate and controversy within modern ecology and systematics (Isaac *et al.* 2004; Moritz and Cicero 2004; Will and Rubinoff 2004).

However, molecular and behavioural investigations still tend to be limited to diagnosing previously unrecognized levels of phylogenetic diversity or testing between existing morphology-based taxonomic hypotheses rather than providing primary data for establishing new species, which still usually requires morphological determinants of taxonomic status. Furthermore, as the Late Quaternary record frequently consists of subfossil rather than fully fossilized material, it also provides a palaeontologically unique opportunity to investigate the genetic status of species that became extinct during recent prehistory, and although degradation of the organic bone fraction occurs frequently in tropical Quaternary deposits, considerable molecular insights have recently been generated into the taxonomy and systematics of many extinct Late Quaternary species. Possibly the most remarkable example of molecular revision of morphology-based taxonomy for extinct species is the recognition through ancient DNA research that the three supposed size-differentiated species of giant moa (*Dinornis struthioides*, *Dinornis novaezealandiae*, and *Dinornis giganteus*) actually represent male and female size morphs displaying extreme reversed sexual size dimorphism, with extensive environmentally controlled size variation in females from different populations, and with the real species boundary existing between North Island and South Island populations (*Dinornis novaezealandiae* and *Dinornis robustus*) (Bunce *et al.* 2003; Huynen *et al.* 2003).

Palaeontological research instead tends to generate taxonomic inflation primarily through alternative processes of taxonomic elevation and overdescription. Many extinct species known only from the Quaternary fossil record have historically been placed in supposedly extinct higher-order taxa, typically monotypic genera (e.g. *Harpagornis*, *Notornis*, *Palaeocorax*, *Wetmoregyps*), which are now interpreted as synonymous with extant genera (see e.g. Livezey 1998; Worthy and Holdaway 2002; Bunce *et al.* 2005; Olson 2007). Many fossil species have also been radically oversplit in the past on the basis of extremely dubious and fine-scale morphological differences that do not stand up to systematic scrutiny; as an extreme example, Rothschild (1907) recognized 37 different species of moa, which have now been reduced to 10 species (Worthy and Holdaway 2002; Bunce *et al.* 2003). The proliferation of dubious fossil taxa is exacerbated by the frequent establishment of extinct species on highly incomplete material. Some of these supposed extinct 'species' are now interpreted as merely locally extirpated populations of surviving species, especially for taxa that formerly occurred on multiple islands (e.g. *Cyclura portoricensis* from Puerto Rico and *Cyclura mattea* from St. Thomas, which are now recognized as being synonymous with the extant but Critically Endangered Anegada Island iguana *Cyclura pinguis*; Pregill 1981). It is easy to speculate that many disjunct populations of living species—for example, the isolated Iberian population of the azure-winged magpie *Cyanopica cyanea*—could have become classified as distinct, extinct species on the basis of fossil material alone (Cooper 2000) had they been locally extirpated in the Late Quaternary. More extreme taxonomic problems have developed on occasion from inaccurate interpretation of historical sources. Several hypothetical species have been formally described in the absence of type material on the basis of erroneous records that were either fictional or actually referred to known species, such as the giant Mauritian rail *Leguatia gigantea* (see Rothschild 1907), and a dodo and a solitaire from Réunion, *Victoriornis imperialis* and *Ornithaptera solitaria*, erected by Hachisuka (1953; see also Olson 2005 and Olson *et al.* 2005a; Fig. 10.3). Several invalid species, notably the supposed Réunion dodo (Hume and Cheke 2004) and the Chatham Island sea eagle *Haliaeetus australis* (in reality a mislabelled bald eagle *Haliaeetus leucocephalus*; Worthy and Holdaway 2002), are also still frequently referred to in the scientific and popular literature, despite having been thoroughly discounted as real taxa. Many other synonyms also continue to be erroneously included in inflated species lists of extinct faunas (e.g. Chatham Island *Pterodroma* petrels; see Cooper and Tennyson 2008).

Although Late Quaternary faunas have been the focus of considerable taxonomic revision in recent decades, it is difficult to assess the extent to which further synonymization remains to be corrected. The conclusions of Alroy (2003) that current name quality varies strongly with body mass for North American fossil mammals (i.e. large-bodied species have been grossly oversplit but are historically older, and therefore more invalid synonyms have already been identified relative to less-studied small-bodied species) are probably also valid for the Late Quaternary record. However, similar problems are not restricted to extinct faunas, and even the taxonomic validity of several living (or supposedly extant) terrestrial megafaunal mammals (e.g. African forest elephants *Loxodonta cyclotis* and *Loxodonta pumilio*: Roca *et al.* 2001; Debruyne *et al.* 2003; kouprey *Bos sauveli*: Galbreath *et al.* 2006;

Figure 10.3 The white dodo of Réunion (copy by Joseph Smit of painting by Pieter Withoos), an invented species probably based on travellers' accounts of the facultatively flightless white Réunion ibis *Threskiornis solitarius*. From Newton (1869); see Hume and Cheke (2004) for further information.

Hassanin and Ropiquet 2007; Vietnamese warty pig *Sus bucculentus*: Robins *et al.* 2006; khting voar or snake-eating cow *Pseudonovibos spiralis*: Brandt *et al.* 2001; Galbreath and Melville 2003) continues to be the subject of debate. Conversely, relatively large numbers of extinct Late Quaternary taxa (e.g. the extensive insular radiation of Lesser Antillean oryzomyine rice rats) still remain undescribed or in open nomenclature, due to inadequate preservation preventing accurate description or identification, or material having been collected by investigators untrained in taxonomic procedure.

10.4 Estimating species richness in pre-human Quaternary ecosystems

It is extremely unlikely that any regional pre-human Late Quaternary vertebrate faunas have yet been fully characterized. For example, New Zealand has one of the most heavily sampled Late Quaternary records in the world (Worthy and Holdaway 2002), but a comparison of the description dates of its extant and extinct bird species reveals fundamental differences in our knowledge of these two avifaunas (Fig. 10.4). The collection curve for living and historically extinct New Zealand bird species (i.e. species described from modern specimens) is almost saturated, with recently erected species representing newly split populations of previously known species (e.g. Tennyson *et al.* 2003). Conversely, whereas the megafaunal and large-bodied components of New Zealand's extinct pre-human fauna (moa, eagles, giant rails, geese) were also described relatively early on during the nineteenth century, new species of smaller birds (waterfowl, petrels, wrens) continue to be described at a relatively constant rate from the Late Quaternary fossil record, making comparison between past and present data difficult even for this well-studied system. Furthermore, pre-human faunas from different island and continental systems are not only consistently more poorly known than extant faunas, but the relative state of knowledge also varies wildly between different geographic regions and taxonomic and ecological groups; geography is probably the most important predictor variable of description date in both modern and Quaternary vertebrate faunas (cf. Collen *et al.* 2004). In comparison to New Zealand, probably only the pre-human avifauna of the Mascarene Islands has been studied as thoroughly, and known taphonomic biases between different Mascarene sites may make predictive faunal analysis possible (Hume 2005; Cheke and Hume 2008). Subfossil material from other regions is typically much more poorly known, posing substantial problems for inter-regional comparison of Quaternary faunas using data sets of comparable quality. Within the remainder of the south-west Pacific Island region, for example, many of the extinct Quaternary taxa known from Fiji and

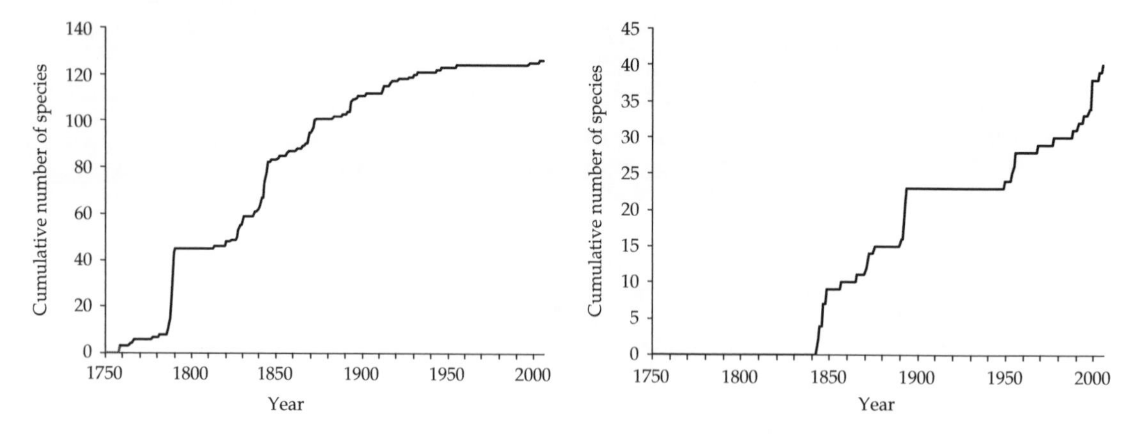

Figure 10.4 Description dates for New Zealand's avifauna (including date of first publications documenting taxa that have not yet been formally described, e.g. Chatham Island penguin *Eudyptes* sp. nov.). Left: extant or historically extinct species. Right: currently recognized extinct species described from the Late Quaternary fossil record. The collection curve for extant and historically extinct species is almost saturated, unlike that for extinct species described from fossil material. Data from Worthy and Holdaway (2002), Bunce *et al.* (2003), Gill (2003), Tennyson *et al.* (2003), and Worthy (2005a).

Vanuatu are large-bodied species only discovered within the last decade (e.g. Worthy and Molnar 1999; Mead *et al.* 2002), and very few subfossil taxa have so far been recorded at all from the Solomon Islands (Steadman 2006b). These problems reflect both variable availability of representative Late Quaternary exposures and variable historical collection effort, making it difficult even to evaluate the status of our current understanding of pre-human Quaternary diversity.

However, whereas considerable attention has been given to the problem of estimating the completeness of the 'deep-time' fossil record (e.g. Foote and Sepkoski 1999; Briggs and Crowther 2001), relatively little attention has been paid to this problem in relation to the youngest Quaternary vertebrate record. The best-known and most frequently cited estimate for the number of unknown extinct species that still remain to be discovered in the pre-human Holocene record is that of Steadman (1995), who attempted to determine the total number of island birds that may have disappeared following the prehistoric expansion of Polynesian settlers across the tropical Pacific (Table 10.1). Four hundred and seventy five landbird and seabird species occur today across the island groups of Near Oceania and Remote Oceania (excluding Hawai'i and New Zealand), and an additional 121 globally extinct species have so far been recorded from the region (Steadman 2006b; see Chapter 4 in this volume). At the time of Steadman's initial estimate, Late Quaternary fossil assemblages containing more than 300 identified bones were only known from seven islands in the Polynesian heartland (Henderson, Ua Huka, Tahuata, Hiva Oa, Huahine, Mangaia, and 'Eua). Each of these assemblages contained approximately 20 or more different bird species (both landbirds and seabirds) that had either become globally extinct or locally extirpated from the island, and none of these regional fossil records was considered to be complete. Using a conservative estimate that an average of 10 bird species or populations had been lost following first human contact from each of tropical Oceania's approximately 800 major islands and atolls, Steadman (1995) suggested that as many as 8000 species and populations had been lost from the region, a figure almost equal to the number of bird species still extant today. This estimate notably included approximately 2000 species of unknown extinct flightless rails. Following a further decade of palaeontological and zooarchaeological data collection across the region, Steadman (2006b) provided a revised 'conservative but crude' estimate for tropical Oceania (exclusive of New Zealand and Hawai'i) of 538–1673 unknown extinct bird species,

Table 10.1 Different estimates of the total number of extinct bird species that remain unknown from the Late Quaternary fossil record of Near Oceania and Remote Oceania. Note that the Pacific Island groups being considered differ slightly between each analysis. See text for further details.

Analytical approach	Pacific bird groups	Predicted number of unknown species	Reference
Simple extrapolation	All species	≈2000	Steadman (1995)
Simple extrapolation	All species	538–1673 (plus 22 New Zealand species)	Steadman (2006b)
Mark–recapture	All landbirds (including non-marine waterbirds)	≈400	Pimm *et al.* (1994, 1995)
Island biogeography	All landbirds, extrapolated from rails, pigeons, and parrots	≤1080	Curnutt and Pimm (2001)

of which 442–1579 were unknown flightless rails, and also suggested that 22 extinct species remain to be discovered from the New Zealand avifauna. The total number of extirpated bird populations across the tropical Pacific may be around 10 times greater, in part because it includes many or most of the region's extant species as well as all of the extinct species. The Late Holocene anthropogenic loss of bird species across the Pacific region may therefore represent the largest vertebrate extinction event ever detected (Steadman 2006b).

The simplicity and biogeographic plausibility of Steadman's estimates have been criticized by several authors (e.g. Bennett and Owens 2002), and lower estimates of the total number of extinct Pacific bird species incorporating alternative extrapolation methods and parameters have also been proposed (Table 10.1). Pimm *et al.* (1994, 1995) used mark–recapture principles to develop their estimates, by dividing regional Pacific avifaunas into four subgroups: those species that were observed or collected by biologists ('skins') versus those that were not ('no skins'), and those species that are known from the recent fossil record ('bones') versus those for which no fossils are known ('no bones'). We are only ignorant of the single subgroup of species for which no skins or bones are known, and the number of these missing species can be estimated as:

No skins and no bones = (skins but no bones × bones but no skins)/skins plus bones

Using this approach, Pimm *et al.* (1994, 1995) made conservative estimates that 21 extinct passerine species and between eight and 12 extinct non-passerine landbird species remain to be discovered in the Late Quaternary fossil record of Hawai'i, and that a further 20–23 species are still missing in the recent fossil record from south-eastern Polynesia, seven species from the Marianas Islands, 24 species from New Caledonia, and a single species from Tonga. However, these authors warned that the limited numbers of missing species predicted from the western Pacific using this approach were likely to be drastic underestimates, because the number of currently endangered and recently extinct bird species is inversely proportional to length of human occupancy on islands across the Pacific region, suggesting that these islands are likely to have lost a larger proportion of their sensitive native species without trace as a result of prehistoric human impacts. As very rough estimates, Fiji, Tonga, and Samoa may therefore have lost 60 or more bird species, and Vanuatu and New Caledonia may each have lost approximately 50 species. Overall, Pimm *et al.* (1994, 1995) suggested that the Pacific bird fossil record was only half complete, and that the original pre-human avifauna consisted of about 800 species. Interestingly, however, this approach suggested that there was only one new bird species still to be discovered from New Zealand's extinct avifauna, but four species have since been described or documented from the recent fossil record. Greater uncertainties are introduced by the taxonomic problems posed by incomplete or poorly preserved subfossil material; many specimens may either represent known extant species or closely related, undescribed extinct taxa (see Steadman 2006b), and incorporating these large amounts of questionable data into predictive extinction models remains a major challenge.

A more refined method of estimation was presented by Curnutt and Pimm (2001), using the theoretical framework of island biogeography to determine the number of endemic species of rails, pigeons, and parrots across Polynesia, Micronesia, central and eastern Melanesia, and New Zealand according to four models (Tables 10.1 and 10.2). Using occurrence data for known extant and extinct species across the Pacific, these authors identified all islands across the region with the potential to maintain viable landbird populations, then extrapolated the numbers of species that could have existed on each individual island (model 1) or island group (model 2) by applying the known maximum species number for each taxon recorded on different island sizes and types. This analysis was also the first such estimate of pre-human Pacific bird diversity to incorporate data on a series of relevant geographical parameters that could further affect species diversity. Subsequent analysis suggested that low-relief island groups were more likely to produce endemic species on single islands, whereas high-relief island groups were more likely to produce endemics shared across the group (model 3), and additional consideration of the effects of changing Holocene sea levels and vulnerability to tsunamis on low-relief island groups (e.g. Tuamotu Archipelago, Tonga) led to further biogeographic revision (model 4). The model assumed to be most realistic (model 4) predicts that only 332 endemic species of all three groups combined formerly occurred across Oceania. Curnutt and Pimm (2001) considered that only approximately one-third of these species were accounted for as living, historically extinct, or fossil species, and concluded that the entire Pacific pre-human landbird fauna contained up to 1620 species.

These three simplistic estimates vary by an order of magnitude, and it remains difficult to assess which assumptions are more valid. Other investigative approaches also remain to be attempted, notably analyses using relative abundances of living and known extinct species in fossil deposits to derive the number of species that have not yet been collected, dependent upon underlying assumptions about sampling distributions (see Pimm *et al.* 1994, 1995), and quantification of the relationship between known Quaternary diversity, perceived faunal differences, and historical collection intensity for different regions. Ongoing fossil discoveries across the tropical Pacific continue to revise our understanding of the region's pre-human avifauna and the quality of the Late Quaternary record, and ultimately our understanding of former levels of diversity is heavily dependent upon appropriate interpretation of the inter-related regional patterns of taxonomy and endemicity, notably whether related island populations are classified as distinct species, subspecies, or as part of a superspecies or allospecies complex. However, as with other extant and recently extinct groups, the taxonomic status of many Pacific bird populations according to biological versus phylogenetic species concepts remains the subject of ongoing debate (e.g. Mayr and Diamond 2001), complicating wider considerations about the magnitude of prehistoric extinctions.

Table 10.2 Total predicted numbers of endemic rails, parrots, and pigeons across Polynesia, Micronesia, central and eastern Melanesia, and New Zealand under four biogeographical models (see text for further details). This study assumes that the smallest Pacific islands that can support viable populations of endemic species are: rails, 6.5 km^2 (*Rallus wakensis*, Wake Island); pigeons, 28 km^2 (*Ptilinopus chalcurus*, Makatea); and parrots, 33.7 km^2 (*Nestor productus*, Norfolk Island). After Curnutt and Pimm (2001).

Pacific bird group	Model 1	Model 2	Model 3	Model 4
Rails	537	145	206	199
Parrots	94	29	38	38
Pigeons	253	64	101	95
Total	884	238	345	332

10.5 Uncertainty of extinction dates and drivers

Even if an extinct Quaternary species can be adequately diagnosed and accurately identified, constraining its extinction chronology and identifying causative extinction driver(s) remains a major challenge to understanding patterns of biodiversity loss in the recent past. In many cases it is impossible even to ascertain whether a species present in the Late Quaternary record died out during the Holocene or the Late Pleistocene, because many sites (especially non-cultural sites) remain undated. In particular, palaeontological samples from tropical environments (e.g. the West Indies) have frequently experienced diagenetic loss of collagen through geologically rapid degradation of the organic fraction of skeletal material under warm, humid conditions, making radiometric analysis impossible (e.g. Turvey *et al.* 2007a). If no direct data are available, estimating whether an extinct species survived into the recent past may therefore require inference from timing of first regional human arrival (which may post-date 'natural' Late Pleistocene extinctions, e.g. the megafaunal rodent *Amblyrhiza inundata* in the West Indies; McFarlane *et al.* 1998) or biostratigraphic association with introduced species in Quaternary deposits (which may be affected by time-averaging through processes such as bioturbation).

Although extinction represents the absence (either locally or globally) of a species, the fossil record instead only documents the presence of a species at a particular stratigraphic interval. Extinction must therefore be inferred rather than directly observed when a species disappears from the fossil or historical record. Even when direct last-occurrence dates can be obtained, these almost certainly precede the actual time of extinction of the target species due to the Signor–Lipps effect (Signor and Lipps 1982), and so must be interpreted as a *terminus post quem* that pre-dates final extinction (see Chapter 9 in this volume for further discussion). This may reflect ecological factors that preclude the likelihood of sampling the last individuals of a species in a particular fossil deposit, such as range fragmentation, uneven patterns of decline, and the persistence of remnant populations in refugia. Preservational bias also influences the likelihood of fossil preservation over time (Peters and Foote 2002). For example, whereas fossil deposits in general become more common in progressively younger strata, Wroe *et al.* (2004) considered that the reverse may in fact be the case for Late Pleistocene Australia, when increasing aridity from *c.*40 000 years BP is predicted to have driven a shift from a depositional to an erosional regime, limiting the formation of new fossil deposits and increasing the rate at which previously formed deposits disappeared; this has obvious implications for interpreting the apparently early disappearance of the Australian megafauna from the Late Quaternary fossil record. On a smaller scale, Worthy (2004a) has suggested that the sudden appearance of South Georgian diving-petrels *Pelecanoides georgicus* in deposits dated to AD 1200–1300 at Mason Bay, New Zealand, does not reflect recent establishment of a petrel colony, but instead poor preservation potential of skeletons in revegetated older dunes into which humic acids may have leached. Although Foote (2000) has noted that overall several studies have shown that most of the apparent variation in extinction rates (and origination rates) is not statistically attributable to independently measured variation in preservation rates, at least in studies from the deep-time Phanerozoic fossil record, the extent to which apparent variation in extinction timing and rate reflects merely variation in preservation rate must be addressed case by case. Determining meaningful last-occurrence dates that bear close correspondence with actual time of extinction remains an ongoing problem for understanding Holocene biodiversity loss, and many examples exist of the often substantial mismatch between observed last occurrence dates in the fossil or zooarchaeological record and independent evidence for much later survival of the target species (e.g. Cooper and Tennyson 2004). Possibly the most remarkable of these is the probable historical-era survival of the Marquesan swamphen *Porphyrio paepae*, which is known only from zooarchaeological remains *c.*700–800 years old, but which is apparently depicted in Paul Gauguin's 1902 painting *Le Sorcier d'Hiva Oa ou le Marquisien à la cape rouge* (Raynal 2002; Steadman 2006b; Fig. 10.5). Recent accelerator mass spectrometry (AMS) radiocarbon dating

Figure 10.5 Probable depiction of the Marquesan swamphen *Porphyrio paepae* being killed by a dog in Paul Gauguin's 1902 painting *Le Sorcier d'Hiva Oa ou le Marquisien à la cape rouge.* ©Musée d'Art Moderne et d'Art Contemporain de la Ville de Liège; reproduced by permission.

of Eurasian lynx *Lynx lynx* material from northern England to 1550±24 years BP, revealing the previously unsuspected survival of this large mammalian predator into the early medieval period in Britain (Hetherington *et al.* 2006), underlines how much we still have to learn about extinction chronologies, even in well-studied Quaternary systems.

Identifying the correct extinction driver(s) also remains difficult even when we know approximately when a given species died out. The timing of first appearance of different putative threat processes, notably first human arrival or the anthropogenic introduction of exotic mammal species, may be as difficult to establish as a species extinction date. This is evidenced by the controversy surrounding the first appearance of the Pacific rat *Rattus exulans* in prehistoric New Zealand, which is usually believed to have arrived with the first Maori colonists in the thirteenth century AD but may have colonized through a supposed earlier fleeting human contact in c.AD 50–150 (Holdaway 1996, 1999a; Brook 2000; Wilmshurst and Higham 2004). Multiple different possible anthropogenic and non-anthropogenic extinction drivers may have been operating around the time of a species disappearance (e.g. Bovy 2007), and these may have acted either alone or in combination to drive a species to extinction. Extinction events are also associated not only with a range of mechanisms but also a range of speeds, and attempts to pinpoint primary extinction drivers through temporal correlation of cause and effect will produce erroneous conclusions if species disappear in gradual sitzkrieg-style events (as opposed to rapid blitzkrieg-style events) as impacted regions slowly pay off their extinction debt of 'doomed' species (Diamond 1989; Tilman *et al.* 1994). Furthermore, some Late Quaternary environmental changes, such as the shift from Late Pleistocene herb-dominated so-called mammoth steppe to Holocene moss/sedge-dominated tundra in Siberia, may in fact represent secondary ecological responses to the extinction of keystone species rather than primary extinction drivers in themselves, further complicating the interpretation of extinction events in the recent fossil record (Willerslev *et al.* 2003; Zimov 2005; see also below).

It is also important to recognize that absence of evidence in the Late Quaternary record for early human hunting of now-extinct species does not necessarily indicate that direct overexploitation was not a significant factor in their disappearance. For example, widespread and intensive hunting of giant tortoises *Cylindraspis* spp. on the Mascarene Islands by Portuguese and Dutch sailors is documented in several historical eye-witness accounts (Strickland and Melville 1848; Bour 1981; Cheke and Hume 2008), but there is negligible direct skeletal evidence for this known major extinction driver in the recent Mascarene subfossil record (Janoo 2005; J. Hume, personal communication). This has significant implications for attempts to reconstruct the extinction dynamics of other Late Quaternary faunas, notably continental megafaunas and the large-bodied mammals of Madagascar, the Mediterranean, and the West Indies, where such absence of evidence for extensive direct human persecution is frequently interpreted as meaningful evidence of absence (Grayson and Meltzer 2002, 2003, 2004).

10.6 Wider-scale ecological reconstructions and questions

Species do not die out in isolation, and both coextinctions and disruptions of mutualistic systems are increasingly being recognized as a major component of biodiversity loss (see Chapter 8 in this volume). Synecological investigation into how Late Quaternary extinction events would have impacted the wider structure and composition of terrestrial ecosystems is therefore of considerable importance. This is particularly significant to assess for mammals, as a range of mammalian ecotypes have the capacity to regulate stable-state shifts in both continental and insular systems (e.g. Croll *et al.* 2005). In particular, large-scale ecosystem changes resulting from the removal of ecologically significant mammalian megaherbivores are implicated in the end-Pleistocene continental mass-extinction event (Owen-Smith 1987). It is therefore necessary to investigate whether the supposedly superior quality of the Holocene fossil record can reveal any other persistent ecological 'ghosts' of now-extinct species, or even indicate the extent to which trophic cascades may have occurred in other systems impacted by severe anthropogenically mediated extinctions. This approach can theoretically reveal the extent to which pre-human ecosystems can be reconstructed, identify the true ecological impacts of introduced vertebrates, and assess whether it is possible to regain any approximation of pre-human ecosystem functions. However, although extinct species undoubtedly interacted with their environments it remains difficult to identify and characterize these interactions given the current resolution provided by the Late Quaternary record. Nearly all such studies have been limited to addressing specific plant–animal interactions involving now-extinct vertebrates (typically inferred fruit-/seed-dispersal mutualisms), and many of these attempts have met with substantial criticism.

Probably the best-known hypotheses concerning Quaternary synecological interactions are those of Temple (1977, 1979) and Janzen and Martin (1982). Following earlier suggestions in the scientific literature, Temple proposed an obligate mutualism between the dodo and the tambalacoque tree *Sideroxylon grandiflorum* (formerly *Calvaria major*), on Mauritius, suggesting that the tree had evolved fruits with specialized, thick-walled endocarps in response to seed predation by dodos, and that these seeds were unable to germinate without first being abraded in an avian gizzard. Temple considered that only 13 tambalacoque individuals, all more than 300 years old, remained on Mauritius by the late twentieth century, suggesting that no tambalacoque seeds had germinated for hundreds of years; but when 17 seeds were fed to turkeys 10 survived and three germinated. However, Temple failed to attempt to germinate any seeds from control fruits not fed to turkeys. In fact, tambalacoque seeds break open along a circumferential suture line, and the population of this tree species on Mauritius actually consists of several hundred individuals, including many young plants which were apparently misidentified by Temple. Its current rarity is instead probably due to the effects of exotic introductions: seed predation by rats, browsing by deer, soil disruption by pigs, and competition by alien plants. The most severe extinction driver is the destruction of unripe fruits by introduced macaques (Witmer and Cheke 1991; Cheke and Hume 2008). A comparable hypothesis by Janzen and Martin (1982) suggested that the neotropical flora contains several tree species with large indehiscent fruits possessing a bony endocarp and fleshy sugar-/oil-/nitrogen-rich pulp (e.g. avocados), similar to fruits dispersed in Africa by large mammals, which have limited dispersal in modern-day ecosystems but which are avidly eaten by introduced ungulates. These authors suggested that such fruits were adapted for dispersal by now-extinct Pleistocene megafaunal mammals such as gomphotheres. However, the 'gomphothere fruits' proposed by Janzen and Martin (1982) in fact display heterogeneous fruit characteristics and are adapted for different kinds of dispersal; many remain abundant today (e.g. *Scheelea* palms) and are still dispersed by native neotropical mammals, and display similar levels of seed/seedling mortality to species adapted for other dispersal mechanisms (Howe 1985). Even the widely accepted suggestion that the divaricating morphology (interlaced wide-angle branches with small widely spaced leaves) present in 18 families and approximately 10% of New Zealand woody plant species represents an adaptation against moa browsing (e.g. Worthy and Holdaway 2002) has been challenged, as this morphology may also provide adaptive benefits against photoinhibition during cold weather (Howell *et al.* 2002; although see Lusk 2002).

Recent attempts at Quaternary synecological analysis have attempted to develop more rigorous research approaches to avoid the problems caused by such casual, ambiguous, or inadequate observations. For example, it has been hypothesized that the Late Quaternary extinction of many tropical Pacific frugivorous/granivorous pigeons and doves has affected the composition of Polynesian forests by limiting dispersal of plant propagules (Steadman 1997a). Meehan *et al.* (2002) investigated the potential ecological implications of the extinction of large-bodied endemic *Ducula* pigeons on Tonga, an island where approximately 79% of native rainforest tree and liana species produce odourless, brightly coloured, dehiscent fruits whose morphology suggests adaptation for bird dispersal, but which has relatively few extant seed dispersers. Quantitative measurement of bill width and gape for extant and extinct Tongan pigeons identified eight apparently bird-dispersed plant species whose fruits exceeded the maximum gape (28 mm) of the island's largest living frugivore, *Ducula pacifica*, but which could have been swallowed by the larger extinct pigeons (maximum gape, 48 mm). Although flying foxes may still occasionally disperse such larger fruit, these plant species now apparently lack their coevolved avian dispersers. Similarly, Riera *et al.* (2002) demonstrated that fruits of the Mediterranean shrub *Cneorum tricoccon*, which remain attached to branches for long periods after ripening unless removed, are removed and dispersed in significantly greater proportions on islands where the inferred native seed dispersers (*Podarcis* lizards) are still present, as compared to islands where the lizards have disappeared following prehistoric introduction of mammalian carnivores. Interestingly, Riera *et al.* (2002) also suggested that this system provides one of the few examples of plant host-switching following native vertebrate extinction, as one of these exotic carnivores, the partially frugivorous pine marten *Martes martes* now acts as a seed disperser for *C. tricoccon*; this may have both generated directional selection pressure for larger seed size and altered the plant's elevational distribution.

However, there remain few other well-supported examples of mutualisms or other ecological interactions involving extinct Late Quaternary species. Developing testable hypotheses to avoid evolutionary 'just so stories' remains an ongoing concern, and without a meaningful understanding of the autecology of extinct species wider synecological reconstructions remain impossible. However, as the thylacine case study presented in the opening section of this chapter suggests, developing accurate data on the adaptations of many unusual extinct Quaternary species for which no obvious modern analogues exist, such as the 'marsupial lion' *Thylacoleo carnifex*, the Hawai'ian moa-nalo (giant geese which possibly occupied the same niche as giant tortoises on other island groups), and the remarkably stocky-bodied emeid moa *Euryapteryx* and *Pachyornis* of New Zealand, is likely to continue to prove elusive (see e.g. Wroe *et al.* 1999). For instance, would studies using ungulates to test hypotheses about the evolution of divaricating plants (Atkinson and Greenwood 1989) be expected to bear any relevance to the ecological effects of moa browsing in pre-human New Zealand? Furthermore, a fundamental paradox exists in trying to understand prehistoric Quaternary mutualisms. Almost nothing remains known about plant species losses before the recent historical period, and so genuine, now-extinct mutualists affected by prehistoric vertebrate extinctions are unlikely to have been documented. However, if plant species from such supposed former mutualisms have been able to survive for thousands of years after the disappearance of the vertebrates with which they were once ecologically associated (e.g. through host-switching or broader ecological tolerances), then they cannot have been tightly co-adapted, and any weak ecological relationship they once shared with now-extinct species will be hard for modern researchers to recognize.

10.7 Case study and conclusions: the Late Quaternary West Indian land mammal fauna

It remains difficult to assess the extent to which pre-human Quaternary faunas and ecosystems still remain to be characterized. Consideration of the current state of our knowledge about the Quaternary land mammals of the West Indies, one of the biogeographic regions severely impacted by anthropogenically driven extinctions during the Holocene, provides some important insights into the magnitude of research that is still required.

Only a relatively small proportion of the 99 extinct land mammal species (including bats) that are recognized herein from Late Quaternary deposits from the Greater and Lesser Antilles and which are interpreted as having died out during the Holocene (see Turvey, this volume) actually represent well-defined taxa. As many as 24 putative extinct West Indian land mammal species have not been formally described and remain in open nomenclature, including seven undescribed species respectively assigned to the genera *Capromys*, *Geocapromys*, *Nesophontes*, *Pteronotus*, and *Thomasomys*, two further undescribed species respectively assigned only to the Hexolobodontinae and the Rodentia, and possibly up to a further 14 undescribed oryzomyine rice rats from the Lesser Antillean archipelago (together with the unpublished '*Ekbletomys hypenemus*' of Ray 1962). Of those species that have been formally described, nine (*Brotomys contractus*, *Galerocnus jaimezi*, *Mormoops magna*, *Mysateles jaumei*, *Paramiocnus riveroi*, *Plagiodontia araeum*, *Puertoricomys corozalus*, *Tainotherium valei*, *Xaymaca fulvopulvis*) have been described on the basis of single fossilized skeletal elements, with no published descriptions available for any other referable specimens. Many further species (e.g. *Antillothrix bernensis*, *Megalomys audreyae*) are also known from extremely limited fossil material, making phylogenetic and ecomorphological interpretation difficult. Recent taxonomic revisions have reduced the number of recognized Quaternary species of Cuban hutias and nesophontid island-shrews (Díaz-Franco 2001; Condis Fernández *et al.* 2005), and further synonyms no doubt remain to be documented within the radiations of extinct *Capromys*, *Mesocapromys*, *Plagiodontia*, and heteropsomyine rodent species; in the absence of formal taxonomic revisions for these groups, it remains essentially up to the arbitrary discretion of different authors to decide how many of these species are probably valid. Conversely, other extinct taxa (*Oryzomys antillarum*, *Desmodus puntajudensis*) have only recently been recognized as distinct at the specific level (Morgan 1993; Suárez 2005).

Wider patterns of mammalian biogeography, ecology, and temporal persistence also remain very incompletely known for the West Indies. Although intra-island mammalian biogeographic regions containing extremely range-restricted endemics have been proposed for some of the geographically heterogeneous larger islands of the Greater Antilles (e.g. *Rhizoplagiodontia lemkei* may have only occurred in the Massif de la Hotte in south-western Hispaniola; Woods 1989), these apparent patterns may alternatively reflect incomplete sampling across the remainder of the species' potential ranges. Independent lines of evidence also suggest that some species had wider inter-island geographic distributions than are currently recognized from fossil data; for example, Ragged Island in the Bahamas was called 'Hutiyakaya' or 'Western Hutia Island' in Ciboney Taíno, a pre-Columbian Arawakan language (Granberry and Vescelius 2004), but no fossil remains of the Bahaman hutia *Geocapromys ingrahami* have yet been reported from this island (Morgan 1989b). Many extinct species, notably large rodents and xenotrichin primates, may be expected to have been important seed and fruit dispersers by analogy with living species (e.g. Asquith *et al.* 1999; Cordeiro and Howe 2001; Chapman 2005), but other than the extinction of the Cuban vampire bat *Desmodus puntajudensis* following the extinction of the island's megalonychid sloths (Suárez 2005), almost nothing remains known about the wider ecological associations of West Indian mammals, or potential ecosystem shifts or trophic cascades following their disappearance; in fact, the quality of the West Indian Quaternary record may be too poor to evaluate such hypotheses. Finally, despite recent attempts to constrain West Indian mammal extinction chronologies, direct radiometric last-occurrence dates are still only available in the published literature for 11 of these extinct mammal species, and even in these cases it remains difficult to identify primary extinction drivers from the complex sequence of pre-Columbian and historical-era human impacts experienced in the Caribbean region (see e.g. Watts 1987; Wilson 2007).

To understand the full extent of human impacts on global systems, it is imperative to consider data from the recent past together with better-resolved data from the present. This new dimension of information has provided invaluable new insights into patterns and processes of evolution, ecology, biogeography, and extinction. However, in many cases the information available from the Quaternary fossil record remains both incomplete and qualitatively different to that provided by neontological research. It is clear, therefore, that palaeontologists, ecologists, and conservationists must all recognize the limitations as well as the opportunities provided by such comparative analyses, and that substantial further research is still required before Quaternary species diversity and ecosystem functioning can be fully quantified.

Acknowledgements

We thank Julian Hume for discussion and suggestions, and three reviewers for providing new ideas and helping to improve an earlier draft of this chapter. Support was provided by a NERC Postdoctoral Fellowship.

CHAPTER 11

Holocene deforestation: a history of human–environmental interactions, climate change, and extinction

Rob Marchant, Simon Brewer, Thompson Webb III, and Samuel T. Turvey

11.1 Introduction

Understanding human–environmental interactions is central for assessing present-day human impacts on forest ecosystems. The process is not unidirectional, however, and a Late Quaternary view of human–environmental interactions shows that the physical and biological environment has influenced the nature and development of human civilizations and their associated environmental impact (Zolitschka *et al.* 2003). Deforestation is a phenomenon that has shaped terrestrial ecosystems extensively throughout the Holocene, both in temperate and tropical regions. In areas where there has been a long history of intensive exploitation of forests for fuel, building material, or purely to clear for agricultural production, formerly widespread forest cover has been heavily impacted leading to little, if any, original forest cover remaining (Fig. 11.1). Pre-industrial human populations across the world frequently had intensive impacts on their local environment, resulting in many forest ecosystems being highly modified. Relatively few plants appear to have become extinct through this process. This may be a remarkable testament to the ability of these species and forest ecosystems to adjust their ranges under a backdrop of climate change and human impacts, or may alternatively reflect our lack of knowledge about recent extinction events in groups other than mammals or birds. As deforestation escalates with continued industrialization and the onset of globalization, the future ability of forest species to adjust will certainly be curtailed, and this past resilience to extinction is not likely to be maintained under future predicted climate states (Thomas *et al.* 2004). With the majority of original forest cover cleared from industrialized areas such as Europe (Fig. 11.1), the geography of deforestation has shifted to the formerly extensive tropical and northern coniferous forests. As human needs for lumber, firewood, and agricultural land grow, forest uses for maintaining clean water supplies, for carbon storage, for soil integrity, and for species preservation will need to be valued and appreciated if natural forest resources are to be preserved.

In this chapter we describe some of the extent of past human impacts on forest ecosystems in Europe, the Americas, Africa, and the Pacific region. We also outline approaches used to reconstruct human impacts and the spatio-temporal signature of these impacts on forests through a series of case studies. Forests are dynamic associations of tree and other plant taxa responding to climate change as well as to human migration, population increase, and associated land-use impacts; hence the structure and composition of forest communities was changing in concert with climate long before any human impact. The time frame of significant human impact, the Holocene (approximately the last 11 500 years), was undoubtedly a diverse period with changes in technology, subsistence, population growth, and ensuing mobility creating new possibilities for social arrangement with associated ecosystem

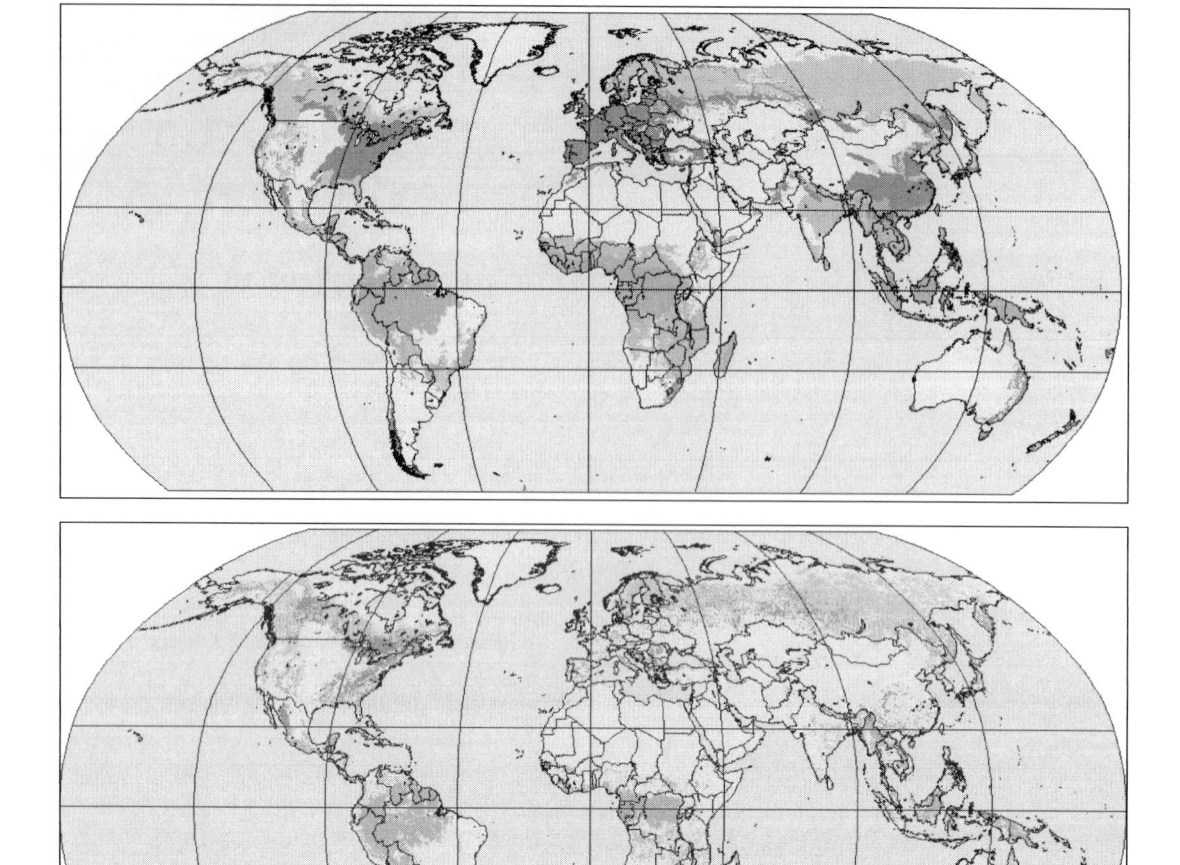

Figure 11.1 Paired images showing the extent of original forest cover in four broad categories prior to extensive human-induced forest clearance under interglacial climate conditions, compared with that of present-day forest cover following extensive deforestation. The four main forest types spreading out from the equator are Tropical Rainforest, Tropical Dry Forest, Temperate Broadleafed and Mixed Forest, and Needleleafed Forest. WCMC-UNEP Cambridge.

impacts. For example, the period between about 8000 and 3000 years BP witnessed temple mounds constructed in Peru, pyramids built in Egypt, and settled agrarian societies established in many areas globally as civilizations rose and fell. We cannot deal with human impacts in isolation from environmental change; indeed, the physical, biological, and climatic environment has influenced the nature and development of human civilizations and associated environmental impacts. Given the evidence of complex spatial and temporal variability in Holocene climate (Zolitschka *et al.* 2003), palaeoecologists question the extent to which human activities were the ultimate cause, as opposed to the proximal trigger, for many palaeoecological changes detected within the sedimentary record. Both archaeologists and palaeoecologists continue to confront the possible causal connections between changes in Holocene environment and human populations around the globe.

11.2 Methods to reconstruct environmental change and human impacts

As seen in Chapter 1 in this volume numerous methods are used to trace the impacts of environmental change on ecosystems. However, establishing cause and effect is difficult, and how human activities may have facilitated in extinction events even more so. Past records from tree rings and sediments show that Holocene drought and flood episodes have been more dramatic and more persistent than any during the instrumental record, which contains only a small sample of the full range of climatic variability within the current interglacial period (Gasse 2002). One good barometer of longer-term climate change comes from sediment records on lake-level fluctuations. Cores of accumulated material within lake basins record changes driven by differences in temperature and precipitation regime within the catchment (Fig. 11.2). Records on past climatic variability show that the only constant in the history of the Earth's climate is change, with the magnitude and direction of change varying depending on the location and nature of the climate system under investigation. With a concentration of research activity within temperate latitudes, the data of both environmental change and human activities are of high quality and quantity and the subsequent understanding much more constrained than within tropical latitudes. Increasingly, a similarly complex picture is emerging from the tropics, particularly with the developing comparison of phase shifts and the timing of impacts. For example, one significant pulse of climate change centred around 4000 years BP is most strongly recorded within the tropics: tropical South America experiencing a change to higher precipitation and/or a shorter dry season, in contrast to tropical Africa, which shifted towards dry conditions, likely due to reduced precipitation, increased evaporation, and/or an extension of the dry season (Marchant and Hooghiemstra 2004).

A range of techniques provide evidence on how cultures impacted their environments. This evidence ranges from direct analysis of changes in past occupation layers revealed by archaeological investigations to documentary, genetic, and linguistic evidence. For example, pottery styles recovered from archaeological sites are used to trace the migrations of different cultural groups (Soper 1971), particularly if pottery styles succeed each other regionally and temporally within the area under investigation. Ancient documents and hieroglyphs provide a rich source of information on past cultures and surrounding environments (Currie and Fairbridge 1985; Fang and Liu 1992). As a caveat, this type of evidence has to be assessed cautiously; in assessing such direct evidence we need to remember that propaganda could have been as rife in ancient populations as in modern politics and business (e.g. advertising). Linguistic analysis also provides insight into past human populations (Nurse 1999; Vansina 1999). Again there are interpretative difficulties; for example, when one notes that languages come into contact with one another, this does not mean that people speaking those languages had to move; languages often differentiate and migrate *in situ* (Nurse 1999).

Based upon the assumption that major changes in activities of humans will alter catchment conditions, information about past human activities is often derived from sedimentary sequences, the impact being commuted to sediment archives. Although this cause and effect is likely to be recorded when population levels are relatively high, evidence for human presence and activity may not be preserved in the sedimentary record, and therefore an absence of human indicators does not mean that humans were absent. Recognizable remains of plants used in agriculture (e.g. bananas, plantains) can also be a useful addition to direct archaeological information. For example, about 4000 years BP domesticated sunflower seeds and cotton pollen appear in Mesoamerican lowlands, marking an expansion in farming which gave rise to one of the New World's first complex societies, the Olmec (Pope *et al.* 2001). Increased demand for land to support food production for a growing population resulted in deforestation. Evidence for this forest clearance comes primarily from four main sources: (1) botanical remains from archaeological sites, (2) vegetation records obtained from perennially wet

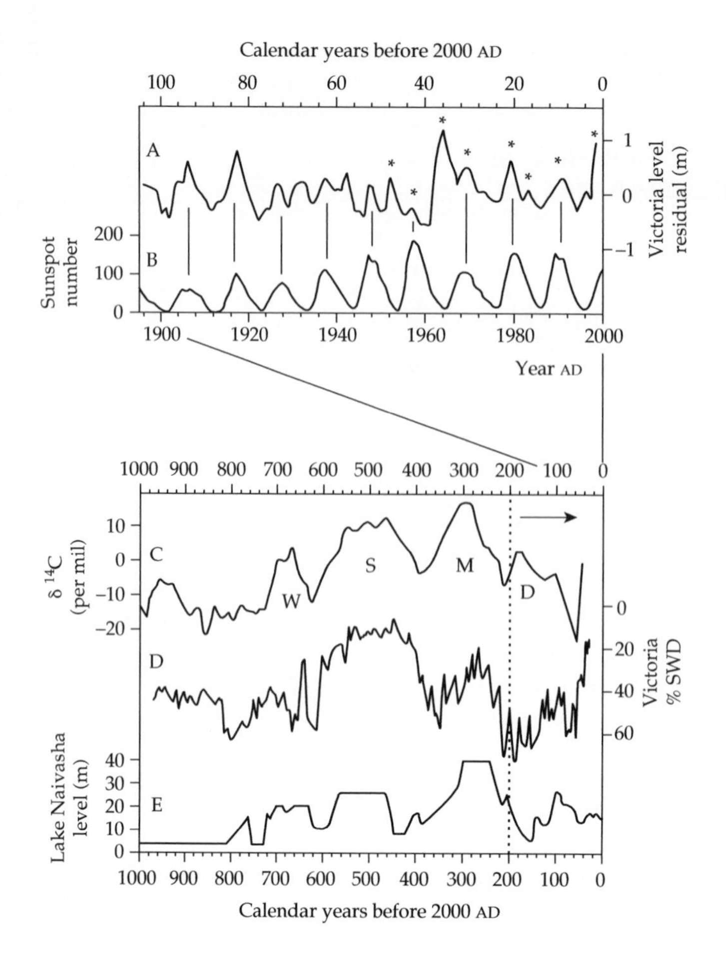

Figure 11.2 Tropical areas are highly responsive to environmental change, as shown from these paired lake-level records from East Africa. (A) Detrended Lake Victoria water levels with asterisks indicating periods when there were outbreaks of Rift Valley Fever (relatively wet periods characterized by high lake level). (B) Number of sunspots, used as a proxy for solar activity. (C) ^{14}C isotope residual series plotted as parts per thousand (per mil); this is used as a proxy for East African climate–sun relationships. Wolf (W), Sporer (S), Maunder (M), and Dalton (D) sunspot minima are marked. (D) Percentage of swallow-water diatoms (SWD) from Lake Victoria sediments accumulated over the past 1000 years. (E) Lake level of Lake Naivasha: present depth is around 20 m with some periods in the past reflecting a 20 m rise or fall (Vershuren *et al.* 2000). The upper record from Lake Victoria (Stager *et al.* 2005) shows that some of the recent high stands of the lake are coincident with high solar activity over the past 100 years.

areas, (3) settlement characteristics, and (4) molecular prints of living crops and their wild ancestors (Piperno and Pearsall 2002). Other information on past human activity comes from the present-day ecological composition of vegetation communities. For example, in north-east Guatemala the patchwork of tropical semi-evergreen forest interspersed with patches of savanna is believed to be a relict of past land-use practices by the Maya (Leyden 1987).

The results from techniques outlined above have contributed much on what we know about the impacts of early farmers, but many gaps remain. As with the palaeoenvironmental data, the amount and quality of archaeological and historical information is skewed towards certain locations and time periods: for every tonne of carefully sifted soil in Egypt, less than a teaspoon has been glanced at in West Africa (Pearce 1999). In spite of these problems, the Holocene is characterized by recognizable cultural developmental stages that allow transitions to centralized, socially structured societies observable today to be reconstructed (Bower and Lubell 1988), and the associated human impact on forests to be assessed. In our brief review, we have concentrated on a series of case studies from Europe, Latin America, Africa, and the Pacific Islands. For North America, we provide a broad-brush description of vegetation changes during the Late Quaternary and compare the magnitude and clarity of human impact before and after settlement by Europeans, which began 500 years ago. The relatively youthful record of major human impact in North America contrasts with the record in other regions, which have a greater longevity of traceable human impacts on the environment.

11.3 Europe

The European region has been extensively studied for changes in both cultural activity and the landscape across the Late Quaternary (e.g. Berglund *et al.* 1996). There is a vast quantity of archaeological data and historical documentation of populations and their activities, and an extensive network of sites from which sedimentary archives are available. We present here an overview of the major changes of the European landscape and the different impacts linked to different stages of its cultural population.

Human activities have been the major factor controlling the structure and composition of European forests over the latter part of the Holocene (Bradshaw and Hannon 2004). Not only have these impacts been widespread in area, but they have also occurred over a long period of time (Moore 2005). For this, we may start with the establishment of forests following the last glacial period, the last large-scale non-anthropogenic restructuring of the European landscape. During the last glacial period, the dry and cold climate led to a widespread distribution of tundra and open steppe-like vegetation in Europe, with temperate trees scattered in small isolated 'refugia', where a milder microclimate existed, often due to local topographical conditions. The end of the glacial period and the concurrent changes to a warmer climate with increased moisture availability led to the expansion of these scattered populations and the recolonization of the continent by temperate tree populations. This recolonization took place surprisingly fast, with the establishment of fairly substantial populations of some species in the north-west of Europe within only a few thousand years of the end of the glacial period. This has been taken as evidence of either the widespread nature of these refugia (Stewart and Lister 2001) and/or an accelerated spread caused by infrequent long-distance dispersal of seeds by animals or birds (Brewer *et al.* 2002). Even prior to the start of intensive human impact on the landscape, the composition of the forests was not fixed, with changes due to ongoing climate change and migrational processes. The early pioneer forests of willow (*Salix*), birch (*Betula*), and pine (*Pinus*) were replaced by a mixture of oak (*Quercus*), hazel (*Corylus*), and alder (*Alnus*) with varying amounts of lime (*Tilia*), elm (*Ulmus*), and ash (*Fraxinus*). These mixed oak forests would eventually become occupied by beech (*Fagus*) and spruce (*Abies*).

In the first part of the Holocene, prior to significant human impact, the vegetation of lowland Europe was originally believed to have been closed canopy deciduous forests (Peterken 1996; Bradshaw and Mitchell 1999). This preconception was questioned by Vera (2000), who suggested that the landscape of Europe would instead have been more open, consisting of a dynamic mosaic of four vegetation types: open parkland, regenerating scrub, groves of deciduous trees, and a 'break-up' type as the canopy of the grove opens out due to the death of trees and returns to the first vegetation type (see also Kirby 2003). The open nature of this landscape would have been maintained by grazing pressure from large herbivores (Vera 2000) as well as fire (Svenning 2002). Indeed, a much greater diversity of herbivores existed across lowland Europe in the early Holocene than the present, including deer, elk *Alces alces*, aurochs *Bos primigenius*, bison *Bison bonasus*, wild horse *Equus ferus*, wild boar *Sus scrofa*, and beaver *Castor fiber* (Bradshaw *et al.* 2003; Mitchell 2004). The main evidence in support of this theory was the continual presence and relatively high values of oak and hazel in pollen records covering the Holocene. Both of these taxa are light-demanding, and their persistence would require the existence of openings in the forest.

More recently, the validity of this theory has been questioned (Moore 2005). Mitchell (2004) compared forest development in Ireland with that in the rest of north-west Europe. Ireland provides an ideal test case, as, due to the early Holocene isolation of the island from mainland Europe, the available evidence indicates that the number of large herbivores was much reduced in both diversity and density, with only wild boar, red deer *Cervus elaphus*, and possibly elk present for some of the interval (Woodman *et al.* 1997). Despite this substantial faunal difference, no significant difference was found between Holocene vegetation development in Ireland and the continent, where a greater diversity of herbivore species and ecotypes was present. Although large mammalian keystone herbivores can attain saturation densities at which they may radically transform vegetation structure and composition (Owen-Smith 1987), there is debate as to whether the European herbivore population was sufficiently dense or diverse to have the impact described by Vera (2000). First, the number of herbivore species in Europe was reduced by more than half when compared to the preceding interglacial period (Bradshaw *et al.* 2003). Second, the scarcity of remains means that it is difficult to establish robust estimates of past herbivore density.

The exact structure and composition of these forests remains therefore under debate. There is, however, a general agreement that the role of herbivores, and therefore the degree of openness of the forest canopies, has been underestimated. A more likely scenario would lie between these two extremes of open or closed canopy, with herbivores maintaining openings in the forest caused by fire, flooding of soils, or wind (Bradshaw and Hannon 2004).

11.3.1 Anthropogenic influence

The second part of the Holocene in Europe is characterized by significant human influence on the vegetation. The start of these impacts is concurrent with the spread of Neolithic agriculture, beginning around 8000 years ago in the south-east, but occurring later (around 5500 years ago) in the north and west (Roberts *et al.* 2004; Moore 2005; Fig. 11.3). Humans have had a number of different effects on forests, not all of which have been destructive. These fall broadly into two types. The first category may be considered as those impacts that result in a loss of trees. This includes the slash-and-burn clearance of trees to obtain land for agriculture, but also selective felling, thinning, burning, litter collection, and grazing of domestic stock (Bradshaw and Hannon 2004). A second category includes those actions that have altered the structure and/or composition of the forest. This includes coppicing and pollarding, where trees are kept in a form to facilitate the harvest of young growth and provide a continual source of wood (Rackham 1980). Forest composition has also been greatly influenced by the introduction of exotic species. This is not a new phenomenon, and was first recorded by the Romans, who introduced a number of species to provide a supply of food, such as walnut *Juglans regia*, chestnut *Castanea sativa*, and vine *Vitis vinifera* (Bradshaw 2004). When these impacts are reduced, however, the forest is able to rapidly re-expand into former areas. This has been notable in Europe during times of war or widespread disease, such as the Black Death (AD 1347–1353) (Williams 2000), but it has also been a feature of much of the last century, as discussed below.

11.3.2 Chronology of impact

The Neolithic period marks the beginning of significant human impact on European forests. As agricultural practices spread from the south-east into the rest of the continent (Roberts *et al.* 2004), open land for grazing and for crops became increasingly needed. Early settlers used slash-and-burn techniques, where the natural vegetation was cut then burned to clear the land for farming. Clearances obtained in this way were then farmed for a few years until the fertility of the soil declined. This practice gave rise to a so-called shifting agriculture and allowed open land to revert to forest after the crops had been gathered. Recent work has shown that the impacts during the Neolithic were more varied than previously thought (Williams 2003), and as time progressed there were a greater number of long-term settlements. These were mainly on

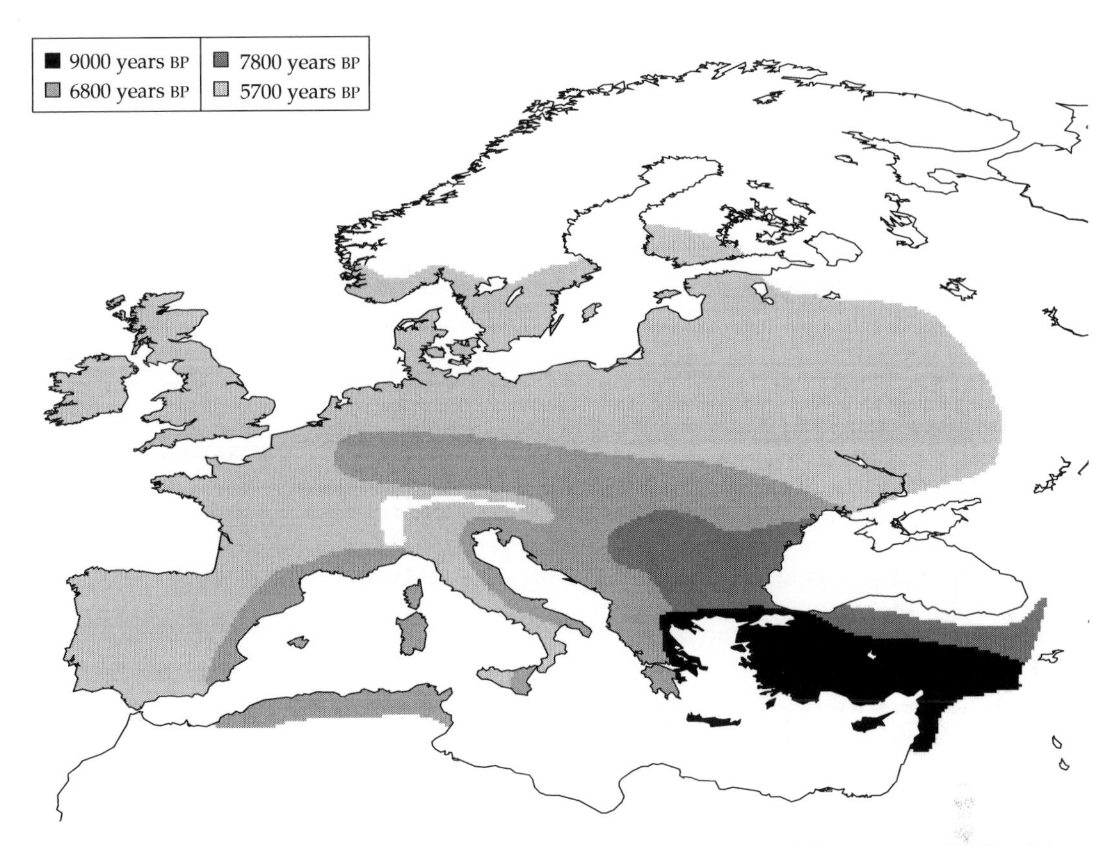

Figure 11.3 Synthetic map of the spread of Neolithic agriculture in Europe. Ages are given as years before present. Modified from Roberts (1998).

loess deposits in the Pannonian Basin and in the lowlands of Central Europe. These regions were fertile and easy to work, and the existing forest cover was opened up to allow cultivation.

These early impacts would have changed the existing forests in a number of ways (Delcourt 1987b). By increasing the frequency of disturbances through grazing and clearing, the degree of openness of the forest increased. The existence of semi-permanent or permanent settlements led to areas being maintained in an open state for crops or meadows, leading to the invasion and spread of weed species. The selective use of individual plant species would have resulted in a change in forest composition. Finally, the distributional limits of the forests would have been altered. Little evidence has been found for the transplantation of species during this period, with the possible exception of species that were used for food, for example stone pine *Pinus pinea* (Bradshaw 2004). However, openings may have helped the establishment of other forest species, for example of lime and elm in Scotland (Tipping *et al.* 1999).

From the start of the Bronze Age (*c.*5000 years BP), a notable shift occurred, as the colonization of north-west Europe became increasingly widespread. The model of shifting agriculture was now entirely abandoned, which further increased the stress on the landscape by limiting any regeneration (Fanta 2005). In addition to clearing for agriculture and grazing, there was an increasing demand for wood for construction, and as fuel for both domestic use and the production of metals. The period of Classical Antiquity that followed the

Bronze Age, which lasted from approximately 3000 to 1000 years BP, was a period of rapid development of human societies, most notable in the northern Mediterranean, including the period of the classical Greek and Roman empires. The demands of a growing population, the move towards urbanism, and the need for boats for trading led to more intensive use of woodland products with further impacts on forests in the south of Europe (Williams 2000). Overgrazing and fire prevented the regrowth of these forests, resulting in a scrub-like maquis or garrigue.

The Middle Ages have been described as the moment that humans changed from being a part of the landscape to exploiting the landscape (Williams 2003). Rapid growth in population density led to a shift in focus from the south to the north of Europe. This expansion occurred in two ways. During the early part of this period land that was previously considered unusable was developed or reclaimed for farming use (Williams 2000). This change was made possible through the development of the plough, which allowed the use of heavier, clay soils for agriculture, and other tools that resulted in a more efficient clearance of the forests remaining in these areas. Later, expansion took place into previously unsettled regions in the centre and east of the continent, notably between the eleventh and thirteenth centuries AD. The rapidity of this expansion was, at least in part, due to the increasing organization of the exploitation of landscape, as more of the region came under some form of political control. A further outcome of the greater population density was a move away from pastoral toward more arable farming, requiring the availability of larger areas of cleared land for an equivalent population. By the end of the twelfth century, the extent of the change in landscape was sufficient for the introduction of the first forestry laws controlling the exploitation and rights of use of the remaining forests.

Following this period of rapid expansion, historical records show increasing occurrences of crop failure and a general reduction in agricultural activities. This was partly caused by the transition to the Little Ice Age, a period of unstable climate with a trend towards lower temperatures (Lamb 1977). By the thirteenth century, the climate was increasingly unsuitable for agriculture, leading to a shortage of supplies and the abandonment of farmland. The decrease in cultural activities was accelerated by the spread of the Black Death throughout Europe, which is believed to have killed approximately one-third of the population. Where agricultural activities continued to exist, there was a return to pastoralism, due to the reduction in the quantity of available labour for crop growing (van Hoof *et al.* 2005). Detailed studies of past vegetation change have shown that the re-establishment of forest on abandoned land may have taken as little as 50 years (Stebich *et al.* 2005; van Hoof *et al.* 2005). This regression was, however, short-lived and there was a return to the practices of land clearance and widespread crop growing from the start of the sixteenth century.

The period following the Industrial Revolution led to an important migration from the country to the cities in Europe. The decline in rural populations lowered the pressure on the landscape, and over the last few decades this has resulted in an increase in forest cover across Europe (Williams 2000; Bradshaw 2004). This reforestation has also been affected by commercial planting and a reduction in grazing, due to a change in agricultural methods towards more intensive but less widespread land use (Bradshaw and Hannon 2004). The changes during this period are therefore quite different from the preceding periods, with an increase in forest cover that results from anthropogenic processes. This is not without implications for the forest, as the transfer of seeds for plantations will have affected the phylogeographical structure of species (Bradshaw 2004).

11.4 North America

The North American record for human impacts on the vegetation picks up more or less where the account of agricultural practices and settlement patterns in Europe leaves off. Within the past 500 years, European settlement in North America initiated the period in sedimentary records with unambiguous widespread evidence for large-scale and major human impacts on North American forests and vegetation (Bernabo and Webb 1977; McAndrews 1988; Williams *et al.* 2004). Before the arrival of Europeans, the record is rather scarce, but much changed after their arrival with the clearing of forests and prairies for farming and settlement. The extent and magnitude of the changes to the vegetation far exceed any changes attributed to Native Americans during the Holocene, even though they practised agriculture in many regions of the continent. The first appearance of Native Americans in large numbers 12000 or more years ago, however, coincides with and probably contributed to the extinction of many large mammal species (Martin and Steadman 1999; Barnosky *et al.* 2004). Both human invasions of North America, first from the west by ancestors of Native Americans and later from the east by Europeans, therefore resulted in significant environmental impacts within a relatively short time of their arrival: the first invasion contributed to mammal extinctions during a time of dramatic climate and vegetation changes, whereas the second invasion resulted in a major transformation of the North American landscape (Flannery 2001). That transformation continues today at an accelerated rate.

Studies estimating the rates of change in pollen records in eastern North America show that changes in the last 500 years are large enough to match those in previous times of large climatic change during the past 21000 years (Bernabo and Webb 1977; Jacobson *et al.* 1987; Shuman *et al.* 2005), and maps of the changes in pollen abundance over the past 500 years show that the spatial scale of the changes match those induced earlier by climate (Bernabo and Webb 1977; Williams *et al.* 2004). The recent disturbance by European-derived practices is therefore clear-cut and unprecedented in the palaeoecological records from regions south of the boreal forest in eastern North America. Increased abundance of *Ambrosia* and *Rumex* pollen and decreased abundance of tree pollen create a well-pronounced European settlement horizon at the tops of long cores from lakes and mires (McAndrews 1988) and marks the time when large areas of forest were cleared and cultivated field and pastures allowed annual weeds to grow in abundance.

What is a mystery is that so little direct evidence of human impact exists during the bulk of the Holocene from 11000 to 500 years ago. This was a time when the cultures of Native Americans developed and changed, their populations grew into the millions, and they settled throughout the continent. By the Late Holocene, they had settled in the high Arctic, domesticated plants, and begun making pottery. The latter two developments qualify them as Neolithic peoples. Why then does their low level of forest destruction and environmental pollution stand in such contrast to what Europeans and their descendents have produced? The palaeoenvironmental record of Native American activity during this time is similar to the record in Europe during Mesolithic times. Evidence for human activity can be found, but it is scarce in space and time and generally small in magnitude and in area impacted (McAndrews 1988). For example, relatively large increases in *Ambrosia* pollen percentages or occurrence of potential cultigens at archaeological sites or in river valleys during Woodland to Mississippian times (1500–500 years BP) do not carry over to nearby upland sites (Delcourt 1987b; McLauchlan 2003), whereas today almost all sites in agricultural regions record big increases in the amount of *Ambrosia* pollen and decreases in tree pollen (Webb 1973; McAndrews 1988; Russell *et al.* 1993).

Part of the answer to the mystery is that Native Americans may have cleared and/or used relatively small areas for agriculture (Spector 1993) and thus did not disturb the landscape in general beyond the level of natural disturbance within the pollen-source region for most lakes. The presence of Native Americans is therefore hard to detect against the noise of blow-downs, fires, and disease that were always prevalent within the area contributing pollen to lakes. We have plenty of evidence that the pollen record from lakes and mires is sensitive enough to record human activity; but, unless human disturbance passes the threshold for disturbance occurring naturally on the landscape, no clear evidence for their activities may be noted in sites with relatively large pollen-source regions. The level of clearance and disturbance by European-style practices of agriculture was well above the threshold, in stark contrast to the practices of Native Americans.

Another part of the answer can be that sites have not been studied in regions with the densest human populations. Cores from lakes near sites where large populations of Native Americans may have lived might yield more telling results, but such areas are yet to be identified and studied. Evidence of Mayan disturbance is clear in pollen diagrams from lakes in the Yucatán Peninsula (McAndrews 1988) and also on the landscape (Leyden 1987; see also section on Latin America, below). On the other hand, sampling of lakes on Cape Cod, where relatively dense populations of Native Americans may once have lived (Thoreau 1951), has so far yielded almost no direct evidence for their activities (Parshall and Foster 2002).

Day (1953) published a much-cited paper attributing widespread burning of the forests in the north-eastern USA to Native Americans. He based his evidence on written accounts by early settlers, but a critical review of these accounts by Russell (1983) showed the evidence to be equivocal as to the amount of clearing. Many of the accounts were from men who were promoting settlement in North America and must be judged as forms of advertising rather than accurate records. Nevertheless, Russell (1997) cited historical records of early colonists witnessing large fires. We are a long way from deciding how much of the fire damage in prehistoric times was caused naturally in comparison with that set by humans. Because charcoal evidence in sediments shows fires in many ecosystems (Clark 1990), those researchers and authors who attribute a large role to Native Americans in setting fires (Williams 2000) can make claims for widespread impacts by humans on the composition and structure of forests, savannas, and prairies. What can appear to be a strong connection between changes in charcoal and changes in certain pollen taxa, for example *Castanea*, at single sites (Delcourt and Delcourt 1998) can also, when mapped at a continental scale, appear to be well correlated with climate changes (Webb 1988; Williams *et al.* 2004). One piece of evidence that counters some of the impacts attributed to Native American burning is the westward movement of forest into prairie during the past 3000 and even 6000 years in the upper Midwest while the population of Native Americans was increasing in that region (Webb 1988; Williams *et al.* 2004). Evidence of rising lake levels in this region shows that climate was the driving force for much of this broad regional change (Bartlein *et al.* 1984; Shuman *et al.* 2002), even though some studies show evidence for water level decreases in central Minnesota in the past 1000 years (Umbanhower 2004). Unpacking what is fire-caused from what is climate-caused or what are human-set fires from those that had natural ignition sources remains a challenge. The significance of pre-European fire management in determining landscape-level vegetation composition has also been debated extensively for other continental regions, notably Australia (Flannery 1994; Bowman 2001).

Another form of human impact on the vegetation has been the unintentional spread of disease to selected tree species, especially recently from other continents. Both the chestnut blight on *Castanea dentata* starting in 1910 and the Dutch elm disease on *Ulmus americana* from the 1950s onward are examples of exotic pathogens that have drastically changed the composition of the canopy in many eastern deciduous forests (Patterson and Backman 1988). Other pests are also now rampant among North American trees, including the woolly adelgid *Adelges tsugae* on hemlock *Tsuga canadensis*. The frequency of such attacks on tree species has gone up drastically since prehistoric times. Before European settlement during the Holocene, only one epidemic among trees species is evident, among hemlocks in eastern North America. Mid-Holocene pollen records from New England and the Maritime Provinces across to the hemlock's western limit in northern Wisconsin show an abrupt synchronous decline in the abundance of *Tsuga* pollen, which 40 selected sites show as having an average age of 5500 years BP with a standard deviation of 380 years (see Bennett and Fuller 2002, who neatly updated Webb 1982). One recent consequence of global trade and travel is the spread of such diseases from regions where the tree species are resistant to regions where the species are susceptible. The combination for new diseases in a near-future time of rapid climate change on a highly disrupted landscape could lead to rapid population declines and perhaps to extinction for some plant species.

11.5 Latin America

This section will initially concentrate on three areas of Latin America (lowland tropical forest, Peruvian Andes, and the Mayan lowlands of Central America) where there are contrasting impacts of past human–forest interactions. Knowledge of pre-Hispanic civilizations and their impacts on, and interactions with, the environment comes mostly from archaeological excavations. The spatial distribution of these sites is quite biased towards the Andes; this stems partly from the range of relatively easily accessible sites in this region and partly from the former relatively high human population densities in the Andes. The lack of information from the lowlands means that the impact of early populations remains largely unknown and controversial (Bahn 1993). Whether Amazonian populations were derived from migrants passing through Central America or up the Amazon River is still open to debate, although it is likely that a series of influxes have taken place. Speakers of Macro-Arawakan languages are believed, on the evidence of lexicostatistics and archaeology, to have colonized most of the vast Central Amazon floodplain about 5000 years BP (Lathrap 1970). There has been a running debate about whether people could survive at high population densities within lowland Amazonia, as the forested environment provides scarce resources for human subsistence; although there are abundant prey species, these occur at relatively low densities and reliance on bush meat invariably necessitates a nomadic lifestyle and relatively small populations. In addition, relatively poor soils and a profusion of biting insects can make for an uncomfortable life (Roosevelt *et al.* 1996; Pringle 1998). However, evidence for substantial prehistoric human occupation of the lowland tropical forests of Amazonia is increasingly becoming more abundant, and in some cases it is well organized in regional sequences (Gnecco 1999). A good example is the extensive fish traps and floodplain management along the Rio Negro (Silva *et al.* 2007). Archaeological investigations demonstrate that human occupation from Colombian Amazonia date back to 9000 years BP (Gnecco 1999). As with Africa, it is likely that ecotonal areas were particularly important sites for these early populations, such sites being characterized by multiple environments that would increase the range of foodstuffs available to sustain the resident and growing populations. The localized impact of these early populations on their environment could have been significant. Evidence suggests that transformation of moist forest to secondary growth forest ecosystems and eventually savanna was initiated by Amerindians through the frequent use of fire used for clearing lands for maize *Zea mays* cultivation (Cavelier *et al.* 1998). Cultivation of *Z. mays* is recorded at several sites in the Colombian savanna from the middle Holocene (Berrio *et al.* 2002). Further north, *Z. mays* and squash (likely to be *Curcurbita moschata*), manioc (*Manihot*), yams (*Dicorea*), and arrowroot (*Maranta*) residues on milling stones dated to 6000 years BP provide direct evidence for the early transition from foraging to food production in lowland neotropical forest (Piperno and Pearsall 2002). From 3000 years BP dried cobs of maize and decorated gourd fragments directly document prehistoric agriculture within a lowland tropical rainforest (Roosevelt *et al.* 1996). Naturalized species such as cashew *Anacardium occidentale* appear to have spread into lowland forest areas following increased forest clearance and/or reduced precipitation (Roosevelt *et al.* 1996). Thus, archaeological and enthnobotanical evidence indicates that the lowland rainforests of Central America and Amazonia, once thought to be virgin, were at least in some parts settled, cut, burned, and cultivated repeatedly during the Holocene (Roosevelt *et al.* 1996; Willis *et al.* 2004). However, these impacts would have been relatively minor; as with Africa it was not until the colonial period, and subsequent commoditization of forest resources, that significant forest clearance started. Deforestation developed rapidly, particularly with the growth of plantation forestry (for rubber production), demonstrating the significant role that cash crops play in deforestation and land cover change. The vast clearance of millions of hectares of lowland rainforest paved the way for alternative land use such as ranching and sugar cane plantations.

Andean civilizations offer exceptional insights into human–environmental interactions; for millennia humans have dramatically changed the forests and landscapes of the high Andes as agricultural and social structures developed and adapted to the ecology of the area (Thompson *et al.* 1988). For example, prehistoric Peru was characterized by a wide range of economies including both hunter-gatherers and pastoralists with sophisticated agricultural communities (Headland and Reid 1989). Andean civilizations developed in a range of different environments; for example, in the Jauja-Huancayo basin of Central Peru there is a long history of semi-nomadic camelid pastoralism dating back 7000 years or more, this economy being well suited to the harsh high-altitude ecosystem (Browman 1974). Nomadic pastoralists associated with hunting and gathering were supplanted about 3800 years BP by sedentary agricultural villages (Browman 1974). The presence of crop taxa such as *Chenopodium quinoa*, *C. pallidicaule*, and agricultural reeds is recorded from 4200 years BP from the Cuzco region, Peru (Chepstow-Lusty *et al.* 2003). By developing irrigated agriculture in harsh environments to sustain increasingly large populations (Binford *et al.* 1997) these agricultural systems would have significantly impacted the surrounding montane forest, so much so that the altitudinal position of the Andean forest line remains open to debate given the longevity of forest clearance in these marginal lands (Wille *et al.* 2001). One of the main climatic mechanisms to impact the Peruvian resource base, particularly through moisture variability, is changeable activity of the El Niño Southern Oscillation (ENSO); these events were absent or significantly reduced between 8800 and 5800 years BP (Sandwiess 2003). The mid-Holocene onset of ENSO and the associated impact on coastal resources, as recorded in archaeological deposits along the Peruvian coast, may have facilitated a number of cultural transitions and subsequent impact on forest resources, with greater connections to highland societies (Núñez *et al.* 2002). A possible stimulus to the growth of trade between maritime hunter-gatherer communities and those operating inland would be a sudden disruption of resource provisioning brought on by a series of ENSO events. The change in ENSO frequency was coincident with the abandonment of the monumental temples that had been built along the central and northern Peruvian coast (Keefer *et al.* 1998; Sandweiss 2003). Concomitant shifts to increased organizational complexity and social integration also took place in the high Andes, with further impact on forest cover (Dillehay *et al.* 2003). For example, intensive cultivation in raised fields, a system inaugurated by the pre-Inca civilizations with connections to coastal cultures, developed around Lake Titicaca and resulted in extensive forest clearance (Kolata 1986; Binford *et al.* 1997).

Another example of close cultural–environmental inter-relationships, and extensive impacts on the forested environment, comes from the lowland area of Yucatán and Guatemala, settled by the Maya. Today, north-east Guatemala is predominantly tropical seasonal forest interspersed with patches of savanna, with this mosaic reflecting Maya land use and demonstrating the importance of past populations in influencing present-day forest structure. Similarly to the Peruvian case study, this area has a wealth of archaeological information that enables past cultures, their economic base, and environmental impacts to be reconstructed; intensive studies have been driven partly by the richness of the artefacts left behind by the Maya and partly because there was large-scale collapse of the population around AD 1100, taking place over a relatively short time (50–100 years) (Leyden 1987). The Mayan civilization was established from about 2000 years BP, with the Classical Period lasting to 850 years BP; this culture culminated in human population that developed an intricate system of agriculture and ritual and social cohesion. Exponential population growth through the Classical Period led to large settlements, increased monoculture, and consequent expansion into the less fertile Yucatán Peninsula (Hsu 2000). Such a large human population would have had a significant impact on regional forest resources; it may have been this reliance on increasingly marginal lands that led to localized collapse (Leyden 1987; Marchant *et al.* 2004). Many reconstructions of the collapse of the Classical Mayan political, social, and economic systems around 850 years BP emphasize anthropogenically induced failure through interstate warfare and subsequent decline in fertility of agricultural areas (Santley *et al.* 1986). Other researchers have suggested that the collapse was not as catastrophic as is believed, and the magnitude of population decline was not uniform throughout the Mayan region (Rice and Rice 1984). The debate as to the nature and duration of the catalyst(s) that precipitated the rapid population decline continues, in part because human responses appear different in different locations. However, there is a consensus that several years of unstable climate would have been a contributory factor required to undermine an otherwise stable agricultural system (Gunn *et al.* 1995), particularly in more mesic locations. Although this is likely to have been of similar duration and magnitude across the region, the 'on-the-ground' human response would have been quite diverse. The Maya appear to have been able to adapt to environmental changes in some areas; thus, the popular model of human-induced destruction of the complex society is not readily applicable throughout the Mayan territory (Inomata 1997). The impact of environmental change, particularly the proposed shift to drier environmental conditions, would have been most strongly recorded where there were intensive land-use practices and/or population pressure was high (Dunning *et al.* 1998).

11.6 Africa

Given the range of cultures and longevity of African archaeology there is a large amount of information to reconstruct past human impacts. However, this information is quite skewed temporally and spatially, and significant gaps remain in our understanding. This section will concentrate on the spread of agriculture primarily driven by Bantu migration, initially focusing on West Africa and the Sahara before moving into the Nile Valley and the interlacustrine region of East Africa.

Across equatorial Africa one of the main transformations in land cover is associated with the Bantu populations and their migrations, from the population cradle in what is now Cameroon and Nigeria (Schwartz 1992), and as such deserves particular attention in this chapter. Understanding the impact of the Bantu is not a new phenomenon. The 1850s German linguist Wilhelm Bleek discovered a relationship between a large number of languages widely distributed over southern Africa; he called these Bantu after the Zulu term *aba-ntu* (meaning 'men'). Despite such longevity of study, the expansion of the Bantu still remains one of the most controversial agricultural expansions (Diamond 2002); there is also continuing debate as to the direction and timing of the Bantu migration, and the character of resources exploited. What initiated population movement from the Bantu homeland in north-west Africa is also difficult to decipher. Increased population growth and environmental change were probably contributory factors; increased aridity and/or greater seasonality facilitated the rapid spread of agriculture as Bantu-speaking people migrated (Schwartz 1992). Although there is a generally accepted diffusion of Bantu influence, this may have been concentrated along a number of so-called migratory pathways (Fig. 11.4). Following migratory routes, it appears that the first Bantu migrants moved in among, and locally replaced, established groups of hunter-gatherer populations (Zogning *et al.* 1997; Neumann 2005). It is difficult to reconstruct how these migrants were received by the populations that they encountered; increasingly it appears that Bantu-speaking peoples were borrowers rather than innovators, acquiring knowledge of ironworking, seed agriculture, and animal husbandry from contact with their non-Bantu-speaking neighbours (Childs and Herbert 2005). Whatever the timing, nature, and direction of this migration, passage was rapid. The Bantu probably followed river courses or dry ridges within the intervening forest, this not being so dense during this period of relative climatic aridity (Schwartz 1992; Marchant and Hoogheimstra 2004).

One of the most dramatic impacts of changing environmental conditions on human populations and subsequent local extinctions is shown by the severe droughts that have punctuated the Holocene in equatorial Africa. Evidence for a previously much more verdant Saharan landscape is blatant. The distribution of giraffe bones within an early Neolithic context along the Egypt–Sudan border records a previously much more mesic environment (Pachur and Roper 1984). Numerous former Saharan lakes contain faunal remains of white rhinoceros *Ceratotherium simum*, camel *Camelus thomasi*, long-horned buffalo *Pelorovis antiquus*, and warthog *Phacochoerus aethiopicus* (Messerli and Winiger 1992). Evidence of previously much more extensive human populations is also to be found: the fairly widespread distribution of rock paintings depicting hippopotamus,

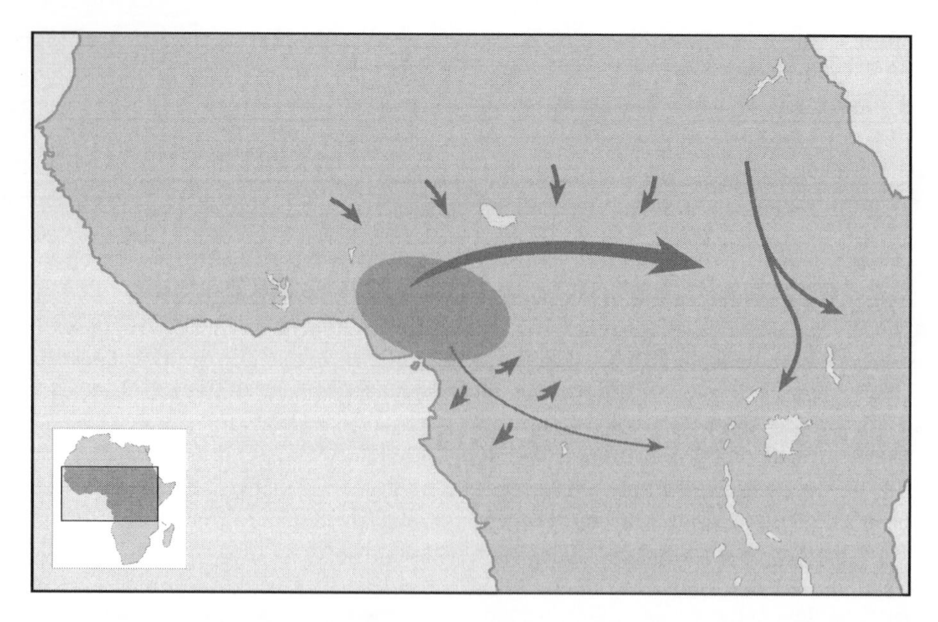

Figure 11.4 Bantu migration routes from the homeland in north-western Africa. Routes are likely to have passed along the Atlantic coastal margins and around and through the Zaire basin. On the northern limit this would have traversed the southern limit of the expanding Sahara, then south down the Nile Valley towards the interlacustrine highlands of East Africa.

elephant, and numerous other savanna species provide direct evidence of permanent water and catchments characterized by mesic savanna and show humans as pastoralists with herds of cattle (Nicoll 2004). The resource base was not just concentrated on pastoralism, fishing, and hunting; archaeological investigations from the Egyptian Sahara have unearthed settled houses with hearths and cooking holes, associated with the remains of some 40 plant species and carbonized seeds of sorghum and millet radiocarbon dated to approximately 8000 years BP (Wendorf *et al.* 1992). Much of the desert became rapidly uninhabitable as the Great Sand Sea was reactivated from the mid Holocene (Besler 2000; Gasse 2002). Increasing aridity is thought to have forced food-producing pastoral populations from the Sahara southward (Stahl 1986; Salzmann and Waller 1998). Rapid aridification can be linked to an Exodus-type event in which humans migrated out of the Saharan and Arabian deserts towards reliable watering sites (Nicoll 2001); one of the main foci of this migration would have been within the Nile Valley.

Population influx into the Nile Valley would have resulted in extensive modification of the land, notably the complete removal of riparian forests and localized clearance of adjacent seasonally flooded forests through an increasingly complex series of drainage and water redistribution schemes. The case of Egypt is an interesting one because it provides an example where the cause-and-effect relationship is complicated. The Nile catchment covers some 30° of latitude, thus encompassing a complex range of environments (Niemi and Smith 1999). Any immediate consequences of environmental change, for example reduced agricultural productivity in Egypt as a result of lower flood levels following reduced rainfall, would be amplified by other factors such as political instability and problems of food distribution (Hassan 2000). Much of the early agricultural transformations along the Nile System focused on agrarian developments and was accompanied by political transformation that took control from villages to provincial states, preparing Egypt for unification into a nation state (Hassan 2000). Historical records show that the Old Kingdom continued successfully until

4160 years BP when it quite suddenly collapsed into anarchy (Wunderlich 1989). Failure of floodwaters to provide nourishment to field systems led to starvation, military weakness, and political instability with the weakened central authority being replaced by local control (Catto and Catto 2004). Geological data from sediment cores collected in the Nile Delta (Stanley *et al.* 2003) substantiate the hypothesis that a very acute drying trend resulted in the failure of the Nile flow; such a change could have played a major role in the demise of the Old Kingdom in Egypt. Thus, the Egyptian Nile provides one of the best examples where climate, hydrological flux, and human development are closely intertwined. Further to the south, a chain of lakes surrounding Wadi Howar supported sedentary hunter-fisher communities from 7300 years BP and pastoralists from 6000 years BP onwards (Neumann 2005), with further increase in pastoral activity about 4500 years BP (Hoelzmann *et al.* 2001). Extensive herds of cattle, goats, and sheep would have had a severe impact on the predominantly savanna woodlands, producing a very open forest structure common today throughout much of the extensive African savanna. The impact of these changes was not only on the immediate area of forest; the cultural and economic importance of the West Nubian lakes region would have acted as a conduit funnelling people south along the Nile Valley towards the interlacustrine region of central Africa.

The interlacustrine region has long been a locus of cultural and socio-economic changes and a major contact zone between varied agricultural practices. This importance has in part been due to the range of environments, mountains, and lakes supporting a diversity of plants, wildlife, and agricultural practices. The transformation to an agro-pastoral lifestyle in the interlacustrine area is thought to be associated with the arrival of the Bantu, rather than independent domestication (Schoenbrun 1993). It is likely that this is part of a general southward movement of pastoralists from north-west Africa after 4500 years BP in a complex, so-called bow-wave migration form (Bower and Lubell 1988). Regional agricultural transformation would have been accompanied by a range of technologies that would have made the pioneers highly effective at modifying land, in particular clearing forest for agricultural production and supporting a developing iron industry. The East African Iron Age is first recorded in the Lake Victoria area around 2500 years BP, where it is thought to have been introduced by Bantu speakers funnelled into the interlacustrine area from the west and north (Pearl and Dickson 2004). Although the regional pattern of spread is controversial, the last few thousand years have seen extensive modification of the forested landscape to support an increasingly settled population practising mixed agriculture. To explain the changing regional vegetation composition and distribution a series of forcing mechanisms (ecological, climatic, and human) need to be invoked. A good example of a 'typical model' of agricultural transition can be seen within south-west Uganda, where some 40 years of palaeoecological research in the Rukiga Highlands (Hamilton and Taylor 1986; Taylor 1990) have led to a good understanding of regional land transformation. It must be stressed that such a 'typical model' is not applicable elsewhere, but it does contain common elements of incoming agriculturalists with associated iron technology and crops, resident hunter-gatherer populations, extensive forest clearance, and development of dense rural populations, with resultant present-day forest remnants restricted to protected areas with sharply defined boundaries (Fig. 11.5). Earliest forest clearance, and replacement by open vegetation characterized by ruderal taxa, is recorded at the highest altitudes from approximately 2200 years BP. For areas that maintained forest cover, a transition to a more open, and possibly drier, form of forest is apparent from approximately 700 years BP; this significant spread in forest clearance to lower altitudes appears to have been in response to a growing agricultural population (Marchant 2007).

As a caveat to such a picture of anthropogenically driven change, increases in aridity or seasonality may also have played a role in changing forest composition. More recently, the driving force in determining forest composition was one of human activity, such as the activities of pitsawyers; gaps originally opened up in the forest appear to be maintained by the activities of elephants and other, more vigorous grazing animals. Despite these major changes in regional land use,

Figure 11.5 The extensively deforested Rukiga Highlands of Uganda following successive and organized expansion of agriculture over the past 2000 years. Formally extensive montane rainforest is now restricted to isolated remnants that are often the focus for protected area status such as Bwindi-Impenetrable Forest National Park (shown).

a large remnant of extensive forest has been maintained: Bwindi-Impenetrable Forest National Park. Why this particular area of forest 'survived' is unknown, although it is likely that there has been some degree of protection afforded by the (until recently) resident indigenous BaTwa population against the colonizing agriculturists (Marchant 2007). For example, one possibility is that a precursor of Bwindi-Impenetrable Forest corresponded to some form of disputed border between territories; the associated economic and political instability would have placed severe limitations on the development of sedentary agriculture. Such a border may relate to the highly centralized societies in the interlacustrine region that became established during the early part of the present millennium (Taylor *et al.* 2000). Indeed, it is possible that any forests which were intact at the time of state formation and located towards the outer limits of a kingdom's sphere of influence were 'protected' as a natural deterrent to potential invaders (*cf.* the last Central European wilderness, The Great Wilderness in former East Prussia, which functioned as a buffer zone between warring countries and provided an unintentional refuge for aurochs, bison, and wild horse until c.AD 1500; see van Vuure 2005). An additional facet to this is that there may have been some degree of protection imparted by the BaTwa against the colonizing agriculturists; although the degree of this protection cannot be quantified, one prominent BaTwa (Bandusya) lamented that his forefathers were once mortal enemies of the pioneering agriculturists, keeping all strangers out of the forest (Kingdon 1990). Whatever the result of the cultural contacts between different groups in the Rukiga Highlands, we can only suggest what may have occurred as modern analogues are unknown. What is quite obvious from the palaeoecological records is that agricultural land use has rapidly spread since approximately 2000 years BP with some protection being afforded to Bwindi-Impenetrable Forest; this protection has continued to present day, now under the guise of National Park legislation.

11.7 Remote Oceania and other island systems

Deforestation was by no means limited to continental regions before the modern era, and islands have experienced some of the worst levels of habitat destruction caused by the activities of early human colonists, in addition to other substantial impacts on their native biotas (see Chapter 2 in this volume). Following the massive-scale dispersal of south-east Asian seafarers across Remote

Oceania (eastern Melanesia and all of Polynesia and Micronesia) from 3600 to 3000 years BP, the native forests of these islands underwent substantial anthropogenic modification. Although there remains tremendous variation in the extent and composition of remaining forest cover between different Pacific islands, original patterns of vegetation are obscured almost everywhere by deforestation and replacement with non-native species and native early successional species, and the composition of an undisturbed forest anywhere in the Pacific region is now uncertain (Steadman 2006b). Native and quasi-native forests are now typically confined to high-elevation areas, and even these are infiltrated by non-native woody plants. Almost all native lowland forest has been disturbed or removed from several islands (e.g. Rarotonga), and on other islands (e.g. Yap) even upland vegetation is wholly or partially anthropogenic. Much of this habitat degradation is the result of active slash-and-burn deforestation by early settlers, particularly for purposes of agricultural clearing; large areas have been degraded by fire and replaced by pyrophitic plant communities dominated by ferns, grasses, and sedges (e.g. *Pandanus* savanna; Fijian *talasiqa* vegetation). The Pacific palynological record typically shows a relative abundance of tree pollen through the Holocene until an abrupt drop associated with first human arrival, when tree pollen declines and is replaced by an increase of grass/sedge pollen and fern spores and the first appearance of charcoal (Steadman 2006b).

However, palaeoenvironmental studies on different Hawai'ian islands have been able to demonstrate that forest decline preceded regional human presence by up to 400 years (as evidenced by a steep decline of native forest pollen before an increase in charcoal particles in sediment cores), after human colonization of the archipelago but before rapid human population growth and expansion. This apparently anomalous pattern probably instead reflects the effects of the Pacific rat *Rattus exulans*, which was introduced across the Pacific region by the first colonists, and presumably underwent a population explosion and rapid radiation throughout the islands ahead of human settlers (Athens *et al.* 2002). Although Pacific rats have been extensively implicated in predating native insular terrestrial vertebrates and invertebrates, plant materials are their primary food; they can penetrate hard, thick seed cases and have been shown to strongly depress seedlings by destroying seeds, underground stems, and leaves, thus halting forest regeneration (Hunt 2007). The increased persistence of native vegetation communities in upland areas may therefore not only reflect the fact that these regions are too steep to cultivate, but also the lowland elevation range of the Pacific rat (Athens *et al.* 2002). The insular vegetation communities of Remote Oceania, which evolved in the absence of native mammalian herbivores and seed predators, were therefore highly vulnerable to the introduction of exotic mammals as well as to the direct effects of human colonists, although the relative importance of each factor in determining the patterns of deforestation observed today across the Pacific region remains to be established.

The most extreme example of ecological devastation in Remote Oceania is that of Easter Island or Rapa Nui, a small (170 km^2) isolated island in the south-east Pacific, which has lost all of its native forest following first human settlement around AD 1200 (Hunt and Lipo 2006). The complete destruction of Easter Island's forests, its perceived association with overpopulation and reckless overexploitation to fuel construction of the island's giant *moai* statues, and the subsequent collapse of the island's society, have been widely promoted as a cautionary example of 'ecocide' (e.g. Diamond 2005); however, the veracity of this paradigm has been the subject of considerable recent debate, and has been influenced by different models of the timing and dynamics of human colonization and forest loss (see Hunt 2007). The island is believed to have supported a relatively simple forest ecosystem dominated by an estimated 16 million *Paschalococos* palms before human arrival, but also containing over 20 other mesophytic forest trees and shrubs (Bork and Mieth 2003; Hunt 2007). Human arrival was followed rapidly by decreasing organic content and a massive influx of charcoal in sediment cores, representing elevated levels of soil erosion and forest fire; these impacts appear to have been abrupt and widespread, with rapid deforestation beginning within the first generations (<100 years) following colonization. However, studies of carbonized plant remains indicate that a wide range of mesic forest vegetation persisted from AD 1300–1650, with at least 10 forest taxa surviving into the last 200–300 years, although after AD 1650 herbaceous materials replaced ligneous charcoal as a fuel source; early European visitors also reported seeing remaining tracts of woodland in AD 1722, and the decaying remains of large trees more than a century later (Orliac 2000; Hunt 2007). The presence of abundant rat-gnawed *Paschalococos* palm endocarps indicates that Pacific rats may again have been a major factor in the disappearance of the island's forests, and it has been estimated that the rat population may have temporarily exceeded 3.1 million animals (Dransfield *et al.* 1984; Hunt 2007). Our understanding of the relative impacts of rats, human forest felling, and fire on Easter Island's native vegetation remains incomplete. However, it is important to note that the subsequent collapse of the island's human population resulted largely from contact with Europeans (i.e. genocide rather than ecocide; Hunt 2007).

The complex pattern of remaining native forest cover and the underlying environmental determinants of the ecological response to human colonization across Remote Oceania have been investigated through comparative analysis by Rolett and Diamond (2004). Statistical decreases in deforestation and/or forest replacement across the Pacific region are associated with increased rainfall, elevation, area, volcanic ash fallout, Asian dust transport, and *makatea* terrain (uplifted reef), whereas increases are associated with latitude, age, and isolation. Forests on small, dry, old, nutrient-poor, low-lying, remote, subtropical rather than tropical islands are therefore less likely to have been able to withstand human environmental impacts, as they would have had lower growth/recovery rates and increased vulnerability to fire, with few inaccessible refugia in which forest remnants could persist. This approach highlights the ecological fragility of Easter Island, and provides a quantitative framework within which it is easier to understand how early colonists managed to destroy the entirety of the island's forests.

Most other island systems across the world have also been severely impacted by intensive and continuing deforestation following first human colonization, combined with further adverse effects on forest regeneration from the presence of introduced ungulates and other mammals. However, less information is often available to assess the relative impacts of prehistoric and European colonists on native forests, because it remains difficult to reconstruct pre-European population densities and environmental pressures, levels of remaining forest cover at the time of European arrival, and indeed the magnitude of ongoing present-day tropical forest loss and changes in land cover for many regions. Most notably, Madagascar has experienced extremely severe deforestation, habitat degradation and erosion following human arrival in the Late Holocene, as a result of rapidly growing human populations and the use of fire as a farming practice. However, current estimates for remaining forest cover on Madagascar vary by at least five-fold (Agarwal *et al.* 2005), and ongoing debate over pre-settlement forest cover and historical ecology continues to present problems for understanding the true extent of human impacts on the island (e.g. Kull and Fairbairn 2000). Similarly, many islands in the West Indies that are almost completely devoid of closed-canopy forests today (e.g. the Bahamas) or have experienced extreme levels of deforestation in the recent historical period (e.g. Puerto Rico) are known to have been heavily forested at the time of European arrival on the basis of early historical records (Watts 1987; Keegan 1992). However, considerable controversy surrounds the maximum size of the pre-Columbian human population of the West Indies, with estimates varying by more than an order of magnitude, confusing the understanding of levels of habitat alteration in the region over the Holocene (Watts 1987). Although extensive historical-era land clearance and timber export records are available for many of the West Indian islands (Watts 1987), little regional analysis has been conducted into quantitative patterns of deforestation over time across the Caribbean.

11.8 Relationship between prehistoric deforestation and extinction

Human-caused deforestation is widely and uncontroversially recognized as one of the primary extinction drivers in the current global mass extinction: it is part of the 'Evil Quartet' as defined by Diamond (1984b). The IUCN (2007) listed 4901 animal and plant species as being threatened by deforestation or associated habitat loss or degradation (conversion to crops, wood plantations, and/or non-timber plantations), and 5659 tree species are either threatened or extinct in the wild as a result of human activities. Sixty-six tree species are also recognized as having become extinct during the historical period (in practice from the eighteenth century onwards, starting with the disappearance of *Trochetiopsis melanoxylon* on St. Helena after 1771); 40 of these extinction events have occurred on oceanic-type islands, notably in the Hawai'ian archipelago, matching the geographical pattern of extinction shown by mammals, birds, and other animal groups during the Holocene (see Chapters 2–4 in this volume). Direct causes of historical tree extinction have rarely been identified, but forest clearance and overexploitation for wood products are recognized as extinction drivers for some species (IUCN 2007).

However, although humans have adversely impacted the structure and distribution of forest systems on both continents and islands throughout the Holocene, prehistoric plant extinctions since the middle Pleistocene remain largely undocumented, and little information is available to assess Late Quaternary tree species extinctions before the historical period. Relatively few plant macrofossils are preserved even in the recent fossil record, and many preservational environments which have been a major source of Late Quaternary vertebrate fossil material (e.g. cave sites) may only have the potential to preserve endocarps or other robust plant structures. Most knowledge of Late Quaternary vegetational and floristic change comes from fossil pollen preserved in lake and wetland sediments, but species-level (and often even genus-level) differentiation is rarely possible from the palynological record, and this 'taxonomic smoothing' may mask genuine extinction events (Jackson and Weng 1999). So far only one end-Pleistocene global extinction of a tree species has been identified, that of the spruce *Picea critchfieldii* in eastern North America (Jackson and Weng 1999). There is almost as little evidence for tree species extinctions during the prehistoric Holocene, which again probably reflects preservational incompleteness rather than a genuine lack of global extinction events. The best-documented example is the extinction of the Easter Island palm *Paschalococos disperta*, an endemic genus which was probably most closely related to the Chilean 'wine palm' *Jubaea chilensis*. It remains difficult to interpret whether the disappearance of trees from the pollen record on other Pacific Islands following prehistoric human arrival (e.g. *Pritchardia* palms on Atiu and Mangaia, Cook Islands; see Steadman 2006b) represent local extirpations of widespread species, or global extinctions of undescribed range-restricted species. Indeed, recorded plant extinction rates are lowest for areas with the longest history of large-scale human colonization, suggesting that most human-caused prehistoric Late Quaternary extinctions in this group remain unknown (Greuter 1995).

There is, unfortunately, similarly little straightforward evidence allowing modern-day researchers to assess the impacts of earlier Holocene deforestation on animal populations. In part, this is because the direct effects of deforestation in terms of habitat loss are likely to become compounded with other anthropogenic threat processes, notably the increased penetration of human hunters and exotic mammalian predators into fragmented forest environments (Diamond 1984b; Pimm 1996). For example, roe deer *Capreolus capreolus* disappeared from England and southern Scotland during the seventeenth and eighteenth centuries, around the time when tree cover reached its historical minimum in Britain; however, medieval accounts indicate that deer became the victim of increased poaching facilitated by the regional reduction of forested habitats, and after the late Middle Ages the species could also be legally hunted by private land-owners (Yalden 1999). This synergistic effect complicates the identification of primary extinction drivers even in relatively well-studied modern-day systems (although see Owens and Bennett 2000). Levels of habitat specialization or generalism also remain largely unknown (and potentially unknowable) even for well-documented extinct Holocene vertebrates, thus further complicating our understanding of the absolute impact of forest loss on different species. Habitat autecology has only been meaningfully reconstructed for extinct species with rich and well-studied Late Quaternary records, notably the extinct moa (Dinornithiformes) of New Zealand, for which distinct closed-canopy forest species (e.g. *Anomalopteryx didiformis*) and open-canopy mosaic forest-savanna species (e.g. *Emeus crassus*, *Pachyornis elephantopus*) can be differentiated (Worthy 1990; Worthy and Holdaway 2002). Indeed, it has been suggested that partial prehistoric deforestation may have created suitable habitat for native vertebrate species, for example certain *Ducula* pigeons and rails in the Pacific region, which were adapted for life among forest patches (Steadman 2006b). The difficulty of retrospective interpretation of potentially complex extinction events, typically on the basis of little or no quantitative ecological data, has meant that the relative significance of deforestation, overexploitation, and/or exotic predators as primary extinction drivers has been debated even for many well-known historical-era vertebrate extinctions; for example, ivory-billed woodpecker *Campephilus principalis* (Snyder 2007), Stephens Island wren *Traversia lyalli* (Worthy and Holdaway 2002; Galbreath and Brown 2004), and passenger pigeon *Ectopistes migratorius* (Pimm and Askins 1995). Lack of contemporaneity between the timing of threat processes and the eventual disappearance of a species in the prehistoric Holocene, due to staggered or protracted declines ('sitzkrieg' extinction events, *sensu* Diamond 1989) and 'extinction debts' (Tilman *et al.* 1994; Loehle and Li 1996), or lack of temporal constraints to identify the relative timing of each event, can further complicate the identification of causality. Furthermore, whereas the significance of other anthropogenic factors, notably exotic introductions, in driving historical extinctions has been the subject of quantitative analysis (Blackburn *et al.* 2004; see also Chapter 12 in this volume), there has to date been little statistical investigation of the link between patterns of Holocene deforestation and extinction. The importance of prehistoric and early historical-era deforestation in the extinctions of both continental and insular animal species therefore remains the subject of continued scientific debate (e.g. Didham *et al.* 2005).

11.9 Future forests

Global climates are changing, and the nature of this change is projected to continue rapidly even if the most extreme abatement scenarios are implemented (IPCC 2007). The rate and direction (from warm to warmer) of predicted climatic change are expected to cause major biodiversity loss (Thomas *et al.* 2004) for which new conservation paradigms must be established (Hannah *et al.* 2002; Roberts *et al.* 2002). In Europe, changes in vegetation cover under future climate scenarios have been investigated using a framework that links modern distributions of different plant species to climate parameters. The future distributions of these species are then projected under expected future climate scenarios. Such predictions are subject to a number of uncertainties, arising from the variety of possible climate scenarios, different modelling methodologies, and the assessment of presence/absence thresholds. The results do, however, show a number of consistent changes (Thuiller *et al.* 2004). The future vegetation in a given region will be affected by two actions: species loss due to a change towards unsuitable climatic conditions and species gain due to the invasion of exotics. Depending on the scenario used the results show that these factors will lead to marked changes in specific composition in Europe of between 45 and 65% over the next 80 years. Of the species that are currently present, 10–15% will become vulnerable, the same number will become endangered, and approximately 5% will become critically endangered. Regionally, central Spain and France and all mountainous areas are expected to suffer the highest levels of species loss. In contrast, the southern Mediterranean and the eastern Pannonian Basin will be less affected, as many of the species that occur in these regions are adapted to the dry and warm conditions that are expected in the future. These changes are expected to have widespread repercussions for human society, altering the possible goods and services that are obtained from the landscape, including crops and wood products, but also the role that the landscape plays in cultural identity.

However, these results do not take into account continuing human activities and environmental impacts, nor the limits on a species' ability to migrate. Many forest habitats have become fragmented by the activities of developing and expanding human populations, with much of the original forest cover removed (Fig. 11.1), leaving only a series of isolated forest islands (Fig. 11.5). This fragmentation will curtail the ability of forest taxa to migrate to new environments as they have under past climate shifts and human impacts. As seen from the case studies of North America and Amazonia, the rapid expansion of Europeans over the past 500 years and introduction of new land-use regimes and exotic crops and diseases together mean that the likelihood of future extinctions is a stark reality. To properly assess the future, models must be constructed and applied that fully integrate skills of archaeologists, ecologists, and modellers to understand ecosystem response to environmental change and human impacts. This research should help mitigate against a future of deforestation and the potential extinction of forest taxa.

11.9.1 Implications of understanding the past: developing an integrated perspective

A primary aim of the chapter has been to demonstrate the longevity of interactions between forested ecosystems and human impacts within a backdrop of environmental change; here we discuss the implications of applying this understanding to predicted future ecosystem changes. Remaining forests throughout the world are influenced by past human hands, a testament to the long history of human impacts. As discussed, there are a variety of methods to investigate past ecosystem dynamics and human interactions that each provides information down a particular path; however, it is necessary to weave together these different strands of information of ecosystem response to climate change to produce an understanding that is greater than the sum of the constituent parts. As shown for all the areas used as case studies in this chapter, the information, and subsequent understanding, is biased towards certain locations. Further to developing new results from crucial areas that are lacking, there is a pressing need to develop more complete methodologies that integrate past, present, and future perspectives on ecosystem dynamics, particularly those that can move from specific sites to local and regional scales (Marchant and Hooghiemstra 2001). As more information becomes available on the nature of past changes it is apparent that climatic change was not a constant process, temporally or spatially, with certain areas experiencing greater reductions in temperature and precipitation than others; it is imperative to understand the full range of natural variability and how this may project to the future. As we reach this situation, studies on past forest dynamics in response to climate change and human impacts can be considered by conservation biologists to impart effective long-term management strategies under future climate-change scenarios (Willis *et al.* 2004).

Acknowledgements

All authors are grateful to Fraser Mitchell who initiated this project but was unable to develop his contribution due to ill health, from which he has made a full recovery. Jack Williams is thanked for insightful comments on an early version of the manuscript. RM is grateful to financial support from award MEXT-CT-2004-517098 from the Marie Curie Excellence programme, FP6, and the Global Land Project (http://www.globallandproject.org/). SB is grateful for financial support for project DECVEG from the European Science Foundation (ESF) under the EUROCORES Programme EuroCLIMATE, through contract no. ERAS-CT-2003–980409 of the European Commission, DG Research, FP6.

CHAPTER 12

The shape of things to come: non-native mammalian predators and the fate of island bird diversity

Julie L. Lockwood, Tim M. Blackburn, Phillip Cassey, and Julian D. Olden

12.1 Introduction

During the Holocene, human populations have dramatically re-shaped regional biodiversity patterns via overexploitation, habitat alteration, and the purposeful introduction of non-native species (Steadman and Martin 2003; Kirch 2005; Pascal and Lorvelec 2005). These human-induced changes predate European colonization and the subsequent effects of the Industrial and Green Revolutions. Evidence for prehistoric species extinctions and invasions comes principally from palaeontology and archaeology, and so it has proven difficult unambiguously to identify the causative agents for such events (Kirch 2005). However, species extinctions and invasions have continued well into historical times, albeit at an apparently accelerating rate (Gaston *et al.* 2003; Pimm *et al.* 2006; Wonham and Pachepsky 2006), and for the latter events we have better, if still incomplete, information on likely drivers.

Oceanic islands and their bird inhabitants provide a wealth of information where changes to diversity across the Holocene have been intensively studied (Blackburn and Gaston 2005). Ecological and archaeological research tells us that birds endemic to islands have been especially vulnerable to human occupancy regardless of the time frame considered. Species on oceanic islands tend to be more prone to extinction than those on continents because of the smaller land areas and population sizes, low rates of increase, and naïveté to predators (Diamond 1985). Past extinctions coincided with the initial arrival of humans to these islands and later colonization by European cultures (Johnson and Stattersfield 1991; Duncan *et al.* 2002). Concurrent with human colonization have also been an astonishing number of non-native bird introductions, both intentional and accidental (Case 1996; Blackburn and Duncan 2001; Kirch 2005).

Despite our relatively good knowledge of the fate of island avifaunas, there are still unanswered questions as to what is driving rates and spatial patterns in bird invasions and extinctions. Four plausible explanations have been forwarded for why island bird populations are highly imperiled and often become extinct. First, the characteristics of disappearing species provide evidence that island bird populations have declined precipitously in the face of hunting by human colonists (Diamond 1984a; Milberg and Tyrberg 1993; Holdaway 1999b; Duncan *et al.* 2002, 2003; Roff and Roff 2003; Duncan and Blackburn 2004). Notably, the large-bodied and flightless species that would have made prime food items for human hunters seem differentially to have disappeared from many islands (but see Duncan *et al.* 2002). Second, multiple lines of evidence suggest that island bird populations have declined in the face of growing conversion of native habitat into agricultural and urban land use (e.g. Anderson 2001; Duncan *et al.* 2002). Island bird populations typically have small geographic ranges, and thus any change to available habitat has an inordinate influence on probability of extinction (Manne *et al.* 1999; Simberloff 2000; Sekercioglu *et al.* 2004). Third, numerous case studies show that island bird populations have suffered from the substantial modification of habitat by non-native mammalian herbivores like rabbits and pigs (Kirch 2005). These herbivores represent novel elements to island ecosystems and thus their consumption of plants and disturbance effects caused by trampling tends to shift habitat from one of dominance by native plants to one primarily of browse-resistant, hardy non-natives (Simberloff 2000). Finally, there is considerable evidence that island bird populations have declined after the introduction of one or more non-native reptilian (Savidge 1987) or mammalian (Atkinson 1985; Holdaway 1999b; Roff and Roff 2003; Duncan and Blackburn 2004) predators. Island birds have typically evolved in the absence of such predators, and thus do not have the behavioural plasticity or life-history traits necessary to counteract their effects (Courchamp *et al.* 2003).

Despite this wealth of knowledge, we do not know the relative importance of the various different factors in driving bird extinctions across oceanic islands, or whether a particular factor predominates across islands and oceans. This knowledge gap is paralleled by a lack of understanding of what drives the success or failure of birds that have been introduced to oceanic islands either accidentally or purposefully by humans. Establishment by exotic birds may depend on characteristics of the species themselves, on features of the non-native island environment, or on idiosyncrasies of the introduction process (Duncan *et al.* 2003). While information on species-level traits is relatively easy to come by, and is the same wherever the species is introduced, location-level and event-level characteristics are generally harder to quantify. However, location-level characteristics are comparably easy to define for simplified islands relative to complex continental locations, while historical records of introductions to many island locations are of relatively high quality (especially New Zealand; Thomson 1922). Thus, introductions of non-native birds to islands represent an ideal opportunity to assess the comparative influence of different drivers of the introduction process.

Here we review recent research on bird extinctions and invasions on oceanic islands which indicates that the presence of non-native predatory mammals is a primary cause of both events. We also explore how these invasions and extinctions have served to re-shape patterns of diversity across entire suites of oceanic islands. The change in the similarity of bird assemblages across islands in the wake of such large numbers of invasions and extinctions has rarely been considered, but there is growing evidence that such modifications may be a key outcome of human-induced environmental change. Since invasion and extinction rates are predicted to accelerate in the future, we end by considering what the past can tell us about the future of island bird biodiversity.

12.2 Island bird extinctions

Fossil evidence indicates that hundreds to thousands of bird species went extinct after their first contact with human cultures during the Holocene (Steadman 1995, 2006b; Blackburn and Gaston 2005; see also Chapter 4 in this volume). On tropical Pacific islands, for example, Quaternary fossil records suggest that many thousands of bird populations (especially seabirds such as petrels, shearwaters, boobies, and terns, and landbirds such as megapodes, rails, pigeons, parrots, and passerines) and perhaps as many as 2000 avian species (especially rails) were lost after the arrival of people (Steadman 1995; see also Chapter 10 in this volume). Furthermore, Johnson and Stattersfield (1991) calculated that over 90% of bird extinctions that have occurred during historical times have occurred on islands, even though less than a fifth of the world's bird species are restricted to islands. Not all islands suffered the same number of extinctions, however. Some oceanic islands have lost well over half of all known bird species that once occurred there, but most islands have witnessed only a few native species extinctions (Pimm *et al.* 1994; Biber 2002; Blackburn *et al.* 2004; Blackburn and Gaston 2005). This variation is typically attributed to island-specific attributes such as total area and habitat complexity, date of human colonization, and degree of isolation. Based on well-known biogeographical principles, we should expect lower proportions of extinct birds on islands with greater habitat heterogeneity, larger area, more recent human colonization, and less isolation from continents (Blackburn and Gaston 2005).

Given the sheer number of case studies showing that non-native mammalian predators and herbivores contribute to bird extinctions from islands, Blackburn *et al.* (2004) suggested that numbers of such species should predict the magnitude of native bird extinction events more generally. To gain a comprehensive understanding of the relative role of non-native mammals versus other factors for causing avian extinctions, Blackburn *et al.* (2004) compiled information on avian assemblages across 220 oceanic islands. The authors used historical information to derive a list of avian species that were extirpated from each island since European colonization, and additionally collated information on island area, isolation, maximum elevation, date of first human colonization, and numbers of non-native mammalian herbivore species, non-native mammalian predator species, and non-native bird species. Each island was grouped with others in its own archipelago, and all archipelagos were categorized as to whether they are located in the Caribbean or the Atlantic, Indian, or Pacific oceans.

Islands varied from zero to 12 non-native mammalian predator species established (Blackburn *et al.* 2004). Non-native predator richness on each island accounted for consistent and substantial amounts of variation in extinction probability: the more non-native mammalian predator species that have established on an island, the higher the proportion of the native avifauna that had been extirpated (Fig. 12.1). The best statistical model for the variables considered by Blackburn *et al.* (2004) also included a negative effect of island area, and a positive effect of island elevation: extinction probability was higher on smaller islands and islands with greater elevational ranges. The influence of island area concurs with ecological theory positing that smaller islands support smaller populations of native bird species, which are thus more prone to extinction (MacArthur and Wilson 1967). By contrast, an explanation for the positive relationship between elevation and extinction probability is less clear. One possible reason is that habitat is further subdivided on elevationally diverse islands, so that already small populations on these islands are even smaller.

The results of Blackburn *et al.* (2004) indicate that the presence of non-native mammalian predator species is related to extinction probability in birds across a large array of islands. This study was the first to illustrate the widespread and consistent effect of predator richness. However, their results still beg the question of what are the specific mechanisms by which these new predators are causing extinction? Ecological experience and theory suggests that there are three possible reasons why more mammalian predator species might be associated with more bird extinctions. First, there could be a lottery (sampling) effect whereby the continual introduction of mammalian predators will increase the chances of a particularly devastating predator

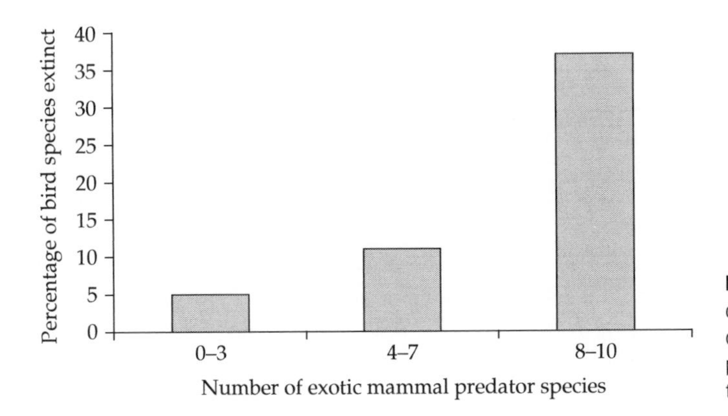

Figure 12.1 The relationship between the number of exotic mammal predator species introduced to oceanic islands around the world and the average percentage of native bird species on those islands that subsequently went extinct.

being released (e.g. cats, rats, or mongoose). Second, the number of mammalian predator species may not be as important as the variety of functional predatory forms this set of species encompasses. Thus, as the number of species increases an island will begin to add functional groups, which for predatory mammals includes species that differ in time of peak predatory activity, hunting behaviour, and/or prey choice. The greater the functional diversity, the greater the variety of birds adversely affected by predation. Third, oceanic islands are typically naturally depauperate of mammalian predators, and thus the continual introduction of non-native mammal species may simply raise the species richness of predators on islands. Even if these predators only occasionally take birds as prey, they will serve to increase predation pressure.

To assess these possibilities, Blackburn *et al.* (2004) assessed the ability of the presence/absence of known bird predators, such as cats and rats, and of the number of mammal orders, to explain variance in avian extinction probability across islands. The latter variable was included as an index of the number of functional groups introduced. However, the inclusion of these variables did not increase the variance explained in extinction probability when compared simply to the absolute number of predator species. Since taxonomic identity is a crude measure of the functional roles of different predatory mammals, Blackburn *et al.* (2005) derived a more nuanced measure of functional roles for predators called Functional Diversity (FD) that captures the differences between species in life history, activity schedule, diet breadth, and predatory behaviour. Interestingly, this metric still did not explain more across-island variability in extinction probability than did absolute numbers of non-native mammalian predator species. This suggested that there is limited 'functional redundancy' among predators, whereby the addition of predatory mammal species with similar roles to species already in the exotic assemblage decreases the impact of subsequent introductions.

Taken as a whole, these results suggest that non-native mammalian predator diversity may be a primary driving force behind the historical loss of birds on oceanic islands. This result extends the case-study evidence for such impacts on oceanic birds, and indicates that this effect explains more variation across islands than traditional explanations such as time of human colonization, area, and habitat complexity. Furthermore, the impact of non-native mammalian predators does not appear to be due to the eventual catastrophic arrival of the worst of the worst, or ameliorated at all by functional redundancy among predators, but rather is one of incremental increases in impact on island bird populations with every new mammalian predator species established.

In fact, Blackburn *et al.* (2005) provided evidence that the effect of non-native mammalian predators is not linear, but quadratic. A quadratic relationship indicates the opposite of functional redundancy: rather, that every newly established mammalian predator has an incrementally more negative effect on native birds than the predator that arrived before. Such a pattern, whereby a collection of non-native species causes more of an impact together than expected based on their individual impacts, is a component of the invasional meltdown scenario of Simberloff and Von Holle (1999). The work of Blackburn *et al.* (2005) is one of the only examples where a meltdown could be clearly linked to extinction events, and it is also the first time that such a scenario has been suggested for the loss of native birds on oceanic islands.

Given its implications, it is helpful to consider how such facilitation may occur. Blackburn *et al.* (2005) provided two potential explanations. First, some non-native mammalian predators could serve as prey for other non-native mammalian predators, thus allowing a larger community of predators on the island to coexist than if the predators were supported on native prey items alone. In addition to the simple build-up of numbers of potential predator species, the establishment of extensive predator–prey networks within ecosystems that did not previously harbour them can lead to hyperpredation: the relative increase in predation experienced by native prey when alternative non-native prey is available to their predator (Courchamp *et al.* 2000). These alternative prey items can serve to sustain the predator through times of low food availability (e.g. winter), thereby allowing the predator to realise larger densities than without the non-native prey (Courchamp *et al.* 2000). Larger prey densities then increase the predation pressure on the often behaviourally naïve native birds. Anecdotal accounts of how the build-up of predator–prey networks on islands adversely affected native birds populations (Courchamp *et al.* 2003) include a well-known example from New Zealand. Here, non-native rodents increase in periodic 'masting' years (approximately 3–5 year cycles), when southern beech trees *Nothofagus* spp. produce relatively large seed crops. Non-native mustelids then prey on the rodents, and so also increase in number. These mustelids then prey on native birds and their nests, more so when rodent numbers subsequently crash in non-masting years, leading to predictable predation risk among native birds (O'Donnell and Phillipson 1996; see also White and King 2006).

A second possible explanation for facilitation in exotic mammalian predator assemblages is that the predators may alter the habitat in such a way that it favours later arrivals (Mack and D'Antonio 1998). This is the typical explanation for facilitative effects between non-native species in general (Simberloff and Von Holle 1999). On oceanic islands such effects may occur following the establishment of particular mammalian predatory species such as rats, which are known to have shifted island habitats dramatically after their arrival through predation of the seeds and fruits of native plants (e.g. Kirch 1996). This altered habitat structure may be more conducive for the establishment of later-arriving non-native species.

12.3 Establishment of introduced bird populations

Another feature of islands is that people have inordinately favoured them as places to release non-native birds (Blackburn and Duncan 2001). Of the 1378 introduction events (i.e. occasions when individuals of a non-native bird species were released) evaluated by Blackburn and Duncan (2001), 953 were to islands even though islands represent less than 3% of the Earth's ice-free surface. Most of these island locations were in the Pacific Ocean, but there were also substantial releases on islands in both the Caribbean and the Indian Ocean. The reasons for introducing birds to islands vary, but it has largely been driven by the desires of early European colonists to populate the islands with useful and known birds from their home countries (Long 1981; Lever 2005). Birds continue to be released and become established on islands, usually accidentally, due to the ubiquity of the trade in pet birds (Lever 2005).

Although not often considered within the context of oceanic islands alone, there are a variety of potential reasons why many non-native birds establish on some islands but only a few on others (Duncan *et al.* 2003). Of these explanations, some are the same as those for differences between islands in extinction probability such as island area and habitat complexity (Case 1996). We should expect more non-native birds on larger islands and those with greater habitat heterogeneity. Others have suggested that competition between native and non-native birds will prohibit the species from the latter category establishing themselves (Elton 1958; Keitt and Marquet 1996). A common explanation for differences across islands in the number of established non-native birds is that some islands have a greater proportion of their area disturbed by humans. This 'disturbance' includes habitat conversion from native plant-dominated ecosystems to those composed entirely of non-native plant species, and the severe habitat fragmentation (native- or non-native-dominated) associated with development.

Cassey *et al.* (2005b) suggested that the number of potential predator species may also reduce the likelihood that non-native bird species will establish. The mechanism is nearly identical to that for the influence of predators on extinction probability. Birds, or their nests, are very vulnerable to depredation and this effect is so strong that it may have shaped the evolutionary history of the entire class (Bennett and Owens 2002). Thus, islands with many predators may provide such a hostile environment that very few bird species introduced to that island will be able to establish a self-sustaining population in the face of predatory losses. The effects of predators are likely to be especially important soon after individuals are released or escaped, as the population at this point is likely to be low and its persistence highly vulnerable to the loss of individuals to predators.

The vulnerability of small populations to predators, and indeed to a variety of environmental and demographic stochastic effects, suggests that non-native birds should be more likely to establish when the number of founding individuals is large. When large numbers of individuals are released in a single introduction attempt, or if there are several releases of the same species to one location, the probability that the incipient non-native population that is created will establish a self-sustaining population increases (Cassey *et al.* 2004; Lockwood *et al.* 2005, 2007). Introduction effort (or propagule pressure) has been shown to be a primary determinant of the successful establishment of introduced birds, generally to the exclusion of hypotheses that suggest a primary role for the environment or species interactions (Forsyth and Duncan 2001; Cassey *et al.* 2005a).

Cassey *et al.* (2005b) evaluated the ability of a variety of factors to explain variation in number of established non-native birds across 41 of the 220 islands considered by Blackburn *et al.* (2004, 2005). This set of 41 islands is the subset for which they could assemble information on the key variable of introduction effort (Cassey *et al.* 2005a). As expected, they found that introduction effort explained most of the variance in avian establishment success between islands. There was also evidence that island area, human population size, number of native bird species, number of non-native mammalian herbivore species, and number of non-native mammalian predator species each explained some fraction of the variation across islands. When Cassey *et al.* (2005b) considered the effect of all variables together in a single model, however, introduction effort and number of non-native mammalian predator species were the only variables that explained significant and independent variation in bird introduction success across islands.

These results provide evidence that *all* birds on oceanic islands are subject to the negative effects of non-native mammalian predators. Further, the work of Cassey *et al.* (2005b) is the first time that non-native predators have been implicated as playing a deciding factor in non-native bird establishment success (Duncan *et al.* 2003). It may be that mammalian predators prevent non-native birds from establishing only on oceanic islands. Mammalian predator abundance and density can be very high on islands as typically there are no natural checks on their populations in the form of larger predators or disease, and non-native birds may be subject to the same synergistic effects of mammalian predators as native birds. It is also possible that there is a general effect of mammalian predators on probability of non-native bird establishment, but this has yet to be addressed empirically. No matter the generality of the result, the combined work of Blackburn *et al.* (2004, 2005) and Cassey *et al.* (2005b) depicts a remarkably consistent picture that non-native mammalian predators have played a critical role in determining the current composition of avian assemblages on oceanic islands through their influence on patterns of species invasions and extinctions.

12.4 Changes in species richness

Given all this activity on islands, one might expect bird richness to have been profoundly changed in the historical period. However, there is a remarkable lack of information on how the forces of extinction and invasion have served to re-shape natural avian diversity on islands into the patterns we observe today (see also Case 1996). If we think of islands as model systems for exploring how human-induced global changes affect biodiversity (Vitousek 2002), then gaining a clear understanding of how invasions and extinctions have combined to create a new template of island biogeography should provide important insights into the fate of the rest of the world's diversity.

A common metric for evaluating the conservation priority of particular events is the change in species richness that they produce (Fleishman *et al.* 2005). Species richness is seen as a determinant of the successful delivery of ecosystem services (Hooper *et al.* 2005), as a suitable proxy measure for the health of ecosystems (e.g. Morley and Karr 2002), and as a property of nature that has intrinsic value (Wilson 1984). Typically changes in richness are tallied using only native species extinctions; however, Hobbs and Mooney (1998) have suggested that a full accounting of richness changes must include the enrichment effects of non-native species. Thus, in this section we explore changes in species richness by simply tallying the number of non-native birds that have established, and the number of native populations that have become extinct, across all 220 oceanic islands considered in Blackburn *et al.* (2004).

The maximum number of species lost across all islands is 21, which occurred on Réunion Island in the Mascarene archipelago. The maximum number of bird species introduced to an island is 37, which has occurred twice: once on the island of Oahu in the Hawai'ian archipelago and again on the North Island of New Zealand (although not the same 37 species). If we do the full accounting of extinctions and introductions across all 220 islands, we find that most islands have not substantially increased or decreased in richness (Fig. 12.2). In other words, on the whole, non-native bird introductions have tended to balance native bird extinctions on a per-island basis. This result concurs with a previous study of island avifaunas worldwide (Sax *et al.* 2002) and suggests that local species richness often does not change even in the face of considerable species extinctions as long as the enrichment effect of non-native species is included in the calculations. However, there are clear outliers across this distribution, and our accounting indicates that the gains in richness can sometimes be double the losses. The maximum decrease in overall richness occurs on Mangere, an island in the Chatham archipelago (Pacific), which has decreased in richness by 12 species. The maximum increase in overall species number occurs on Oahu, which gained a total of 30 species after tallying the number of established non-native birds.

12.5 Who are these birds?

While empirical evidence suggests that human activities have led to increases in local species richness (Sax and Gaines 2003), the recent debate on the role of biodiversity in maintaining ecosystem function illustrates that species composition (meaning the types of species present and their ecological role) influences ecological processes to a much greater extent than merely the total number of species (e.g. Kinzig *et al.* 2002). Sekercioglu (2006) has made a strong case that birds provide substantial ecological services and thus the extinction, or decline in population size, of key species will have long-lasting implications for ecosystem regulation and support. Olden *et al.* (2004) went on to argue that if the species that make up these local biotas are the same across regions, collectively these systems may suffer decreased ecosystem functioning, stability, and resistance in the face of environmental change. Indeed, there is a robust argument to be made that people experience and value biodiversity not as a function of the absolute numbers of species they are seeing, but as the relative uniqueness of the species they are seeing (Cassey *et al.* 2005c; Olden *et al.* 2005). All of this argues that the more profound change to avian diversity on oceanic islands is not altered local species richness (positive or negative), but in how similar (or differentiated) bird assemblages on each island have become relative to each other. The increasing similarity of biotas across space has been called biotic homogenization (McKinney and Lockwood 1999).

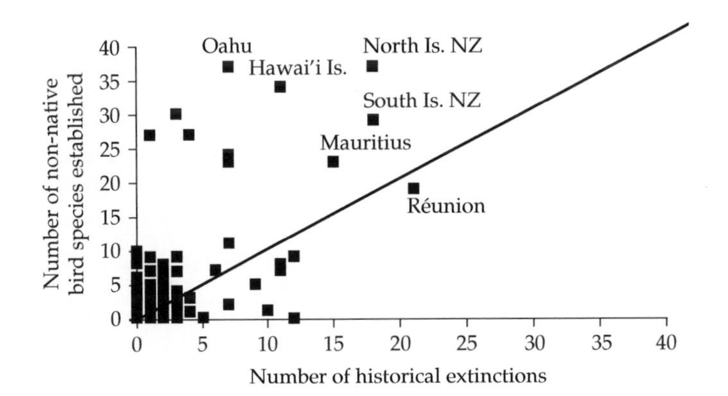

Figure 12.2 Number of native bird extinctions relative to the number of non-native bird species established across the 220 oceanic islands considered in Blackburn *et al.* (2004). Island bird richness did not change in the majority of cases; however, there were many more examples of increases in species richness than decreases.

We have good reason to expect that oceanic islands have become homogenized through time as they have tended to gain a common set of non-native species and have suffered the loss of unique, native species. Blackburn *et al.* (2004) suggested that non-native mammalian predators have differentially driven endemics to local extinction. Most studies of the loss of island birds centre on the loss of unique species such as the suite of endemic birds lost from New Zealand or the Hawai'ian Islands. On the other side of the ledger, it is easy to point to the now ubiquitous house sparrow *Passer domesticus*, common myna *Acridotheres tristis*, or barred ground dove *Geopelia striata* as evidence that the set of non-native populations that establish on islands are probably of the same species. For example, Lockwood (2006) shows that the six main Hawai'ian islands have greatly increased in passerine bird similarity relative to one another since the arrival of Polynesians even though passerine species richness on each island has either not changed, or only dropped slightly.

Nevertheless, simulations of the homogenization process (e.g. Olden and Poff 2003; Cassey *et al.* 2006) combined with the few empirical examples of how similarity has changed through time (reviewed in Olden 2006) show that invasions and extinctions can combine to produce a variety of changes in community similarity across locations, including both biotic homogenization and increased distinctiveness between local biotas (i.e. biotic differentiation). Whether assemblages become homogenized or differentiated depends on several factors, including the spatial scale of analyses, the taxa considered, and the identity of the species invading or becoming extinct (Olden and Poff 2003). It is theoretically possible for assemblages to differentiate or homogenize after having had only species invade or only species become extinct (Olden 2006).

These previous investigations suggest that it is well worth tracking the identity of the birds lost or gained on the islands considered by Blackburn *et al.* (2004). Furthermore, the available information on island bird assemblages allows an investigation into two aspects of homogenization that previously have been unexplored. First, we can examine how bird assemblage similarity has changed at two distinct spatial scales: between islands in the same archipelago, and across islands that are not in the same archipelago but are in the same ocean basin (here also regarding the Caribbean as a distinct ocean-level biogeographical unit). Based on previous work, we should expect to see differing degrees of change in similarity across these two scales, with perhaps more evidence of homogenization at the larger scale. Second, we can explore a range of potential drivers of observed changes in similarity of island bird assemblages. The list of potential drivers can include all variables evaluated in reference to explaining variation in the proportion of an island's avifauna that was extirpated or introduced (see above; Blackburn *et al.* 2004; Cassey *et al.* 2005b). Several of these drivers have been found to be important in changing the across-assemblage similarity of other taxa such as freshwater fish. In particular, there is good evidence that human colonization history should correlate with the degree of homogenization (e.g. Olden *et al.* 2006; Marchetti *et al.* 2006). Given the role that non-native mammalian predators have played in bird invasions and extinctions on the islands we consider, we may also expect to see predator richness explain changes in across-island similarity.

12.5.1 Historical patterns of bird assemblage similarity across islands

Cassey *et al.* (2007) evaluated the change in bird assemblage similarity for 152 of the 220 islands originally analysed by Blackburn *et al.* (2004). This group of islands is the subset for which there were established populations of non-native birds, and for which extinct species could be identified to species-level. They used the Bray–Curtis Index as their metric for evaluating the compositional similarity of avian assemblages between any two islands. The Bray–Curtis score is well established in the community ecology literature, in part because it is easy to interpret. A score of zero indicates that the two assemblages under consideration share no species in common, and a score of 100 indicates the opposite, where all species are shared (Legendre and Legendre 1998).

Using the historical species lists for each island, Cassey *et al.* (2007) partitioned the initial assemblage

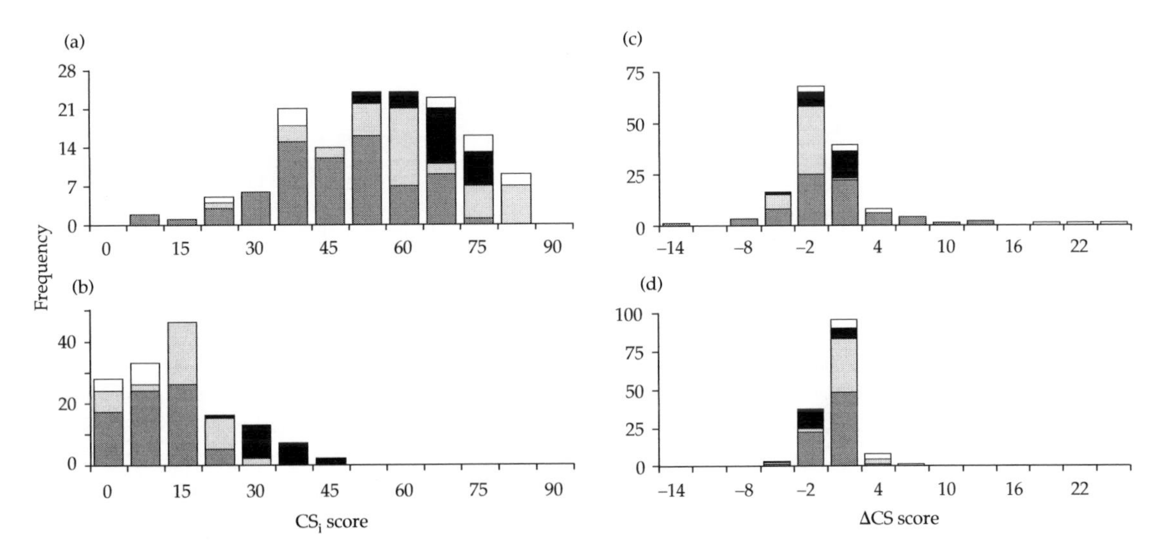

Figure 12.3 Frequency histograms of initial community similarity pairwise scores (CS_i) computed between (a) islands in the same archipelago prior to post-European colonization invasions or extinctions and (b) islands in different archipelagos prior to post-European colonization invasions or extinctions. Frequency histograms of change in community similarity scores (ΔCS) between (c) islands within the same archipelago after avian invasion and extinctions have occurred and (d) islands in the same ocean. Islands in the Atlantic Ocean are denoted by light grey, islands in the Caribbean by black, islands in the Pacific Ocean by dark grey, and islands in the Indian Ocean by white. Redrawn from Cassey *et al.* (2007).

similarity (CS_i) between two scales (Fig. 12.3). The first scale measures how similar all islands are to all other islands within their archipelago, and the second measures the similarity of each island to all other islands outside the archipelago but within the same ocean basin. The end result is two measures of assemblage similarity, across nested spatial scales, which can be directly compared.

Initial assemblage similarity (CS_i) varies between spatial scales such that islands within the same archipelago are much more similar on average than are islands between archipelagos (Fig. 12.3). CS_i only takes values less than 0.50 (50% similarity) when it is calculated between islands within different archipelagos, whereas CS_i spans nearly the full range of potential values (0–100%) for comparisons of islands within archipelagos. The tendency for CS_i scores to be larger for islands within archipelagos than between archipelagos holds regardless of the ocean basin in which the islands are located. However, there was a difference in CS_i scores between ocean basins within each spatial scale. Initial assemblage similarity is higher at both spatial scales for islands in the Caribbean and the Atlantic Ocean, and lower at both spatial scales for islands within the Indian Ocean and Pacific Ocean (Cassey *et al.* 2007).

The reasons for this pattern are fairly obvious if we consider the evolutionary history of oceanic birds. Many archipelagos harbour unique assemblages of bird species because they evolved *in situ*, and these species are more likely to be shared between islands within this archipelago but not across archipelagos, as islands and island groups emerged from the sea at very different rates through geological time. Similarly, non-endemic avian species tend to occupy all islands in the group, rather than just one, because of their inherent vagility. This vagility, however, has limitations. Most native avian species are not likely to occupy many archipelagos within a single ocean, as these tend to be separated by distances that many species, even in a vagile group like birds, cannot cross. It is noteworthy that Cassey *et al.* (2007) still detected a clear signature of this evolutionary history in their analyses even though there were many avian extinction events due to first human contact that were not able to be incorporated (i.e. the data set has already passed through a prehistoric extinction filter; e.g. Balmford 1996).

12.5.2 Changes in similarity after bird invasions and extinctions

Cassey *et al.* (2007) recalculated these two similarity measures for the current avian assemblages on these islands (CS_c), which may have changed from the initial assemblages through either or both of the processes of native extirpation and non-native introduction following human colonization of the islands. Calculating CS_c allowed them to assess the change in assemblage similarity ($\Delta CS = CS_c - CS_i$) that has resulted from any avian extirpations and introductions that might have occurred. They then used ΔCS to examine how similarity has changed at the two different scales, and to compare the direction of change in the different components within these scales. If ΔCS is positive the island has (on average) become more similar to the other locales after avian invasions and extirpations; that is, taxonomic homogenization. If the ΔCS is negative the island has (on average) become less similar to the other islands; that is, taxonomic differentiation.

Cassey *et al.* (2007) found that approximately half of the island values of ΔCS indicated no change in assemblage similarity through time although there are clear outliers in this distribution (Fig. 12.3). The maximum positive increase in similarity of 24% was for Rodrigues Island, indicating that this island has become much more similar to others in the Indian Ocean. The maximum negative value of ΔCS was −13% for the island of Guam, indicating that this island has greatly differentiated in composition from its sister islands within the Marianas archipelago.

Of more interest is that Cassey *et al.* (2007) found that islands within each ocean behave in a unique manner, and that the degree of change in similarity observed is distinctly scale-dependent. Islands in the Mascarene and Comoros groups in the Indian Ocean are homogenizing no matter the scale at which changes are measured, whereas the similarity of islands in the southern Indian Ocean has not changed. In the Caribbean (with the exception of Puerto Rico), some islands have differentiated and others have homogenized with a slight tendency to becoming more differentiated between archipelagos. More variable ΔCS values can also be seen among the islands within the Pacific Ocean. When ΔCS is calculated within the Atlantic, islands within their archipelago have all clearly differentiated. In contrast, when these same islands are compared to others outside their archipelago they have become, on average, more homogenous (Fig. 12.3).

Observing such unique changes across oceans indicates that the (remaining) avian evolutionary history of these islands has been overlain by human cultural history to produce location-specific changes in the identity of the current assemblage. For example, islands in the Atlantic Ocean tend to differentiate in relation to other islands within their own archipelago, but homogenize when compared to islands outside their archipelago. Such changes in similarity suggest that a pattern of the introduction of common non-natives and/or extirpation of endemic natives is driving an increase in similarity between archipelagos, while the introduction of different sets of non-natives (and perhaps the extirpation of widespread native species) is increasing differentiation within archipelagos. Similar mechanisms have also been shown for driving patterns of taxonomic homogenization and differentiation of fish faunas across different spatial scales in the USA (Marchetti *et al.* 2001; Olden and Poff 2004).

The islands in the Caribbean and Pacific Ocean show no clear pattern towards being either predominantly homogenized or differentiated at either spatial scale. Instead, some islands seem to have been drastically altered by invasions and extinctions and others left largely unchanged. For example, in the Pacific, the Hawai'ian Islands have had a plethora of avian invasions and extinctions that have driven up between-island similarity (see also Lockwood 2006), while at the same time these changes have made the islands much more distinct from the other islands in the Pacific. The same is true for the islands of New Zealand, although to a lesser extent. On Guam the large number of very well-known extirpations (Savidge 1987) has strongly differentiated the island from others in the Marianas archipelago. In the Caribbean, only

Puerto Rico has seen equivalent numbers of bird invasions to New Zealand and Hawai'i (although only half or fewer extinctions). Because Puerto Rico has had similar numbers of extirpations as the other Caribbean islands within and outside its own archipelago, while hosting such large numbers of invaders, it has greatly differentiated in composition at both spatial scales. Most other islands in the Pacific and Caribbean have had many fewer invasions and extinctions and the identity of these species appears somewhat idiosyncratic, such that there is no clear overall pattern in either homogenization or differentiation. Whether this lack of pattern reflects truth, or is perhaps a function of an extinction filter applied to these islands by their earlier colonization by humans (Biber 2002; Blackburn *et al.* 2004) relative to those in the Indian and Atlantic Oceans, remains to be seen.

A key question that remains to be explored in the context of changes in island bird community similarity is whether the location- and scale-dependence of the patterns observed are essentially idiosyncratic consequences of the specific identities of the species introduced and/or going extinct, or whether the apparent idiosyncrasies mask consistent underlying drivers. If the latter, then since patterns in homogenization subsume both the process of species invasion and extinction, we can imagine that any of the biogeographical or ecological variables that were explored in the work of Blackburn *et al.* (2004, 2005) or Cassey *et al.* (2005b) could correlate strongly to changes in avifaunal similarity between oceanic islands. Given the evident influence of non-native mammalian predator species richness on both native bird extinction and exotic bird establishment, the hypothesis of an additional role of predators in driving changes in island bird community similarity would seem ripe for testing.

12.6 What does this history tell us about the shape of things to come?

The idea that the transport and release of non-native species will continue relatively unabated into the future has prompted several authors to term the coming era in Earth's history the Homogocene (Rosenzweig 2001). The world envisioned is one where the barriers to species dispersal are largely broken down by human actions, resulting in a massive anthropogenic blender of biological mixing, creating a 'New Pangaea' (Olden 2006). There is no doubt that species are being moved around the world at an unprecedented rate and these species are finding themselves in places they could never have established under their own steam (Perrings *et al.* 2005). It is also depressingly clear that the scale of human enterprise is large enough to greatly accelerate the pace at which species are driven extinct, either globally or locally. What is not so clear is what these diversity convulsions will do to the distribution of the species that remain.

Biotic homogenization is a complex process that subsumes several mechanisms of change in species distribution. These changes in distribution are overlain on a complex biogeographical history. Even though these 'old' biogeographical patterns have certainly been altered to some extent by the effects of prehistoric and historical human colonization, they still persist and continue to set the foundation for large-scale biodiversity patterns. In addition, non-native species invasions are not always so monospecific. Human desires for different species are fickle and thus change through time. This change is reflected in the set of species that are purposefully moved around the world as non-natives. Similarly, human commerce patterns change as world economic conditions and relationships evolve and thus the set of species that hitchhike into non-native locales, as well as the distribution of these locales, will naturally change as well. The end result is often an idiosyncratic pattern of non-native species introductions, with a few species making it to almost everywhere and many more making it to only a few new places.

Another aspect of the changing human influence on non-native species introductions plays a dramatic role in how island avifaunas look today. Mammalian predators have driven many birds extinct from islands, apparently even causing synergistic meltdowns whereby many more native birds become extinct than expected based on the absolute number of predators present. Mammalian predators also appear to play a dominant role in determining how many non-native bird species will establish on these islands. It is unusual that a single biological interaction (here predation) can so strongly influence the composition of an entire suite of species (island birds). In addition, they may yet be shown to affect relative changes in composition.

Will the trends we have identified continue into the future? In some senses, undoubtedly yes. Increasing numbers of bird species are becoming threatened with extinction, as indexed by the growth in the length of the IUCN Red List over recent years (Baillie *et al.* 2004). In addition, simulations of future extinction rates suggest that a further 28–56% of bird species restricted to oceanic islands are likely to become extinct (Sekercioglu *et al.* 2004). Increasing numbers of exotic species are being added to avifaunas, as accidental or deliberate releases by bird traders and owners take over from the planned activities of officially sanctioned naturalization societies as the main sources for immigrants (Blackburn *et al.* 2009). We can expect avifaunas on islands (and indeed other faunas and floras on continents and islands) to continue to change in composition as a result.

In other senses, however, future changes in island avifaunas will probably differ from those observed to date. Mammalian predator species have been shown to have a smaller influence on the suite of island birds currently threatened with extinction (Blackburn *et al.* 2004). This is consistent with a filter effect (Pimm *et al.* 1995; Balmford 1996; Biber 2002; Blackburn *et al.* 2004), as most species susceptible to the current assemblages of exotic predators have presumably already been driven extinct (Blackburn *et al.* 2004), albeit that the apparently synergistic effect of predator species additions (Blackburn *et al.* 2005) means that more should not be sanctioned. The current major threat to island birds is habitat destruction, as it is for most imperiled species worldwide. This may eventually switch to climate change. We can anticipate that such changes will alter the characteristics of species disappearing from islands, as well as the islands most imperiled, and so perhaps also the pattern of loss. The set of non-native birds that will be introduced to islands in the future will likely continue to be somewhat idiosyncratic. There are a few species that are ubiquitous in the pet bird trade, but which species make it to what islands and which of these will establish themselves is an inherently complex process. It will all depend on the vagaries of the international trade in birds, the whims of human demand, and the laws that governments pass to protect their native species.

In the face of all these changes, how can we best explore the New Biogeography created in the wake of human actions? We finish by suggesting some rules of thumb.

First, it is imperative to consider the existing biogeography of the region under consideration when examining contemporary patterns of biodiversity. Although it is entertaining to think of the outcome of creating a new Pangaea, most regions will retain a non-trivial amount of their original identity and thus the relevance of standard information on biogeographical distributions will remain important into the future.

Second, there is a distinct and highly relevant biogeography of human influence on species distributions. Non-native species are distributed by and large in the places where humans wanted them to be, or where humans took the products in or on which the species hitchhiked (Lockwood 2005). This is true for the non-native birds that directly contribute to the calculus of island bird homogenization, as well as the non-native mammalian predators that contribute to variation in native bird extinction and non-native species establishment. Although exotic species have in many cases spread widely from the point of introduction (Lockwood *et al.* 2007), current evidence suggests that it is the geographic patterns of release events that largely determines the geographic extent of non-native species and spatial patterns of richness (Blackburn *et al.* 2008).

Third, in studying the New Biogeography, we must consider changes in identity and not just in number of species. Whereas it is conceptually and analytically harder to express diversity changes in ways that incorporate quality as well as quantity, it will be essential to do so fully to understand human effects on the environment. For example, we do not understand how the change in the identity of species within an assemblage affects ecosystem functions, resilience of communities in the face of disturbances such as climate change, or the evolutionary pathway of the species within the assemblage (Olden 2006). Research on the effect of losses (or gains) in species richness does not provide insight into these issues. Furthermore, the observed increases in local species richness do not generally map on to changes in the similarity of those assemblages (Olden 2006).

Fourth, we should not let apparent idiosyncrasies in diversity changes divert us from the search for underlying regularities in the processes driving those changes. How natural biogeography interacts with that of human influence, as determined by the identities and numbers of species concerned, will undoubtedly be hard to understand. We have already seen the variety of changes in the richness and similarity of island bird assemblages that these processes can generate. It is only by understanding the processes that drive them can we hope to control, and perhaps one day ameliorate, their effects.

CHAPTER 13

The Quaternary fossil record as a source of data for evidence-based conservation: is the past the key to the future?

John R. Stewart

13.1 Introduction

The debate regarding whether modern climate change is taking place due to anthropogenic carbon emissions is apparently now settled. The debate must now move to how organisms are likely to respond in advance of any conservation measures that may be employed to limit the negative effects of climate change.

It has been recognized that there is a need for conservation practices to be based on properly assessed evidence rather than anecdote or even popular myths (Sutherland *et al.* 2004). A parallel was drawn between the medical sciences 30 years ago where procedures and therapies were not necessarily evaluated. Since that time the use of evidence-based practice has transformed the medicine that is routinely taught and practised. A plea was made by Sutherland *et al.* (2004) for a similar revolution in the conservation sciences. Sutherland (2002) has also drawn attention to the work of palaeoecologist J.-C. Svenning (2002) whose work on European landscapes before agriculture has questioned the perception that there was extensive dense woodland in Europe and North America during the earlier Holocene. The latter builds on what Vera (2000), and others (for references see Svenning 2002), have suggested regarding the role of factors such as large herbivores in opening up the landscape. Quaternary palaeoecology can therefore be required to establish what is 'natural', and is thus an important form of evidence on which to base conservation measures (Sutherland 2002). Clearly the debate among palaeoecologists is relevant to those actively conserving species and landscapes today.

The application of palaeoecological data to the understanding of how environments are likely to change in the future has existed for some time, with initiatives such as that by Huntley *et al.* (1997). More recently others, such as Willis and Birks (2006), Willis *et al.* (2007), Lyman (2006), and Botkin *et al.* (2007), have considered the use of palaeoecological data to inform biodiversity conservation practices. A generally ignored but very important body of work is that by Graham (1988, 1997) who has cautioned against the simple use of palaeoecological data in future predictions due to the individualistic response of species to climate change. More alarming still, however, is research which suggests from modelling experiments that climates themselves will change into novel systems not previously seen, and that some present climates will disappear due to factors such as CO_2 levels that exceed those which have existed for the past 650 000 years (Williams *et al.* 2007). Clearly this will have major effects on the response of biotas to such changes in the future and could limit the use of palaeoecology for predicting the future. However, despite questions over the applicability of the past to the future, it is one of the few empirical sources of information we have (and maybe the only such source of information) and so cannot be ignored.

The aim of this chapter is to review some of the attempts at using the past to predict the future, as well as a discussion of some important issues that need to be addressed and incorporated into models of the future. The particular issues that have been highlighted are the related subjects of refugia, population extinction, and the existence of non-analogue communities in the Quaternary. The first of these is important as differently adapted taxa have different refugial areas; because species are often at their most vulnerable when they are in their restricted refugial populations rather than in their optimally distributed populations, this correspondingly suggests that different geographical areas may need to be recognized as of core conservation importance. The theme of population extinctions is a related subject, and a discussion is presented here to consider the increasing evidence for population extinctions in the Quaternary that have been demonstrated using ancient DNA techniques. These micro-extinctions are evidence of environmental stress on species and should be considered in addition to species extinctions in order to better understand the whole extinction process. Finally, the existence of non-analogue communities in the Quaternary is considered in relation to the manner in which species distributions change. The fact that species respond, rather than ecological communities as a whole, is an important factor that requires consideration by conservation practitioners. Such responses imply that communities will not simply move northwards as units with global warming, but will break up and form new communities.

13.2 Responses to climate change

Lister (1997) described five responses by vertebrates to environmental change during the Quaternary, including: (1) behavioural accommodation, (2) distributional shifts, (3) ecophenotypic modification (non-genetic), (4) evolution (genetic), and (5) extinction. Of these responses, behavioural accommodation is not relevant to plants and is very difficult to recognize in the fossil record. This leaves four general response modes to environmental change by organisms that can be identified in the past. However, while these responses are listed as distinct, it is becoming increasingly clear that extinction and range shifts are part of the process continuum of an organism's population dynamics. It is generally accepted that distributional change either involves population increase with an area-occupied increase or population decrease (local extinction) with an area-occupied decrease. When all populations of a species are locally extinct, the species as a whole is extinct. Therefore, responses 2 and 5 simply describe different levels of population size and so these responses are more similar than generally portrayed. Another problem exists in distinguishing ecophenotypic change from evolution, as the morphological evidence may be identical. For example, Lister (1997) cited research on domestic pigs where different individuals from the same litter were brought up at different temperatures to produce widely varying bodyforms (Weaver and Ingram 1969). These are exactly the kinds of changes in bodyform seen in Quaternary mammals throughout time with changing climates and interpreted as representing evolution. Therefore, distinguishing between responses 3 and 4 will be difficult. Another confounding factor in interpretations of the fossil record is that such morphological changes may instead involve extinctions followed by immigrations of other populations (see below). Recent analytical developments, such as ancient DNA studies, will hopefully aid distinguishing between the various responses which have taken place (see below). This is an important consideration when the significance of past extinctions and ecological responses are considered in relation to conservation biology, and signifies that the whole spectrum of past ecological responses by organisms to environmental change need to be included in any inferential application to the future. A similar point was made by Guthrie (1990a) in relation to interpretations of the causes for the megafaunal extinctions of the Late Pleistocene. He noted that the Late Pleistocene saw a faunal revolution involving many ecological changes, which he suggested argued for a climate-led cause of the extinctions rather than the alternative human overkill hypothesis. He believed that too much focus on the extinctions themselves has obscured the fact that other events had taken place.

Listed among the phenomena which took place during the end of the Pleistocene were rapid evolutionary changes, fractionation of biotic communities, and enormous reductions in distributional ranges. It can be argued that many or all of these changes reflect population size changes, because distribution reductions involve a retreat by local population extinction, community fractionation is a more complex version of distribution change (see below), and rapid evolution often involves extinction followed by immigration of allopatrically distributed different, but congeneric, populations.

What the above implies is that although extinction is generally considered to be a negative phenomenon it is also a natural process and indeed the expected ultimate fate of all species (Willis *et al.* 2007, and see below). Neither of these considerations signify that extinction is an acceptable outcome when humans are involved, but it does imply that extinction, as a process, is inevitable and natural. Therefore, we need to understand past extinctions, and their patterns and causes, to assess what we should accept and what we should be alarmed by. The same applies to all response types described by Lister (1997).

13.3 Evidence-based conservation

As described above, Sutherland *et al.* (2004) have made the case for evidence-based conservation and have pointed to palaeoecology as a potential source of evidence. Others have made a similar case for the use of such data although most former calls to arms have come from palaeoecologists themselves, so an advocate from conservation biology will hopefully prove more influential.

A former initiative, that by Huntley *et al.* (1997), is a published proceedings of a workshop which aimed to assess the degree to which different organisms respond to climate change by geographical range change versus evolutionary change, and which included workers studying a range of different types of organisms and using various methods. It was hoped that by bringing together Quaternary scientists specializing in organisms as varied as plants, molluscs, beetles, birds, and mammals, an assessment could be made of the abilities of different organisms to respond to varying rates of change. Included among the conclusions to the workshop were that it was unlikely that organisms would be able to adapt fast enough to deal with forecast global warming, and that population fragmentation and reduction, and species- and subspecific-level extinctions, were likely. It was also suggested that spatial scales of response would vary, and that more complex shifts in population size and extent would take place within landscapes. The individualistic response, particularly emphasized by Graham (1997), was also considered important as rendering predictions difficult to make. Perhaps most importantly, a caution was made regarding the lack of basic biological knowledge of many organisms which will inevitably hinder any ability to predict how they might respond to climate change (Huntley *et al.* 1997).

A significant contribution to the use of the Quaternary fossil record for understanding the processes affecting biodiversity in the light of climate change is that by Botkin *et al.* (2007). In this instance focus was given to the phenomenon he described as the 'Quaternary conundrum', whereby it was identified that comparatively little extinction occurred during past Quaternary climatic changes when much more has been predicted to accompany impending global warming. Explanations suggested for this conundrum include the possibility that climate changes were different in the past from that predicted for the future, that unaccounted-for evolutionary and ecological processes allow species persistence, and that other factors mediated rates of extinction (Botkin *et al.* 2007). To these explanations must be added the fact that different species appear to have different rates of speciation/extinction, which in turn probably depend on the nature of the fossil evidence available for each species, and how that evidence relates to the distinction between modern species of the taxonomic group in question.

Willis and Birks (2006) and Willis *et al.* (2007) recently reviewed the use of long-term ecological studies to biodiversity conservation. Willis *et al.* (2007) highlighted themes such as extinction, identifying regions of greatest diversity and threat, climate change, and biological invasion. Within that framework was included a perspective which is rarely considered, where extinctions are acknowledged as an inevitable and 'natural' process, and it was suggested that species at the end of their evolutionary lifespan should be identified to make the best use of the finite resources available to conservation biologists to best preserve 'evolutionary potential' (Willis *et al.* 2007). However, whether such identifications are possible, or even theoretically meaningful, remains unclear. In Willis and Birks (2006) considerations are given to the use of palaeoecological data to inform biodiversity conservation practices and describe a number of areas in which such data can contribute, including biological invasions, wildfires, climate change, and the determination of natural variability.

The subject of biological invasions is one which is a concern to many (Lyman 2006; Willis and Birks 2006; Willis *et al.* 2007; see also Chapter 12 in this volume) and clearly the verification of whether a species is native to a particular geographic area can be achieved by using the fossil and archaeological records. Lyman (2006) discussed exotic and native taxa and further dealt with the subject of distinguishing between invasive taxa from recolonizing taxa. He provided an example based on zooarchaeological remains of two fur seal species, the Guadalupe fur seal *Arctocephalus townsendi* and the northern fur seal *Callorhinus ursinus*, whose ranges on the west coast of the USA included more northern areas in the past. Clearly, if these species were to spread to these areas today they would be recolonizing rather than invading these northern coastal reaches. In fact there is evidence that the northern fur seal did so during the late twentieth century (Lyman 2006).

13.3.1 Biological invasion case study: eagle owl in Britain

A recent example of a potential biological invasion is that of the eagle owl *Bubo bubo* in Britain (Stewart 2007a). The species has recently started breeding in northern England, and has been the cause of some controversy because several authors have questioned whether it previously occurred in Britain and whether it can be considered part of the original native avifauna (Turk 2004; Dennis 2005). A key concern with the newly arrived eagle owls is that they may predate other rare raptors (hobbies *Falco subbuteo* and hen harriers *Circus cyaneus*) in the area. The fossil and more recent zooarchaeological records of eagle owl for Britain are clearly of relevance to the debate. Holocene records are necessary to confirm that the birds belong to the present interglacial fauna of the British Isles. However, post-glacial records (later than the Last Glacial Maximum (LGM), which ended *c.*16 000 years BP) are also relevant, as this is the time by which the species is likely to have arrived. Taxa like the eagle owl with temperate latitude distributions in Europe today are less likely to have undergone range reductions after the end of the Pleistocene than cold-adapted taxa such as reindeer *Rangifer tarandus*. Eagle owl remains, if they were found from the Late Glacial of the British Isles, may therefore represent the same population, as is evidenced by remains from the Holocene.

The absence of the eagle owl from Britain in historical times has been thought to be due to persecution (Mikkola 1983), although this is difficult to prove. Until recently, the species has occupied most of Europe, except the extreme north-west, where many other raptors (e.g. peregrine falcon *Falco peregrinus*, common buzzard *Buteo buteo*, red kite *Milvus milvus*, white-tailed eagle *Haliaeetus albicilla*, and marsh harrier *Circus aeruginosus*) have certainly been affected by persecution (e.g. Whitlock 1952; Brown 1976; O'Connor 1993; Yalden 2007). Indeed, the wider phenomenon of human disturbance has affected a whole suite of organisms in Britain and elsewhere (Stewart 2004). Protective legislation has allowed many of these birds to recover to some degree and recolonize parts of their former ranges. The recent breeding of eagle owl in Britain may have stemmed from the release of captive birds; alternatively, it may be the result of natural recolonization from mainland Europe, where the population has both increased and spread westwards into Belgium and The Netherlands (Dennis 2005). The change in protective legislation concerning the eagle owl in Europe is likely to be partly responsible for its range expansion on the mainland, and it is certainly plausible that the birds in Britain arrived naturally.

An examination of the fossil record suggests that eagle owls, or at least a species from the genus *Bubo* very closely allied with modern eagle owl *B. bubo*, have been present in Britain for up to 700 000 years, through to the end of the last ice age glaciation c.10 000 years ago and into the Holocene. It is, however, important to re-examine the material on which the written record is based. For example, one problem with previous studies is that the osteological similarity between snowy owl *Bubo scandiaca* (see Sangster *et al.* 2004 for generic-level assignment of this species) and eagle owl has been overlooked, so that eagle owl records should not be accepted unless it can be shown convincingly that snowy owl has been considered and eliminated. The latest confirmed find of eagle owl in Britain is known to be from the early Holocene, associated with Mesolithic archaeology (Demen's Dale) (Stewart 2007a). Furthermore, the eagle owl has probably been a native species in Britain for a long time, and possibly through different climatic regimes with different habitats (see Table 13.1). It is possible that this native status was not continuous and was punctuated by the extreme conditions of the most severe glacial periods. However, at most other times, including the warmer parts of glacial episodes, the climate and environment of the British Isles would probably have been suitable for the species.

A careful examination of the fossil record of the eagle owl from the Quaternary of Britain, and particularly the last 16 000 years including the early Holocene record, shows that the species is indeed a native to the country. This suggests that given protection from persecution, the species is not unlikely to recover to occupy this part of its natural range. This is a separate argument to that regarding the mode by which the species recovers its range, whether introduced by humans legitimately or otherwise. However, the precedent set by the unofficial reintroduction of the goshawk *Accipiter gentilis* into Britain (Hodder and Bullock 1997) is relevant here, as there appears to be no serious suggestion that this species should be eliminated.

The cause of species invasions can, however, be elusive, as is demonstrated by the explanations of perhaps the best-known invasion of Europe in living memory, that of the collared dove *Streptopelia decaocto* (Fisher 1966; Rocha-Camarero and Hildago de Truios 2002). In this case the species first reached north-west Europe from south-east Europe and south-west Asia during the 1950s. No satisfactory explanation has yet been proposed that adequately describes why it increased in range so far and so fast, spreading at least 1900 km in 20 years (Fisher 1966) and still spreading into parts of southern Europe (Rocha-Camarero and Hildago de Truios 2002). The reasons given for the spread include changes in nesting behaviour, climate change, changes in agricultural policy and farming practices, or explosive expansion caused by genetic drift (Rocha-Camarero and Hildago de Truios 2002).

A similar large-scale population movement of a commensal bird seems to have taken place during the deeper past when the house sparrow *Passer domesticus* spread in a westerly direction across Europe during late prehistory (Ericson *et al.* 1997). The explanation for the spread of the house sparrow is even less clear; it was formerly assumed to be associated with the human-driven spread of chickens across Europe from Asia, but it may instead be linked to the arrival of the domestic horse in the region. These examples demonstrate that even if an invasion can be ascribed a cause, such phenomena can be difficult to adequately explain mechanistically. The house sparrow and collared dove examples are also of interest as they suggest a link between agricultural practices and bird dispersal. It could be further speculated that when the European landscape changed dramatically as a result of the advent of cereal cultivation during the prehistoric Holocene, this new 'grassland' provided an agricultural steppe for open habitat birds such as grey partridge *Perdix perdix*, corncrake *Crex crex*, skylark *Alauda arvensis*, linnet *Carduelis cannabina*, and buntings (*Emberiza*, *Miliaria*). Indeed, some of these bird populations, if not species, may even have moved west with Neolithic agriculture across Europe, while others expanded their existing populations as this habitat spread. If this were true, it may be important to consider how this relates to the fact that many of these birds have experienced recent population declines in response to modern changes in farming practices (Robinson and Sutherland 2002).

A plea that Quaternary fossil and archaeological records of birds should be made available to conservation scientists was made in an article on the history of wetland birds in Western Europe (Stewart 2001, 2004). The British Ornithologists' Union had been

Table 13.1 The fossil record of the genus *Bubo* in Britain. From Stewart (2007a).

No.	Site	Dates	Age	Climate	Skeletal element and identification [identified by] with notes	Revised taxon
1.	Forest Bed, East Runton, Norfolk	c.1.7 million–700 000 years BP	Pastonian	Temperate	Distal right tarsometatarsus [ID Harrison, confirmed by Stewart]	*Bubo* (*bubo*)
2.	Unit 4c, Boxgrove, West Sussex	c.500 000 years BP	Late Cromerian Complex	Temperate	Left coracoid fragment, *Bubo* cf. *bubo*. [ID Stewart]	*Bubo* (*bubo*)
3.	Lower Loam, Swanscombe, Kent	c.400 000 years BP (although TL dates: 202 000±15 000 and 228 000±23 000)	Hoxnian: Pollen zone Ho II	Temperate	L/R carpometacarpus fragment [ID Harrison, confirmed by Stewart]	*Bubo* (*bubo*)
4.	Tornewton Cave, Devon	c.200 000–10 000 years BP	Middle or Late Pleistocene	Glacial or interglacial	Ungual phalanx [ID Harrison, unconfirmed by Stewart]	Undet. large predatory bird
5.	Chelm's Combe Shelter, Cheddar, Somerset	^{14}C dates: 10 190±130 and 10 910±110 years BP	Devensian: Dryas 3	Glacial	Distal right carpometacarpus [ID Harrison, reassigned Stewart]	*Bubo* (*bubo*) or *B.* (*scandiaca*)
6.	Ossom's Cave, Staffordshire	^{14}C dates: 10 190±130, 10 780±70, and 10 600±140 years BP	Devensian: Dryas 3	Glacial	Element? [ID Bramwell, not seen]	Not seen
7.	'Derbyshire Peak Caves'	c.110 000–10 000 years BP	Devensian	Glacial?	Element? [ID Bramwell, not seen]	Not seen
8.	Kent's Cavern, Devon	c.125 000–10 000 years BP	Late Pleistocene	Glacial?	Element? [ID ?, not seen] Record of *B. bubo* in Tyrberg (1998) from Kent's Cavern; may refer to the specimen below	Not seen
9.	Kent's Cavern, Devon	c.110 000–10 000 years BP	Devensian	Glacial?	Complete right tarsometatarsus, *Bubo scandiaca* [ID Harrison, confirmed by Stewart]	*Bubo* (*scandiaca*)
10.	Langwith Basset Cave, Derbyshire	c.125 000 years BP to present	Late Pleistocene/ Holocene	Glacial or temperate	Element? [ID Bramwell?, not seen]	Not seen
11.	Merlin's Cave (Wye Valley Cave), Herefordshire	c.15 000–10 000 years BP	Devensian: Late Glacial?	Glacial?	Element? [ID Newton, not seen], may have been lost in WWII	Not seen
12.	Demen's Dale, Derbyshire	c.10 000–5500 years BP	Early Holocene (Mesolithic)	Temperate?	Right tarsometatarsus [ID A. Hazelwood and others, confirmed from photograph by Stewart]	*Bubo* (*bubo*)
13.	Meare Lake Village, Somerset	c.700 BC to AD 43	Holocene (Iron Age)	Temperate	Two portions of ulnae [ID D. Bate but may be unreliable; D. Yalden, personal communication]	Not seen

TL, thermoluminescence date.

interested for some time in the use of the older bird records to establish baselines of native status. As a result of the controversy over the status of the eagle owl and its corresponding conservation challenge the British Ornithologists' Union established a new section (Category F subcommittee) of the official list of native British bird taxa (British Ornithologists' Union Records Committee 2007). This list includes all records of birds dated to before AD 1800 (the start date of the British list of observed live birds). It is divided into the F1 subcategory (including birds from *c.*16 000 years BP until AD 1800, and including species records based on skeletal remains and on documentary records only) and the F2 subcategory (including all records from *c.*16 000 years BP back to *c.*700 000 years BP). This initiative will hopefully be of help if any further controversies exist over native status of birds in Britain.

13.4 The role of refugia in the survival of species

The idea that the geographical ranges of species have changed through time in response to past climate change has existed since the time of Darwin (1859). However, it was not until the twentieth century, with a better understanding of the ice ages and after the advent of the Modern Synthesis of evolutionary biology, that the concept of the Glacial Refugium Hypothesis began to dominate studies of ice age biogeography (Mengel 1964; MacPherson 1965; Holder *et al.* 1999). This concept recognizes the cold glacial phases of the Earth's Milankovitch cycles as being the primary forcers of northern-latitude temperate and boreal animal and plant population break-up (e.g. Hewitt 1996). It also leads to the general assumption that many organisms were pushed southwards as the glaciated north became inhospitable to most species. However, the dominant view of biogeographic change over the last glacial–interglacial climatic cycle (*c.*30 000 years) has been challenged, and the different current perspectives have varying implications for the conservation status of today's animal and plant populations. Nevertheless, refugia themselves are important for the long-term survival of species, as extinction is often the result of a complete loss of refugia.

Phylogeographic studies have recently confirmed many of the assumptions regarding southern refugia, with modern haplotype distributions suggesting that different organisms migrated out of various European peninsular and other southern refugia at the end of the last glaciation (Hewitt 1996, 1999, 2000; Taberlet *et al.* 1998). This has, however, been complicated by the suggestion that cryptic northern refugia had existed in the Late Pleistocene for temperate organisms (Stewart and Lister 2001; Stewart 2003; Willis and van Andel 2004). It has also been claimed that the peninsular refugia were areas of endemism rather than refugia (Bilton *et al.* 1998). The cryptic northern refugium hypothesis has found significant further support, with phylogeographic studies yielding evidence for northern refugia among various temperate organisms including small mammals (Wójcik *et al.* 2002; Jaarola and Searle 2003; Deffontaine *et al.* 2005; Kotlìk *et al.* 2006), reptiles (Carlsson 2003; Ursenbacher *et al.* 2006), snails (Haase and Bisenberger 2003; Pfenninger *et al.* 2003), sedges (Tyler 2002a, 2002b), and ferns (Trewick *et al.* 2002). Furthermore, the brown bear *Ursus arctos*, one of the key taxa used by Hewitt (1996) to suggest that the southern peninsular areas of Europe were the dominant glacial refugia in the Late Pleistocene, has been shown by ancient DNA analysis to also have had populations further north during this time to allow gene flow across southern Europe. (Valdiosera *et al.* 2007). Recent research has also shown that many boreal organisms also continued to live at high latitudes in areas such as Beringia even during the glacial maxima (Holder *et al.* 1999; Tremblay and Schoen 1999; Fedorov and Stenseth 2001; Loehr *et al.* 2005; Pruett and Winker 2005; Anderson *et al.* 2006).

Increasingly, however, it has become clear that the Glacial Refugium Hypothesis was missing the interglacial perspective. Animals that are now confined to northern latitudes, and in some cases to mountain ranges in the south, are now occupying refugia during the Holocene interglacial (Stewart and Lister 2001; Dalén *et al.* 2005). This additional dimension does not completely negate the existence of glacial refugia for such cold-adapted taxa and particularly during glacial maxima such as the LGM. However, cold-adapted boreal taxa are geographically restricted to refugia during interglacial intervals, whereas during the majority of glacial intervals they are at their maximum distribution (Musil 1985; Tyrberg 1991b; Stewart *et al.* 2003). This signifies that phylogeographic (and biogeographic) studies of taxa such as the rock sandpiper *Calidris ptilocnemis* in Beringia (Pruett and Winker 2005) may need further interpretation (Stewart and Dalén 2008). Pruett and Winker (2005) found that this species was variable with partitioned genetic diversity which did not, however, reflect the different subspecies that have been described. They concluded that this diversity had evolved in multiple refugia in Beringia and that these refugia were in operation during the last glaciation or between 117000 and 10000 years BP. However, because it is likely that cold-adapted taxa such as the rock sandpiper had more extensive and more southwards-extending populations during the last and previous glacials, they may have been found more extensively across the North American and Asian continents. The populations seen today are the refugial populations likely to have resulted from a major contraction along the coasts of both continents. From an evolutionary, and indeed conservation, perspective geographic regions that have been continuously inhabited during both glacials and interglacials are of special importance. These are the 'long-term refugia' or 'true refugia' where genetic diversity evolves, and between which there is sufficient time for strong population divergence to evolve, suggesting that subspecies differentiation may not have taken place during the last glaciation. In the editorial comment on the dialogue of Pruett and Winker (2005, 2008) with Stewart and Dalén (2008), Barnosky (2008) made the point that it is of great importance to conservation biology to distinguish between when species are restricted to refugia as opposed to being at their most extensively distributed. Clearly taxa that are in refugia are at their most restricted geographically and hence are more at risk of extinction.

The number of types of refugia that are being invoked to explain the various phylogeographic patterns seen in animals and plants is growing, although they remain to be adequately defined and distinguished. Clearly the locations of refugia are important to conservation biology, as these areas have long-term conservation significance. These locations may also be unexpected, as seen from a study of hairy wood ants *Formica lugubris*, the Irish populations of which have been shown to be native and likely to have survived in ice-free areas during the last part of the Pleistocene. These ants are therefore very vulnerable to extinction because of low genetic variation, which is made worse by the inherent low diversity present in social insects (Mäki-Petäys and Breen 2007).

The timing of subspecies differentiation is also important, as it seems clear that if—as suggested from the reinterpretation of the rock sandpiper and other studies (Stewart and Dalén 2008)—differentiation took place during the Holocene then subspecies may not all be 'incipient species' and so are not necessarily of long-term conservation importance. The populations or subspecies which should be prioritized are the ones living in the long-term or true refugia, as these taxa will contribute to the genetic diversity available through Milankovitch cycles.

13.5 Range contractions and population extinctions

As has been indicated above, the species range contractions caused by climatic perturbations of the Quaternary probably represent the extinction of the populations that lived in the areas that became vacated. This has been suspected for some time (Stewart *et al.* 2003), and was recently confirmed in a study of the former wider distribution of the arctic fox *Alopex lagopus* in Europe (Dalén *et al.* 2007). The more southerly distributed populations of this species during the Late Pleistocene appear to have been made up of extinct haplotypes, which demonstrates the distinctiveness of these arctic fox fossils from their modern, northern counterparts. It remains to be seen, but seems likely, that other species whose ranges contracted after the Pleistocene also did so by population die-off rather than actual movement. The latter applies, for instance, to species that lived over much of the mid latitudes of north-west Europe, and are either now restricted, like the arctic fox, to northern latitudes (e.g. reindeer, Norway lemming *Lemmus lemmus*, and collared lemming *Dicrostonyx torquatus*) and/or to lower-latitude montane areas such as the Alps (e.g. ptarmigan *Lagopus muta*, and arctic hare *Lepus timidus*) and eastern areas of the Palaearctic (e.g. sousliks *Citelus* spp., and saiga *Saiga tatarica*). The same applies to many, if not most, other Pleistocene geographical range contractions, including population movements at the onset of glacials where temperate- and warm-adapted taxa contracted south, east, or into more northern cryptic refugia.

Regional extinctions have rarely received attention compared to global extinctions. This is no doubt because they are not easily recognized in the morphology of fossils, and even if they are recognized they are not often considered to be greatly significant in Quaternary palaeontology because of their subspecific or population-level status. An exception to this is that of the northern European water vole *Arvicola terrestris* during the middle Pleistocene of northern Europe. In this instance there was an apparent reversal in the lineage's evolutionary trends. *A. terrestris* apparently underwent an evolutionary reversal in the morphology of its molars, which is probably explained by an immigration of a more 'primitive' population from southern Europe (van Kolfschoten 1990; Lister 1993). The population preceding the apparent reversal is likely to have been replaced either because it became extinct or because it moved. It may be that an extinction is more likely because whole populations do not generally become displaced, but instead retreat by population extirpation *in situ* (see above). This further demonstrates that extinction is not merely a completely separate phenomenon from distribution change, and that it is also involved in allopatric evolution.

A recent example of a morphologically distinct population which has been shown to be genetically distinct is that of the Late Pleistocene wolves of eastern Beringia by Leonard *et al.* (2007). In this case the extinct Beringian wolf population was shown to have a different skull morphology from both contemporary southern wolves from La Brea, and from modern Alaskan and non-Alaskan wolves. Furthermore, these distinctive wolves were interpreted from morphology and stable isotope evidence to have been adapted to killing and/or scavenging large (megafaunal) prey. Finally, ancient DNA analysis showed that these wolves are now extinct, presumably as a result of the loss of their megafaunal prey base (Leonard *et al.* 2007).

The majority of morphologically distinctive past populations of still-extant animal species have not been shown to be extinct, although this should be (and is being) tested by the use of ancient DNA methods. For example, the consistently larger-bodied populations of both species of *Lagopus* (red/willow grouse *Lagopus lagopus*, and ptarmigan *L. muta*) in the Late Pleistocene, for which the richer carrying capacity of the steppe tundra was invoked as an explanation (Stewart 1999, 2007b), may well prove to be extinct. These larger birds may not be ancestral to the smaller individuals of those species today and so do not represent the intraspecific evolution of those lineages. Instead they probably represent extinctions of the Pleistocene lineages that were then replaced by smaller intraspecific populations of those species from other geographical regions (i.e. allopatric evolution). This alternative could cause a reinterpretation of the conclusions of Smith and Betancourt (2006), who reported size change in woodrats (*Neotoma*) in the USA over the last 40000 years and suggested that *in situ* body-size change was a likely response of woodrats to future climate change. If these findings instead represent extinctions and immigration events, the future response of woodrats would therefore rely instead on the viability of suitably adapted populations elsewhere to expand into vacant territories left by locally extinct populations. Clearly the two interpretations have different implications for the adaptability of woodrats to human-caused climate change.

Research on ancient mitochondrial DNA, such as that by Barnes *et al.* (2002) on brown bear, Shapiro *et al.* (2004) on bison, and MacPhee *et al.* (2005) on musk ox, has led to a growing awareness of Late Quaternary subspecific-level extinctions. These three studies have showed that the genetic diversity in these species was hugely reduced at the LGM, probably due to the environmental changes that accompanied this extreme global cooling event. This further demonstrates, in a direct manner, the importance of climate and environment on genetic diversity through extinction. There are suggestions that the climatic deterioration of the LGM also caused extensive environmental depletion, including population declines in smaller carnivores and other animals (Stewart *et al.* 2003; Stewart 2005) and even drove the retreat of mammoths *Mammuthus primigenius* from Europe (Stuart *et al.* 2002). Population declines, sometimes accompanying geographical range reduction, are also increasingly apparent in the Late Pleistocene as a result of climate change. For example, new findings from ancient DNA of neanderthals *Homo neanderthalensis* have shown that the extreme cold phase during Marine Isotope Stage (MIS) 4 (74000–60000 years BP) caused a reduction of genetic diversity prior to their eventual complete disappearance in MIS3 (approximately 60000–25000 years BP; Orlando *et al.* 2006a).

These examples demonstrate that, whether accompanying geographical range reduction or not, there is a growing body of evidence that population reductions and extinctions have taken place during the Late Quaternary in addition to species-level extinctions. This shows that species are dynamic entities whose overall genetic make-up responds to climate change. It further suggests that an over-emphasis on species-level extinctions has caused a lack of consideration of intraspecific-level extinction, which has in turn hampered interpretation of the causes of species extinctions in the past. This may in part explain the so-called Quaternary conundrum, because the coeval reductions in genetic diversity in many taxa, including that in smaller mammals such as arctic fox, are unlikely to have been caused by agents other than climate. If Palaeolithic humans did not cause the extinctions in smaller mammal populations they may not have had a role in the global extinctions of megafaunal taxa. It should also be remembered that species extinctions generally take place through range contractions into smaller isolated populations (Ceballos and Ehrlich 2002; Lister and Bahn 2007).

13.6 The lack of uniformitarianism: non-analogue ecological communities in the Quaternary

Willis and Birks (2006) have described how organisms are likely to respond to climate change in the future based on past responses observed in the fossil record. This valuable contribution is an important counter to the general lack of high-profile attention that palaeoecological research is receiving in conservation research. However, while there may be data (e.g. Grayson 2000) suggesting that taxa are predictable using uniformitarianism (i.e. modern taxonomic distributions and their corresponding climatic variables can inform the responses of taxa to past climatic variation), there is also a wealth of information demonstrating that there are many non-conformist species in this regard (Graham 1988, 1997; Fox 2007; Williams *et al.* 2007). For example, many extant mammal species are known to have occurred sympatrically in the past but have completely allopatric distributions today (FAUNMAP Working Group 1996). These disharmonious ecological mixes or 'non-analogue communities' are not exceptional in Pleistocene palaeontology, but instead are the rule (Huntley 1991; Roy *et al.* 1995), and are not restricted to mammals but instead have been documented in almost all major taxonomic groups during the Quaternary. This includes vascular plants (Bell 1969; Huntley 1991; Kullman 1998, 2002; Jackson and Williams 2004), diatoms (Gasse *et al.* 1997), dinoflagellates (Head 1998), foraminifera (Wittaker in Bates *et al.* 2000), and animals, including mammals (Bramwell 1976; Graham 1986; Graham and Grimm 1990; FAUNMAP Working Group 1996; Stafford *et al.* 1999; Stewart *et al.* 2003; Stewart 2005; Stewart 2008), birds (Bramwell 1984; Emslie 1986; Brasso and Emslie 2006), beetles (Coope and Angus 1975; Coope 2000), terrestrial molluscs (Kerney 1963; Preece 1997; Preece and Bridgland 1998), marine molluscs (Roy *et al.* 1995), and polychaetes (Sanfilippo 1998).

Different explanations for these non-analogue communities have been offered by various authors (Kerney 1963; Coope and Angus 1975; Bramwell 1984; Graham and Grimm 1990; FAUNMAP Working Group 1996; Stafford *et al.* 1999). One explanation lies in the fact that species respond individually to environmental change (the Gleasonian model; Gleason 1926) and not as part of communities (the Clementsian model; Clements 1904) (Graham 1985, 1986; FAUNMAP Working Group 1996). The individual response of species has long been used to explain the different vegetational elements that recombine during the successive climatic episodes of the Quaternary (West 1980; Prentice 1986; Webb

1986). Phylogeographers have recognized the same polarized hypotheses and have called these the concerted (community) versus independent (individual species) responses to environmental fluctuations (Sullivan *et al.* 2000). Firm support for the model of individualistic species responses was provided by Taberlet *et al.* (1998) based on comparative European phylogeography. Another explanation is that the non-analogue component species of disharmonious communities have poorly understood modern geographical ranges; that is, the observed differences between past and present community assemblages reflect differences between the real and actualized niches of different species (Kerney 1963). However, the main alternative explanation for non-analogue assemblages is that they are mixtures of material of different ages. For example, the non-analogue coleopteran assemblages from the Late Pleistocene have been explained as a mixture of material dating from distinct intervals with different climates (e.g. Coope 2000). Although post-depositional mixing is a real possibility, two factors suggest that many or most non-analogue communities are likely to be genuine: the phenomenon is so well known across a wide range of taxa living in a diverse range of environments, with their fossils being preserved in very different deposits with different accumulation histories; and radiometric dating has demonstrated the contemporaneity of some non-analogue species, for example Late Pleistocene mammals (Stafford *et al.* 1999).

Although non-analogue ecological communities are commonly observed in the Quaternary fossil record of many taxonomic groups of animals and plants, the organisms with which this phenomenon is perhaps most closely associated are mammals. Species from the terminal Pleistocene (*c.*60 000–10 000 years BP) are maybe the best understood in relation to dramatic climatic and environmental upheavals reflecting Milankovitch- and sub-Milankovitch-scale perturbations. The main ecological controversy for this interval concerns the ecology of the vegetational environment, and there have been substantial disagreements between researchers studying the palynological and mammalian records. Mammalian palaeontologists (e.g. Guthrie 1990a, 1990b) have described the mid-latitudes of the northern hemisphere as having a biome that they called the 'steppe-tundra' or 'mammoth-steppe', described on the basis of its mixture of mammals that are associated today with either more continental climes (steppe) or northern regions (tundra). A classic example of this is the regular co-occurrence of horse *Equus* sp. and reindeer as common elements of Late Pleistocene sites (Musil 1985; Markova *et al.* 1995; Stewart *et al.* 2003). Palynologists, however, have rejected this description of the vegetation (Colinvaux and West 1984; Ritchie 1984). On the basis of pollen analyses they considered that the region was dominated by tundra vegetation, which mammalian palaeontologists cannot reconcile with the large herbivores of this interval (e.g. mammoth, woolly rhinoceros, giant deer, horse, bison) and the carnivores they supported (e.g. lion, hyaena). Palynologists argue that these megafaunal elements were only present in short-lived intervals during which conditions were more optimal, although dating of the faunas has demonstrated that the megafauna were present relatively continuously (Guthrie 1990a).

Several studies of Late Pleistocene plant fossils have been in accordance with the mammalian explanation, such as the analysis of plant species composition from Earith, southern England, by Bell (1969) based on macrofossils (primarily seeds and leaves, in contrast to the palynological focus on genera and families). This study described the region's Late Pleistocene vegetation as being a curious mixture of southern, halophytic, and steppe elements as well as the more expected cold-tolerant plant species, and concluded that either the climate was unlike any analogues available today or the plants were responding to factors other than climate. The ecological associations are in accordance with the mammalian fossil evidence for this interval at other localities, although the beetle fossil samples taken at the same time from exactly the same sampling sites at Earith have subsequently been reported to show that at least two periods of markedly different climatic regimes were involved (Coope 2000). Based on modern climatic associations, the first of these suggested climatic regimes was characterized by a cold and continental climate, and the second climatic regime was characterized by warm summers and a markedly less seasonal climate (Coope 2000). Clearly the interpretations of Bell and Coope cannot both be correct. Either Earith has sediments with organisms with different climatic tolerances because the fossils were physically mixed together (Coope's explanation) and they represent different time episodes, or the assemblage is a genuine non-analogue mixture (Bell's explanation). It is notable that studies of mammals and birds from the similar-aged deposits in England have also yielded anomalous assemblages with many of the same northern, southern, and eastern affinities (Stewart *et al.* 2003; JR Stewart and R Jacobi, unpublished work).

Similar patterns of intermingled faunas and floras are also found in North America, although there is possibly a greater degree of consensus between the different palaeontological disciplines including mammalian, molluscan, coleopteran, and floral research (Graham 1986). It has also been suggested that the non-analogue mammalian communities of the Late Pleistocene ceased to exist at the same time as much of the megafauna went extinct at the end of the Pleistocene (Graham 1986; Semken *et al.* 1998; Tankersley 1999). This would seem to indicate that they are intrinsically tied to ecological characteristics such as the carrying capacity of the environment as a whole, and hence need to be understood in the light of other Quaternary ecological phenomena such as extinctions and evolution (Guthrie 1990a). The patterns of distribution of modern-day biotas in the Holarctic (or in the Holocene as a whole) are largely dictated by the latitudinally zoned nature of the vegetation, which is in turn regulated by the relatively stable climates. In the Late Pleistocene this was apparently not the case, and patchy mosaic-like vegetational assemblages existed instead, as reflected by the disharmonious nature of the described biotas; these ecosystem changes may have been caused by the climatic instability that has been revealed by Greenland ice cores (Lister and Sher 1995).

Non-analogue Quaternary communities are not restricted to terrestrial assemblages. The marine environment also seems to have been characterized by comparable associations of taxa during the Pleistocene, although workers on marine assemblages are not convinced that non-analogue assemblages are genuine. The most extensively studied region is the Eastern Pacific coast of North America (Roy *et al.* 1995, 1996). Non-analogue mollusc communities from this region have been described as thermally anomalous assemblages. The explanation for these faunas is that the assemblages are caused by short-term, high-amplitude climatic 'flickers' (Roy *et al.* 1996); it is suggested that different mollusc species responded close to the resolution limits of the fossil record, and hence the assemblages are actually mixes of organisms that are responding fast to short climatic fluctuations. Roy *et al.* (1996) also cited the fact that Pleistocene climatic changes are not known to be accompanied by significant speciation or extinction events as evidence that these thermally anomalous assemblages are mixtures. Other explanations include that of Zinsmeister (1974), who suggested that temporary ocean currents, not present today, were introducing tropical mollusc larvae which survived to adulthood in areas with cooler waters. However, the alternative explanation that the thermally anomalous mollusc associations are real has not been given a great deal of weight. It is interesting that a different consideration of the phenomenon of rapid climatic fluctuations was made by Lister and Sher (1995), who suggested that this phenomenon would promote the existence of a mosaic-like mix of ecologies as opposed to the latitudinal ecoregions seen under the temporally more stable modern-day climatic regime.

Another perspective on the subject of non-analogue communities is the argument over whether such ecologies existed during the Holocene, or whether they were restricted (at least in large part) to the Pleistocene (FAUNMAP Working Group 1996; Alroy 1999). Notably, Alroy (1999) criticized the link that has been made by other researchers (FAUNMAP Working Group 1996) between the megafaunal extinctions of the Late Pleistocene and the loss of disharmonious assemblages. He suggested that the non-analogue communities of the Late Pleistocene have no bearing on the subject of extinction, and also argued that such non-analogues are very rare and not restricted to the Pleistocene. Alroy (1999) did acknowledge that differences between glacial and interglacial faunas had been noted and attributed to differences in seasonal climates (e.g. Graham 1986) and that this phenomenon was partly responsible for the non-analogue phenomenon. However, he asserted that most mammals 'maintained ecologically stable and predictable distributions that only gradually contracted in the face of Holocene environmental change' (Alroy 1999). Graham (2006) vigorously rejected these conclusions, suggesting that Alroy (1999) used a non-traditional definition of non-analogues which artificially decreased the number of observed examples of such biotas. Indeed, significant changes in mammalian ranges and diversity have been documented accompanying Holocene climate change in the Great Basin of the western USA (Grayson 2000) and found to progress along predictable lines. In this case, an increase in Holocene aridity led to a corresponding decrease in mesic species and an increase in xeric mammals.

In summary, it seems that there were a number of different organisms living in the Pleistocene that were in disharmonious associations from a modern point of view, including taxa that today are not found in sympatry. It seems unlikely that these different non-analogue communities, documented from a range of different sites with a range of depositional histories and having such similarly disharmonious ecological indications, should be entirely the result of post-depositional mixing. This suggests that the disharmonious nature of the ecology of the mid-latitudes of the northern hemisphere in the Late Pleistocene was genuine. The existence of these non-analogue communities has been acknowledged by population geneticists such as Hewitt (1996), who have recognized their importance for evolution. The implications of this phenomenon to conservation biology are that there will be a significant level of unpredictability in environmental responses to climatic change at the community level, and that it may therefore be unwise to attempt to maintain perceived 'natural' conditions when 'natural' may take on as-yet-unseen permutations of community assemblage (see Graham 1988). This in turn questions the degree of importance that should be given to the concept of keystone species discussed by Willis and Birks (2006), and to the 'refuge view' of conservation areas (Graham 1988). Whereas the refuge view assumes that all taxa of a presumed community may be accommodated in a single reserve, each taxon instead needs to be considered separately from a conservation viewpoint; taxa will each have their own individualistic responses to climate change, and these individualistic refugia are not likely to share similar geographical distributions (Graham 1988).

13.7 Conclusion

The discussion above, of various aspects of palaeoecology that are relevant to conservation biology, is one that emphasizes some of the problems that exist. It is hoped that the tone is not so negative as to render the topic apparently redundant. However, it will be imperative that we understand the subject at hand in an interconnected, holistic way if we are to make best use of the lessons that palaeoecology can give us. Broadly, there are two types of information that can be sought from palaeoecology. First, palaeoecological data can provide information about how individual taxa have responded in the past to specific climatic and environmental changes. Second, data from the past can suggest the kinds of processes that are likely to take place in the future at a broader ecological level, but without necessarily allowing access to detailed predictions. Once the dialogue between palaeoecologists and conservation biologists gets underway, the challenge is to develop a flexible definition of 'natural' which allows for unexpected developments. This will be necessary if the limited resources available for conservation are to be deployed in a pragmatic and effective manner.

Acknowledgements

Thanks are due to Sam Turvey for asking me to write this contribution and to Kathy Willis for suggesting that he ask me. I would also like to acknowledge Derek Yalden's help with Table 13.1. Finally, I would like to thank Tabitha Stewart Stacey and Edward Stewart for making me smile.

CHAPTER 14

Holocene extinctions and the loss of feature diversity

Arne Ø. Mooers, Simon J. Goring, Samuel T. Turvey, and Tyler S. Kuhn

14.1 Introduction

Species are hard to define, but under all definitions they are unique. Each species can be considered to possess a set of unique characters that comprise its feature diversity, the diversity that will be lost when that species goes extinct. However, because species differentiate via evolution from other species, they also share features as a function of degree of relationship. So, within placental mammals, all mice share all the characteristics that allow us to call something a mouse, whereas the sole living species of aardvark shares very few basic mammal characteristics with anything: it must be true that the Earth loses less diversity when any single mouse species goes extinct than when the last living aardvark shuffles off its mortal coil. This is one well-established way in which species are not created equal (Vane-Wright *et al.* 1991; Faith 1992).

We can start to formalize this simple idea with the help of Fig. 14.1. It depicts the common pattern of a diversifying phylogenetic tree, but also can represent the pattern of shared feature diversity, with branch lengths representing the number of unique features that have evolved along that lineage. Pruning the tree illustrates how species and groups of species may differ in the number of unique features they have: if we prune distinctive species (such as the aardvark in Fig. 14.1) we will lose more of the tree (more unique features) than if we prune less distinctive species (e.g. the cat). Likewise, if we prune species {dog, cat}, we will lose more of the tree than if we prune {beaver, mouse 2}, because, on the tree, species {mouse 1} preserves much of the feature diversity represented by {beaver, mouse 2}. We can say that each of {beaver, mouse 1, mouse 2} are redundant with respect to the features represented by the rodent lineage that gave rise to them.

The mouse/aardvark comparison and Fig. 14.1 allow us to consider a set of interesting observations about how feature diversity is distributed across biodiversity, and what the impacts of historically and prehistorically recent (i.e. Holocene) extinctions have been on this distribution. First, it is now well known that biodiversity is very unevenly distributed across the tree of life: the fact that at the ordinal level the mammals are approximately 50% rodents (Rodentia) and 0.02% aardvark (Tubulidentata) is typical (reviewed in Purvis and Hector 2000). The few living monotremes are another obvious case within mammals, being only distantly related to the remaining approximately 5500 known Holocene species. In phylogenetic terms, the tree of life is very imbalanced, because clades differ greatly in the net diversification they have experienced. Interestingly, the simplest null models of diversification also lead to imbalance, although they are usually not as extreme as those observed in nature (reviewed in Mooers *et al.* 2007).

The second observation, made most clearly by Nee and May (1997), is that, at least under simple null models of diversification, random loss of species has a mild impact on the total loss of the tree. This is easy to intuit: if loss is random, more

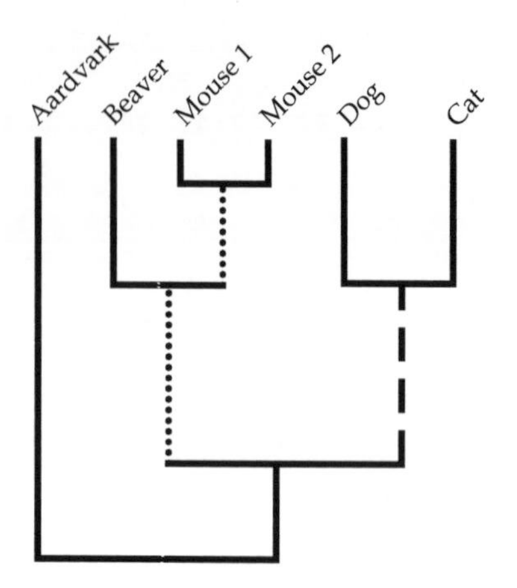

Figure 14.1 Schematic diagram demonstrating clumping of extinction on a phylogenetic tree. Branch lengths represent the number of unique features. The loss of the aardvark would lead to the greatest loss of feature diversity. If both dog and cat go extinct, their shared feature diversity (dashed branch) would also be lost. Extinction of beaver and mouse 2 would lead to less loss, because mouse 1 would represent the shared features of the set (dotted branches).

species will be chosen from the most speciose parts of the tree, which are those parts that contain the most redundancy. From the point of view of conserving the products of evolution in the face of current extinctions, this observation is often cited with some relief (see e.g. Avise 2005). Importantly, Nee and May (1997) highlighted that the actual amount of evolutionary history lost depends strongly on the shape of the tree: the more imbalanced the tree, and the shorter the branches near the root, the more that can be lost (Vazquez and Gittleman 1998; Heard and Mooers 2000). Interestingly, Rauch and Bar-Yam (2004) applied diversification models similar to Nee and May's to make a very different claim: the distribution of genetic redundancy (for them, within species) is highly skewed, such that some lineages (individuals) would be genetically much more distinct than others; that is, *non-random* loss of these lineages would have a large effect on total genetic diversity. This perspective of non-random loss is very important, and the question of the fail-safe of phylogenetic redundancy becomes largely empirical. Heard and Mooers (2000) documented that (1) although non-random loss alone may contribute little to loss overall, (2) non-random loss concentrated within slowly diversifying groups could lead to very large losses.

The Nee and May (1997) result spurred a series of empirical investigations into how much of the tree of life would be lost if current extinction risk projections were predictive: von Euler (2001) considered birds of the world, Purvis *et al.* (2000a) considered both the world's birds and the world's carnivores and primates, Sechrest *et al.* (2002) looked at these latter two groups in the world's biodiversity hotspots, Johnson *et al.* (2002) studied marsupials in Australia, Russell *et al.* (1998) looked at historical and projected extinctions for all mammals and birds (using taxonomies), and Mooers and Atkins (2003) considered at-risk birds in Indonesia. All these studies agreed that projected losses are significantly greater than if extinctions were random, because projected losses are clumped and/or concentrated in species-poor groups. In addition, von Euler (2001) offered evidence that projected extinctions will make the future bird tree even less balanced than it is today. This means that feature diversity will be even less uniformly distributed among living species than at present, increasing the distinctiveness of some remaining lineages.

It might help to contextualize the main observation (non-random loss of feature diversity) by comparison with past extinctions, and this is the focus of this final chapter. For this we must move from time-based model trees to primarily morphology-based taxonomies, and this presents several interesting issues. Phylogenetic trees are often depicted with all the species an equal distance from their common root, so that we can think of branch lengths as equal to time. Elapsed time must be correlated with feature diversity (Crozier 1997). However, we know of no good quantitative test of the strength of this correlation, partly because the very notion of feature diversity is vague. Williams and Gaston (1994) argue clearly that the correlation may be low, and indeed there is no theoretical reason to expect that morphological evolution should proceed at a constant rate (this is the reason why molecular clocks based on neutral genetic mutations are used instead of cumulative phenotypic character-state changes to estimate phylogenetic divergence dates). So, for example, *Amborella trichopoda* seems to be the sister group to all other living flowering plants (Mathews and Donoghue 1999). As such it is a monotypic lineage on a very long branch of the tree of life. However, it is presented in textbooks as classically 'primitive', representing a suite of 'basal' characters rather than a suite of novel ones: time seems to have stood still for this particular lineage. The tuatara *Sphenodon punctatus* is another case. Recent genetic studies (Hay *et al.* 2008) suggest that although morphologically unchanged since the Cretaceous, *S. punctatus* may in fact possess one of the fastest rates of molecular evolution for the mitochondrial DNA control region ever observed, making a strong case that although morphologically 'primitive' it might be inappropriate to label *S. punctatus* as primitive in a phylogenetic context. *A. trichopoda* and *S. punctatus* are only two examples of many such small-taxon number lineages, and considerable quantitative work still needs to be done to define and understand the relationships between feature diversity and phylogenetic diversity.

Furthermore, it is important to bear in mind that there are two different hypotheses to explain the existence of small taxonomic groups today. Some small taxa (e.g. obligate river dolphins, most of which represent monotypic families) have persisted for considerable periods but have not diversified further, presumably as a result of the diminished likelihood of diversification under certain ecological settings (see also Vrba 1984). This long persistence at low species diversity provides no intuitive reason to suspect any increased vulnerability to extinction. Conversely, other small taxa were formerly very species-rich (e.g. sloths; see Kurtén and Anderson 1980), but have experienced disproportionately high levels of extinction. If extinction risk is phylogenetically patterned (see e.g. Purvis *et al.* 2000a), then species from these taxa may be at increased risk of extinction in the future. However, these two different kinds of small taxa are generally not differentiated in analyses of extinction risk, and further research is again required to quantify their relative contributions to present-day feature diversity.

14.2 Holocene extinctions

Here, we use taxonomies as surrogates for phylogenetic trees, and in the case of mammals the recently published supertree for all mammals (Bininda-Emonds *et al.* 2007), to examine how Holocene extinctions were distributed among taxa. We begin by examining taxonomies, a classification system based primarily on morphological differences. Taxonomies may offer a crude compound measure of feature diversity and time, in so far as they can often be interpreted as reflections of underlying phylogeny (especially following the incorporation of cladistic methodology into taxonomy from the mid twentieth century onwards). New taxa are probably more likely to be recognized when groups of individuals or species are phenotypically distinct (Scotland and Sanderson 2003), and so a taxonomy must contain at least some information about how feature diversity is shared among its members. We recognize, however, that until we have a better concept of feature diversity, this argument is weak (e.g. *Amborella trichopoda* had already been assigned to its own order before phylogenetic work was conducted, and it is 'different' because it is primitive, not because it is derived).

Our point of departure is the study by Russell *et al.* (1998) on the taxonomic patterns of recent (post-AD 1600) and projected bird and mammal extinctions. We ask the same two questions these authors did: were Holocene extinctions non-random in that they were concentrated in particular subtaxa, and, if so, were these subtaxa small? The first pattern would point to biological correlates of extinction risk, but would not lead to an appreciable loss of feature diversity (see e.g. fig. 3 in Heard and Mooers 2000), whereas the second pattern could lead to substantial losses of higher-order taxa (see e.g. fig. 4 in Heard and Mooers 2000). Using the recently published mammalian supertree (Bininda-Emonds *et al.* 2007) we examine these same two questions by looking at the change in imbalance from a reconstructed Holocene tree to the pruned-by-extinction current tree. Non-random loss that is not clustered within small taxa might have little effect on the overall balance of the tree (but see Heath *et al.* 2008), while non-random losses clustered within small taxa would increased imbalance, at least until a tipping point is surpassed where monotypic taxa are completely removed from the pruned phylogeny, rendering the tree more balanced.

We made use of the taxonomy and lists of mammals and birds known to have gone extinct in the past 11 000 years presented earlier in this volume (see Chapters 3 and 4 in this volume) cross-checked with the 2007 Red List (IUCN 2007), Wilson and Reeder (2005), and the Systema Naturae 2000 online taxonomic database (Brands 2007) (see Appendix 14.1 for decisions made). This database, which lists many extinct taxa, was also important in checking synonymies for extinct and modern species. Because the taxonomic positions of extinct species are sometimes not fully resolved, the genera and family data sets are not fully nested. Our mammal data set lists as extinct 249/5577 species, 70/1276 genera (with 16/5577 extinct species not assigned to a genus), 9/159 families, and 1/14 orders (the enigmatic aardvark-like Bibymalagasia of Madagascar). Our bird data set lists as extinct 520/10 324 species, 89/2166 genera (with 38/10 324 species not assigned to a genus), 11/204 families, and 2/24 orders (Aepyornithiformes and Dinornithiformes). This list is obviously not complete: for instance, an estimated but undocumented 2000 flightless rail species may have gone extinct between 3500 and 1000 years BP in the South Pacific (Steadman 1995; see Chapter 10 in this volume for further discussion).

Extinct species were added into the mammal supertree manually using the program PhyloWidget (Jordan 2008). The taxonomic position of each extinct species from primary literature and the Systema Naturae 2000 database were used to work out the sister taxa and depth of each node added. Only extinct species whose taxonomic positions were fully resolved were added to the supertree.

A first reasonable question is whether approximately 5% of 5500 mammal (or of 10 000 bird species) is an anomalous number to lose over 11 000 years. The following are some very rough calculations based on simple null models of the average tempo of diversification that might help (for a guide, see e.g. Baldwin and Sanderson 1998; Ricklefs 2003). We start with a constant-rate birth-death model of diversification:

$$n_t = n_0 e^{rt}$$

where $r = b - d$ (b and d refer to instantaneous speciation and extinction rates per lineage) and n_0 is the number of species at time (t) zero. If we set t_0 to 1.66×10^8 years ago (the deepest split in the placental mammal tree; Bininda-Emonds *et al.* 2007) when $n_0 = 2$, and if we set $n_t = 5500$ mammal species (the total number in our data set), then, by rearrangement, we get $r = [\ln (5500) - \ln (2)]/1.66 \times 10^8$ years $= 5 \times 10^{-8}$ years^{-1}. To try to achieve a high enough number of extinctions, we can set $d = 0.9b$ (a high turnover rate sometimes assumed in modelling diversification; see e.g. Magallon and Sanderson 2001); b would be approximately 5×10^{-7} years^{-1} and $d = 4.5 \times 10^{-7}$ years^{-1}. If we now look at any time slice when many lineages are extant, then the overall rate of addition of new species is deterministically $n_0 b e^{rt}$, and so the total (deterministic) number of 'births' can be calculated as

$$\text{births} = n_0 \int_{t=0}^{t=t^*} b e^{rt} dt$$

For $n_0 = 5500$ and $t^* = 11\,000$ years and the b and d estimates above, this is less than 25, and the number of deaths would be less than that; that is, much less than 0.5% of the standing crop. Increasing turnover further to achieve the observed approximately 240 Holocene extinctions, in line with the surely underestimated record number of mammal extinctions, requires that $d = 0.9875b$, but would imply a birth rate of 4×10^{-6} years^{-1}. This in turn demands that the average species is only $b^{-1} = 250\,000$ years old. Given that genetic evidence suggests that new species of vertebrate take on the order of 2 million years to form (Avise *et al.* 1998), and an oft-quoted average species age for mammals is 5 million years (see e.g. Purvis and Hector 2000), this does not seem like a biologically reasonable turnover rate (see also Ricklefs 2003). Calculations for birds lead to similar values. Even including preservation biases, estimates for average species ages taken directly from the fossil record are also on the order of 1 million years (Kemp 1999). Taken together, the evidence suggests that the number of recorded mammal and bird extinctions in the Holocene is unusual and was not compensated for by the production of new species. This is not a controversial conclusion, as it is widely recognized that extinctions have occurred rapidly over ecological rather than evolutionary timescales during the

Late Quaternary. In addition, no Holocene species-level bird or mammal extinctions are considered likely to represent 'natural' (i.e. non-anthropogenic) events (see e.g. Chapter 2 in this volume). These prehistoric Late Quaternary extinctions are interpreted as the beginning of a mass extinction event, comparable to those observed in the Phanerozoic fossil record, and which is ongoing today.

To investigate the taxonomic patterning of these extinctions, we followed the clear procedures outlined in Russell *et al.* (1998). We scored the size n_i of each taxon i (genus or family) and the proportion of species extinct and extant in each. Standard binomial theory allows us to produce numbers of species one would expect to go extinct in taxa of a given size and so the expected number of taxa of each size that would be lost entirely. If extinction were random, then the observed proportion of extinct species p is our maximum likelihood estimate of the global probability of extinction. For the taxonomically assigned mammals, $p=0.042$, and for taxonomically assigned birds, $p=0.047$. If the number of taxa of each size n is S_n, then the expected number of taxa of S_n that are wholly extinct is just $S_n p^n$. This expectation will have a standard deviation of $[S_n p^n (1-p)^n]^{1/2}$. This pattern of random extinction is presented in the grey bars in Fig. 14.2 and compared with the observed number of extinct taxa. Overall, if we take two standard deviations above the expected value as our guide, roughly twice as many higher taxa than expected under a random extinction scenario have been lost throughout the Holocene (e.g. 70 observed mammal genera extinct compared with a maximum of 24 expected, nine observed compared with one or two expected extinct mammal families, 89 compared with 43 expected extinct bird genera,

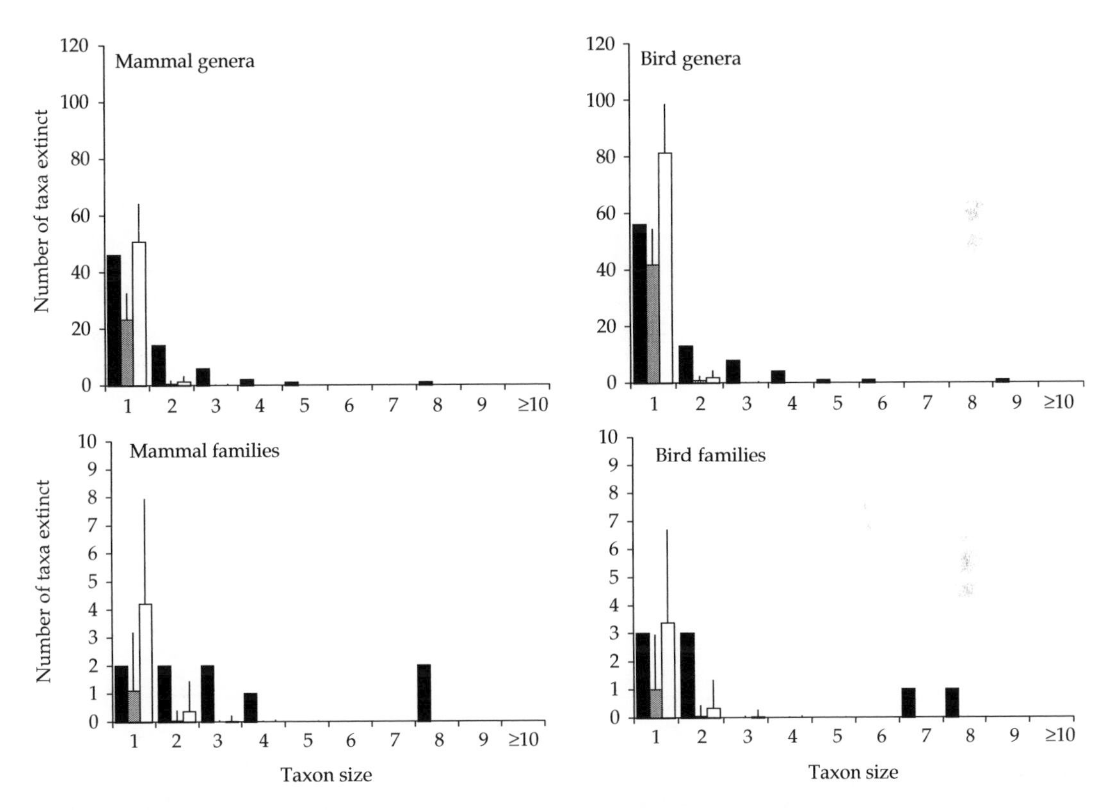

Figure 14.2 The numbers of extinct Holocene taxa of a given size. Dark bars indicate the actual number, grey bars are the expectations if extinction were random, and open bars the expectation under size selectivity (from models presented in Fig. 14.3). Error bars depict two standard deviations under a binomial distribution. Panels are standard for all subsequent figures: top left panel depicts mammal genera; top right panel depicts bird genera; bottom left panel depicts mammal families; and bottom right panel depicts bird families.

and 11 compared with one or two expected extinct bird families).

Russell *et al.* (1998) speak of 'selectivity by size', where probability of extinction is some smooth function of taxon size. Using the standard R-package statistical library (www.r-project.org), and assuming binomial error, we used maximum likelihood to fit our data using the same generalized non-linear model as Russell *et al.* (1998):

$$p_i = \frac{e^{(b_0+b_1\ln(n_i))}}{1-e^{(b_0+b_1\ln(n_i))}}$$

where p_i is the proportion of species extinct in taxon i and n_i is its size, and b_0 and b_1 are fitted parameters. Fig. 14.3 plots the fitted lines of the binomial models of p_i on $\ln(n_i)$ and also the average proportion extinct for taxa of size $\ln(n)$. For both mammal and bird genera, and for bird families, there is a significant effect of taxon size on p_i, such that smaller taxa were more likely to lose species to extinction (see Fig. 14.3 legend for values). Outliers in Fig. 14.3 are individual taxon *sizes* that are ill-fit by the size-selectivity model.

We then used these fitted equations to produce a set of p_n, the probability of extinction for each taxon of size n. This can then be substituted into the simple binomial equations above to produce the expected number of taxa, and associated standard deviations, that would go extinct if there were only

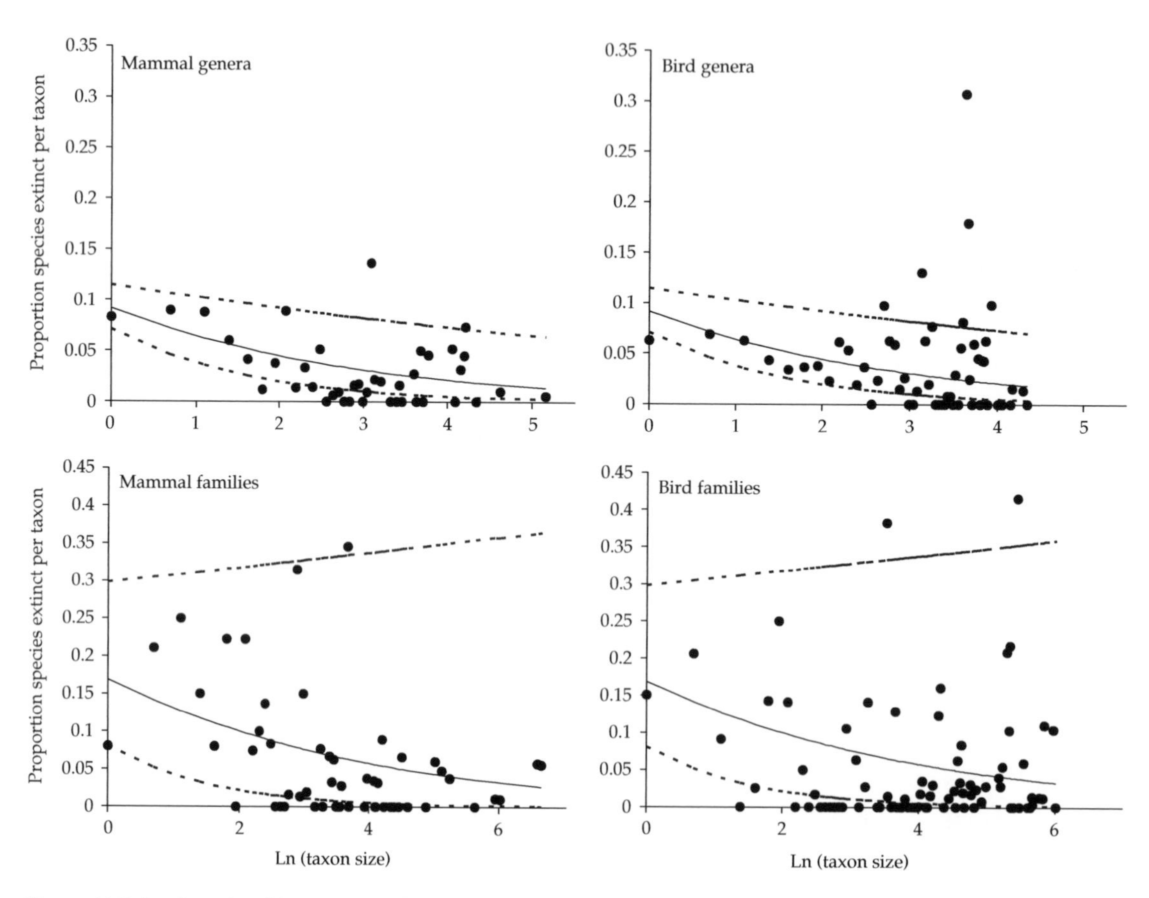

Figure 14.3 Non-linear fits of the proportion of species extinct in a taxon as a function of its size (see equation in the text). The data points are the average proportion of extinct species for taxa of that size. Panels are arranged as in Fig. 14.2. For mammal genera (top left), $b_0=-2.30$ ($P<0.0001$) and $b_1=-0.38$ ($P=0.005$). Corresponding values for the other data sets are as follows: bird genera (top right), $b_0=-2.61$ ($P<0.0001$), $b_1=-0.20$ ($P=0.05$); mammal families (bottom left), $b_0=-1.60$ ($P<0.0001$), $b_1=-0.30$ ($P=0.11$); bird families (bottom right), $b_0=-1.38$ ($P=0.0003$), $b_1=-0.41$ ($P=0.01$).

selectivity by size. These are depicted in the open bars in Fig. 14.2. Taxa sizes that are not well-fit by this equation are those for which there has been some 'selectivity by taxon'; that is, clumping or dispersion of extinction that is not predicted by taxon size. Risk to monotypic taxa is well described by size selectivity. However, for both mammal and bird genera, taxa with two to five members are at higher risk even than expected based on their size, and a few entire larger-sized genera have also been wiped out, consistent with taxon selectivity not predicted by size. The same general patterns hold at the family level: whereas monotypic families are well-fit by the size-selectivity model, taxa of size two, and the three extinct large-bodied families of Paleognathes (Aepyornithidae (elephant birds), Dinornithidae (giant moas), and Emeidae (smaller moas)) that radiated on islands, show selectivity not predicted by taxon size.

Overall, these patterns we report are similar those reported by Russell *et al.* (1998) for historical extinctions and extinctions projected by IUCN Red List data.

What contributes to this clumping of extinctions within taxa? Holocene extinctions and the subset considered 'historical' (post-AD 1600) by Russell *et al.* exhibit a strong taxon-size bias, where smaller taxa are more likely to be affected. Consistent with Russell *et al.*'s IUCN projections, birds show a weaker taxon-size effect than do mammals for genera. Russell *et al.*'s preferred explanation for this difference was that mammals in larger taxa were understudied, leading to a bias. However, this does not seem to hold here; if anything, we might have better data for mammal than for bird Holocene extinctions. Another possibility may be that this pattern reflects the increased dispersal capabilities of birds, which has led to a difference in the taxonomic distribution of island-dwelling birds compared with mammals: bird taxa tend to contain both (more often extinct) island and (more often extant) mainland taxa. In birds, extent of annual dispersal and diversification rate are positively correlated (Phillimore *et al.* 2006). If high-dispersing and more species-rich avian taxa (e.g. rails, pigeons, parrots) contain species more likely to reach islands, this would weaken (or even reverse) a negative relationship between taxon size and extinction probability.

Superficially, patterns of modern extinctions are not overwhelmed by a contrasting pattern if we extend back over the Holocene. Table 14.1 and Fig. 14.3 highlight that there was also a strong taxon-size-independent component of selectivity: many taxa are ill-fit by the taxon-size selectivity curve. We identified individual taxa with too many extinct species as those that have lost at least $(n_i p_n + (2n_i p_n(1-p_n))^{1/2})$, rounded up to the nearest integer, and these are listed in Appendix 14.2.

These two types of taxonomic clumping of extinction can be further disentangled by considering a further quantity, the number of taxa of each size affected (i.e. taxa with at least one extinction; Russell *et al.* 1998). The observed numbers can be compared both with expected numbers under a global p [equal to $S_n(1-(1-p)^n)$] and with the expected numbers under the size-selective model [equal to $S_n(1-(1-p_n)^n)$]. These numbers are presented in Table 14.2 and summarized in Fig. 14.4 alongside the total numbers of entire taxa lost and the total expected. The logic, as presented by Russell *et al.* (1998), is straightforward. Selectivity by taxon size (where species in smaller taxa are more likely to be extinct) results in more taxa lost, but can offer a form of compensation in that fewer taxa are affected. However, extreme clumping in the smallest taxa can have the opposite effect (e.g. if every extinction were in a monotypic genus, then the number of taxa affected would equal the total number of extinctions, much higher than any random expectation; Russell *et al.* 1998). If only large taxa were hit with extinctions, or at least if non-size selectivity is concentrated in relatively larger taxa, then fewer taxa might be expected to be lost, with fewer taxa affected. This might be considered the ideal if we are interested in the preservation of evolutionary feature diversity. All four groups show a consistent pattern: size selectivity means more entire taxa are lost than expected, but selectivity in some larger-sized taxa means that there are fewer taxa affected than one would expect from the size-effect model alone. In other words, many of the taxa that are outliers in Fig. 14.3 by virtue of having no recorded extinctions are the result of extinctions being clumped elsewhere. This also brings up a critical issue of scale: whether clumped extinction leads to the overall loss of feature diversity depends on how

Table 14.1 Number of recorded Holocene extinct 'groups' per group size for mammal and bird genera and families. Predicted values based on size-selective models.

Group		Group size	Number of groups	p_n	Actual	Predicted	2 SD
Mammals	Genus	1	555	0.091	46	50.628	13.566
		2	234	0.072	14	1.198	2.033
		3	110	0.062	6	0.026	0.294
		4	92	0.056	2	0.001	0.053
		5	53	0.052	1	0.000	0.008
		6	28	0.048	0	0.000	0.001
		7	19	0.046	0	0.000	0.000
		8	31	0.043	1	0.000	0.000
		9	25	0.042	0	0.000	0.000
		10+	129	0.028	0	0.000	0.000
	Family	1	25	0.168	2	4.207	3.741
		2	19	0.141	2	0.379	1.057
		3	8	0.127	2	0.016	0.209
		4	10	0.118	1	0.002	0.068
		5	5	0.111	0	0.000	0.014
		6	3	0.106	0	0.000	0.003
		7	6	0.101	0	0.000	0.001
		8	9	0.098	2	0.000	0.000
		9	3	0.095	0	0.000	0.000
		10+	71	0.061	0	0.000	0.000
Birds	Genus	1	896	0.068	60	61.106	15.092
		2	347	0.060	12	1.247	2.099
		3	211	0.056	7	0.036	0.349
		4	151	0.053	5	0.001	0.061
		5	95	0.050	2	0.000	0.010
		6	63	0.049	1	0.000	0.002
		7	57	0.047	1	0.000	0.000
		8	35	0.046	0	0.000	0.000
		9	46	0.045	1	0.000	0.000
		10+	265	0.036	0	0.000	0.000
	Family	1	21	0.200	4	4.206	3.668
		2	18	0.158	4	0.451	1.130
		3	13	0.137	0	0.034	0.294
		4	7	0.124	0	0.002	0.062
		5	8	0.114	0	0.000	0.018
		6	6	0.107	0	0.000	0.004
		7	4	0.101	1	0.000	0.001
		8	8	0.096	1	0.000	0.000
		9	2	0.092	0	0.000	0.000
		10+	117	0.044	1	0.000	0.000

those features are distributed among taxa through the tree. Our implicit interpretation here is that more diversity is distributed among families than among genera within families. Null models of diversification and feature evolution may offer a guide here, but more empirical work is also needed.

We augment these analyses with a preliminary look of the change in the shape of the mammal supertree (Bininda-Emonds *et al.* 2007) following Holocene extinctions. We made use of a measure of tree shape, I_w, which allows the inclusion of polytomies (Fusco and Cronk 1995, modified by

Table 14.2 Number of groups of mammal and bird genera and families with one or more recorded extinct Holocene species. Predicted values based on size-selective models.

Group		Group size	Number of groups	p_n	Actual	Predicted	2 SD
Mammals	Genus	1	0.091	555	46	50.628	13.566
		2	0.138	234	28	32.290	3.639
		3	0.175	110	15	19.201	1.147
		4	0.205	92	10	18.900	0.511
		5	0.233	53	6	12.323	0.196
		6	0.257	28	1	7.189	0.074
		7	0.279	19	1	5.298	0.032
		8	0.299	31	4	9.274	0.022
		9	0.318	25	2	7.950	0.010
		10+	0.560	129	24	1.809	0.000
	Family	1	0.168	25	2	4.207	3.741
		2	0.262	19	6	4.985	1.687
		3	0.335	8	2	2.678	0.594
		4	0.394	10	2	3.942	0.361
		5	0.445	5	1	2.224	0.136
		6	0.488	3	1	1.465	0.054
		7	0.527	6	0	3.162	0.038
		8	0.561	9	2	5.051	0.022
		9	0.592	3	2	1.775	0.006
		10+	0.895	71	30	1.358	0.000
Birds	Genus	1	0.068	896	60	61.106	15.092
		2	0.116	347	34	40.355	3.829
		3	0.158	211	19	33.252	1.405
		4	0.194	151	11	29.369	0.603
		5	0.228	95	9	21.671	0.254
		6	0.259	63	5	16.325	0.112
		7	0.288	57	7	16.413	0.059
		8	0.315	35	7	11.021	0.026
		9	0.340	46	10	15.650	0.016
		10+	0.670	265	61	2.950	0.000
	Family	1	0.200	21	4	4.206	3.668
		2	0.291	18	5	5.246	1.752
		3	0.358	13	3	4.649	0.794
		4	0.410	7	0	2.872	0.310
		5	0.454	8	1	3.632	0.173
		6	0.492	6	1	2.949	0.076
		7	0.524	4	1	2.097	0.031
		8	0.553	8	2	4.425	0.021
		9	0.579	2	0	1.158	0.005
		10+	0.907	117	51	1.348	0.000

Purvis *et al.* 2002). This measure has an expectation of 0.5 under the simplest null model of random diversification, and approaches 1.0 as trees become more imbalanced. We compared the change in shape from the loss of 233 extinct species that we could place on the mammal supertree with the expectation if such losses were taxonomically random.

Not surprisingly, the Holocene mammalian phylogeny appears imbalanced (I_w=0.633). When extinct taxa are pruned from the tree imbalance increases (I_w=0.645), and this increase is significant

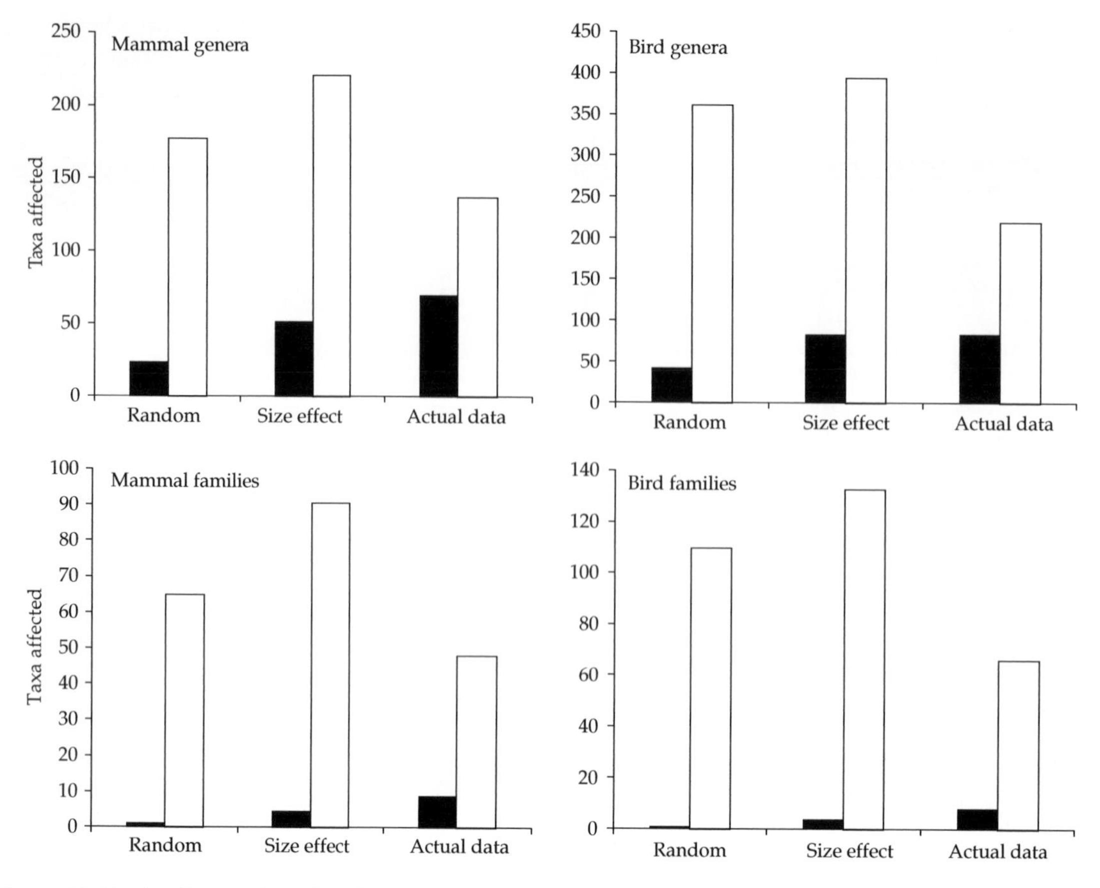

Figure 14.4 In closed bars are the total number of taxa expected to have been lost during the Holocene under taxonomically random extinction [sum across sizes n of $n \times p^n$]; the total number under size-selective extinction [sum across sizes n of $n \times p_n^{\,n}$], and the observed total number. Open bars depict the total number of taxa with at least one extinction during the Holocene under random extinction [sum across sizes n of $n(1-(1-p)^n]$, the total under the size selectivity model [sum across sizes n of $n(1-(1-p_n)^n)$], and the observed total. Panels are arranged as in Figs 14.2 and 14.3, with genera above and mammals to the left.

(P<0.05) relative to the null expectation of random loss [$E(I_w)=0.628$, SD=0.008, $n=1000$ bootstraps]. In agreement with the taxonomic work, the increase in imbalance is most likely due to non-random losses from species-poor clades over the course of the Holocene.

The explanation for the non-random loss of species from small taxa during the Holocene is most certainly the island effect. First, small-range species have experienced the great majority of post-glacial anthropogenic extinctions (for passerine birds, see Manne *et al.* 1999), and island species tend to have small ranges (usually by virtue of geographic isolation of insular populations). Indeed, 207/249 of the extinct mammal species and 495/520 of the extinct bird species were either endemic to islands or geographically restricted to islands or island systems by the start of the Holocene. The three extinct orders were also each endemic to one of two island systems (Madagascar and New Zealand). Island species tend to be placed in their own taxonomic groups, perhaps due to divergent selection imposed by island habitats, or by predetermined biases in taxonomic classification. It is important also to recognize that island systems can act both as refugia for ancient, typically species-poor lineages (e.g. tuatara) and also as centres of evolutionary radiation of more recent colonists (e.g. drepanidine honeycreepers and *Drosophila* in the Hawai'ian archipelago). However, both sets of species appear to be similarly vulnerable to anthropogenic impacts, as they have typically evolved in the absence of many native predators (notably mammalian predators). It remains unclear whether there is any relationship between relative age and size of island taxa and their vulnerability to extinction. Following on from the work reported here, it would be interesting to compare patterns of taxon size selectivity and extinction in the earlier Late Pleistocene extinctions, which had a continental rather than insular focus but which also clearly impacted many mammals that survive today only as small taxa; however, this is beyond the scope of the present volume.

14.3 Holocene effects on present-day distributions

Humans have had a significant impact on global biodiversity during the Holocene. The extinctions recorded here and projected extinctions based on IUCN Red List data (Russell *et al.* 1998) have had, and likely will lead to, a greater than random loss of feature diversity, at least for birds and mammals (the best-studied taxonomic groups for which we currently possess the most meaningful data). In addition, past taxon selectivity could produce present-day small, at-risk taxa. For example, the two living members of the family Elephantidae (*Elephas maximus* and *Loxodonta africana*) that express the unique feature diversity of the entire classical order Proboscidea are the remnants of a clade with at least 10 members that still survived after the Last Glacial Maximum in the terminal Pleistocene. These two remaining species are also at fairly high risk of extinction: *L. africana* is classified as Near Threatened and *E. maximus* is classified as Endangered by IUCN (2008).

Is it possible that past anthropogenic extinction has created extreme skew in present-day feature diversity? Eleven of the top 100 most evolutionarily distinctive mammals as measured by Isaac *et al.* (2007) have had close (confamilial) relatives lost in the Holocene, increasing their taxonomic and phenotypic isolation. Besides the two elephant species, the list includes the two remaining hippopotamus species *Hexaprotodon liberiensis* and *Hippopotamus amphibius*, the aye aye *Daubentonia madagascarensis*, the greater bilby *Macrotis lagotis*, the two living solenodons *Solenodon cubanus* and *Solenodon paradoxus*, the dugong *Dugong dugon*, the steppe pika *Ochotona pusilla*, and the New Zealand lesser short-tailed bat *Mystacina tuberculata*. Disconcertingly, 10 of these 11 species are also listed as globally threatened by the IUCN (2008). A combination of evolutionary distinctiveness and global threat is encapsulated in the EDGE of Existence programme administered by the Zoological Society of London. Nine of these evolutionarily distinct species are on the Society's top 100 EDGE mammal conservation priority list (and the missing two, *H. amphibius* and *M. lagotis*, are ranked 137th and 130th respectively). It may be that many of the species now listed by this innovative programme may be current priorities as a result of human-precipitated extinctions of close relatives within the last 11 000 years. However, connections among past losses, current threat, and current distinctiveness would need to be investigated more formally. So, while many of the top 100 EDGE species are insular, the Spearman rank correlation between threat status and evolutionary distinctiveness, as measured by Isaac *et al.* (2007), while significantly positive, has very low explanatory power ($\rho = 0.05$, $P = 0.001$, $n = 4507$).

In conclusion, not only have Holocene mammal and bird extinctions occurred at a significantly elevated rate, but taxa containing disproportionately few species are both disproportionately threatened with extinction today (Russell *et al.* 1998) and have also experienced elevated rates of species loss over the past 11 000 years (our results) as well as farther back in time (McKinney 1997). We end this short chapter by noting that it is not immediately obvious how to evaluate the importance of the non-random loss of feature diversity through the Holocene that we document here. If the same number of extinctions had taken place but they had been random with respect to the tree of life, would the world be better off? Implicit in the research agenda that looks at the loss of feature diversity through extinction is the idea that species are of different value: to quote George Orwell, 'all animals are equal, but some animals are more equal than others'. But it need not be true from first principles that more feature diversity as we are measuring it is much better in any ecologically or evolutionarily meaningful way. For example, the species-richness/ecosystem function debate (Cardinale *et al.* 2006; Worm *et al.* 2006) is ongoing. We need more work on the relationship between feature diversity and phylogenetic diversity. We also need more work on the use and non-use values of each.

Acknowledgements

AOM, TSK and SJG thank NSERC Canada, and AOM thanks the Institute for Advanced Study, Berlin, for financial support. We are grateful to IRMACS and to the fab*-lab, both at SFU, for stimulating environments, and to Sally Otto, Walter Jetz, and Gareth Russell for important intellectual input of various kinds. We thank the two anonymous reviewers, one who made us go back to the mammal supertree, and Jonathan Davies, who directed us to PhyloWidget.

Appendix 14.1 Notes on the taxonomic database used

The data sets used here are from Chapter 3 (for mammals) and Chapter 4 (for birds). Revisions and exclusions are listed below for mammals and birds. Revisions to the taxonomies were only made when extant taxa listed in the IUCN 2007 Red List were affected by the taxonomic discrepancies between data sets.

A. Mammals
Taxa excluded from generic analysis
Rodentia
Cricetidae
Oryzomyini gen. et sp. indet., spp. 1–13
Muridae
Melomys/Pogonomelomys spp. nov. A–B

Taxa excluded from all analysis
Rodentia
Family Indet.
Tainotherium valei
Rodentia? gen. et sp. nov.
Chiroptera
Family Indet.
Boryptera alba

B. Birds
Taxonomic conflicts
Columbiformes
Columbidae
Pezophaps solitaria, *Raphus cucullatus*: listed as Raphidae by IUCN (2007)
Gruiformes
Rallidae
Atlantisia elpenor: transferred to *Atlantisia* from *Mundia* to be consistent with IUCN (2007)
Passeriformes
Fringillidae
Loxops sagittirostris: listed as *Hemignathus sagittirostris* by IUCN (2007)
Psittirostra kona: listed as *Chloridops kona* by IUCN (2007)
Struthioniformes
Casuariidae
Dromaius ater, *D. baudinianus*: all *Dromaius* species listed as Dromaiidae by IUCN (2007)

Taxa excluded from generic analysis
Anseriformes
Anatidae
aff. *Anas* undescribed species
aff. *Tadorna* undescribed sp.
cf. *Dendrocygna* undescribed sp.
Anatidae undescribed sp.
Anatidae "supernumerary Oahu goose"
Charadriiformes
Scolopacidae
Coenocorypha? undescribed sp.
Columbiformes
Columbidae
cf. *Alectroenas* undescribed sp.
Gallicolumba? *norfolciensis*
"Raperia" godmanae
undescribed gen. et sp.
undescribed gen. et spp. A-C
Falconiformes
Falconidae
Falconidae undescribed small sp.
Galliformes
Megapodiidae
Megapodiidae undescribed sp.

Gruiformes
Rallidae
cf. *Dryolimnas* undescribed sp.
"*Fulica*" *podagrica*
cf. *Gallinula* sp.
cf. *Porzana* undescribed sp.
Rallidae undescribed sp.
Rallidae undescribed spp. A–D
Passeriformes
Campephagidae
cf. *Lalage* sp.
Meliphagidae
cf. *Chaetoptila* undescribed sp.
Passeridae
Foudia belloni: formally described after our analysis was conducted; previously referred to in the literature as *Foudia*? undescribed sp.
Sylviidae
cf. *Cettia* sp.
Timaliidae
Timaliinae undescribed gen. et sp.
Turdidae
Turdidae undescribed sp.
Zosteropidae
Zosteropidae undescribed spp. 1–2
Psittaciformes
Psittacidae
"*Necropsittacus*" *borbonicus*
cf. Psittacidae undescribed spp. 1–2
Psittacidae undescribed sp.
Strigiformes
Tytonidae
Tyto? *letocarti*

Taxa excluded from all analysis
Aves incertae sedis
"*Aquila*" *simurgh*
Passeriformes incertae sedis
"*Turdus*" *ulientensis*
aff. *Carduelis* undescribed species
Passeriformes undescribed slender-billed species
Passeriformes undescribed very small species

Appendix 14.2 List of taxa that have undergone more extinctions than expected

Families listed in bold have suffered higher than expected numbers of extinct species. Each taxon is listed with the number of extinct species/total number of species in that taxon. §, Extinct order; ‡, extinct family; †, extinct genus.

A. Mammals	Genus	Family
Artiodactyla		
Bovidae		
Bubalus	2/6	
Hippopotamidae		4/6
Hippopotamus	3/4	
§Bibymalagasia		
‡Plesiorycteropodidae		2/2
†Plesiorycteropus	2/2	
Carnivora		
Canidae		
†Dusicyon	2/2	
Chiroptera		
Phyllostomidae		
Desmodus	3/4	
Phyllops	2/3	
Pteropodidae		
Pteropus	6/67	
Dasyuromorphia		
‡Thylacinidae		2/2
†Thylacinus	2/2	
Diprotodontia		
Macropodidae		
Lagorchestes	2/4	
Pilosa		
Megalonychidae		16/18
†Acratocnus	4/4	
†Megalocnus	2/2	
†Neocnus	5/5	
†Parocnus	2/2	
Primates		
‡Archaeolemuridae		3/3
†Archaeolemur	2/2	
Lemuridae		
†Pachylemur	2/2	
‡Megaladapidae		3/3
†Megaladapis	3/3	

Table continued

A. Mammals	Genus	Family		Genus	Family
†Palaeopropithecidae		8/8	†gen. 1	3/3	
†*Mesopropithecus*	3/3		†gen. 2	3/3	
†*Palaeopropithecus*	3/3		†gen. 4	2/2	
			Notomys	5/10	
Rodentia			*Rattus*	5/68	
Capromyidae		27/39			
Capromys	6/8		Soricomorpha		
Geocapromys	5/7		**†Nesophontidae**		8/8
†*Isolobodon*	2/2		†*Nesophontes*	8/8	
Mesocapromys	8/12		Solenodontidae		
Plagiodontia	2/3		*Solenodon*	2/4	
Cricetidae		41/710	Soricidae		
†*Megalomys*	4/4		†*Asoriculus*	3/3	
†*Megaoryzomys*	2/2				
Neotoma	3/22				
Nesoryzomys	6/8				
Echimyidae					
†*Boromys*	2/2				
†*Brotomys*	2/2				
Gliridae					
Eliomys	2/5				
†Heptaxodontidae		4/4			
Muridae		42/754			
†*Canariomys*	2/2				
†*Coryphomys*	2/2				

B. Birds	Genus	Family		Genus	Family
§Aepyornithiformes			**Ardeidae**		12/75
†Aepyornithidae		7/7	*Nycticorax*	8/10	
†*Aepyornis*	4/4		**Falconidae**		9/73
†*Mullerornis*	3/3		*Caracara*	3/5	
			Milvago	2/4	
Anseriformes			**Procellariidae**		17/96
Anatidae		41/197	*Pterodroma*	7/38	
Alopochen	3/4		*Puffinus*	7/26	
Anas	6/48		**Scolopacidae**		14/102
Branta	4/10		*Coenocorypha*	5/7	
†*Cnemiornis*	2/2		†*Prosobonia*	6/6	
Cygnus	3/9		Threskiornithidae		
†*Thambetochen*	2/2		†*Apteribis*	3/3	
Ciconiiformes			Columbiformes		
Accipitridae		15/255	**Columbidae**		38/345
Aquila	3/15		*Alectroenas*	2/5	
†*Titanohierax*	2/2		*Ducula*	5/42	

Table continued

B. Birds	Genus	Family
Gallicolumba	6/24	
Craciformes		
Megapodiidae		13/34
Megapodius	11/23	
Cuculiformes		
Cuculidae		
Coua	3/12	
§Dinornithiformes		
‡Dinornithidae		2/2
†*Dinornis*	2/2	
‡Emeidae		8/8
†*Eurapteryx*	2/2	
†*Pachyornis*	3/3	
Strigiformes		
Caprimulgidae		
Siphonorhis	2/3	
Strigidae		21/204
Athene	4/7	
†*Gallistrix*	4/4	
†*Mascarenotus*	3/3	
†*Ornimegalonyx*	2/2	
Tytonidae		11/26
Tyto	10/23	
Struthioniformes		
Casuariidae		
Dromaius	2/3	
Galliformes		
Phasianidae		
Coturnix	3/11	
Gruiformes		
‡Aptornithidae		2/2
†*Aptornis*	2/2	
Rallidae		96/231
†*Aphanapteryx*	2/2	
Atlantisia	2/3	
Fulica	4/15	
Gallirallus	28/38	
†*Nesotrochis*	3/3	
†*Pareudiastes*	2/2	
Porphyrio	10/15	
Porzana	26/39	
Rallus	3/12	
Gruidae		
Grus	4/17	
Passeriformes		
Acanthisittidae		4/6
Corvidae		
Corvus	8/51	
Fringillidae		45/207
†*Aidemedia*	3/3	
†*Akialoa*	9/9	
†*Chloridops*	3/3	
†*Ciridops*	4/4	
†*Drepanis*	2/2	
Loxoides	2/3	
†*Rhodacanthis*	4/4	
Telespiza	4/6	
†*Vangulifer*	2/2	
†*Xestospiza*	2/2	
Meliphagidae		
†*Chaetoptila*	3/3	
†*Moho*	5/5	
Monarchidae		
Pomarea	4/9	
Turdidae		
Myadestes	3/14	
‡Turnagridae		2/2
†*Turnagra*	2/2	
Sturnidae		
Aplonis	5/26	
Psittacidae		41/394
Amazona	6/37	
Ara	8/16	
Eclectus	2/3	
Nestor	2/4	
Psittacula	4/17	
Vini	2/7	

References

Abell RA, Olson DM, Dinerstein E, Hurley PT, Diggs JT, Eichbaum W, Walters S, Wettengel W, Allnutt T, Loucks CJ, and Hedao P (2000) *Freshwater Ecoregions of North America: a Conservation Assessment*. Island Press, Washington DC.

Adam PJ and Garcia GG (2003) New information on the natural history, distribution, and skull size of the extinct (?) West Indian monk seal, *Monachus tropicalis*. *Marine Mammal Science* **19**, 297–317.

Agarwal DK, Silander Jr JA, Gelfand AE, Dewar RE, and Mickelson Jr JG (2005) Tropical deforestation in Madagascar: analysis using hierarchical, spatially explicit, Bayesian regression models. *Ecological Modelling* **185**, 105–131.

Agnolin FL (2006) Presencia de *Ciconia maltha* (Aves, Ciconiidae) en el Pleistoceno Inferior-Medio del Valle de Tarija, Bolivia. *Revista Española de Paleontología* **21**, 39–41.

Ahlstedt SA and McDonough TA (1993) Quantitative evaluation of commercial mussel populations in the Tennessee River portion of Wheeler Reservoir, Alabama. In Cummings KS, Buchanan AC, and Koch LM, eds, *Conservation and Management of Freshwater Mussels. Proceedings of a UMRCC symposium*, pp. 38–49. Upper Mississippi River Conservation Committee, Rock Island, IL.

Ahlstedt SA and McDonough TA (1995–1996) Summary of pre-operational monitoring of the mussel fauna in the upper Chickamauga reservoir (Tennessee River) in the vicinity of TVA's Watts Bar nuclear plant, 1983–1993. *Walkerana* **8**, 107–126.

Ainley DG, Telfer TC, and Reynolds MH (1997) Townsend's and Newell's Shearwater (*Puffinus auricularis*). In Poole A and Gill F, eds, *The Birds of North America*, no. 297. The Academy of Natural Sciences, Philadelphia, and The American Ornithologists' Union, Washington DC.

Ainley DG, Hobson KA, Crosta X, Rau GH, Wassenaar LI, and Augustinus PC (2006) Holocene variation in the Antarctic coastal food web: linking δD and δ^{13}C in snow petrel diet and marine sediments. *Marine Ecology Progress Series* **306**, 31–40.

Alcover JA (1989) Les aus marines fossils de les Pitiüses en el context de la Mediterrània. In López-Jurado C, ed., *Aves Marinas*, pp. 33–44. Actas de la IV Reunión del Grupo Ibérico de Aves Marinas. GOB, Palma de Mallorca.

Alcover JA and Florit F (1987) Una nueva especie de *Carduelis* (Fringillidae) de La Palma. *Vieraea* **17**, 75–86.

Alcover JA and McMinn M (1995) Fossil birds from the Canary Islands. *Courier Forschungsinstitut Senckenberg* **181**, 207–213.

Alcover JA, Campillo X, Macias M, and Sans A (1998) Mammal species of the world: additional data on insular mammals. *American Museum Novitates* **3248**, 1–29.

Alcover JA, Bover P, and Seguí B (1999a) Una aproximació a paleoecologia de les illes. *Monografies de la Societat d'Història Natural de les Balears* **6**, 169–204.

Alcover JA, Seguí B, and Bover P (1999b) Extinctions and local disappearances of vertebrates in the western Mediterranean Islands. In MacPhee RDE, ed., *Extinctions in Near Time: Causes, Contexts, and Consequences*, pp. 165–188. Kluwer Academic/Plenum, New York.

Alcover JA, McMinn M, and Seguí B (2005) Fossil rails (Gruiformes: Rallidae) from the Balearic Islands. *Monografies de la Societat d'Història Natural de les Balears* **12**, 9–16.

Allen GM (1942) Extinct and vanishing mammals of the Western Hemisphere with the marine species of all the oceans. *American Committee for International Wild Life Protection, Special Publication* **11**, 1–620.

Allen KR (1980) *Conservation and Management of Whales*. University of Washington Press, Seattle.

Allen MS and Steadman DW (1990) Excavations at the Ureia site, Aitukaki, Cook Islands: preliminary results. *Archaeology in Oceania* **25**, 24–37.

Allen RC and Keay I (2006) Bowhead whales in the eastern Arctic, 1611–1911: population reconstruction with historical whaling records. *Environment and History* **12**, 89–113.

Alley RB (2000) The Younger Dryas cold interval as viewed from central Greenland. *Quaternary Science Reviews* **19**, 213–226.

Alongi DM (2002) Present state and future of the world's mangrove forests. *Environmental Conservation* **29**, 331–349.

Alroy J (1999) Putting North America's end-Pleistocene megafaunal extinction in context: large-scale analyses of spatial patterns, extinction rates and size distributions. In MacPhee RDE, ed., *Extinctions in Near Time: Causes, Contexts, and Consequences*, pp. 105–143. Kluwer Academic/Plenum, New York.

Alroy J (2001) A multispecies overkill simulation of the end-Pleistocene megafaunal mass extinction. *Science* **292**, 1893–1896.

Alroy J (2003) Taxonomic inflation and body mass distributions in North American fossil mammals. *Journal of Mammalogy* **84**, 431–443.

Altizer S, Nunn CL, and Lindenfors P (2007) Do threatened hosts have fewer parasites? A comparative study in primates. *Journal of Animal Ecology* **76**, 304–314.

Álvarez-Castañeda ST and Cortés-Calva P (2003) *Peromyscus pembertoni*. *Mammalian Species* **734**, 1–2.

Álvarez-Castañeda ST and Ortega-Rubio A (2003) Current status of rodents on islands in the Gulf of California. *Biological Conservation* **109**, 157–163.

Anderson A (2001) No meat on that beautiful shore: the prehistoric abandonment of subtropical Polynesian islands. *International Journal of Osteoarchaeology* **11**, 14–23.

Anderson A, Holdaway RN, and Jacomb C (2000) Less is moa. *Science* **289**, 1472–1474.

Anderson DE, Goudie AS, and Parker AG (2007) *Global Environments Through the Quaternary: Exploring Environmental Change*. Oxford University Press, Oxford.

Anderson LL, Hu F, Nelson DM, Petit RJ, and Paige KN (2006) Ice-age endurance: DNA evidence of a white spruce refugium in Alaska. *Proceedings of the National Academy of Sciences USA* **13**, 12447–12450.

Anderson P (1995) Competition, predation, and the evolution and extinction of Steller's sea cow, *Hydrodamalis gigas*. *Marine Mammal Science* **11**, 391–394.

Anderson RM and May RM (1991) *Infectious Diseases of Humans: Dynamics and Control*. Oxford University Press, New York.

Anderson WB and Polis GA (1999) Nutrient fluxes from water to land: seabirds affect plant nutrient status on Gulf of California islands. *Oecologia* **118**, 324–332.

Andrew NL, Bene C, Hall SJ, Allison EH, Heck S, and Ratner BD (2007) Diagnosis and management of small-scale fisheries in developing countries. *Fish and Fisheries* **8**, 227–240.

Andrews CW (1897) On some fossil remains of carinate birds from central Madagascar. *Ibis* **(7)3**, 343–359.

Andrews P (1990) *Owls, Caves and Fossils*. Natural History Museum Publications, London.

Andriaholinirina VN, Fausser J-L, Roos C, Zinner D, Thalmann U, Rabarivola C, Ravoarimanana I, Ganzhorn JU, Meier B, Hilgartner R *et al.* (2006) Molecular phylogeny and taxonomic revision of the sportive lemurs (*Lepilemur*, Primates). *BMC Evolutionary Biology* **6**, 17.

Anon (1990) Computers help to hunt the Tasmanian tiger. *New Scientist* 10 March, 24.

Anthony JL and Downing JA (2001) Exploitation trajectory of a declining fauna: a century of freshwater mussel fisheries in North America. *Canadian Journal of Fisheries and Aquatic Sciences* **58**, 2071–2090.

Antonelis GA, Baker JD, Johanos TC, Braun RC, and Hartling AL (2006) Hawaiian monk seal (*Monachus schauinslandi*): status and conservation issues. *Atoll Research Bulletin* **543**, 75–101.

Antunes MT (2006) The *zebro* (Equidae) and its extinction in Portugal, with an Appendix on the noun *zebro* and the modern "zebra". In Mashkour M, ed., *Equids in Time and Space*, pp. 210–235. Oxbow Books, Oxford.

Aplin K and Pasveer J (2005) Mammals and other vertebrates from late Quaternary archaeological sites on Pulau Kobroor, Aru Islands, eastern Indonesia. In O'Connor S, Spriggs M, and Veth P, eds, *The Archaeology of the Aru Islands, Eastern Indonesia*, pp. 41–62. Pandanus Books, Canberra.

Aplin KP, Pasveer JM, and Boles WE (1999) Late Quaternary vertebrates from the Bird's Head Peninsula, Irian Jaya, Indonesia, including descriptions of two previously unknown marsupial species. *Records of the Western Australian Museum, Supplement* **57**, 351–387.

Araujo AGM, Neves WA, and Piló LB (2004) Vegetation changes and megafaunal extinctions in South America: comments on de Vivo and Carmignotto (2004). *Journal of Biogeography* **31**, 2039–2040.

Arbogast R-M, Jacomet S, Magny M, and Schibler J (2006) The significance of climate fluctuations for lake level changes and shifts in subsistence economy during the late Neolithic (4300–2400 B.C.) in central Europe. *Vegetation History and Archaeobotany* **15**, 403–418.

Arbuckle KE and Downing JA (2002) Freshwater mussel abundance and species richness: GIS relationships with watershed land use and geology. *Canadian Journal of Fisheries and Aquatic Sciences* **59**, 310–316.

Archer M and Baynes A (1972) Prehistoric mammal faunas from two small caves in the extreme south-west of Western Australia. *Journal of the Royal Society of Western Australia* **55**, 80–89.

Arcos JM and Oro D (2002) Significance of fisheries discards for a threatened Mediterranean seabird, the Balearic shearwater *Puffinus mauretanicus*. *Marine Ecology Progress Series* **239**, 209–220.

Aronson RB, Macintyre IG, Precht WF, Murdoch TJT, and Wapnick CM (2002) The expanding scale of species turnover events on coral reefs in Belize. *Ecological Monographs* **72**, 233–249.

Arredondo O (1958) Aves gigantes de nuestro pasado prehistorico. *El Cartero Cubano*, **17(7)**, 10–12.

Arredondo O (1970) Nueva especie de ave pleistocénica del orden Accipitriformes (Accipitridae) y nuevo género para las Antillas. *Ciencias, series 4 (Ciencias Biologicas)* **8**, 1–19.

Arredondo O (1971) Nuevo genero y especie de ave fósil (Accipitriformes: Vulturidae) del pleistoceno de Cuba. *Memoria de la Sociedad de Ciencias Naturales La Salle* **31(90)**, 309–323.

Arredondo O (1972a) Nueva especie de ave fósil (Strigiformes: Tytonidae) del pleistoceno superior de Cuba. *Boletin de la Sociedad Venezolana de Ciencias Naturales* **29(122/123)**, 415–431.

Arredondo O (1972b) Especie nueva de lechuza gigante (Strigiformes: Tytonidae) del pleistoceno cubano. *Boletin de la Sociedad Venezolana de Ciencias Naturales* **30(124/125)**, 129–140.

Arredondo O (1982) Los strigiformes fosiles del Pleistoceno Cubano. *Sociedad Venezolana de Ciencias Naturales Boletin* **37**, 33–55.

Arredondo O and Arredondo C (2002a) Nuevos género y especie de ave fósil (Falconiformes: Accipitridae) del Cuaternario de Cuba. *Poeyana* **470–475**, 9–14.

Arredondo O and Arredondo C (2002b) Nuevos género y especie de ave fósil (Falconiformes: Teratornithidae) del Pleistoceno de Cuba. *Poeyana* **470–475**, 15–21.

Arredondo O and Olson SL (1976) The great predatory birds of the Pleistocene of Cuba. *Smithsonian Contributions to Paleobiology* **27**, 169–187.

Arredondo O and Olson SL (1994) A new species of owl of the genus *Bubo* from the Pleistocene of Cuba (Aves: Strigiformes). *Proceedings of the Biological Society of Washington* **107**, 436–444.

Ashmole NP (1963) The extinct avifauna of St. Helena Island. *Ibis* **103b**, 390–408.

Ashmole NP and Ashmole MJ (1997) The land fauna of Ascension Island: new data from caves and lava flows, and a reconstruction of the prehistoric ecosystem. *Journal of Biogeography* **24**, 549–589.

Asquith NM, Terborgh J, Arnold AE, and Riveros CM (1999) The fruits the agoutis ate: *Hymenaea courbaril* seed fate when its disperser is absent. *Journal of Tropical Ecology* **15**, 229–235.

Athens JS, Tuggle HD, Ward JV, and Welch DJ (2002) Avifaunal extinctions, vegetation change, and Polynesian impacts in prehistoric Hawai'i. *Archaeology in Oceania* **37**, 57–78.

Atkinson IAE (1985) The spread of commensal species of *Rattus* to oceanic islands and their effect on island avifaunas. In Moors PJ, ed., *Conservation of Island Birds: Case Studies for the Management of Threatened Island Species*, pp. 35–81. Technical Publication no. 3. International Council for Bird Preservation, Cambridge.

Atkinson IAE and Bell BD (1973) Offshore and outlying islands. In Williams GR, ed., *The Natural History of New Zealand*, pp. 372–392. AH & AW Reed, Wellington.

Atkinson IAE and Greenwood RM (1989) Relationships between moas and plants. *New Zealand Journal of Ecology* **12** (suppl), 67–96.

Augusto de Freitas H, Pessenda LCR, Aravena R, Gouveia SEM, Ribeiro A de S, and Boulet R (2001) Late Quaternary vegetation dynamics in the southern Amazon Basin inferred from carbon isotopes in soil organic matter. *Quaternary Research* **55**, 39–46.

Auler AS, Piló LB, Smart PL, Wang X, Hoffmann D, Richards DA, Edwards RL, Neves WA, and Hai C (2006) U-series dating and taphonomy of Quaternary vertebrates from Brazilian caves. *Palaeogeography, Palaeoclimatology, Palaeoecology* **240**, 508–522.

Avise JC (2005) Phylogenetic units and currencies above and below the species level. In Purvis A, Gittleman JL, and Brooks T, eds, *Phylogeny and Conservation*, pp. 77–100. Cambridge University Press, Cambridge.

Avise JC, Walker D, and Johns GC (1998) Speciation durations and Pleistocene effects on vertebrate phylogeography. *Proceedings of the Royal Society of London Series B Biological Sciences* **265**, 1707–1712.

Bachmayer F and Symeonidis NK (1975) Eigenartige Abspaltungen von Stosszähnen der Zwergelefanten aus der Höhle "Charkadio" auf der Insel Tilos—Artefacte? *Annales Géologiques des Pays Helléniques* **26**, 320–323.

Bachmayer F, Symeonidis NK, Seemann R, and Zapfe H (1976) Die Ausgrabungen in der Zwergelefantenhöle "Charkadio" auf der Insel Tilos (Dodekanes, Griechenland) in den Jahren 1974 und 1975. *Annalen des Naturhistorischen Museums Wien* **80**, 113–144.

Baffa O, Brunetti A, Karmann I, and Neto CMD (2000) ESR dating of a toxodon tooth from a Brazilian karstic cave. *Applied Radiation and Isotopes* **52**, 1345–1349.

Bahn PG (1993) 50,000-year-old Americans of Pedra Furada. *Nature* **362**, 114–115.

Bailey G (2007) Postglacial adaptations. In Elias S, ed., *Encyclopedia of Quaternary science*, pp. 145–152. Elsevier, Oxford.

Baillie JEM, Hilton-Taylor C, and Stuart SN (2004) *2004 IUCN Red List of Threatened Species: a Global Species Assessment*. IUCN, Gland and Cambridge.

Baino R, Serena F, Ragonese S, Rey J, and Rinelli P (2001) Catch composition and abundance of elasmobranchs based on the MEDITS program. *Rapports de la Commission Internationale pour l'Exploration Scientifique de la Mer Mediterranee* **36**, 234.

Baird RF (1991) The dingo as a possible factor in the disappearance of *Gallinula mortierii* from the Australian mainland. *The Emu* **91**, 121–122.

Baker CS and Clapham PJ (2004) Modelling the past and future of whales and whaling. *Trends in Ecology & Evolution* **19**, 365–371.

Baker P, Baker H, and Seto N (1997) Tristram's storm petrel (*Oceanodroma tristrami*) on Midway: a probable breeding record. *'Elepaio* **57**, 30.

Baldwin BG and Sanderson MJ (1998) Age and rate of diversification of the Hawaiian silversword alliance (Compositae). *Proceedings of the National Academy of Sciences USA* **95**, 9402–9406.

Balmford A (1996) Extinction filters and current resilience: the significance of past selection pressures for conservation biology. *Trends in Ecology & Evolution* **11**, 193–196.

Balouet J-C and Alibert E (1990) *Extinct Species of the World*. Baron's Educational Series, New York.

Balouet J-C and Olson SL (1987) A new extinct species of giant pigeon (Columbidae: *Ducula*) from archeological deposits on Wallis (Uvea) Island, South Pacific. *Proceedings of the Biological Society of Washington* **100**, 769–775.

Balouet J-C and Olson SL (1989) Fossil birds from late Quaternary deposits in New Caledonia. *Smithsonian Contributions to Zoology* **469**, 1–38.

Bancroft WJ, Hill D, and Roberts JD (2004) A new method for calculating volume of excavated burrows: the geomorphic impact of wedge-tailed shearwater burrows on Rottnest Island. *Functional Ecology* **18**, 752–759.

Bancroft WJ, Roberts JD, and Garkaklis MJ (2005) Burrowing seabirds drive decreased diversity and structural complexity, and increased productivity in insular-vegetation communities. *Australian Journal of Botany* **53**, 231–241.

Barel CDN, Dorit R, Greenwood PH, Fryer G, Hughes N, Jackson PBN, Kawanabe H, Lowe-McConnell RH, Nagoshi M, Ribbink AJ *et al.* (1985) Destruction of fisheries in Africa's lakes. *Nature* **315**, 19–20.

Barnes I, Matheus P, Shapiro B, Jensen D, and Cooper A (2002) Dynamics of Pleistocene population extinctions in Beringian brown bears. *Science* **295**, 2267–2270.

Barnosky AD (2008) Climatic change, refugia, and biodiversity: where do we go from here? An editorial comment. *Climatic Change* **86**, 29–32.

Barnosky AD, Koch PL, Feranec RS, Wing SL, and Shabel AB (2004) Assessing the causes of Late Pleistocene extinctions on the continents. *Science* **306**, 70–75.

Barrett JH, Nicholson RA, and Cerón-Carrasco R (1999) Archaeo-ichthyological evidence for long-term socioeconomic trends in northern Scotland: 3500 BC to AD 1500. *Journal of Archaeological Science* **26**, 353–388.

Barrett JH, Locker AM, and Roberts CM (2004a) 'Dark age economics' revisited: the English fish bone evidence AD 600–1600. *Antiquity* **78**, 618–636.

Barrett JH, Locker AM, and Roberts CM (2004b) The origins of intensive marine fishing in medieval Europe: the English evidence. *Proceedings of the Royal Society of London Series B Biological Sciences* **271**, 2417–2421.

Bartlein PJ, Webb III T, and Fleri EC (1984) Holocene climatic change in the northern Midwest: pollen-derived estimates. *Quaternary Research* **22**, 361–374.

Baruch U and Bottema S (1999) A new pollen diagram from Lake Hula. Vegetational, climatic and anthropogenic implications. In Kawanabe H, Coulter GW, and Roosevelt AC, eds, *Ancient Lakes: their Cultural and Biological Diversity*, pp. 75–86. Kenobi Productions, Ghent.

Bate DMA (1916) On a small collection of vertebrate remains from the Har Dalam cavern, Malta, with note on a new species of the genus *Cygnus*. *Proceedings of the Zoological Society of London* **1916**, 421–430.

Bates MR, Bates CR, Gibbard PL, Macphail RI, Owen FJ, Parfitt SA, Preece RC, Roberts MB, Robinson JE, Whittaker JE, and Wilkinson KN (2000) Late Middle Pleistocene deposits at Norton Farm on the West Sussex coastal plain, southern England. *Journal of Quaternary Science* **15**, 61–89.

Baum DA (1998) Individuality and the existence of species through time. *Systematic Biology* **47**, 641–653.

Baum JK and Myers RA (2004) Shifting baselines and the decline of pelagic sharks in the Gulf of Mexico. *Ecology Letters* **7**, 135–145.

Baum JK, Myers RA, Kehler DG, Worm B, Harley SJ, and Doherty PA (2003) Collapse and conservation of shark populations in the northwest Atlantic. *Science* **299**, 389–392.

Beaumont PB (1990) Wonderwerk Cave. In Beaumont PB and Morris D, eds, *Guide to Archaeological Sites in the Northern Cape*, pp. 101–134. McGregor Museum, Kimberley.

Beebee TJC, Buckley J, Evans I, Foster JP, Gent AH, Gleed-Owen CP, Kelly G, Rowe G, Snell C, Wycherley JT, and Zeisset I (2005) Neglected native or undesirable alien? Resolution of a conservation dilemma concerning the pool frog *Rana lessonae*. *Biodiversity and Conservation* **14**, 1607–1626.

Behrensmeyer AK (1991) Terrestrial vertebrate accumulations. In Allison PA and Briggs DEG, eds, *Taphonomy: Releasing the Data Locked in the Fossil Record*, pp. 291–335. Plenum Press, New York.

Bell BD (1978) The Big South Cape rat irruption. In Dingwall PR, Atkinson IAE, and Hay C, eds, *The Ecology and Control of Rodents in New Zealand Nature Reserves*, pp. 33–40. Department of Lands and Survey Information Series 4. Government Printer, Wellington.

Bell GF (1969) The occurrence of southern, steppe and halophytic elements in the Weichselian (Last Glacial) floras of southern Britain. *New Phytologist* **68**, 913–922.

Bell M and Walker MJC (2005) *Late Quaternary Environmental Change: Physical and Human Perspectives.* Pearson Education, Harlow.

Bell M, Bell BD, and Bell EA (2005) Translocation of fluttering shearwater (*Puffinus gavia*) chicks to create a new colony. *Notornis* **52**, 11–15.

Bellwood DR, Hoey AS, and Choat JH (2003) Limited functional redundancy in high diversity systems: resilience and ecosystem function on coral reefs. *Ecology Letters* **6**, 281–285.

Bellwood P (2005) *First Farmers: the Origins of Agricultural Societies.* Blackwell, Oxford.

Benke AC (1990) A perspective on America's vanishing streams. *Journal of the North American Benthological Society* **9**, 77–88.

Bennett KD and Fuller JL (2002) Determining the age of the mid-Holocene *Tsuga canadensis* (hemlock) decline, eastern North America. *The Holocene* **12**, 421–429.

Bennett PM and Owens IPF (2002) *Evolutionary Ecology of Birds: Life Histories, Mating Systems and Extinction.* Oxford University Press, Oxford.

Bent AC (1922) Life histories of North American petrels and pelicans and their allies. *United States National Museum Bulletin* **121**, 1–343.

Benton MJ (1994) Palaeontological data and identifying mass extinctions. *Trends in Ecology & Evolution* **9**, 181–185.

Benzi V, Abbazzi L, Bartolomei P, Esposito M, Fassò C, Fonzo O, Giampieri R, Murgia F, and Reyss J-L (2007) Radiocarbon and U-series dating of the endemic deer *Praemegaceros cazioti* (Depéret) from "Grotta Juntu", Sardinia. *Journal of Archaeological Science* **34**, 790–794.

Berger J, Stacey PB, Bellis L, and Johnson MP (2001) A mammalian predator-prey imbalance: grizzly bear and wolf extinction affect avian Neotropical migrants. *Ecological Applications* **11**, 947–960.

Berglund BE, Birks HJB, Ralska-Jasiewiczowa M, and Wright HE (eds) (1996) *Palaeoecological Events During the Last 15000 years.* Wiley, Chichester.

Berman WD and Tonni EP (1987) *Canis* (*Dusicyon*) *avus* Burmeister, 1864 (Carnivora, Canidae) en el Pleistoceno tardío y Holoceno de la provincial de Buenos Aires. Aspectos sistemáticos y bioestratigráficos relacionados. *Ameghiniana* **24**, 245–250.

Bernabo JC and Webb III T (1977) Changing patterns in the Holocene pollen record from northeastern North America: a mapped summary. *Quaternary Research* **8**, 64–96.

Berovides Alvarez V and Comas González A (1991) The critical condition of hutias in Cuba. *Oryx* **25**, 206–208.

Berrio JC, Hooghiemstra H, Behling H, Botero P, and van der Borg K (2002) Late Quaternary savanna history of the Colombian Llanos Orientales from Laguna Chenevo and Mozambique: a transect synthesis. *The Holocene* **12**, 35–48.

Besler H (2000) Modern and palaeo-modelling in the Great Sand Sea of Egypt (initial results from the Cologne Cooperative Research project 389). *Global and Planetary Change* **26**, 13–24.

Biber E (2002) Patterns of endemic extinctions among island bird species. *Ecography* **25**, 661–676.

Biesmeijer JC, Roberts SPM, Reemer M, Ohlemuller R, Edwards M, Peeters T, Schaffers AP, Potts SG, Kleukers R, Thomas CD *et al.* (2006) Parallel declines in pollinators and insect-pollinated plants in Britain and the Netherlands. *Science* **313**, 351–354.

Bilton DT, Mirol PM, Mascheretti S, Fredga K, Zima J, and Searle JB (1998) Mediterranean Europe as an area of endemism for small mammals rather than a source for northwards postglacial colonisation. *Proceedings of the Royal Society of London Series B Biological Sciences* **265**, 1219–1226.

Binford MW, Kolata AL, Brenner M, Janusek JW, Seddon MT, Abbott M, and Curtis JH (1997) Climate variation and the rise and fall of an Andean civilisation. *Quaternary Research* **47**, 235–248.

Bininda-Emonds ORP, Cardillo M, Jones KE, MacPhee RDE, Beck RMD, Grenyer R, Price SA, Vos RA, Gittleman JL, and Purvis A (2007) The delayed rise of present-day mammals. *Nature* **446**, 507–512.

BirdLife International (2000) *Threatened Birds of the World.* Lynx Edicions and BirdLife International, Barcelona and Cambridge.

BirdLife International (2004) *Threatened Birds of the World 2004.* BirdLife International, Cambridge.

BirdLife International (2007a) Species factsheet: *Pseudobulweria becki.* www.birdlife.org.

BirdLife International (2007b) Species factsheet: *Pterodroma brevipes.* www.birdlife.org.

Birks HJB (2003) Quantitative palaeoenvironmental reconstructions from Holocene biological data. In Mackay AW, Battarbee RW, Birks HJB, and Oldfield F, eds, *Global Change in the Holocene*, pp. 328–241. Arnold, London.

Birks HH and Birks HJB (2006) Multiproxy studies in palaeolimnology. *Vegetation History and Archaeobotany* **15**, 235–251.

Björck S, Walker MJC, Cwynar LC, Johnsen S, Knudsen K-L, Lowe JJ, Wohlfarth B, and INTIMATE members (1998) An event stratigraphy for the Last Termination in the north Atlantic region based on the Greenland ice-core record: a proposal by the INTIMATE group. *Journal of Quaternary Science* **13**, 283–292.

Björck S, Muscheler R, Kromer B, Andresen CS, Heinemeier J, Johnsen SJ, Conley D, Koç N, Spurk M, and Veski S (2001) High-resolution analyses of an early Holocene climate event may imply decreased solar forcing as an important climate trigger. *Geology* **29**, 1107–1110.

Blackburn TM and Duncan RP (2001) Establishment patterns of exotic birds are constrained by non-random patterns in introduction. *Journal of Biogeography* **28**, 927–939.

Blackburn TM and Gaston KJ (2005) *Macroecology: Concepts and Consequences.* Blackwell Scientific Press, Oxford.

Blackburn TM, Cassey P, Duncan RP, Evans KL, and Gaston KJ (2004) Avian extinction and mammalian introductions on oceanic islands. *Science* **305**, 1955–1958.

Blackburn TM, Petchey OL, Cassey P, and Gaston KJ (2005) Functional diversity of mammalian predators and extinction in island birds. *Ecology* **86**, 2916–2923.

Blackburn TM, Cassey P, and Lockwood JL (2008) The island biogeography of exotic bird species. *Global Ecology and Biogeography* **17**, 246–251.

Blackburn, T.M., Lockwood, J.L., and Cassey, P. (2009) *Avian Invaders. The Ecology and Evolution of Exotic Birds.* Oxford University Press, Oxford (in press).

Blanqui LA (1885) *Critique Sociale.* Published by the author, Paris.

Blight LK and Burger AE (1997) Occurrence of plastic particles in seabirds from the eastern North Pacific. *Marine Pollution Bulletin* **34**, 323–325.

Blockley SPE, Donahue RE, and Pollard AM (2000) Radiocarbon calibration and Late Glacial occupation in northwest Europe. *Antiquity* **74**, 112–121.

Blockstein DE (1998) Lyme disease and the passenger pigeon? *Science* **279**, 5358.

Bocherens H, Michaux J, Garcia Talavera F, and Van der Plicht J (2006) Extinction of endemic vertebrates on islands: the case of the giant rat *Canariomys bravoi* (Mammalia, Rodentia) on Tenerife (Canary Islands, Spain). *Comptes Rendus Palevol* **5**, 885–891.

Bocquet-Appel J-P and Demars P-Y (2000) Population kinetics in the Upper Palaeolithic in western Europe. *Journal of Archaeological Science* **27**, 551–570.

Boersma PD and Silva MC (2001) Fork-tailed storm-petrel (*Oceanodroma furcata*). In Poole A and Gill F, eds, *The Birds of North America*, no. 569. The Academy of Natural Sciences, Philadelphia, and The American Ornithologists' Union, Washington DC.

Boessenkool S, Austin JJ, Worthy TH, Scofield P, Cooper A, Seddon PJ, and Waters JM (2009) Relict or colonizer? Extinction and range expansion of penguins in southern New Zealand. *Proceedings of the Royal Society of London Series B Biological Sciences* **276**, 815–821.

Bogan AE (1990) Stability of recent unionid (Mollusca: Bivalvia) communities over the past 6000 years. In Miller III W, ed., *Paleocommunity Temporal Dynamics: the Long-term Development of Multispecies Assemblies*, pp. 112–136. Special Publication 5. The Paleontological Society, Knoxville.

Bogan AE (1993) Freshwater bivalve extinctions (Mollusca: Unionida): a search for causes. *American Zoologist* **33**, 599–609.

Bond G, Kromer B, Beer J, Muscheler R, Evans MN, Showers W, Hoffmann S, Lotti-Bond R, Hajdas I, and Bonani G (2001) Persistent solar influence on North Atlantic climate during the Holocene. *Science* **294**, 2130–2136.

Bond WJ (1994) Do mutualisms matter? Assessing the impact of pollinator and dispersal disruption on plant extinction. *Philosophical Transactions of the Royal Society of London Series B* **344**, 83–90.

Bond WJ (1995) Assessing the risk of plant extinction due to pollinator and disperser failure. In Lawton JH and May RM, eds, *Extinction Rates*, pp. 131–146. Oxford University Press, Oxford.

Bond WJ and Slingsby P (1984) Collapse of an ant-plant mutualism: the Argentine ant (*Iridomyrmex humilis*) and myrmecochorous Proteaceae. *Ecology* **65**, 1031–1037.

Böpple JF and Coker RE (1912) *Mussel Resources of the Holston and Clinch River, Tennessee and Virginia.* Document no. 765, pp. 3–13. United States Bureau of Fisheries, Washington DC.

Bork H-R and Mieth A (2003) The key role of *Jubaea* palm trees in the history of Rapa Nui: a provocative interpretation. *Rapa Nui Journal* **17**, 119–121.

Borrero LA, Zárate M, Miotti L, and Massone M (1998) The Pleistocene-Holocene transition and human occupations in the Southern Cone of South America. *Quaternary International* **49/50**, 191–199.

Botkin DB, Saxe H, Araújo MB, Betts R, Bradshaw RHW, Cedhagen T, Chesson P, Dawson TP, Etterson JR, Faith DP *et al.* (2007) Forecasting the effects of global warming on biodiversity. *BioScience* **57**, 227–235.

Bour R (1981) Histoire de la tortue terrestre de Bourbon. *Bulletin de l'Académie de l'île de la Réunion* **25**, 98–147.

Bourne WRP, Ashmole NP, and Simmons KEL (2003) A new subfossil night heron and a new genus for the extinct rail from Ascension Island, central tropical Atlantic Ocean. *Ardea* **91**, 45–51.

Bover P and Alcover JA (2003) Understanding Late Quaternary extinctions: the case of *Myotragus balearicus* (Bate, 1909). *Journal of Biogeography* **30**, 771–781.

Bover P and Ramis D (2005) Requiem for *Myotragus balearicus* domestication. *Monografies de la Societat d'Història Natural de les Balears* **12**, 73–84.

Bover P and Alcover JA (2008) Extinction of the autochthonous small mammals of Mallorca (Gymnesic Islands, western Mediterranean) and its ecological consequences. *Journal of Biogeography* **35**, 1112–1122.

Bovy KM (2007) Global human impacts or climate change?: explaining the sooty shearwater decline at the Minard site, Washington State, USA. *Journal of Archaeological Science* **34**, 1087–1097.

Bower J and Lubell D (eds) (1988) *Prehistoric Cultures and Environments in the Late Quaternary of Africa*. BAR International Series 1653, Oxford.

Bowman DMJS (2001) Future eating and country keeping: what role has environmental history in the management of biodiversity? *Journal of Biogeography* **28**, 549–564.

Boye P, Hutterer R, López-Martínez N, and Michaux J (1992) A reconstruction of the lava mouse (*Malpaisomys insularis*), an extinct rodent of the Canary Islands. *Zeitschrift für Säugetierkunde* **57**, 29–38.

Bradley RS (2003) Climate forcing during the Holocene. In Mackay AW, Battarbee RW, Birks HJB, and Oldfield F, eds, *Global Change in the Holocene*, pp. 10–19. Arnold, London.

Bradshaw RHW (2004) Past anthropogenic influence on European forests and some possible genetic consequences. *Forest Ecology and Management* **197**, 203–212.

Bradshaw RHW and Mitchell FJG (1999) The palaeoecological approach to reconstructing former grazing-vegetation interactions. *Forest Ecology and Management* **120**, 3–12.

Bradshaw RHW and Hannon GE (2004) The Holocene structure of north-west European forest induced from palaeoecological data. In Honnay O, Verheyen K, Bossuyt B, and Hermy M, eds, *Forest Biodiversity: Lessons from History for Conservation*, pp. 11–25. CABI Publishing, Oxford.

Bradshaw RHW, Hannon GE, and Lister AM (2003) A long-term perspective on ungulate-vegetation interactions. *Forest Ecology and Management* **181**, 267–280.

Bramwell D (1976) The vertebrate fauna at Wetton Mill Rock Shelter. In Kelly JH, ed., *The Excavation of Wetton Mill Rock Shelter, Manifold Valley, Staffs*, pp. 40–51. City Museum and Art Gallery, Stoke on Trent.

Bramwell D (1984) The birds of Britain: when did they arrive? In Gilbertson DD and Jenkinson RDS, eds, *In the Shadow of Extinction: a Quaternary Archaeology and Palaeoecology of the Lake, Fissures and Smaller Caves at Cresswell Crags SSSI*, pp. 89–99. Department of Prehistory and Archaeology, University of Sheffield.

Brander K (1981) Disappearance of common skate *Raja batis* from Irish Sea. *Nature* **290**, 48–49.

Brander K (2007) Global fish production and climate change. *Proceedings of the National Academy of Sciences USA* **104**, 19709–19714.

Brands SJ (2007) *Systema Naturae 2000*. The Taxonomicon, Universal Taxonomic Services, Amsterdam.

Brandt JH, Dioli M, Hassanin A, Melville RA, Olson LE, Seveau A, and Timm RM (2001) Debate on the authenticity of *Pseudonovibos spiralis* as a new species of wild bovid from Vietnam and Cambodia. *Journal of Zoology* **255**, 437–444.

Brasso RL and Emslie SD (2006) Two new Late Pleistocene avifaunas from New Mexico. *The Condor* **108**, 721–730.

Breitburg E (1986) Paleoenvironmental exploitation strategies: the faunal data. In PA Cridlebaugh, ed., *Penitentiary Branch: a Late Archaic Cumberland River Shell Midden in Middle Tennessee*, pp. 87–125. Report of Investigations 4, Division of Archaeology, Tennessee Department of Environment and Conservation, Nashville.

Brewer S, Cheddadi R, de Beaulieu J-L, and Reille M (2002) The spread of deciduous *Quercus* throughout Europe since the last glacial period. *Forest Ecology and Management* **156**, 27–48.

Briggs DEG and Crowther PR (2001) *Palaeobiology II*. Blackwell Science, Oxford.

Briggs JM, Spielmann KA, Schaafsma H, Kintigh KW, Kruse M, Morehouse K, and Schollmeyer K (2006) Why ecology needs archaeologists and archaeology needs ecologists. *Frontiers in Ecology and the Environment* **4**, 180–188.

Brim Box J and Mossa J (1999) Sediment, land use, and freshwater mussels: prospects and problems. *Journal of the North American Benthological Society* **18**, 99–117.

Brim Box J and Williams JD (2000) Unionid mollusks of the Apalachicola Basin in Alabama, Florida, and Georgia. *Bulletin of the Alabama Museum of Natural History* **21**, 1–143.

British Ornithologists' Union Records Committee (2007) British Ornithologists' Union Records Committee: 35th Report (April 2007). *Ibis* **149**, 652–654.

Brodkorb P (1959) Pleistocene birds from New Providence Island, Bahamas. *Bulletin of the Florida State Museum, Biological Sciences* **4**, 349–371.

Brodkorb P (1963a) Birds from the Upper Cretaceous of Wyoming. In Sibley CG, ed., *Proceedings of the 13th International Ornithological Congress, Ithaca, New York*, pp. 50–70. American Ornithologists' Union, Baton Rouge.

Brodkorb P (1963b) Catalogue of fossil birds. Part 1 (Archaeopterygiformes through Ardeiformes). *Bulletin of the Florida State Museum, Biological Sciences Series* **7**, 179–293.

Brodkorb P (1965) Fossil birds from Barbados, West Indies. *Journal of the Barbados Museum and Historical Society* **31**, 3–10.

Bronk Ramsey C (1995) Radiocarbon calibration and analysis of stratigraphy: The OxCal program. *Radiocarbon* **37**, 425–430.

Bronk Ramsey C (2001) Development of the radiocarbon calibration program OxCal. *Radiocarbon* **43**, 355–363.

Bronstein JL, Dieckmann U, and Ferriére R (2004) Coevolutionary dynamics and the conservation of mutualisms. In Ferriére R, Dieckmann U, and Couvet D, eds, *Evolutionary Conservation Biology*, pp. 305–326. Cambridge University Press, Cambridge.

Broodbank C (2006) The origins and early development of Mediterranean maritime activity. *Journal of Mediterranean Archaeology* **19**, 199–230.

Brook FJ (2000) Prehistoric predation of the landsnail *Placostylus ambagiosus* Suter (Stylommatophora: Bulimulidae), and evidence for the timing of establishment of rats in northernmost New Zealand. *Journal of the Royal Society of New Zealand* **30**, 227–241.

Brooke M de L (1987) *The birds of the Juan Fernández Islands,Chile.* International Council for Bird Preservation, Fauna and Flora Preservation Society, Report 16. World Wildlife Fund, Cambridge.

Brooke M de L (1990) *The Manx Shearwater.* T & AD Poyser, London.

Brooke M de L (2004) *Albatrosses and Petrels across the World.* Oxford University Press, Oxford.

Broughton JM (2004) Prehistoric human impacts on California birds: evidence from the Emeryville Shellmound avifauna. *Ornithological Monographs* **56**, 1–90.

Browman DL (1974) Pastoral nomadism in the Andes. *Current Anthropologist* **15**, 188–196.

Brown AJV and Verhagen BT (1985) Two *Antidorcas bondi* individuals from the Late Stone Age site of Kruger Cave 35/83, Olifantsnek, Rustenberg District, South Africa. *South African Journal of Science* **81**, 102.

Brown DM, Brenneman RA, Koepfli K-P, Pollinger JP, Mila B, Georgiadis NJ, Louis Jr EE, Grether GF, Jacobs DK, and Wayne RK (2007) Extensive population genetic structure in the giraffe. *BMC Biology* **5**, 57.

Brown L (1976) *British Birds of Prey.* New Naturalist, Collins, London.

Bruno JF and O'Connor MI (2005) Cascading effects of predator diversity and omnivory in a marine food web. *Ecology Letters* **8**, 1048–1056.

Bryant PJ (1995) Dating remains of gray whales from the eastern North Atlantic. *Journal of Mammalogy* **76**, 857–861.

Buck CE and Bard E (2007) A calendar chronology for Pleistocene mammoth and horse extinction in North America based on Bayesian radiocarbon calibration. *Quaternary Science Reviews* **26**, 2031–2035.

Bulte EH, Horan RD, and Shogren JF (2003) Is the Tasmanian tiger extinct? A biological-economic re-evaluation. *Ecological Economics* **45**, 271–279.

Bunce M, Worthy TH, Ford T, Hoppitt W, Willerslev E, Drummond A, and Cooper A (2003) Extreme reversed sexual size dimorphism in the extinct New Zealand moa *Dinornis. Nature* **425**, 172–175.

Bunce M, Szulkin M, Lerner HRL, Barnes I, Shapiro B, Cooper A, and Holdaway RN (2005) Ancient DNA provides new insights into the evolutionary history of New Zealand's extinct giant eagle. *PLoS Biology* **3**, 44–46.

Bunin JS and Jamieson IG (1995) New approaches toward a better understanding of the decline of takahe (*Porphyrio mantelli*) in New Zealand. *Conservation Biology* **9**, 100–106.

Burbidge AA, Johnson KA, Fuller PJ, and Southgate RJ (1988) Aboriginal knowledge of the mammals of the central deserts of Australia. *Australian Wildlife Research* **15**, 9–39.

Burckhardt R (1893) Über *Aepyornis. Palæontologische Abhandlungen* **2**, 127–145.

Burness GP, Diamond J, and Flannery T (2001) Dinosaurs, dragons, and dwarfs: the evolution of maximal body size. *Proceedings of the National Academy of Sciences USA* **98**, 14518–14523.

Burney DA (1999) Rates, patterns, and processes of landscape transformation and extinction in Madagascar. In MacPhee RDE, ed., *Extinctions in Near Time: Causes, Contexts, and Consequences*, pp. 145–164. Kluwer Academic/Plenum, New York.

Burney DA and Ramilisonina (1998) The kilopilopitsofy, kidoky and bokyboky: accounts of strange animals from Belo-sur-Mer, Madagascar, and the megafaunal "extinction window". *American Anthropologist* **100**, 957–966.

Burney DA and Flannery TF (2005) Fifty millennia of catastrophic extinctions after human contact. *Trends in Ecology & Evolution* **20**, 395–401.

Burney DA, Burney LP, and MacPhee RDE (1994) Holocene charcoal stratigraphy from Laguna Tortuguero, Puerto Rico, and the timing of human arrival on the island. *Journal of Archaeological Science* **21**, 273–281.

Burney DA, James HF, Burney LP, Olson SL, Kikuchi W, Wagner WL, Burney M, McCloskey D, Kikuchi D, Grady FV *et al.* (2001) Fossil evidence for a diverse biota from Kaua'i and its transformation since human arrival. *Ecological Monographs* **74**, 615–641.

Burney DA, Robinson GS, and Burney LP (2003) *Sporormiella* and the late Holocene extinctions in Madagascar. *Proceedings of the National Academy of Sciences USA* **100**, 10800–10805.

Burney DA, Burney LP, Godfrey LR, Jungers WL, Goodman SM, Wright HT, and Jull AJT (2004) A chronology for late prehistoric Madagascar. *Journal of Human Evolution* **47**, 25–63.

Bush MB (2003) Holocene climates of the lowland tropical forests. In Mackay AW, Battarbee RW, Birks HJB, and Oldfield F, eds, *Global Change in the Holocene*, pp. 384–395. Arnold, London.

Bush MB, Stute M, Ledru M-P, Behling H, Colinvaux PA, de Oliveira PE, Grimm EC, Hooghiemstra H, Haberle S, Leyden BW *et al.* (2001) Paleotemperature estimates for the lowland Americas between 30°S and 30°N at the last glacial maximum. In Markgraf V, ed., *Interhemispheric Climate Linkages: Present and Past Interhemispheric Climate Linkages in the Americas and their Societal Effects*, pp. 293–306. Academic Press, New York.

Butchart SHM, Stattersfield AJ, Bennun LA, Shutes SM, Akçakaya HR, Baillie JEM, Stuart SN, Hilton-Taylor C, and Mace GM (2004) Measuring global trends in the status of biodiversity: Red List indices for birds. *PLoS Biology* **2**, e383.

Butchart SHM, Stattersfield AJ, and Brooks TM (2006) Going or gone: defining 'Possibly Extinct' species to give a truer picture of recent extinctions. *Bulletin of the British Ornithologists' Club* **126A**, 7–24.

Butler PJ (1992) Parrots, pressures, people and pride. In Beissinger SR and Snyder NFR, eds, *New World Parrots in Crisis*, pp. 23–46. Smithsonian Institution Press, Washington DC.

Byrd GV, Sincock JL, Telfer TC, Moriarty DI, and Brady BG (1984) A cross-fostering experiment with Newell's race of Manx shearwater. *Journal of Wildlife Management* **48**, 163–168.

Caddy JF and Agnew DJ (2004) An overview of recent global experience with recovery plans for depleted marine resources and suggested guidelines for recovery planning. *Reviews in Fish Biology and Fisheries* **14**, 43–112.

Calviño-Cancela M, Dunn RR, van Etten E, and Lamont BB (2006) Emus as non-standard seed dispersers and their potential for long-distance dispersal. *Ecography* **29**, 632–640.

Campbell RR (1988) Status of the sea mink, *Mustela macrodon*, in Canada. *Canadian Field-Naturalist* **102**, 304–306.

Carcaillet C, Almquist H, Asnong H, Bradshaw RHW, Carrión JS, Gaillard M-J, Gajewski K, Haas JN, Haberle SG, Hadorn P *et al.* (2002) Holocene biomass burning and global dynamics of the carbon cycle. *Chemosphere* **49**, 845–863.

Carder N, Reitz EJ, and Crock JG (2007) Fish communities and populations during the post-Saladoid period (AD 600/800–1500), Anguilla, Lesser Antilles. *Journal of Archaeological Science* **34**, 588–599.

Cardillo M, Mace GM, Jones KE, Bielby J, Bininda-Emonds ORP, Sechrest W, Orme CDL, and Purvis A (2005) Multiple causes of high extinction risk in large mammal species. *Science* **309**, 1239–1241.

Cardinale BJ, Srivastava DS, Duffy JE, Wright JP, Downing AL, Sankaran M, and Jouseau C (2006) Effects of biodiversity on the functioning of trophic groups and ecosystems. *Nature* **443**, 989–992.

Carleton MD and Olson SL (1999) Amerigo Vespucci and the rat of Fernando de Noronha: a new genus and species of Rodentia (Muridae: Sigmodontinae) from a volcanic island off Brazil's continental shelf. *American Museum Novitates* **3256**, 1–59.

Carlson LA (1999) *Aftermath of a Feast: Human Colonization of the Southern Bahamian Archipelago and its Effects on the Indigenous Fauna.* PhD thesis, University of Florida.

Carlsson M (2003) *Phylogeography of the adder, Vipera berus.* Comprehensive Summaries of Uppsala Dissertations from the Faculty of Sciences and Technology 849. Acta Universitatis Upsaliensis, Uppsala.

Carlton JT (1993) Neoextinctions of marine invertebrates. *American Zoologist* **33**, 499–509.

Carlton JT, Vermeij GJ, Lindberg DR, Carlton DA, and Dudley EC (1991) The first historical extinction of a marine invertebrate in an ocean basin: the demise of the eelgrass limpet *Lottia alveus*. *Biological Bulletin* **180**, 72–80.

Carlton JT, Geller JB, Reaka-Kudla ML, and Norse EA (1999) Historical extinctions in the sea. *Annual Review of Ecology and Systematics* **30**, 525–538.

Caro T (2007) The Pleistocene re-wilding gambit. *Trends in Ecology & Evolution* **22**, 281–283.

Carr SG and Robinson AC (1997) The present status and distribution of the desert rat-kangaroo *Caloprymnus campestris* (Marsupialia: Potoroidae). *South Australian Naturalist* **72**, 4–27.

Carreiro-Silva M and McClanahan TR (2001) Echinoid bioerosion and herbivory on Kenyan coral reefs: the role of protection from fishing. *Journal of Experimental Marine Biology and Ecology* **262**, 133–153.

Cartelle C and Hartwig WC (1996) A new extinct primate among the Pleistocene megafauna of Bahia, Brazil. *Proceedings of the National Academy of Sciences USA* **93**, 6405–6409.

Carter I, Newbery P, Grice P, and Hughes J (2008) The role of reintroductions in conserving British birds. *British Bird* **101**, 2–25.

Carter SK, VanBlaricom GR, and Allen BL (2007) Testing the generality of the trophic cascade paradigm for sea otters: a case study with kelp forests in northern Washington, USA. *Hydrobiologia* **579**, 233–249.

Case TJ (1996) Global patterns in the establishment and distribution of exotic birds. *Biological Conservation* **78**, 69–96.

Cassels, RJS, Jones KL, Walton A, and Worthy TH (1988) Late prehistoric subsistence practices at Parewanui,

Lower Rangitikei River, New Zealand. *New Zealand Journal of Archaeology* **10**, 109–128.

Cassey P (2001) Determining variation in the success of New Zealand land birds. *Global Ecology & Biogeography* **10**, 161–172.

Cassey P, Blackburn TM, Sol D, Duncan RP, and Lockwood J (2004) Introduction effort and establishment success in birds. *Proceedings of the Royal Society of London Series B Biological Sciences* **271**, S405–S408.

Cassey P, Blackburn TM, Duncan RP, and Lockwood JL (2005a) Lessons from the establishment of exotic species: a meta-analytical case study using birds. *Journal of Animal Ecology* **74**, 250–258.

Cassey P, Blackburn TM, Duncan RP, and Gaston KJ (2005b) Causes of exotic bird establishment across oceanic islands *Proceedings of the Royal Society of London Series B Biological Sciences* **272**, 2059–2063.

Cassey P, Blackburn TM, and Duncan RP (2005c) Concerning invasive species: a reply to Brown and Sax. *Austral Ecology* **30**, 465–480.

Cassey P, Blackburn TM, Lockwood JL, and Sax DF (2006) A stochastic model for integrating changes in species richness similarity across spatial scales. *Oikos* **115**, 207–218.

Cassey P, Lockwood JL, Blackburn TM, and Olden JD (2007) Spatial scale and evolutionary history determine the degree of taxonomic homogenization across island bird assemblages. *Diversity and Distributions* **13**, 458–466.

Castillo C, Martín-Gonzalez E, and Coello JJ (2001) Small vertebrate taphonomy of La Cueva del Llano, a volcanic cave on Fuerteventura (Canary Islands, Spain). Palaeoecological implications. *Palaeogeography, Palaeoclimatology, Palaeoecology* **166**, 277–291.

Catto N and Catto G (2004) Climate change, communities, and civilisations: driving force, supporting player, or background noise. *Quaternary International* **123**, 7–10.

Cavanagh R and Dulvy NK (2004) Disappearing from the depths: sharks on the Red List. In Baillie JEM, Hilton-Taylor C, and Stuart S, eds, *2004 IUCN Red List of Threatened Species: a Global Species Assessment*, pp. 21–22. IUCN, Gland and Cambridge.

Cavelier J, Aide TM, Santos C, Eusse AM, and Dupuy JM (1998) The savannization of moist forests in the Sierra Nevada de Santa Marta, Colombia. *Journal of Biogeography* **25**, 901–912.

Ceballos G and Ehrlich PR (2002) Mammal population losses and the extinction crisis. *Science* **296**, 904–907.

Cetti F (1777) *Appendice alla storia naturale dei quadrupedi di Sardegna*. Giuseppe Piattoli, Sassari.

Challies CN (1975) Feral pigs (*Sus scrofa*) on Auckland Island: status, and effects on vegetation and nesting sea birds. *New Zealand Journal of Zoology* **2**, 479–490.

Chandler RM (1990) Fossil birds of the San Diego Formation, late Pliocene, Blancan, San Diego County, California, Part 2. Recent advances in the study of Neogene fossil birds. *Ornithological Monographs* **44**, 73–171.

Channell R and Lomolino MV (2000a) Dynamic biogeography and conservation of endangered species. *Nature* **403**, 84–86.

Channell R and Lomolino MV (2000b) Trajectories to extinction: spatial dynamics of the contraction of geographical ranges. *Journal of Biogeography* **27**, 169–179.

Chapman CA (2005) Primate seed dispersal: coevolution and conservation implications. *Evolutionary Anthropology* **4**, 74–82.

Chapman J, Delcourt PA, Cridlebaugh PA, Shea AB, and Delcourt HR (1982) Man-land interactions: 10,000 years of American Indian impact on native ecosystems in the lower Little Tennessee River valley. *Southeastern Archaeology* **1**, 115–121.

Cheke AS (1987) An ecological history of the Mascarene Islands, with particular reference to extinctions and introductions of land vertebrates. In Diamond AW, ed., *Studies of Mascarene Island Birds*, pp. 5–89. Cambridge University Press, Cambridge.

Cheke AS (2006) Establishing extinction dates – the curious case of the dodo *Raphus cucullatus* and the red hen *Aphanapteryx bonasia*. *Ibis* **148**, 155–158.

Cheke AS and Hume JP (2008) *Lost Land of the Dodo: an Ecological History of Mauritius, Réunion & Rodrigues*. A & C Black, London.

Chepstow-Lusty AJ, Frogley MR, Bauer BS, Bush M, and Herrera AT (2003) A late Holocene record of arid events from the Cuzco region, Peru. *Journal of Quaternary Science* **18**, 491–502.

Cherel Y, Ridoux V, Weimerskirch H, Tveraa T, and Chastel O (2001) Capelin (*Mallotus villosus*) as an important food source for northern fulmars (*Fulmarus glacialis*) breeding at Bjørnøya (Bear Island), Barents Sea. *ICES Journal of Marine Science* **58**, 355–361.

Childe VG (1952) *New Light on the Most Ancient East*. Routledge & Kegan Paul, London.

Childs TS and Herbert EW (2005) Metallurgy and its consequences. In Stahl AB, ed., *African Archaeology*, pp. 276–301. Blackwell Publishing, Oxford.

Chinsany-Turan A (2005) *The Microstructure of Dinosaur Bone: Deciphering Biology with Fine-Scale Techniques*. Johns Hopkins University Press, Baltimore.

Christensen V, Guénette S, Heymans JJ, Walters CJ, Watson R, Dirk Zeller D, and Pauly D (2003) Hundred-year decline of North Atlantic predatory fishes. *Fish and Fisheries* **4**, 1–25.

Christian CE (2001) Consequences of a biological invasion reveal the importance of mutualism for plant communities. *Nature* **413**, 635–639.

Church J, Gregory J, Huybrechts P, Kuhn M, Lambeck K, Nhuan M, Qin D, and Woodworth P (2001) Changes in sea level. In Houghton J, Ding Y, Griggs D, Noguer M, van der Linden P, Dai X, Maskell K, and Johnson C, eds, *Climate Change 2001: the Scientific Basis*, pp. 639–693. Cambridge University Press, Cambridge.

Cibois A, Thibault J-C, and Pasquet E (2004) Biogeography of eastern Polynesian monarchs (*Pomarea*): an endemic genus close to extinction. *The Condor* **106**, 837–851.

Clare EL, Lim BK, Engstrom MD, Eger LJ, and Hebert PDN (2007) DNA barcoding of Neotropical bats: species identification and discovery within Guyana. *Molecular Ecology Notes* **7**, 184–190.

Clark JS (1990) Fire and climate change during the last 750-yr in northwestern Minnesota. *Ecological Monographs* **60**, 135–159.

Clark JS, Fastie C, Hurtt G, Jackson ST, Johnson C, King GA, Lewis M, Lynch J, Pacala S, Prenctice C *et al.* (1998) Reid's paradox of rapid plant migration. Dispersal theory and interpretation of paleoecological record. *BioScience* **48**, 13–24.

Clark JS, Lewis M, and Horvath L (2001) Invasion by extremes: population spread with variation in dispersal and reproduction. *American Naturalist* **157**, 537–554.

Clarke AH (1981) The tribe Alasmidontini (Unionidae: Anodontinae), Part I: *Pegias*, *Alasmidonta*, and *Arcidens*. *Smithsonian Contributions to Zoology* **326**, 1–101.

Clarke R and Schulz M (2005) Land-based observations of seabirds off sub-Antarctic Macquarie Island during 2002 and 2003. *Marine Ornithology* **33**, 7–17.

Clarke SC (2004) Understanding pressures on fishery resources through trade statistics: a pilot study of four products in the Chinese dried seafood market. *Fish and Fisheries* **5**, 53–74.

Clarke SC, Magnussen JE, Abercrombie DL, McAllister MK, and Shivji MS (2006a) Identification of shark species composition and proportion in the Hong Kong shark fin market based on molecular genetics and trade records. *Conservation Biology* **20**, 201–211.

Clarke SC, McAllister MK, Milner-Gulland EJ, Kirkwood GP, Michielsens CGJ, Agnew DJ, Pikitch EK, Nakano H, and Shivji MS (2006b) Global estimates of shark catches using trade records from commercial markets. *Ecology Letters* **9**, 1115–1126.

Claussen M, Kubatzki C, Brovkin V, Ganopolski A, Hoelzmann P, and Pachur HJ (1999) Simulation of an abrupt change in Saharan vegetation at the end of the mid-Holocene. *Geophysical Research Letters* **24**, 2037–2040.

Claussen M, Brovkin V, Calov R, Ganopolski A, and Kubatzki C (2005) Did humankind prevent a Holocene glaciation? *Climatic Change* **69**, 409–417.

Clayton DH and Price RD (1999) Taxonomy of New World Columbicola (Phthiraptera: Philopteridae) from the Columbiformes (Aves), with descriptions of five new species. *Annals of the Entomological Society of America* **92**, 675–685.

Clements FE (1904) *Plant Succession: an Analysis of the Development of Vegetation*. Carnegie Institute, Washington DC.

Clements FE and Long FL (1923) *Experimental Pollination: an Outline of the Ecology of Flowers and Insects*. Carnegie Institution of Washington, Washington DC.

Clench WJ (1926) Some notes and a list of shells of Rio, Kentucky. *The Nautilus* **38**, 7–12, 65–67.

Coard R and Chamberlain AT (1999) The nature and timing of faunal change in the British Isles across the Pleistocene/Holocene transition. *The Holocene* **9**, 372–376.

Cohen JE, Small C, Mellinger A, Gallup J, and Sachs J (1997) Estimates of coastal populations. *Science* **278**, 1211–1212.

Coker RE (1919) *Fresh-water Mussels and Mussel Industries of the United States*. Document no. 865. United States Bureau of Fisheries, Washington DC.

Colinvaux PA and West FH (1984) The Beringian ecosystem. *Quarterly Review of Archaeology* **5**, 10–16.

Colinvaux PA and de Oliveira PE (2001) Amazon plant diversity and climate through the Cenozoic. *Palaeogeography, Palaeoclimatology, Palaeoecology* **166**, 51–63.

Collar NJ (1998) Extinction by assumption; or, the Romeo Error on Cebu. *Oryx* **32**, 239–244.

Collar NJ, Crosby MJ, and Stattersfield AJ (1994) *Birds to Watch 2: the World List of Threatened Birds*. BirdLife International, Cambridge.

Collen B, Purvis A, and Gittleman JL (2004) Biological correlates of description date in carnivores and primates. *Global Ecology and Biogeography* **13**, 459–467.

Condis Fernández MM, Jiménez Vásquez O, and Arredondo C (2005) Revisión taxonómica del género *Nesophontes* (Insectivora: Nesophontidae) en Cuba: análisis de los caracteres diagnóstico. *Monografies de la Societat d'Història Natural de les Balears* **12**, 95–100.

Cook JM and Rasplus JY (2003) Mutualists with attitude: coevolving fig wasps and figs *Trends in Ecology & Evolution* **18**, 325–325.

Cooke P (1980) Optimal linear estimation of bounds of random variables. *Biometrika* **67**, 257–258.

Cooke R (2001) Prehistoric nearshore and littoral fishing in the eastern tropical Pacific: an ichthyological evaluation. *Journal of World Prehistory* **6**, 1–49.

Coope GR (2000) Middle Devensian (Weichselian) coleopteran assemblages from Earith, Cambridge (UK) and their bearing on the interpretation of 'Full Glacial' floras and faunas. *Journal of Quaternary Science* **15**, 779–788.

Coope GR and Angus RB (1975) An ecological study of a temperate interlude in the middle of the last glaciation, based on fossil Coleoptera from Isleworth, Middlesex. *Journal of Animal Ecology* **44**, 365–391.

Cooper A, Lalueza-Fox C, Anderson S, Rambaut A, Austin J, and Ward R (2001) Complete mitochondrial genome sequences of two extinct moas clarify ratite evolution. *Nature* **409**, 704–707.

Cooper JH (2000) First fossil record of azure-winged magpie *Cyanopica cyanus* in Europe. *Ibis* **142**, 150–151.

Cooper JH and Tennyson AJD (2004) New evidence on the life and death of Hawkins' rail (*Diaphorapteryx hawkinsi*): Moriori accounts recorded by Sigvard Dannefaerd and Alexander Shand. *Notornis* **51**, 212–216.

Cooper JH and Tennyson AJD (2008) Wrecks and residents: the fossil gadfly petrels (*Pterodroma* spp.) of the Chatham Islands, New Zealand. *Oryctos* **7**, 227–248.

Corbett L (1995) *The Dingo in Australia and Asia*. University of New South Wales Press, Sydney.

Cordeiro NJ and Howe HF (2001) Low recruitment of trees dispersed by animals in African forest fragments. *Conservation Biology* **15**, 1733–1741.

Cortés E (1999) Standardized diet compositions and trophic levels of sharks. *ICES Journal of Marine Science* **56**, 707–717.

Cortés-Calva P, Álvarez-Castañeda ST, and Yensen E (2001a) *Neotoma anthonyi*. *Mammalian Species* **663**, 1–3.

Cortés-Calva P, Yensen E, and Álvarez-Castañeda ST (2001b) *Neotoma martinensis*. *Mammalian Species* **657**, 1–3.

Côté IM and Reynolds JD (2006) *Coral Reef Conservation*. Cambridge University Press, Cambridge.

Courchamp F, Langlais M, and Sugihara G (2000) Rabbits killing birds: modeling the hyperpredation process. *Journal of Animal Ecology* **69**, 154–164.

Courchamp F, Chapuis J-L, and Pascal M (2003) Mammal invaders on islands: impact, control and control impact. *Biological Reviews* **78**, 347–383.

Cowie RH (2001) Invertebrate invasions on Pacific islands and the replacement of unique native faunas: a synthesis of the land and freshwater snails. *Biological Invasions* **3**, 119–136.

Cowles GS (1994) A new genus, three new species and two new records of extinct Holocene birds from Réunion Island, Indian Ocean. *Geobios* **27**, 87–93.

Crock JG (2000) *Interisland Interaction and the Development of Chiefdoms in the Eastern Caribbean*. PhD thesis, University of Pittsburgh.

Croft DA, Heaney LR, Flynn JJ, and Bautista AP (2006) Fossil remains of a new, diminutive *Bubalus* (Artiodactyla: Bovidae: Bovini) from Cebu Island, Philippines. *Journal of Mammalogy* **87**, 1037–1051.

Croll DA, Maron JL, Estes JA, Danner EM, and Byrd GV (2005) Introduced predators transform subarctic islands from grassland to tundra. *Science* **307**, 1959–1961.

Crooks KR and Soulé ME (1999) Mesopredator release and avifaunal extinctions in a fragmented system. *Nature* **400**, 563–566.

Crossen KJ, Yesner DR, Veltre DW, and Graham RW (2005) 5,700-year-old mammoth remains from the Pribilof Islands, Alaska: last outpost of North American megafauna. *Geological Society of America Abstracts with Programs* **37**, 463.

Crowder LB, Hopkins-Murphy SR, and Royle JA (1995) Effects of turtle excluder devices (TEDs) on loggerhead sea turtle strandings with implications for conservation. *Copeia* **1995**, 773–779.

Crowley TJ (2000) Causes of climate change over the past 1000 years. *Science* **289**, 270–277.

Croxall JP and Prince PA (1994) Dead or alive, night or day: how do albatrosses catch squid? *Antarctic Science* **6**, 155–162.

Crozier RH (1997) Preserving the information content of species: genetic diversity, phylogeny, and conservation worth. *Annual Review of Ecology and Systematics* **28**, 243–268.

Crutzen PJ (2002) Geology of mankind. *Nature* **415**, 23.

Cruz I and Elkin D (2003) Structural bone density of the lesser rhea (*Pterocnemia pennata*) (Aves: Rheidae). Taphonomic and archaeological implications. *Journal of Archaeological Science* **30**, 37–44.

Cruz JB and Cruz F (1987) Conservation of the dark-rumped petrel *Pterodroma phaeopygia* in the Galapagos Islands, Ecuador. *Biological Conservation* **42**, 303–311.

Cumbaa SL (1986) Archaeological evidence of the 16th century Basque right whale fishery in Labrador. *Reports of the International Whaling Commission, Special Issue* **10**, 187–190.

Cummings KS, Mayer CA, and Page LM (1988) Survey of the freshwater mussels (Mollusca: Unionidae) of the Wabash River drainage. Phase II: upper and middle Wabash River. *Illinois Natural History Survey Technical Report* **1988(8)**, 1–47.

Curnutt J and Pimm S (2001) How many bird species in Hawai'i and the central Pacific before first contact? *Studies in Avian Biology* **22**, 15–30.

Currie RG and Fairbridge RW (1985) Periodic 18.6 year and cyclic 11-year induced drought and flood in northeastern China and some global implications. *Quaternary Science Reviews* **4**, 109–134.

Dabert J, Mironov SV, and Proctor H (2006) A new species of the feather mite genus *Titanolichus* Gaud & Atyeo, 1996 (Acari: Astigmata: Pterolichidae) from the endangered orange-bellied parrot *Neophema chrysogaster* (Aves: Psittaciformes) from Australia. *Australian Journal of Entomology* **45**, 206–214.

Dalén L, Fuglei E, Hersteinsson P, Kapel CMO, Roth JD, Samelius G, Tannerfeldt M, and Angerbjörn A (2005) Population history and genetic structure of a circumpolar species: the arctic fox. *Biological Journal of the Linnean Society* **84**, 79–89.

Dalén L, Nyström V, Valdiosera C, Germonpré M, Sablin M, Turner E, Angerbjörn A, Arsuaga JL, and Götherström A (2007) Ancient DNA reveals lack of postglacial habitat tracking in the arctic fox. *Proceedings of the National Academy of Sciences USA* **104**, 6726–6729.

Dalton R (2005) Ornithologists stunned by bird collector's deceit. *Nature* **437**, 302–303.

Dalzell P (1998) The role of archaeological and cultural-historical records in long-range coastal fisheries resources management strategies and policies in the Pacific Islands. *Ocean and Coastal Management* **40**, 237–252.

Danielson F, Sørenson MK, Olwig MF, Selvam VPF, Burgess ND, Hiraishi T, Karunagran VM, Rasmussen MS, Hansen LB, Quarto A, and Suryadiputra N (2005) The Asian tsunami: a protective role for coastal vegetation. *Science* **310**, 643.

Dansgaard W, Johnsen SJ, Clausen HB, Dahl-Jensen D, Gundestrup NS, Hammer CU, Hvidberg CS, Steffensen JP, Sveinbjornsdottir AE, Jouzel J, and Bond G (1993) Evidence for general instability of past climate from a 250-kyr ice-core record. *Nature* **364**, 218–20.

Darwin C (1859) *On the Origin of Species by Means of Natural Selection, or the Preservation of Favoured Races in the Struggle for Life.* John Murray, London.

Davis MB and Shaw RG (2001) Range shifts and adaptive responses to Quaternary climate change. *Science* **292**, 673–679.

Dawson L (1982) Taxonomic status of fossil thylacines (*Thylacinus*, Thylacinidae, Marsupialia) from late Quaternary deposits in eastern Australia. In Archer M, ed., *Carnivorous Marsupials*, pp. 527–536. Royal Zoological Society of New South Wales, Mosman.

Day GM (1953) The Indian as an ecological factor in the northeastern forest. *Ecology* **34**, 329–346.

Debruyne R, Van Holt A, Barriel V, and Tassy P (2003) Status of the so-called African pygmy elephant (*Loxodonta pumilio* (Noack 1906)): phylogeny of cytochrome *b* and mitochondrial control region sequences. *Comptes Rendus Biologies* **326**, 687–697.

de Castro F and Bolker B (2005) Mechanisms of disease-induced extinction. *Ecology Letters* **8**, 117–126.

Decher J and Abedi-Lartey M (2002) *Small Mammal Zoogeography and Diversity in West African Forest Remnants.* Unpublished report.

Deffontaine V, Libois R, Kotlík P, Ommer R, Nieberding C, Pradis E, Searle JB, and Michaux JR (2005) Beyond the Mediterranean peninsulas: evidence of central European glacial refugia for temperate forest mammal species, the bank vole (*Clethrionomys glareolus*). *Molecular Ecology* **14**, 1727–1739.

de Flacourt E (1658) *Histoire de la grande isle de Madagascar.* J. Hénault, Paris.

DEFRA (2004) *Securing the Benefits: the Joint UK Response to the Prime Minister's Strategy Unit Net Benefits Report on the Future of the Fishing Industry in the UK.* Department for Environment, Food and Rural Affairs, London.

de la Mare WK and Kerry KR (1994) Population dynamics of the wandering albatross (*Diomedea exulans*) on Macquarie Island and the effects of mortality from longline fishing. *Polar Biology* **14**, 231–241.

Delcourt H (1987a) The impact of prehistoric agriculture and land occupation on natural vegetation. *Ecology* **68**, 341–346.

Delcourt HR (1987b) The impact of pre-Columbian agriculture and land occupation on natural vegetation. *Trends in Ecology & Evolution* **2**, 39–44.

Delcourt HR (1997) Pre-Columbian Native American use of fire on southern Appalachian landscapes. *Conservation Biology* **11**, 1010–1014.

Delcourt PA and Delcourt HR (1998) The influence of prehistoric human-set fires on oak-chestnut forests in the southern Appalachians. *Castanea* **63**, 337–345.

Delcourt PA and Delcourt HR (2004) *Prehistoric Native Americans and Ecological Change: Human Ecosystems in Eastern North America since the Pleistocene.* Cambridge University Press, New York.

Delgado CL, Wada N, Rosegrant MW, Meijer S, and Ahmed M (2003) *Fish to 2020 Supply and Demand in Changing Global Markets.* International Food Policy Research Institute (IFPRI) and Worldfish Center. WorldFish Center, Washington DC.

Del Hoyo J, Elliott A, and Sargatal L (eds) (1994) *Handbook of the Birds of the World. Vol. 2. New World Vultures to Guineafowl.* Lynx Edicions, Barcelona.

del Monte-Luna P, Lluch-Belda D, Serviere-Zaragoza E, Carmona R, Reyes-Bonilla H, Aurioles-Gamboa D, Castro-Aguirre JL, Próo SAGD, Trujillo-Millán O, and Brook BW (2007) Marine extinctions revisited. *Fish and Fisheries* **8**, 107–122.

DeLord J (2007) The nature of extinction. *Studies in History and Philosophy of Biological and Biomedical Sciences* **38**, 656–667.

Dennis R (2005) The eagle owl has landed. *BBC Wildlife* **23(13)**, 24–29.

Deredec A and Courchamp P (2003) Extinction thresholds in host-parasite dynamics. *Annales Zoologici Fennici* **40**, 115–130.

Derenne P and Mougin J-L (1976) Les procellariiformes à nidification hypogée de l'île aux Cochons (Archipelego Crozet, 46°06'S. 50°14'E). *Comité National Français pour les Recherches Antarctiques* **40**, 149–175.

Dermitzakis MD and Sondaar PY (1978) The importance of fossil mammals in reconstructing palaeogeography with special reference to the Pleistocene Aegean Archipelago. *Annales Géologiques des Pays Helléniques* **46**, 808–840.

de Smet K and Smith TR (2001) Algeria. In Mallon DP and Kingswood SP, compilers, *Antelopes. Part IV. North Africa, the Middle East and Asia. Global Survey and Regional Action Plans.* IUCN, Gland and Cambridge.

Desse J and Desse-Berset N (1993) Pêche et surpêche en Méditerranée: le témoignage des os. In Desse D and Audoin-Rouzeau F, eds, *Expoitation des animaux sauvages a travers le temps*, pp. 332–333. Recontres Internationales d'Archéologie de l'Histoire d'Antibes, APDCA, Antibes.

Devine JA, Baker KD, and Haedrich RL (2006) Deep-sea fishes qualify as endangered. *Nature* **439**, 29.

de Vivo M and Carmignotto AP (2004) Holocene vegetation change and the mammal faunas of South America and Africa. *Journal of Biogeography* **31**, 943–957.

de Waal MS (1996) *The Petite Riviere excavations, La Desirade, French West Indies: Fieldwork Report and Subsistence Studies for a Pre-Columbian Site with Late Saladoid and Post-Saladoid Components.* MA thesis, University of Leiden.

Dewar RE and Richard AF (2007) Evolution in the hypervariable environment of Madagascar. *Proceedings of the National Academy of Sciences USA* **104**, 13723–13727.

Diamond JM (1984a) Historic extinctions: a Rosetta Stone for understanding preshistoric extinctions. In PS Martin and RG Klein, eds, *Quaternary Extinctions: a Prehistoric Revolution*, pp. 824–862. University of Arizona Press, Tucson and London.

Diamond JM (1984b) "Normal" extinctions of isolated populations. In Nitecki MH, ed., *Extinctions*, pp. 191–246. University of Chicago Press, Chicago.

Diamond JM (1985) Population processes in island birds: immigration, extinction and fluctuation. *International Council for Bird Preservation Technical Publication* **3**, 17–21.

Diamond JM (1987) Extant unless proven extinct? Or, extinct unless proven extant? *Conservation Biology* **1**, 77–79.

Diamond JM (1989) Quaternary megafaunal extinctions: variations on a Theme by Paganini. *Journal of Archaeological Science* **16**, 167–175.

Diamond J (2002) Evolution, consequences and future of plant domestication. *Nature* **418**, 700–706.

Diamond J (2005) *Collapse: How Societies Choose to Fail or Succeed.* Viking, New York.

Díaz MM, Flores DA, and Barquez RM (2002) A new species of gracile mouse opossum, genus *Gracilinanus* (Didelphimorphia: Didelphidae) from Argentina. *Journal of Mammalogy* **83**, 824–833.

Diaz RJ and Rosenberg R (1995) Marine benthic hypoxia: a review of its ecological effects and the behavioural responses of benthic macrofauna. *Oceanography and Marine Biology: an Annual Review* **33**, 245–03.

Díaz-Franco S (2001) Situación taxonómica de *Geocapromys megas* (Rodentia: Capromyidae). *Caribbean Journal of Science* **37**, 72–80.

Didham RK, Ewers RM, and Gemmell NJ (2005) Comment on "Avian extinction and mammalian introductions on oceanic islands". *Science* **307**, 1412a.

Dillehay TD (1999) The late Pleistocene cultures of South America. *Evolutionary Anthropology* **7**, 206–216.

Dillehay TD, Rossen J, Maggard G, Stackelbeck K, and Netherly P (2003) Localization and possible social aggregation in the Late Pleistocene and early Holocene on the north coast of Peru. *Quaternary International* **109**, 3–11.

Dobson A and Foufopoulos J (2000) Emerging infectious pathogens of wildlife. *Philosophical Transactions of the Royal Society of London B* **356**, 219–244.

Domning DP, Thomason J, and Corbett DG (2007) Steller's sea cow in the Aleutian Islands. *Marine Mammal Science* **23**, 976–983.

Donaldson TJ and Dulvy NK (2004) Threatened fishes of the world: *Bolbometopon muricatum* (Valenciennes, 1840) (Scaridae). *Environmental Biology of Fishes* **70**, 373.

Donlan CJ, Tershy BR, Keitt BS, Wood B, Sanchez JA, Weinstein A, Croll DA, and Alguilar JL (2000) Island conservation action in northwest Mexico. In Browne DR, Mitchell KL, and Chaney HW, eds, *Proceedings of the Fifth California Islands Symposium*, pp. 330–338. Santa Barbara Museum of Natural History, Santa Barbara.

Donlan J, Greene HW, Berger J, Bock CE, Bock JH, Burney DA, Estes JA, Foreman D, Martin PS, Roemer GW *et al.* (2005) Re-wilding North America. *Nature* **436**, 913–914.

Douglas MSV, Smol JP, Savelle JM, and Blais JM (2004) Prehistoric Inuit whalers affected Arctic freshwater ecosystems. *Proceedings of the National Academy of Sciences USA* **101**, 1613–1617.

Dowler RC, Caroll DS, and Edwards CW (2000) Rediscovery of rodents (Genus *Nesoryzomys*) considered extinct in the Galápagos Islands. *Oryx* **34**, 109–117.

Dransfield J, Flenley JR, King SM, Harkness DD, and Rapu S (1984) A recently extinct palm from Easter Island. *Nature* **312**, 750–752.

Droxler AW, Poore RZ, and Burckle LH (2003) *Earth's Climate and Orbital Eccentricity: the Marine Isotope Stage 11 Question.* Geophysical Monograph 137, pp. 1–240. American Geophysical Union, Washington DC.

Duarte CM, Marba N, and Holmer M (2007) Rapid domestication of marine species. *Science* **316**, 382–383.

Duckworth JW, Salter RE, and Khounboline K (1999) *Wildlife in Lao DPR: 1999 status report.* IUCN, Vientiane.

Duff, R. (1950) *The Moa-hunter Period of Maori Culture.* R.E. Owen, Wellington.

Dulvy NK and Polunin NVC (2004) Using informal knowledge to infer human-induced rarity of a conspicuous reef fish. *Animal Conservation* **7**, 365–374.

Dulvy NK, Metcalfe JD, Glanville J, Pawson MG, and Reynolds JD (2000) Fishery stability, local extinctions and shifts in community structure in skates. *Conservation Biology* **14**, 283–293.

Dulvy NK, Sadovy Y, and Reynolds JD (2003) Extinction vulnerability in marine populations. *Fish and Fisheries* **4**, 25–64.

Dulvy NK, Freckleton RP, and Polunin NVC (2004) Coral reef cascades and the indirect effects of predator removal by exploitation. *Ecology Letters* **7**, 410–416.

Dulvy NK, Baum JK, Clarke S, Compagno LVJ, Cortés E, Domingo A, Fordham S, Fowler S, Francis MP, Gibson C *et al.* (2008) You can swim but you can't hide: the global status and conservation of oceanic pelagic sharks. *Aquatic Conservation—Marine and Freshwater Ecosystems* doi: 10.1002/aqc.975.

Duncan RP and Blackburn TM (2004) Extinction and endemism in the New Zealand avifauna. *Global Ecology and Biogeography* **13**, 509–517.

Duncan RP and Forsyth DM (2006) Modelling population persistence on islands: mammal introductions in the New Zealand archipelago. *Proceedings of the Royal Society of London Series B Biological Sciences* **273**, 2969–2975.

Duncan RP, Blackburn TM, and Worthy TH (2002) Prehistoric bird extinctions and human hunting. *Proceedings of the Royal Society of London Series B Biological Sciences* **269**, 517–521.

Duncan RP, Blackburn TM, and Sol D (2003) The ecology of avian introductions. *Annual Review of Ecology and Systematics* **34**, 71–98.

Dunn RR (2002) On parasites lost. *Wild Earth* **12**, 28–31.

Dunn RR (2005) Modern insect extinctions, the neglected majority. *Conservation Biology* **19**, 1030–1036.

Dunning NP, Rue D, Beach T, Covich A, and Traverse A (1998) Human environmental interactions in a tropical watershed: the palaeoecology of Laguna Tamarindito, El Peten, Guatemala. *Journal of Field Archaeology* **25**, 139–151.

Dynesius M and Nilsson C (1994) Fragmentation and flow regulation of river systems in the northern third of the world. *Science* **266**, 753–762.

Ebenman B and Jonsson T (2005) Using community viability analysis to identify fragile systems and keystone species. *Trends in Ecology & Evolution* **20**, 568–575.

Edgar GJ, Samson CR, and Barrett NS (2005) Species extinction in the marine environment: Tasmania as a regional example of overlooked losses in biodiversity. *Conservation Biology* **19**, 1294–1300.

EJF (2004) *Farming the Sea, Costing the Earth.* Environmental Justice Foundation, London.

Ekdahl EJ, Teranes JL, Guilderson TP, Turton CL, McAndrews JH, Wittkop CA, and Stoermer EF (2004) Prehistorical record of cultural eutrophication from Crawford Lake, Canada. *Geology* **32**, 745–748.

Eldredge N (1979) Cladism and common sense. In Cracraft J and Eldredge N, eds, *Phylogenetic Analysis and Paleontology*, pp. 165–198. Columbia University Press, New York.

Ellis JC (2005) Marine birds on land: a review of plant biomass, species richness, and community composition in seabird colonies. *Plant Ecology* **181**, 227–241.

Ellis JC, Farina JM, and Witman JD (2006) Nutrient transfer from sea to land: the case of gulls and cormorants in the Gulf of Maine. *Journal of Animal Ecology* **75**, 565–574.

Ellsworth JW and McComb BC (2003) Potential effects of passenger pigeon flocks on the structure and composition of presettlement forests of eastern North America. *Conservation Biology* **17**, 1548–1558.

Elton C (1958) *The Ecology of Invasions by Animals and Plants.* Methuen, London.

Emerson KC (1964) *Checklist of the Mallophaga of North America (North of Mexico). Part I. Suborder Ischnocera.* Desert Test Center, Dugway Proving Ground, Dugway, Utah.

Emmons LH (1999) A new genus and species of abrocomid rodent from Peru (Rodentia: Abrocomidae). *American Museum Novitates* **3279**, 1–14.

Emslie SD (1986) The Late Pleistocene (Rancholabrean) avifauna of Little Box Elder Cave, Wyoming. *Contributions to Geology, University of Wyoming* **23(2)**, 63–82.

Eno NC, Clark RA, and Sanderson WG (1997) *Non-Native Marine Species in British Waters: a Review and Directory.* Joint Nature Conservation Committee, Peterborough.

Ericson PGP (1987) Interpretations of archaeological bird remains: a taphonomic approach. *Journal of Archaeological Science* **14**, 65–75.

Ericson PGP and Tyrberg T (2004) The early history of the Swedish avifauna: a review of the subfossil record and early written sources. *Kungliga Vitterhets Historie och Antikvitets Akademiens Handlingar, Antikvariska Serien* **45**, 1–349.

Ericson PGP, Tyrberg T, Kjellberg AS, Jonsson L, and Ullén I (1997) The earliest record of the house sparrow (*Passer domesticus*) in northern Europe. *Journal of Archaeological Science* **24**, 183–191.

Erskine PD, Bergstrom DM, Schmidt S, Stewart GR, Tweedie CE, and Shaw JD (1998) Subantarctic Macquarie Island—a model ecosystem for studying animal-derived nitrogen sources using 15N natural abundance. *Oecologia* **117**, 187–193.

Eshleman JA, Malhi RS, and Smith DG (2003) Mitochondrial DNA studies of Native Americans: conceptions and misconceptions of the population prehistory of the Americas. *Evolutionary Anthropology* **12**, 7–18.

Essington TE, Beaudreau AH, and Wiedenmann J (2006) Fishing through marine food webs. *Proceedings of the National Academy of Sciences USA* **103**, 3171–3175.

Estes JA (1998) Killer whale predation on otters: linking oceanic and nearshore ecosystems. *Science* **282**, 473–476.

Estes JA, Danner EM, Doak DF, Konar B, Springer AM, Steinberg PD, Tinker MT, and Williams TM (2004) Complex trophic interactions in kelp forest ecosystems. *Bulletin of Marine Science* **74**, 621–638.

Etheridge R (1889) Lord Howe Island (General Zoology). *Australian Museum Memoirs* **2**, 2–42.

Etnier DA and Starnes WC (1993) *The Fishes of Tennessee.* University of Tennessee Press, Knoxville.

Everett WT and Anderson DW (1991) *Status and Conservation of the Breeding Seabirds on Offshore Pacific Islands of Baja California and the Gulf of California. Seabird Status and Conservation: a Supplement.* International Council for Bird Preservation Technical Publication no. 11, pp. 115–139. J.P. Croxall, Cambridge.

Faith DP (1992) Conservation evaluation and phylogenetic diversity. *Biological Conservation* **61**, 1–10.

Falla RA (1954) A new rail from cave deposits in the North Island of New Zealand. *Records of the Auckland Institute and Museum* **4**, 241–244.

Fang J and Liu G (1992) Relationship between climate change and the nomad southward migration in eastern Asia. *Climate Change* **20**, 151–169.

Fanta J (2005) Forests and forest environments. In Koster EA, ed., *The Physical Geography of Western Europe*, pp. 331–351. Oxford University Press, Oxford.

FAO (2004a) *Report of the Expert Consultation on Interactions Between Sea Turtles and Fisheries within an Ecosystem Context.* Food and Agriculture Organization, Rome.

FAO (2004b) *The State of World Fisheries and Aquaculture 2004.* Food and Agriculture Organization, Rome.

FAO (2007) *The State of World Fisheries and Aquaculture 2006.* Food and Agriculture Organization, Rome.

FAOSTAT (2004) *Statistical Databases of the Food and Agriculture Organization.* Food and Agriculture Organization, Rome.

FAUNMAP Working Group (1996) Spatial response of mammals to Late Quaternary environmental fluctuations. *Science* **272**, 1601–1606.

Faure M, Guérin C, and Parenti F (1999) The Holocene megafauna from the Toca do Serrote do Artur (São Raimundo Nonato archaeological area, Piauí, Brazil). *Comptes Rendus de l'Académie des Sciences de Paris, Série II, Sciences de la Terre et des Planètes* **329**, 443–448.

Fedorov VB and Stenseth NC (2001) Glacial survival of the Norwegian lemming (*Lemmus lemmus*) in Scandinavia: inference from mitochondrial DNA variation. *Proceedings of the Royal Society of London Series B Biological Sciences* **268**, 809–814.

Feduccia A (1996) *The Origin and Evolution of Birds.* Yale University Press, New Haven.

Feduccia A and McPherson B (1993) A petrel-like bird from the late Eocene of Louisiana: earliest record of the order Procellariiformes. *Proceedings of the Biological Society of Washington* **106**, 749–751.

Ferreira CEL, Gasparini JL, Carvalho-Filho A, and Floeter SR (2005) A recently extinct parrotfish species from Brazil. *Coral Reefs* **24**, 128.

Ferretti DF, Miller JB, White JWC, Etheridge DM, Lassey KR, Lowe DC, MacFarling Meure CM, Dreier MF, Trudinger CM, van Ommen TD, and Langenfelds RL (2005) Unexpected changes to the global methane budget over the past 2000 years. *Science* **309**, 1714–1717.

Ficcarelli G, Coltorti M, Moreno-Espinosa M, Pieruccini PL, Rook L, and Torre D (2003) A model for the Holocene extinction of the mammal megafauna in Ecuador. *Journal of South American Earth Sciences* **15**, 835–845.

Firestone RB, West A, Kennett JP, Becker L, Bunch TE, Revay ZS, Schultz PH, Belgya T, Kennett DJ, Erlandson JM *et al.* (2007) Evidence for an extraterrestrial impact 12,900 years ago that contributed to the megafaunal extinctions and the Younger Dryas cooling. *Proceedings of the National Academy of Sciences USA* **104**, 16016–16021.

Fischer K (1968) Ein flugunfähiger Kranich aus dem Pleistozän von Cuba. *Monatschrift für Ornithologie und Vivariekunde (Ausgabe A "Der Falke")* **15**, 270–271.

Fischer K (1985) Ein albatrosartiger Vogel (*Diomedeoides minimus* nov. gen., nov. sp., Diomedeoididae nov. fam., Procellariiformes) aus dem Mitteloligozän bei Leipzig (DDR). *Mitteilungen aus dem Zoologischen Museum in Berlin* **61** (suppl: Annalen für Ornithologie 9), 113–118.

Fischer K and Stephan B (1971a) Ein flugunfähiger Kranich (*Grus cubensis* n. sp.) aus dem Pleistozän von Kuba: Eine Osteologie der Familie der Kraniche (Gruidae). *Wissenschaftliche Zeitschrift der Humboldt-Universität zu Berlin, Mathematisch-Naturwissenschaftliche Reihe* **20**, 541–592.

Fischer K and Stephan B (1971b) Weitere Vogelreste aus dem Pleistozän der Pio-Domingo-Höhle in Kuba. *Wissenschaftliche Zeitschrift der Humboldt-Universität zu Berlin, Mathemathisch-Naturwissenschaftliche Reihe* **20**, 593–607.

Fisher J (1952) *The Fulmar.* Collins, London.

Fisher J (1966) *The Shell Bird Book.* Ebury Press, London.

Fisher J and Lockley R (1954) *Seabirds: an Introduction to the Natural History of the Seabirds of the North Atlantic.* Collins, London.

Fitzpatrick JW, Lammertink M, Luneau MD, Gallagher TW, Harrison BR, Sparling GM, Rosenberg KV, Rohrbaugh RW, Swarthout ECH, Wrege PH *et al.* (2005) Ivory-billed woodpecker (*Campephilus principalis*) persists in continental North America. *Science* **308**, 1460–1462.

Flannery TF (1994) *The Future Eaters.* Reed Books, Melbourne.

Flannery T (1995a) *Mammals of the South-west Pacific and Moluccan Islands.* Australian Museum/Reed New Holland, Sydney.

Flannery T (1995b) *Mammals of New Guinea.* Australian Museum/Reed New Holland, Sydney.

Flannery TF (2001) *The Eternal Frontier: an Ecological History of North America and its People.* Atlantic Monthly Press, New York.

Flannery TF and Wickler S (1990) Quaternary murids (Rodentia: Muridae) from Buka Island, Papua New Guinea, with descriptions of two new species. *Australian Mammalogy* **13**, 127–139.

Flannery TF and White JP (1991) Animal translocation. Zoogeography of New Ireland mammals. *National Geographic Research and Exploration* **7**, 96–113.

Flannery T and Schouten P (2001) *A Gap in Nature: Discovering the World's Extinct Animals.* William Heinemann, London.

Flannery TF, Bellwood P, White JP, Moore A, Boeadi, and Nitihaminoto G (1995) Fossil marsupials (Macropodidae, Peroryctidae) and other mammals of Holocene age from Halmahera, North Moluccas, Indonesia. *Alcheringa* **19**, 17–25.

Flannery TF, Bellwood P, White JP, Ennis T, Irwin G, Schubert K, and Balasubramaniam S (1999) Mammals from Holocene archaeological deposits on Gebe and Morotai Islands, northern Moluccas, Indonesia. *Australian Mammalogy* **20**, 391–400.

Fleishman E, Thomson JR, MacNally R, Murphy DD, and Fay JP (2005) Using indicator species to predict species richness of multiple taxonomic groups. *Conservation Biology* **19**, 1125–1137.

Flemming C and MacPhee RDE (1999) Redetermination of holotype of *Isolobodon portoricensis* (Rodentia, Capromyidae), with notes on recent mammalian extinctions in Puerto Rico. *American Museum Novitates* **3278**, 1–11.

Florit X, Mourer-Chauviré C, and Alcover JA (1989) Els ocells pleistocenics d'Es Pouas, Eivissa. Nota preliminar. *Butlleti de la Institució Catalana d'Història Natural (Seccio Geol.)* **56**, 35–46.

Foote M (2000) Origination and extinction components of taxonomic diversity: general problems. In Erwin DH and Wing SL, eds, *Deep Time: Paleobiology's Perspective,* pp. 74–102. Allen Press, Lawrence, Kansas.

Foote M and Raup DM (1996) Fossil preservation and the stratigraphic ranges of taxa. *Paleobiology* **22**, 121–140.

Foote M and Sepkoski Jr JJ (1999) Absolute measures of the completeness of the fossil record. *Nature* **398**, 415–417.

Forcada J, Hammond PS, and Aguilar A (1999) Status of the Mediterranean monk seal *Monachus monachus* in the western Sahara and the implications of a mass mortality event. *Marine Ecology Progress Series* **188**, 249–261.

Forsyth DM and Duncan RP (2001) Propagule size and the relative success of exotic ungulate and bird introductions in New Zealand. *American Naturalist* **157**, 583–595.

Fox D (2007) Back to the no-analog future? *Science* **316**, 823–825.

Frank KT, Petrie B, Choi JS, and Leggett WC (2005) Trophic cascades in a formerly cod-dominated ecosystem. *Science* **308**, 1621–1623.

Freeman MC, Irwin ER, Burkhead NM, Freeman BJ, and Bart Jr HL (2005) Status and conservation of the fish fauna of the Alabama River system. In Rinne JN, Hughes RM, and Calamusso B, eds, *Historical Changes in Large River Fish Assemblages of the Americas,* pp. 557–585. Symposium 45. American Fisheries Society, Bethesda.

Freiwald A, Wilson JB, and Henrich R (1999) Grounding Pleistocene icebergs shape recent deep-water coral reefs. *Sedimentary Geology* **125**, 1–8.

Frías AI, Berovides V, and Fernández C (1988) Situación actual de la jutita de la tierra *Capromys sanfelipensis* (Rodentia, Mammalia). *Doñana, Acta Vertebrata* **15**, 252–254.

Fritts TH and Rodda GH (1998) The role of introduced species in the degradation of island ecosystems: a case history of Guam. *Annual Review of Ecology and Systematics* **29**, 113–140.

Fukami T, Wardle DA, Bellingham PJ, Mulder CPH, Towns DP, Yeates GW, Bonner KI, Durrett MS,

Grant-Hoffman MN, and Williamson WM (2006) Above- and below-ground impacts of introduced predators in seabird-dominated island ecosystems. *Ecology Letters* **9**, 1299–1307.

Furness RW (1988) Predation on ground-nesting seabirds by island populations of red deer *Cervus elaphus* and sheep *Ovis*. *Journal of Zoology* **216**, 565–573.

Furness RW and Camphuysen CJ (1997) Seabirds as monitors of the marine environment. *ICES Journal of Marine Science* **54**, 726–737.

Fusco G and Cronk QCB (1995) A new method for evaluating the shape of large phylogenies. *Journal of Theoretical Biology* **175**, 235–243.

Galbreath GJ and Melville RA (2003) *Pseudonovibos spiralis*: epitaph. *Journal of Zoology* **259**, 169–170.

Galbreath GJ, Mordacq JC, and Weiler FH (2006) Genetically solving a zoological mystery: was the kouprey (*Bos sauveli*) a feral hybrid? *Journal of Zoology* **270**, 561–564.

Galbreath R and Brown D (2004) The tale of the lighthouse-keeper's cat: discovery and extinction of the Stephens Island wren (*Traversia lyalli*). *Notornis* **51**, 193–200.

Gallagher T (2005) *The Grail Bird*. Houghton Mifflin, Boston and New York.

Gamble C, Davies W, Pettitt P, Hazelwood L, and Richards M (2005) The archaeological and genetic foundations of the European population during the Late Glacial: implications for 'agricultural thinking'. *Cambridge Archaeological Journal* **15**, 193–223.

García LV, Marañón T, Ojeda F, Clemente L, and Redondo R (2002) Seagull influence on soil properties, chenopod shrub distribution, and leaf nutrient status in semi-arid Mediterranean islands. *Oikos* **98**, 75–86.

Garcia SM and Newton C (1995) *Current Situation, Trends and Prospects in World Capture Fisheries*. Fisheries Technical Paper. Food and Agriculture Organization, Rome.

Garner JT and McGregor SW (2001) Current status of freshwater mussels (Unionidae, Margaritiferidae) in the Muscle Shoals area of Tennessee River in Alabama (Muscle Shoals revisited again). *American Malacological Bulletin* **16**, 155–170.

Garthe S and Scherp B (2003) Utilization of discards and offal from commercial fisheries by seabirds in the Baltic Sea. *ICES Journal of Marine Science* **60**, 980–989.

Gaskell J (2000) *Who Killed the Great Auk*? Oxford University Press, Oxford.

Gasse F (2002) Diatom-inferred salinity and carbonated oxygen isotopes in Holocene water bodies of the western Sahara and Sahel (Africa). *Quaternary Science Reviews* **21**, 737–767.

Gasse F, Barker P, Gell PA, Fritz SC, and Chalié F (1997) Diatom-inferred salinity in palaeolakes: an indirect tracer of climate change. *Quaternary Science Reviews* **16**, 547–563.

Gaston KJ, Blackburn TM, and Goldewijk KK (2003) Habitat conversion and global avian biodiversity loss. *Proceedings of the Royal Society of London Series B Biological Sciences* **270**, 1293–1300.

Gautier A and Muzzolini A (1991) The life and times of the giant buffalo alias *Bubalus/Homoioceras/Pelorovis antiquus* in North Africa. *Archaeozoologia* **4**, 39–92.

Gentry A, Clutton-Brock J, and Groves CP (2004) The naming of wild animal species and their domestic derivatives. *Journal of Archaeological Science* **31**, 645–651.

Geoffroy Saint-Hilaire I (1851) Note sur des ossements et des oeufs trouvés à Madagascar, dans des alluvions modernes, et provenant d'un oiseau gigantesque. *Comptes Rendus Hebdomadaires des Séances de l'Académie des Sciences (Paris)* **32**, 101–107.

Gibson JR (1999) *De bestis marinis*: Steller's sea cow and Russian expansion from Siberia to America, 1741–1768. In Bolkhovitinov NN, ed., *Русская Америка 1799–1867*, pp. 24–44. Institut Vseobshchey Istorii RAN, Moscow.

Giffin JG (2003) *Pu'u Wa'awa'a Biological Assessment*. Report, State of Hawaii, Department of Natural Resources, Division of Forestry and Wildlife.

Gilbert MTP, Jenkins DL, Götherstrom A, Naveran N, Sanchez JJ, Hofreiter M, Thomsen PF, Binladen J, Higham TFG, Yohe II RM *et al.* (2008) DNA from pre-Clovis human coprolites in Oregon, North America. *Science* **320**, 786–789.

Gill BJ (2003) Osteometry and systematics of the extinct New Zealand ravens (Aves: Corvidae: *Corvus*). *Journal of Systematic Palaeontology* **1**, 43–58.

Gillies RR, Brim Box J, Symanzik J, and Rodemaker EJ (2003) Effects of urbanization on the aquatic fauna of the Line Creek watershed, Atlanta—a satellite perspective. *Remote Sensing of Environment* **86**, 411–422.

Gilman E, Ellison J, and Coleman R (2007) Assessment of mangrove response to projected relative sea-level rise and recent historical reconstruction of shoreline position. *Environmental Monitoring and Assessment* **124**, 105–130.

Gilpin ME and Soulé ME (1986) Minimum viable populations: processes of species extinction. In Soulé ME, ed., *Conservation Biology*, pp. 19–34. Sinauer Associates, Sunderland.

Gleason HA (1926) The individualistic concept of the plant association. *Bulletin of the Torrey Botanical Club* **53**, 1–20.

Gnecco C (1999) An archaeological perspective on the Pleistocene/Holocene boundary in northern South America. *Quaternary International* **53**, 3–9.

Goebal T, Waters MR, and O'Rourke DH (2008) The Late Pleistocene dispersal of modern humans in the Americas. *Science* **319**, 1497–1501.

Gog J, Woodroffe R, and Swinton J (2002) Disease in endangered metapopulations: the importance of alternative hosts. *Proceedings of the Royal Society of London Series B Biological Sciences* **269**, 671–676.

Gommery D, Tombomiadana S, Valentin F, Ramanivosoa B, and Bezoma R (2004) New discovery in the northwest of Madagascar and geographical distribution of the different species of *Palaeopropithecus*. *Annales de Paléontologie* **90**, 279–286.

Gompper ME and Williams ES (1998) Parasite conservation and the black-footed ferret recovery program. *Conservation Biology* **12**, 730–732.

Gonzalez S, Kitchener AC, and Lister AM (2000) Survival of the Irish elk into the Holocene. *Nature* **405**, 753–754.

Goodman SM (1994) Description of a new species of subfossil eagle from Madagascar: *Stephanoaetus* (Aves: Falconiformes) from the deposits of Ampasambazimba. *Proceedings of the Biological Society of Washington* **107**, 421–428.

Goodman SM (1996) Description of a new species of subfossil lapwing (Aves, Charadriiformes, Charadriidae, Vanellinae) from Madagascar. *Bulletin du Muséum National d'Histoire Naturelle (C)* **18**, 607–614.

Goodman SM (2000) A description of a new species of *Brachypteracias* (family Brachypteraciidae) from the Holocene of Madagascar. *Ostrich* **71**, 318–322.

Goodman SM and Ravoavy F (1993) Identification of bird subfossils from cave surface deposits at Anjohibe, Madagascar, with a description of a new giant coua (Cuculidae: Couinae). *Proceedings of the Biological Society of Washington* **106**, 24–33.

Goodman SM and Rakotozafy LMA (1995) Evidence for the existence of two species of *Aquila* on Madagascar during the Quaternary. *Geobios* **28**, 241–246.

Goodman SM and Rakotozafy LMA (1997) Subfossil birds from coastal sites in western and southwestern Madagascar: a paleoenvironmental reconstruction. In Goodman SM and Patterson BD, eds, *Natural Change and Human Impact in Madagascar*, pp. 257–279. Smithsonian Institution Press, Washington and London.

Goodman SM, Ganzhorn JU, and Rakotondravony D (2003) Introduction to the mammals. In Goodman SM and Benstead JP, eds, *The Natural History of Madagascar*, pp. 1159–1186. University of Chicago Press, Chicago.

Goodman SM, Rasoloarison RM, and Ganzhorn JU (2004) On the specific identification of subfossil *Cryptoprocta* (Mammalia, Carnivora) from Madagascar. *Zoosystema* **26**, 129–143.

Goodman SM, Vasey N, and Burney DA (2007) Description of a new species of subfossil shrew tenrec (Afrosoricida: Tenrecidae: *Microgale*) from cave deposits in southeastern Madagascar. *Proceedings of the Biological Society of Washington* **120**, 367–376.

Gosden C and Robertson N (1991) Models for Matenkupkum: interpreting a late Pleistocene site from southern New Ireland, Papua New Guinea. In Allen J and Gosden C, eds, *The Report of the Lapita Homeland Project*, pp. 20–45. Occasional Papers no. 20. Department of Prehistory, Research School of Pacific Studies, Australian National University, Canberra.

Gotelli NJ and Entsminger GL (2001) *EcoSim: Null Models Software for Ecology. Version 7.0*. Acquired Intelligence & Kesey-Bear. http://homepages.together.net/~gentsmin/ecosim.htm.

Gove AD, Dunn RR, and Majer JD (2007) A keystone ant species promotes seed dispersal in a "diffuse" mutualism. *Oecologia* **153**, 687–697.

Graf DL and Cummings KS (2007) Review of the systematics and global diversity of freshwater mussel species (Bivalvia: Unionoida). *Journal of Molluscan Studies* **73**, 291–314.

Graham RW (1985) Diversity and community structure of the Late Pleistocene mammal fauna of North America. *Acta Zoologica Fennica* **170**, 181–192.

Graham RW (1986) Response of mammalian communities to environmental changes during the late Quaternary. In Diamond J and Case TJ, eds, *Community Ecology*, pp. 300–313. Harper and Row, New York.

Graham RW (1988) The role of climatic change in the design of biological reserves: the palaeoecological perspective for conservation biology. *Conservation Biology* **2**, 391–394.

Graham RW (1993) Processes of time-averaging in the terrestrial vertebrate record. In Kidwell SM and Behrensmeyer AK, eds, *Taphonomic Approaches to Time Resolution in Fossil Assemblages*, pp. 102–124. The Paleontological Society, Knoxville.

Graham RW (1997) The special response of mammals to Quaternary climate changes. In Huntley B, Cramer W, Morgan AV, Prentice HC, and Allen JRM, eds, *Past and Future Rapid Environmental Change: the Spatial and Evolutionary Responses of Terrestrial Biota*. Springer, Berlin.

Graham RW (2006) Fallacies of the disharmonious index and relevance of Quaternary non-analogue mammal faunas for future environment change. In: Biotic Response to Global Environmental Change: Analogs for the Future of Life on Earth, Philadelphia Annual Meeting, 22–25 October 2006. *Geological Society of America Abstracts with Programs* **38(7)**, 118.

Graham RW and Grimm EC (1990) Effects of global climate change on the patterns of terrestrial biological communities. *Trends in Ecology & Evolution* **5**, 289–292.

Granberry J and Vescelius GS (2004) *Languages of the pre-Columbian Antilles*. University of Alabama Press, Tuscaloosa.

Grattan J (2006) Aspects of Armageddon: an exploration of the role of volcanic eruptions in human history and civilisation. *Quaternary International* **151**, 10–18.

Graves GR and Olson SL (1987) *Chlorostilbon bracei* Lawrence, an extinct species of hummingbird from New Providence Island, Bahamas. *The Auk* **104**, 296–302.

Grayson DK (1984) Nineteenth-century explanations of Pleistocene extinctions: a review and analysis. In Martin PS and Klein RG, eds, *Quaternary Extinctions: a Prehistoric Revolution*, pp. 5–39. University of Arizona Press, Tucson.

Grayson DK (1989) Bone transport, bone destruction, and reverse utility curves. *Journal of Archaeological Science* **16**, 643–652.

Grayson DK (2000) Mammalian responses to middle Holocene climatic change in the Great Basin of the western United States. *Journal of Biogeography* **27**, 181–192.

Grayson DK (2001) The archaeological record of human impacts on animal populations. *Journal of World Prehistory* **15**, 1–68.

Grayson DK (2005) A brief history of Great Basin pikas. *Journal of Biogeography* **32**, 2103–2111.

Grayson DK and Meltzer DJ (2002) Clovis hunting and large mammal extinction: a critical review of the evidence. *Journal of World Prehistory* **16**, 313–359.

Grayson DK and Meltzer DJ (2003) A requiem for North American overkill. *Journal of Archaeological Science* **30**, 585–593.

Grayson DK and Meltzer DJ (2004) North American overkill continued? *Journal of Archaeological Science* **31**, 133–136.

Green EP, Mumby PJ, Edwards AJ, and Clark CD (1996) A review of remote sensing for the assessment and management of tropical coastal resources. *Coastal Management* **24**, 1–40.

Greenway JC (1967) *Extinct and Vanishing Birds of the World*. American Committee for International Wild Life Protection, Special Publication no 13, 2nd edn. Dover Publications, New York.

Greuter W (1995) Extinctions in Mediterranean areas. In Lawton JH and May RM, eds, *Extinction Rates*, pp. 88–97. Oxford University Press, Oxford.

Grieve S (1885) *The Great Auk, or Garefowl (Alca impennis Linn.) its History, Archaeology and Remains*. Th. C. Jack, London.

Groves CP and Flannery TF (1994) A revision of the genus *Uromys* Peters, 1867 (Muridae: Mammalia) with descriptions of two new species. *Records of the Australian Museum* **46**, 145–169.

Guiler ER (1985) *Thylacine: the Tragedy of the Tasmanian Tiger*. Oxford University Press, Melbourne.

Guiot J, de Beaulueu JL, Cheddadi R, David F, Ponel P, and Reille M (1993) The climate in Western Europe during the last Glacial/Interglacial cycle derived from pollen and insect remains. *Palaeogeography, Palaeoclimatology, Palaeoecology* **103**, 73–93.

Gulland JA (1974) Antarctic whaling. In Gulland JA, ed., *The Management of Marine Fisheries*, pp. 10–37. Scientechnica, Bristol.

Gummer H (2003) *Chick Translocation as a Method of Establishing New Surface-nesting Seabird Colonies: a Review*. Internal Series 150. Department of Conservation Science, Wellington.

Gunn JD, Folan WJ, and Robichaux HR (1995) A landscape analysis of the Candelaria watershed in Mexico: insights into paleoclimates affecting upland horticulture in the southern Yucatán peninsula semi-karst. *Geoarchaeology* **10**, 3–42.

Günther A and Newton E (1879) The extinct birds of Rodriguez. *Philosophical Transactions of the Royal Society of London* **168** (extra vol.), 423–437.

Guthrie DA (1980) Analysis of avifaunal and bat remains from midden sites on San Miguel Island. In Power DM, ed., *The California Islands: Proceedings of a Multidisciplinary Symposium*, pp. 689–702. Santa Barbara Museum of Natural History, Santa Barbara.

Guthrie DA (1993) New information on the prehistoric fauna of San Miguel Island, California. In Hochberg FG, ed., *Third California Islands Symposium: Recent Advances in Research in the California Islands*, pp. 405–416. Santa Barbara Museum of Natural History, Santa Barbara.

Guthrie DA (2005) Distribution and provenance of fossil avifauna on San Miguel Island. In Garcelon DK and Schwemm CA, eds, *Proceedings of the Sixth California Islands Symposium*, NPS Technical Publication CHIS-05-01, pp. 35–42.

Guthrie DA, Thomas HW, and Kennedy GL (2002) A new species of extinct Late Pleistocene puffin (Aves: Alcidae) from the southern California Channel Islands. In Browne DR, Mitchell KL, and Chaney HW, eds, *Proceedings of the Fifth California Islands Symposium*, pp. 525–530.

Guthrie RD (1990a) Late Pleistocene faunal revolution – a new perspective on the extinction debate. In Agenbroad LD, Mead JI, and Nelson LW, eds, *Megafauna and Man: Discovery of America's Heartland*, pp. 42–53. North Arizona Press, Flagstaff.

Guthrie RD (1990b) *Frozen Fauna of the Mammoth Steppe: the Story of Blue Babe.* The University of Chicago Press, London.

Guthrie RD (2003) Rapid body size decline in Alaskan Pleistocene horses before extinction. *Nature* **426**, 169–171.

Guthrie RD (2004) Radiocarbon evidence of mid-Holocene mammoths stranded on an Alaskan Bering Sea island. *Nature* **429**, 746–749.

Haag WR (2002) *Spatial, Temporal, and Taxonomic Variation in Population Dynamics and Community Structure of Freshwater Mussels.* PhD thesis, University of Mississippi.

Haag WR (2004a) Flat pigtoe *Pleurobema marshalli.* In Mirarchi R, Garner JT, Mettee MF, and O'Neil PE, eds, *Alabama Wildlife. Volume 2. Imperiled Aquatic Mollusks and Fishes,* p. 30. The University of Alabama Press, Tuscaloosa.

Haag WR (2004b) Black clubshell *Pleurobema curtum.* In Mirarchi R, Garner JT, Mettee MF, and O'Neil PE, eds, *Alabama Wildlife. Volume 2. Imperiled Aquatic Mollusks and Fishes,* p. 29. The University of Alabama Press, Tuscaloosa.

Haag WR (2004c) Alabama spike *Elliptio arca.* In Mirarchi R, Garner JT, Mettee MF, and O'Neil PE, eds, *Alabama Wildlife. Volume 2. Imperiled Aquatic Mollusks and Fishes,* p. 44. The University of Alabama Press, Tuscaloosa.

Haag WR and Warren Jr ML (1997) Host fishes and reproductive biology of 6 freshwater mussel species from the Mobile Basin, USA. *Journal of the North American Benthological Society* **16**, 576–585.

Haag WR and Warren Jr ML (2003) Host fishes and infection strategies of freshwater mussels in large Mobile Basin streams, USA. *Journal of the North American Benthological Society* **22**, 78–91.

Haag WR and Warren Jr ML (2008) Effects of severe drought on freshwater mussel assemblages. *Transactions of the American Fisheries Society* **137**, 1165–1178.

Haase M and Bisenberger A (2003) Allozymic differentiation in the land snail *Arianta arbustorum* (Stylommatophora, Helicidae): historical inferences. *Journal of Zoological Systematics and Evolutionary Research* **41**, 175–185.

Hachisuka M (1953) *The Dodo and Kindred Birds.* Witherby, London.

Haffer J (1969) Speciation in Amazonian forest birds. *Science* **165**, 131–137.

Hall P and Wang JZ (1999) Estimating the end point of a probability distribution using minimum-distance methods. *Bernoulli* **5**, 177–189.

Hall-Spencer J, Allain V, and Fossa JH (2002) Trawling damage to northeast Atlantic ancient coral reefs. *Proceedings of the Royal Society of London Series B Biological Sciences* **269**, 507–511.

Hambler C, Hambler H, and Ewing J (1985) Some observations on *Nesillas aldabranus,* the endangered brush warbler of Aldabra Atoll, with hypotheses on its distribution. *Atoll Research Bulletin* **290**, 1–19.

Hamel PB (1986) *Bachman's Warbler: a Species in Peril.* Smithsonian Institution Press, Washington DC.

Hamilton AC, and Taylor DM (1986) Mire sediments in East Africa. In Frostick LE, Renault RW, Reid I, and Tiercelin JJ, eds, *Sedimentation in the African Rifts,* pp. 211–217. Special Publication 25. Geological Society, London.

Hamilton R (2003) The role of indigenous knowledge in depleting a limited resource—a case study of the bumphead parrotfish (*Bolbometopon muricatum*) artisanal fishery in Roviana Lagoon, western Province, Solomon Islands. In Haggan N, Brignall C, and Wood L, eds, *Putting Fishers' Knowledge to Work,* pp. 68–77. Fisheries Centre Research Reports. University of British Columbia, Vancouver.

Hannah L, Midgley GF, and Miller D (2002) Climate change-integrated conservation strategies. *Global Ecology and Biogeography* **11**, 485–495.

Hansen JE (2007) Scientific reticence and sea level rise. *Environmental Research Letters* **2**, 024002.

Hanski I (1998) Metapopulation dynamics. *Nature* **396**, 41–49.

Hanski I and Ovaskainen O (2002) Extinction debt at extinction threshold. *Conservation Biology* **16**, 666–673.

Harding J, Hawke D, Holdaway RN, and Winterbourn MJ (2004) Incorporation of marine-derived nutrients from petrel breeding colonies into stream food webs. *Freshwater Biology* **49**, 576–586.

Harrison CJO and Cowles GS (1977) The extinct large cranes of the north-west Palearctic. *Journal of Archaeological Science* **4**, 25–27.

Harrison CJO and Walker CA (1978) Pleistocene bird remains from Aldabra Atoll, Indian Ocean. *Journal of Natural History* **12**, 7–14.

Harrow G, Hawke DJ, and Holdaway RN (2006) Surface soil chemistry at an alpine procellariid breeding colony in New Zealand, and comparison with a lowland site. *New Zealand Journal of Zoology* **33**, 165–174.

Hartfield PD and Rummel RG (1985) Freshwater mussels (Unionidae) of the Big Black River, Mississippi. *The Nautilus* **99**, 116–119.

Hartwig WC and Cartelle C (1996) A complete skeleton of the giant South American primate *Protopithecus. Nature* **381**, 307–311.

Hasegawa H (1984) Status and conservation of seabirds in Japan, with special attention to the short-tailed albatross. In Croxall JP, Evans PGH, and Schreiber RW, eds, *Status and Conservation of the World's Seabirds*, pp. 487–500. Technical Publication 2. International Council for Bird Preservation, Cambridge.

Hasegawa H (1991) The status of seabirds and the assessment on the influence of feral goats. In *Reports on the Influence of the Feral Goats on Native Animals and Plants in the Bonin Islands*, pp. 85–100. Japan Wildlife Research Center, Tokyo.

Hasegawa H and DeGange AR (1982) The short-tailed albatross, *Diomedea albatrus*, its status, distribution and natural history. *American Birds* **36**, 806–814.

Hassan FA (2000) Environmental perception and human responses: history and prehistory. In McIntosh RJ, Tainter JA, and McIntosh SK, eds, *The Way the Wind Blows: Climate, History and Human Action*, pp. 121–140. Columbia University Press, New York.

Hassanin A and Ropiquet A (2007) Resolving a zoological mystery: the kouprey is a real species. *Proceedings of the Royal Society of London Series B Biological Sciences* **274**, 2849–2856.

Haviser J (1997) Settlement strategies in the Early Ceramic Age. In Wilson SM, ed., *The Indigenous People of the Caribbean*, pp. 59–69. University of Florida Press, Gainesville.

Havlik ME and Stansbery DH (1978) The naiad mollusks of the Mississippi River in the vicinity of Prairie Du Chien, Wisconsin. *Bulletin of the American Malacological Union, Inc.* **1977**, 9–12.

Hawkins JP, Roberts CM, and Clark V (2000) The threatened status of restricted-range coral reef fish species. *Animal Conservation* **3**, 81–88.

Hay JM, Subramanian S, Millar CD, Mohandesan E, and Lambert DM (2008) Rapid molecular evolution in a living fossil. *Trends in Genetics* **24**, 106–109.

Haynes G (2002) The catastrophic extinction of North American mammoths and mastodonts. *World Archaeology* **33**, 391–416.

Head MJ (1998) Marine environmental change in the Pliocene and early Pleistocene of eastern England: the dinoflagellate evidence reviewed. *Mededelingen Nederlands Instituut voor Toegepaste Geowetenschappen TNO* **60**, 199–226.

Headland TN and Reid L (1989) Hunter-gatherers and their neighbours from prehistory to the present. *Current Anthropology* **30**, 43–66.

Heard SB and Mooers AØ (2000) Phylogenetically patterned speciation rates and extinction risks change the loss of evolutionary history during extinctions. *Proceedings of the Royal Society of London Series B Biological Sciences* **267**, 613–620.

Hearty PJ, James HF, and Olson SL (2005) The geological context of Middle Pleistocene crater lake deposits and fossil birds at Ulupau, Oahu, Hawaiian Islands. *Monografies de la Societat d'Història Natural de les Balears* **12**, 113–128.

Heath TA, Zwickl DJ, Kim J, and Hillis DM (2008) Taxon sampling affects inferences of macroevolutionary processes from phylogenetic trees. *Systematic Biology* **57**, 160–166.

Heberle G (2004) Reports of alleged thylacine sightings in Western Australia. *Conservation Science of Western Australia* **5**, 1–5.

Hebert PDN, Cywinska A, Ball SL, and deWaard JR (2003) Biological identifications through DNA barcodes. *Proceedings of the Royal Society of London Series B Biological Sciences* **270**, 313–321.

Heiri O, Tinner W, and Lotter A (2004) Evidence for cooler European summers during periods of changing meltwater flux to the North Atlantic. *Proceedings of the National Academy of Sciences* **101**, 15285–15288.

Heithaus MR, Frid A, Wirsing AJ, and Worm B (2008) Predicting ecological consequences of marine top predator declines. *Trends in Ecology & Evolution* **23**, 202–210.

Helgen KM, Helgen LE, and Wilson DE (2009) Pacific flying foxes (Mammalia: Chroptera): two new species of *Pteropus* from Samoa, probably extinct. *American Museum Novitates* (in press).

Hershkovitz P (1998) Report on some sigmodontine rodents collected in southeastern Brazil with descriptions of a new genus and six species. *Bonner zoologische Beiträge* **47**, 193–256.

Hetherington DA, Lord TC, and Jacobi RM (2006) New evidence for the occurrence of Eurasian lynx (*Lynx lynx*) in medieval Britain. *Journal of Quaternary Science* **21**, 3–8.

Hewitt G (1996) Some genetic consequences of ice ages, and their role in divergence and speciation. *Biological Journal of the Linnean Society* **58**, 247–276.

Hewitt GM (1999) Post-glacial recolonization of European biota. *Biological Journal of the Linnean Society* **68**, 87–112.

Hewitt G (2000) The genetic legacy of the Quaternary ice ages. *Nature* **405**, 907–913.

Heyning JE and Thacker C (1999) Phylogenies, temporal data, and negative evidence. *Science* **285**, 1179.

Hilborn R, Quinn TP, Schindler DE, and Rogers DE (2003) Biocomplexity and fisheries sustainability. *Proceedings of the National Academy of Sciences USA* **100**, 6564–6568.

Hiller A, Wand U, Kämpf H, and Stackebrandt W (1988) Occupation of the Antarctic continent by petrels during the past 35 000 years: inferences from a 14C study of stomach oil deposits. *Polar Biology* **9**, 69–77.

Hobbs RJ and Mooney HA (1998) Broadening the extinction debate: population deletions and additions in California and western Australia. *Conservation Biology* **12**, 271–283.

Hobson KA, Drever MC, and Kaiser GW (1999) Norway rats as predators of burrow-nesting seabirds: insights from stable isotope analyses. *Journal of Wildlife Management* **63**, 14–25.

Hoch E (1977) Reflections on prehistoric life at Umm-an-Nar (Trucial Oman) based on faunal remains from the Third Millenium BC. *South Asian Archaeology* **1977**, 589–638.

Hochberg ME, Thomas JA, and Elmes GW (1992) A modelling study of the population dynamics of a large blue butterfly, *Maculinea rebeli*, a parasite of red ant nests. *Journal of Animal Ecology* **61**, 397–409.

Hodder KH and Bullock JM (1997) Translocations of native species in the UK: implications for biodiversity. *Journal of Applied Ecology* **34**, 547–565.

Hoegh-Guldburg O (1999) Climate change, coral bleaching and the future of the world's coral reefs. *Marine and Freshwater Research* **50**, 839–866.

Hoegh-Guldberg O, Mumby PJ, Hooten AJ, Steneck RS, Greenfield P, Gomez E, Harvell CD, Sale PF, Edwards AJ, Caldeira K *et al.* (2007) Coral reefs under rapid climate change and ocean acidification. *Science* **318**, 1737–1742.

Hoelzmann P, Keding B, Berke H, Kröpelin S, and Kruse H-J (2001) Environmental change and archaeology: lake evolution and human occupation in the Eastern Sahara during the Holocene. *Palaeogeography, Palaeoclimatology, Palaeoecology* **169**, 193–217.

Hoffmann RC (1996) Economic development and aquatic ecosystems in medieval Europe. *American Historical Review* **101**, 631–669.

Hogg EH and Morton JK (1983) The effects of nesting gulls on the vegetation and soil of islands in the Great Lakes. *Canadian Journal of Botany* **61**, 3240–3254.

Hoggarth MA, Rice DL, and Lee DM (1995) Discovery of the federally endangered freshwater mussel, *Epioblasma obliquata obliquata* (Rafinesque, 1820) (Unionidae), in Ohio. *Ohio Journal of Science* **95**, 298–299.

Holdaway RN (1989) New Zealand's pre-human avifauna and its vulnerability. *New Zealand Journal of Ecology* **12** (suppl), 11–25.

Holdaway RN (1996) Arrival of rats in New Zealand. *Nature* **384**, 225–226.

Holdaway RN (1999a) A spatio-temporal model for the invasion of the New Zealand archipelago by the Pacific rat *Rattus exulans*. *Journal of the Royal Society of New Zealand* **29**, 91–105.

Holdaway RN (1999b) Introduced predators and avifaunal extinction in New Zealand. In MacPhee RDE, ed., *Extinctions in Near Time: Causes, Contexts, and Consequences*, pp. 189–238. Kluwer Academic/Plenum, New York.

Holdaway RN and Worthy TH (1994) A new fossil species of shearwater *Puffinus* from the late Quaternary of the South Island, New Zealand, and notes on the biogeography and evolution of the *Puffinus gavia* superspecies. *Emu* **94**, 201–215.

Holdaway RN and Worthy TH (1997) A reappraisal of the late Quaternary fossil vertebrates of Pyramid Valley Swamp, North Canterbury, New Zealand. *New Zealand Journal of Zoology* **24**, 69–121.

Holdaway RN and Jacomb C (2000) Rapid extinction of the moas (Aves: Dinornithiformes): model, test and implications. *Science* **287**, 2250–2254.

Holdaway RN and Anderson AJ (2001) Avifauna from the Emily Bay settlement site, Norfolk Island: a preliminary account. *Records of the Australian Museum* suppl 27, 85–100.

Holdaway RN, Jones MD, and Beavan Athfield NR (2002a) Late Holocene extinction of the New Zealand owlet-nightjar *Aegotheles novaezealandiae*. *Journal of The Royal Society of New Zealand* **32**, 653–667.

Holdaway RN, Jones MD, and Beavan Athfield NR (2002b) Late Holocene extinction of Finsch's duck (*Chenonetta finschi*), an endemic, possibly flightless, New Zealand duck. *Journal of the Royal Society of New Zealand* **32**, 629–651.

Holdaway RN, Hawke DJ, Hyatt OM, and Wood GC (2007) Stable isotopic (δ15N, δ13C) analysis of wood in trees growing in past and present colonies of burrow-nesting seabirds in New Zealand. I. δ15N in two species of conifer (Podocarpaceae) from a mainland colony of Westland petrels (*Procellaria westlandica*), Punakaiki, South Island. *Journal of the Royal Society of New Zealand* **37**, 75–84.

Holden MJ (1992) *The Common Fisheries Policy*. Fishing News Books, London.

Holder K, Montgomerie R, and Friesen VL (1999) A test of the glacial refugium hypothesis using patterns of mitochondrial and nuclear DNA sequence variation in rock ptarmigan (*Lagopus mutus*). *Evolution* **53**, 1936–1950.

Honea K (1975) Prehistoric remains on the island of Kythnos. *American Journal of Archaeology* **79**, 277–279.

Hoogland MLP (1996) *In Search of the Native Population of Pre-Columbian Saba (400–1450 AD). Part Two: Settlements in their Natural and Social Environment*. PhD thesis, University of Leiden.

Hooijer DA (1963) Mammalian remains from an Indian site on Curaçao. *Studies on the Fauna of Curaçao and other Caribbean Islands* **14**, 119–122.

Hooijer DA (1965) Note on *Coryphomys bühleri* Schaub, a gigantic murine rodent from Timor. *Israel Journal of Zoology* **14**, 128–133.

Hooijer DA (1966) Fossil mammals of the Netherlands Antilles. *Archives Néerlandaises de Zoologie* **16**, 531–532.

Hooper DU, Chapin FS III, Ewel JJ, Hector A, Inchausti P, Lavorel S, Lawton JH, Lodge DM, Loreau M, Naeem S *et al.* (2005) Effects of biodiversity on ecosystem functioning: a consensus of current knowledge. *Ecological Monographs* **75**, 3–35.

Hope GS, Flannery TF, and Boeadi (1993) A preliminary report of changing Quaternary mammal faunas in subalpine New Guinea. *Quaternary Research* **40**, 117–126.

Hope JH (1981) A new species of *Thylogale* (Marsupialia: Macropodidae) from Mapala rock shelter, Jaya (Carstenz) Mountains, Irian Jaya (western New Guinea), Indonesia. *Records of the Australian Museum* **33**, 369–387.

Hope S (2002) The Mesozoic record of Neornithes (modern birds). In Chiappe LM and Witmer L, eds, *Above the Heads of the Dinosaurs*, pp. 339–388. University of California Press, Berkeley.

Horst GR, Hoagland DB, and Kilpatrick CW (2001) The mongoose in the West Indies: the biogeography and population biology of an introduced species. In Woods CA and Sergile FE, eds, *Biogeography of the West Indies: Patterns and Perspectives*, pp. 409–424. CRC Press, Boca Raton.

Howard H (1935) The Rancho La Brea wood ibis. *The Condor* **37**, 251–253.

Howard H (1964) A fossil owl from Santa Rosa Island, California, with comments on the eared owls of Rancho La Brea. *Bulletin of the Southern California Academy of Sciences* **63**, 27–31.

Howard H (1971) Tertiary birds from Laguna Hills, Orange County, California. *Los Angeles County Museum Contributions to Science* **142**, 1–21.

Howe HF (1985) Gomphothere fruits: a critique. *American Naturalist* **125**, 853–865.

Howell CJ, Kelly D, and Turnbull MH (2002) Moa ghosts exorcised? New Zealand's divaricate shrubs avoid photoinhibition. *Functional Ecology* **16**, 232–240.

Howells RG, Neck RW, and Murray HD (1996) *Freshwater Mussels of Texas*. Texas Parks and Wildlife Department, Inland Fisheries Division, Austin.

Hsu KJ (2000) *Climate and Peoples: a Theory of History*. Orell Fussli, Zurich.

Hubbe A, Hubbe M, and Neves W (2007) Early Holocene survival of megafauna in South America. *Journal of Biogeography* **34**, 1642–1646.

Hudson P, Rizzoli A, Grenfell B, Heesterbeek H, and Dobson A (2002) *The Ecology of Wildlife Diseases*. Oxford University Press, Oxford.

Hughes MH and Parmalee PW (1999) Pre-Columbian and modern freshwater mussel (Mollusca: Bivalvia: Unionoidea) faunas of the Tennessee River: Alabama, Kentucky, and Tennessee. *Regulated Rivers: Research and Management* **15**, 25–42.

Hume JP (2005) Contrasting taphofacies in ocean island settings: the fossil record of Mascarene vertebrates. *Monografies de la Societat d'Història Natural de les Balears* **12**, 129–144.

Hume JP (2007) Reappraisal of the parrots (Aves: Psittacidae) from the Mascarene Islands, with comments on their ecology, morphology, and affinities. *Zootaxa* **1513**, 1–76.

Hume JP and Cheke AS (2004) The white dodo of Réunion Island: unravelling a scientific and historical myth. *Archives of Natural History* **31**, 57–79.

Hume JP and Prŷs-Jones RP (2005) New discoveries from old sources, with reference to the original bird and mammal fauna of the Mascarene Islands, Indian Ocean. *Zoologische Mededelingen* **79**, 85–95.

Hunt TL (2007) Rethinking Easter Island's ecological catastrophe. *Journal of Archaeological Science* **34**, 485–502.

Hunt TL and Lipo CP (2006) Late colonization of Easter Island. *Science* **311**, 1603–1606.

Hunter S and Brooke M de L (1992) The diet of giant petrels *Macronectes* spp. at Marion Island, Southern Indian Ocean. *Colonial Waterbirds* **15**, 56–65.

Huntington CE, Butler RG, and Mauck RA (1996) Leach's storm-petrel (*Oceanodroma leucorhoa*). In Poole A and Gill F, eds, *The Birds of North America*, No. 233. The Academy of Natural Sciences, Philadelphia, and The American Ornithologists' Union, Washington DC.

Huntley B (1991) How plants respond to climate change: Migration rates, individualism and the consequences for plant communities. *Annals of Botany* **67** (suppl 1), 15–22.

Huntley B, Cramer W, Morgan AV, Prentice HC, and Allen JRM (1997) *Past and Future Rapid Environmental Change: the Spatial and Evolutionary Responses of Terrestrial Biota*. Springer, Berlin.

Hutchings JA (1996) Spatial and temporal variation in the density of northern cod and a review of the hypotheses for the stock collapse. *Canadian Journal of Fisheries and Aquatic Sciences* **53**, 943–962.

Hutchings JA (2000) Collapse and recovery of marine fishes. *Nature* **406**, 882–885.

Hutchings JA (2005) Life history consequences of overexploitation to population recovery in northwest Atlantic cod (*Gadus morhua*). *Canadian Journal of Fisheries and Aquatic Sciences* **62**, 824–832.

Hutchings JA and Reynolds JD (2004) Marine fish population collapses: consequences for recovery and extinction risk. *BioScience* **54**, 297–309.

Hutchings JA and Baum JK (2005) Measuring fish biodiversity: temporal changes in abundance, life history and demography. *Philosophical Transactions of the Royal Society, B* **360**, 315–338.

Hutterer R (1994) Shrews of ancient Egypt: biogeographical interpretation of a new species. In Merritt J, Kirkland Jr GL, and Rose RK, eds, *Advances in the Biology of Shrews*, pp. 407–414. Special Publication 18. Carnegie Museum of Natural History, Pittsburgh.

Hutterer R, Lopez-Martínez N, and Michauz J (1988) A new rodent from Quaternary deposits of the Canary Islands and its relationships with Neogene and Recent murids of Europe and Africa. *Palaeovertebrata* **18**, 241–262.

Huynen L, Millar CD, Scofield RP, and Lambert DM (2003) Nuclear DNA sequences detect species limits in ancient moa. *Nature* **425**, 175–178.

ICES (2003) *Report of the Working Group on the Biology and Assessment of Deep-sea Fisheries Resources*. International Council for the Exploration of the Sea, Copenhagen.

ICES (2005) *Guidance on the Application of the Ecosystem Approach to Management of Human Activities in the European Marine Environment*. ICES Cooperative Research Report, Copenhagen.

Igual JM, Forero MG, Gomez T, Orueta JF, and Oro D (2006) Rat control and breeding performance in Cory's shearwater (*Calonectris diomedea*): effects of poisoning effort and habitat features. *Animal Conservation* **9**, 59–65.

Imber MJ (1994) Seabirds recorded at the Chatham Islands, 1960 to May 1993. *Notornis* **41** (suppl), 97–108.

Imbrie J and Kipp NG (1971) A new micropalaeoentological method for quantitative paleoclimatology: application to late Pleistocene Caribbean core V28–238. In Turekian KK, ed., *The Late Cenozoic Glacial Ages*, pp. 77–181. Yale University Press, New Haven.

Imbrie J, Berger A, Boyle EA, Clemens SC, Duffy A, Howard WR, Kukla G, Kutzbach J, Martinson DG, McIntyre A *et al.* (1992) On the structure and origin of major glaciation cycles. 1: Linear responses to Milankovitch forcing. *Paleoceanography* **7**, 701–738.

Inomata T (1997) The last day of a fortified Classic Maya centre: archaeological investigations at Aguateca, Guatemala. *Ancient Mesoamerica* **8**, 337–351.

IPCC (2007) *Climate Change 2007: the Physical Science Basis. Contribution of Working Group I to the Fourth Assessment Report of the Intergovernmental Panel on Climate Change*, Solomon S, Qin D, Manning M, Chen Z, Marquis M, Averyt KB, Tignor M, and Miller HL, eds. Cambridge University Press, Cambridge and New York.

Irwin DE, Bensch S, and Price TD (2001) Speciation in a ring. *Nature* **409**, 333–337.

Isaac NJB, Mallet J, and Mace GM (2004) Taxonomic inflation: its influence on macroecology and conservation. *Trends in Ecology & Evolution* **19**, 464–469.

Isaac NJB, Turvey ST, Collen B, Waterman C, and Baillie JEM (2007) Mammals on the EDGE: conservation priorities based on threat and phylogeny. *PLoS One* **2(3)**, e29.

Islam MA and Macdonald SE (2005) Effects of variable nitrogen fertilization on growth, gas exchange and biomass partitioning in black spruce and tamarack seedlings. *Canadian Journal of Botany* **83**, 1574–1580.

Isom BG and Yokley Jr P (1968) The mussel fauna of Duck River in Tennessee, 1965. *American Midland Naturalist* **80**, 34–42.

Iturralde-Vinent M, MacPhee RDE, Díaz-Franco S, Rojas-Consuegra R, Suárez W, and Lomba A (2000) Las Breas de San Felipe, a Quaternary fossiliferous asphalt seep near Martí (Matanzas Province, Cuba). *Caribbean Journal of Science* **36**, 300–313.

IUCN (2001) *IUCN Red List Categories and Criteria. Version 3.1.* World Conservation Union, Gland, Switzerland.

IUCN (2006) *2006 IUCN Red List of Threatened Species.* www.iucnredlist.org.

IUCN (2007) *2007 IUCN Red List of Threatened Species.* www.iucnredlist.org.

Jaarola M and Searle JB (2003) Phylogeography of field voles (*Microtus agrestis*) in Eurasia inferred from mitochondrial DNA sequences. *Molecular Ecology* **11**, 2613–2621.

Jackson JA (2004) *In Search of the Ivory-billed Woodpecker.* Smithsonian Books, Washington DC.

Jackson JBC (1997) Reefs since Columbus. *Coral Reefs* **16**, S23–S32.

Jackson JBC, Kirby MX, Bergoer WH, Bjorndal KA, Botsford LW, Bourque BJ, Bradbury RH, Cooke R, Erlandson J, Estes JA *et al.* (2001) Historical overfishing and the recent collapse of coastal ecosystems. *Science* **293**, 629–637.

Jackson ST and Overpeck JT (2000) Responses of plant populations and communities to environmental changes of the late Quaternary. *Paleobiology* **26**, 194–220.

Jackson ST and Weng C (1999) Late Quaternary extinction of a tree species in eastern North America. *Proceedings of the National Academy of Sciences USA* **96**, 13847–13852.

Jackson ST and Williams JW (2004) Modern analogs in Quaternary palaeoecology: here today, gone yesterday, gone tomorrow? *Annual Review of Earth and Planetary Sciences* **32**, 495–537.

Jacobson Jr GL, Webb III T, and Grimm EC (1987) Patterns and rates of vegetation change during deglaciation of eastern North America. In Ruddiman WF and Wright Jr HE, eds, *North America and Adjacent Oceans During the Last Deglaciation*, pp. 277–288. Geological Society of America, Boulder.

Jahncke J, Checkley DM, and Hunt GL (2004) Trends in carbon flux to seabirds in the Peruvian upwelling system: effects of wind and fisheries on population regulation. *Fisheries Oceanography* **13**, 208–223.

James HF (1987) A Late Pleistocene avifauna from the Island of Oahu (Hawaiian Islands). In Mourer-Chauviré C, ed., *L'évolution des oiseaux d'après le témoignage des fossiles. Documents des Laboratoires de Geologie, Lyon* **99**, 221–230.

James HF (2004) The osteology and phylogeny of the Hawaiian finch radiation (Fringillidae: Drepanidini), including extinct taxa. *Zoological Journal of the Linnean Society* **141**, 207–255.

James HF and Olson SL (1991) Descriptions of thirty-two new species of birds from the Hawaiian Islands: Part II. Passeriformes. *Ornithological Monographs* **46**, 1–88.

James HF and Olson SL (2003) A giant new species of nukupuu (Fringillidae: Drepanidini, *Hemignathus*) from the island of Hawaii. *The Auk* **120**, 970–981.

James HF and Olson SL (2005) The diversity and biogeography of koa-finches (Drepanidini: *Rhodacanthis*), with descriptions of two new species. *Zoological Journal of the Linnean Society* **144**, 527–541.

James HF and Olson SL (2006) A new species of Hawaiian finch (Drepanidini: *Loxioides*) from Makauwahi Cave, Kauaì. *The Auk* **123**, 335–344.

James HF, Stafford Jr TW, Steadman DW, Olson SL, Martin PS, Jull AJT, and McCoy PC (1987) Radiocarbon dates on bones of extinct birds from Hawaii. *Proceedings of the National Academy of Sciences USA* **84**, 2350–2354.

James HF, Zusi RL, and Olson SL (1989) *Dysmodrepanis munroi* (Fringillidae, Drepanidini), a valid genus and species of Hawaiian finch. *Wilson Bulletin* **101**, 159–179.

Janetski JC (1997) Fremont hunting and resource intensification in the eastern Great Basin. *Journal of Archaeological Science* **24**, 1075–1089.

Janoo A (2005) Discovery of isolated dodo bones [*Raphus cucullatus* (L.), Aves, Columbiformes] from Mauritius cave shelters highlights human predation, with a comment on the status of the family Raphidae Wetmore, 1930. *Annales de Paléontologie* **91**, 167–180.

Janzen DH and Martin PS (1982) Neotropical anachronisms: the fruits the gomphotheres ate. *Science* **215**, 19–27.

Jaramillo-Legorreta A, Rojas-Bracho L, Brownell RL, Read AJ, Reeves RR, Ralls K, and Taylor BL (2007) Saving the vaquita: immediate action, not more data. *Conservation Biology* **21**, 1653–1655.

Jaume D, McMinn M, and Alcover JA (1993) Fossil birds from the Bujero del Silo, La Gomera (Canary Islands) with a description of a new species of quail (Galliformes: Phasianidae). *Boletim do Museu Municipal do Funchal* Suplemento 2, 147–165.

Jefferies RL, Henry HAL, and Abraham KF (2004) Agricultural nutrient subsidies to migratory geese and change in Arctic coastal habitats. In Polis GA, Power ME, and Huxel GR, eds, *Food Webs at the Landscape Level*, pp. 268–283. University of Chicago Press, Chicago.

Jehl JR and Parkes KC (1982) The status of the avifauna of the Revillagigedo Islands, Mexico. *Wilson Bulletin* **94**, 1–19.

Jenkins RE and Burkhead NM (1993) *Freshwater Fishes of Virginia*. American Fisheries Society, Bethesda.

Jennings S (2007) Reporting and advising on the effects of fishing. *Fish and Fisheries* **8**, 269–276.

Jennings S and Blanchard JL (2004) Fish abundance with no fishing: predictions based on macroecological theory. *Journal of Animal Ecology* **73**, 632–642.

Jennings S, Brierley AS, and Walker JW (1994) The inshore fish assemblages of the Galapagos archipelago. *Biological Conservation* **70**, 49–57.

Jiménez Vázquez O (1997) La biáya o bambiáya de los indocubanos. *El Pitirre* **10**, 96–97.

Jiménez Vázquez O (2001) Registros ornitológicos en residuarios de dieta de los aborigenes precerámicos Cubanos. *El Pitirre* **14**, 120–126.

Jiménez Vázquez O, Condis MM, and Elvis García C (2005) Vertebrados post-glaciales en un residuario fósil de *Tyto alba scopoli* (Aves: Tytonidae) en el occidente de Cuba. *Revista Mexicana de Mastozoología* **9**, 85–112.

Johannes RE (1981) *Words of the Lagoon*. University of California, Berkley.

Johannessen S (1993) Farmers of the Late Woodland. In Scarry CM, ed., *Foraging and Farming in the Eastern Woodlands*, pp. 57–77. University Press of Florida, Gainesville.

Johnsen SJ, Dahl-Jensen D, Gundestrup N, Steffensen JP, Clausen HB, Miller H, Masson-Delmotte V, Sveinbjörnsdottir AE, and White J (2001) Oxygen isotope and palaeotemperature records from six Greenland ice-core stations: Camp Century, Dye-3, GRIP, GISP2, Renland and NorthGRIP. *Journal of Quaternary Science* **16**, 299–307.

Johnson CN (2002) Determinants of loss of mammal species during the Late Quaternary 'megafauna' extinctions: life history and ecology, but not body size. *Proceedings of the Royal Society of London Series B Biological Sciences* **269**, 2221–2227.

Johnson C (2006) *Australia's Mammal Extinctions: a 50,000 Year History*. Cambridge University Press, Cambridge.

Johnson CN and Wroe S (2003) Causes of extinction of vertebrates during the Holocene of mainland Australia: arrival of the dingo, or human impact? *The Holocene* **13**, 1009–1016.

Johnson CN, Delean S, and Balmford A (2002) Phylogeny and the selectivity of extinction in Australian marsupials. *Animal Conservation* **5**, 135–142.

Johnson CN, Isaac JL, and Fisher DO (2007) Rarity of a top predator triggers continent-wide collapse of mammal prey: dingoes and marsupials in Australia. *Proceedings of the Royal Society of London Series B Biological Sciences* **274**, 341–346.

Johnson PM (2001) *Habitat Associations and Drought Responses of Freshwater Mussels in the Lower Flint River Basin*. MSc thesis, University of Georgia.

Johnson RI (1978) Systematics and zoogeography of *Plagiola* (=*Dysnomia* =*Epioblasma*), an almost extinct genus of freshwater mussels (Bivalvia: Unionidae) from Middle North America. *Bulletin of the Museum of Comparative Zoology* **148**, 239–321.

Johnson TH and Stattersfield AJ (1991) A global review of island endemic birds. *Ibis* **132**, 167–180.

Johnston DW (1969) The thrushes of Grand Cayman Islands, B.W.I. *The Condor* **71**, 120–128.

Jones CG, Lawton JH, and Shachak M (1994) Organisms as ecosystem engineers. *Oikos* **69**, 373–386.

Jones DN, Dekker RWRJ, and Roselaar CS (1995) *The Megapodes*. Oxford University Press, Oxford.

Jones E (1977) Ecology of the feral cat, *Felis catus* (L.), (Carnivora: Felidae), on Macquarie Island. *Australian Wildlife Research* **4**, 249–262.

Jones HP, Tershy BR, Zavaleta ES, Croll DA, Keitt BS, Finkelstein ME, and Howald GR (2008a) Severity of the effects of invasive rats on seabirds: a global review. *Conservation Biology* **22**, 16–26.

Jones JW, Neves RJ, Ahlstedt SA, and Hallerman EM (2006a) A holistic taxonomic evaluation of two closely related freshwater mussel species, the oyster mussel *Epioblasma capsaeformis* and tan riffleshell *Epioblasma florentina walkeri* (Bivalvia:Unionidae). *Journal of Molluscan Studies* **72**, 267–283.

Jones JW, Hallerman EM, and Neves RJ (2006b) Genetic management guidelines for captive propagation of freshwater mussels (Unionidea). *Journal of Shellfish Research* **25**, 527–535.

Jones ME and Stoddart DM (1998) Reconstruction of the predatory behaviour of the extinct marsupial thylacine (*Thylacinus cynocephalus*). *Journal of Zoology* **246**, 239–246.

Jones T, Ehardt CL, Butynski TM, Davenport TRB, Mpunga NE, Machaga SJ, and De Luca DW (2005) The highland mangabey *Lophocebus kipunji*: a new species of African monkey. *Science* **308**, 1161–1164.

Jones TL, Porcasi JF, Erlandson JM, Dallas Jr H, Wake A, and Schwaderera R (2008b) The protracted Holocene extinction of California's flightless sea duck (*Chendytes lawi*) and its implications for the Pleistocene overkill hypothesis. *Proceedings of the National Academy of Sciences USA* **105**, 4105–4108.

Jordan G (2008) *Phylowidget: an Online Phylogenetics Tool*. www.phylowidget.org.

Kay J (2004) Etienne de Flacourt: *L'Histoire de le Grand Île de Madagascar*. *Curtis's Botanical Magazine* **21**, 251–257.

Kear J and Scarlett RJ (1970) The Auckland Islands merganser. *Wildfowl* **21**, 78–86.

Kearns CA, Inouye DW, and Waser NM (1998) Endangered mutualisms: the conservation of plant-pollinator interactions. *Annual Review of Ecology and Systematics* **29**, 83–112.

Keefer DK, de France SD, Mosely ME, Richardson JB III, Santerlee DR, and Day-Lewis A (1998) Early maritime economy and El Niño at Quebrada Tacahuay, Peru. *Science* **281**, 1833–1835.

Keegan WF (1992) *The People who Discovered Columbus: the Prehistory of the Bahamas*. University Press of Florida, Gainesville.

Keith DA and Burgman MA (2004) The Lazarus effect: can the dynamics of extinct species lists tell us anything about the status of biodiversity? *Biological Conservation* **117**, 41–48.

Keitt TH and Marquet PA (1996) The introduced Hawaiian avifauna reconsidered: evidence for self-organized criticality? *Journal of Theoretical Biology* **182**, 161–167.

Kemp TS (1999) *Fossils and Evolution*. Oxford University Press, Oxford.

Kenyon KW (1975) *The Sea Otter in the Eastern Pacific Ocean*. Dover Publications, New York.

Kepler CB (1967) Polynesian rat predation on nesting Laysan albatrosses and other Pacific seabirds. *The Auk* **84**, 426–430.

Kerney MP (1963) Late-glacial deposits on the chalk of South-East England. *Philosophical Transactions of the Royal Society B* **246**, 203–254.

Kerr KCR, Stoeckle MY, Dove CJ, Weigt LA, Francis CM, and Hebert PDN (2007) Comprehensive DNA barcode coverage of North American birds. *Molecular Ecology Notes* **7**, 535–543.

Kidwell SM and Flessa KW (1995) The quality of the fossil record: populations, species, and communities. *Annual Review of Ecology and Systematics* **26**, 269–299.

King FW (1987) Thirteen milestones on the road to extinction. In Fitter R and Fitter MS, eds, *The Road to Extinction*, pp. 7–18. IUCN, Gland.

Kingdon J (1990) *Island Africa: the Evolution of Africa's Rare Animals and Plants*. Princeton University Press, Princeton.

Kinzig AP, Pacala S, and Tilman D (eds) (2002) *Functional Consequences of Biodiversity: Empirical Progress and*

Theoretical Extensions. Princeton University Press, Princeton.

Kirby KJ (2003) *What Might British Forest-Landscape Driven by Large Herbivores Look Like?* English Nature Report 530. English Nature, Peterborough.

Kirch PV (1996) Late Holocene human-induced modifications to a central Polynesian island ecosystem. *Proceedings of the National Academy of Sciences USA* **93**, 5296–5300.

Kirch PV (2005) Archaeology and global change: the Holocene record. *Annual Review of Environment and Resources* **30**, 409–440.

Kirch PV (2007) Three islands and an archipelago: reciprocal interactions between humans and island ecosystems in Polynesia. *Earth and Environmental Science Transactions of the Royal Society of Edinburgh* **98**, 85–99.

Kirchman JJ and Steadman DW (2005) Rails (Aves: Rallidae: *Gallirallus*) from prehistoric sites in the Kingdom of Tonga, including description of a new species. *Proceedings of the Biological Society of Washington* **118**, 465–477.

Kirchman JJ and Steadman DW (2006a) New species of rails (Aves: Rallidae) from an archaeological site on Huahine, Society Islands. *Pacific Science* **60**, 279–295.

Kirchman JJ and Steadman DW (2006b) Rails (Rallidae: *Gallirallus*) from prehistoric archaeological sites in Western Oceania. *Zootaxa* **1316**, 1–31.

Kirchman JJ and Steadman DW (2007) New species of extinct rails (Aves: Rallidae) from archaeological sites in the Marquesas Islands, French Polynesia. *Pacific Science* **61**, 145–163.

Klein RG (1974) On the taxonomic status, distribution and ecology of the blue antelope, *Hippotragus leucophaeus* (Pallas, 1766). *Annals of the South African Museum* **65**, 99–143.

Klein RG (1984) Mammalian extinctions and Stone Age people in Africa. In Martin PS and Klein RG, eds, *Quaternary Extinctions: a Prehistoric Revolution*, pp. 553–573. Arizona University Press, Tucson.

Klein RG (1994) The long-horned African buffalo (*Pelorovis antiquus*) is an extinct species. *Journal of Archaeological Science* **21**, 725–733.

Klein RG, Avery G, Cruz-Uribe K, Halkett D, Parkington JE, Steele T, Volman TP, and Yates R (2004) The Ysterfontein 1 Middle Stone Age site, South Africa, and early human exploitation of coastal resources. *Proceedings of the National Academy of Sciences USA* **101**, 5708–5715.

Kleypas JA, Buddemeier RW, Archer D, Gattuso J-P, Langdon C, and Opdyke BN (1999) Geochemical consequences of increased atmospheric carbon dioxide on coral reefs. *Science* **284**, 118–120.

Klippel WE, Celmer G, and Purdue JR (1978) The Holocene naiad record at Rodgers Shelter in the western Ozark Highlands of Missouri. *Plains Anthropologist* **23**, 257–271.

Knowlton N (1993) Sibling species in the sea. *Annual Review of Ecology and Systematics* **24**, 189–216.

Koenigswald WV and Van Kolfschoten T (1995) The *Mimonys-Arvicola* boundary and the enamel thickness quotient (SDQ) of *Arvicola* as stratigraphic markers in the Middle Pleistocene. In Turner C, ed., *The Early Middle Pleistocene in Europe*, pp. 211–226. Balkema, Rotterdam.

Koh LP, Dunn RR, Sodhi NS, Colwell RK, Proctor HC, and Smith VS (2004) Species coextinctions and the biodiversity crisis. *Science* **305**, 1632–1634.

Kokita T and Nakazono A (2001) Rapid response of an obligately corallivorous filefish *Oxymonacanthus longirostris* (Monacanthidae) to a mass coral bleaching event. *Coral Reefs* **20**, 155–158.

Kolata AK (1986) The agricultural foundations of the Tiwanaku State: a view from the heartland. *American Antiquity* **51**, 748–762.

Kolb MJ (1994) Ritual activity and chiefly economy at an upland religious site on Maui, Hawai'i. *Journal of Field Archaeology* **21**, 417–436.

Kos AM (2003) Pre-burial taphonomic characterisation of a vertebrate assemblage from a pitfall cave fossil deposit in southeastern Australia. *Journal of Archaeological Science* **30**, 769–779.

Kotlìk P, Deffontain V, Maschertti S, Zima J, Michaux JR, and Searle JB (2006) A northern glacial refugium for bank voles (*Clethrionomys glareolus*). *Proceedings of the National Academy of Sciences USA* **103**, 14860–14864.

Kuang-Ti L (2001) Prehistoric marine marine fishing adaptation in southern Taiwan. *Journal of East Asian Archaeology* **3**, 47–74.

Kull CA and Fairbairn J (2000) Deforestation, erosion, and fire: degradation myths in the environmental history of Madagascar. *Environment and History* **6**, 423–450.

Kullman L (1998) Non-analogous tree flora in the Scandes Mountains, Sweden, during the early Holocene—macrofossil evidence of rapid geographic spread and response to palaeoclimate. *Boreas* **27**, 153–161.

Kullman L (2002) Boreal tree taxa in the central Scandes during the Late-Glacial: implications for Late-Quaternary forest history. *Journal of Biogeography* **29**, 1117–1124.

Kurlansky M (1998) *Cod: a Biography of the Fish that Changed the World.* Jonathan Cape, London.

Kurochkin E (1995) The assemblage of Cretaceous birds in Asia. In Sun A and Wang Y, eds, *Sixth Symposium on Mesozoic Terrestrial Ecosystems and Biota*, pp. 203–208. China Ocean Press, Beijing.

Kurochkin EN (2000) Mesozoic birds of Mongolia and the former USSR. In Benton MJ, Shishkin MA, Unwin DM, and Kurochkin EN, eds, *The Age of Dinosaurs in Russia and Mongolia*, pp. 544–559. Cambridge University Press, Cambridge.

Kurtén B (1956) The status and affinities of *Hyena sinensis* Owen and *Hyena ultima* Matsumoto. *American Museum Novitates* **1764**, 1–48.

Kurtén B (1968) *Pleistocene Mammals of Europe*. Weidenfeld & Nicolson, London.

Kurtén B and Anderson E (1980) *Pleistocene Mammals of North America*. Columbia University Press, New York.

Lamb HH (1977) *Climate: Present, Past and Future. Volume 2. Climatic History and the Future*. Methuen & Co, London.

Lathrap DW (1970) *The Upper Amazon*. Thames and Hudson, London.

Law R and Grey DR (1989) Evolution of yields from populations with age-specific cropping. *Evolutionary Ecology* **3**, 343–359.

Layzer JB, Gordon ME, and Anderson RM (1993) Mussels: the forgotten fauna of regulated rivers. A case study of the Caney Fork River. *Regulated Rivers: Research and Management* **8**, 63–71.

Lee WG and Jamieson IG (eds) (2001) *The Takahe: Fifty Years of Conservation Management and Research*. University of Otago Press, Dunedin.

Lees K, Pitois S, Scott C, Frid C, and Mackinson S (2006) Characterizing regime shifts in the marine environment. *Fish and Fisheries* **7**, 104–127.

LeFebvre MJ (2007) Zooarchaeological analysis of prehistoric vertebrate exploitation at the Grand Bay Site, Carriacou, West Indies. *Coral Reefs* **26**, 931–944.

Legendre P and Legendre L (1998) *Numerical Ecology*, 2nd edn. Elsevier Science BV, Amsterdam.

Leonard JA, Vilá C, Fox-Dobbs K, Koch, PL, Wayne RK, and Van Valkenberg B (2007) Megafaunal extinctions and the disappearance of a specialized wolf ecomorph. *Current Biology* **17**, 1146–1150.

Leopold MF (2005) Ooit broedden er *Pterodroma*'s rond de Noordzee. *Nieuwsbrief Nederlandse Zeevogelgroep* **7**, 3.

Lepiksaar J (1958) Fossilfynd av stormfåglar (Procellariiformes) från Sveriges västkust. *Zoologisk Revy* **4**, 77–85.

Lever C (2005) *Naturalised Birds of the World*. T & AD Poyser, London.

Lewis FT (1944) The passenger pigeon as observed by the Rev. Cotton Mather. *The Auk* **61**, 587–592.

Lewison RL, Crowder LB, Read AJ, and Freeman SA (2004) Understanding impacts of fisheries bycatch on marine megafauna. *Trends in Ecology & Evolution* **19**, 598–604.

Leyden BW (1987) Man and climate in Maya lowlands. *Quaternary Research* **28**, 407–414.

Lister AM (1989) Rapid dwarfing of red deer on Jersey in the Last Interglacial. *Nature* **342**, 539–542.

Lister AM (1993) Patterns of evolution in Quaternary mammal lineages. In Lees DR and Edwards D, eds, *Evolutionary Patterns and Processes*, pp. 71–93. Academic Press, London.

Lister AM (1997) The evolutionary response of vertebrates to Quaternary environmental change. In Huntley B, Cramer W, Prentice AV, and Allen JRM, eds, *Past and Future Rapid Environmental Change: the Spatial and Evolutionary Responses of Terrestrial Biota*. Springer, Berlin.

Lister AM and Sher AV (1995) Ice cores and mammoth extinction. *Nature* **378**, 23–24.

Lister AM and Bahn P (2007) *Mammoths: Giants of the Ice Age*. Frances Lincoln, London.

Livezey BC (1998) A phylogenetic analysis of the Gruiformes (Aves) based on morphological characters, with an emphasis on the rails (Rallidae). *Philosophical Transactions of the Royal Society B* **353**, 2077–2151.

Livingston SD (1989) The taphonomic interpretation of avian skeletal part frequencies. *Journal of Archaeological Science* **16**, 537–547.

Lloyd BD (2001) Advances in New Zealand mammalogy 1990–2000: short-tailed bats. *Journal of the Royal Society of New Zealand* **31**, 59–81.

Lockwood JL (2005) Predicting which species will become invasive: what's taxonomy got to do with it? In Purvis A, Gittleman J, and Brooks T, eds, *Phylogeny and Conservation*, pp. 365–386. Cambridge University Press, Cambridge.

Lockwood JL (2006) Life in a double-hotspot: the transformation of Hawaiian passerine bird diversity following invasion and extinction. *Biological Invasions* **8**, 449–457.

Lockwood JL, Cassey P, and Blackburn TM (2005) The role of propagule pressure in explaining species invasion. *Trends in Ecology & Evolution* **20**, 223–228.

Lockwood JL, Hoopes MF, and Marchetti MP (2007) *Invasion Ecology*. Blackwell Scientific Press, Oxford.

Loehle C and Li B (1996) Habitat destruction and the extinction debt revisited. *Ecological Applications* **6**, 784–789.

Loehr J, Worley K, Grapputo A, Carey J, Veitch A, and Coltman DW (2005) Evidence for cryptic glacial refugia from North American mountain sheep mitochondrial DNA. *Journal of Evolutionary Biology* **19**, 419–430.

Long JL (1981) *Introduced Birds of the World*. David & Charles, London.

Lopinot NH (1992) Spatial and temporal variability in Mississippian subsistence: the archaeobotanical record.

In Woods WI, ed., *Late Prehistoric Agriculture*, pp. 44–94. Illinois Historic Preservation Agency, Springfield.

Lotze HK (2005) Radical changes in the Wadden Sea fauna and flora over the last 2000 years. *Helgoland Marine Research* **59**, 71–83.

Lotze HK, Lenihan HS, Bourque BJ, Bradbury RH, Cooke RG, Kay MC, Kidwell SM, Kirby MX, Peterson CH, and Jackson JBC (2006) Depletion, degradation, and recovery potential of estuaries and coastal seas. *Science* **312**, 1806–1809.

Louchart A (2002) Les oiseaux du Pléistocène de Corse et de quelques localités Sardes. Ecologie, évolution, biogéographie et extinctions. *Documents des Laboratoires de Geologie, Lyon* **155**, 1–287.

Louchart A (2004) An extinct large thrush (Aves: Turdidae) from the Late Quaternary of Mediterranean Europe. *Neues Jahrbuch für Geologie und Paläontologie, Abhandlungen* **233**, 275–296.

Louchart A (2005) Integrating the fossil record in the study of insular body size evolution. *Monografies de la Societat d'Història Natural de les Balears* **12**, 155–174.

Louchart A, Bedetti C, and Pavia M (2005) A new species of eagle (Aves: Accipitridae) close to the steppe eagle, from the Pleistocene of Corsica and Sardinia, France and Italy. *Palaeontographica (A)* **272(5–6)**, 121–148.

Louis EE, Engberg SE, Lei R, Geng H, Sommer JA, Randriamampionona R, Randriamanana JC, Zaonarivelo JR, Andriantompohavana R, Randria G *et al.* (2006) Molecular and morphological analyses of the sportive lemurs (family Megaladapidae: genus *Lepilemur*) reveals 11 previously unrecognised species. *Special Publications, Museum of Texas Tech University* **49**, 1–47.

Loutre MF and Berger A (2003) Marine Isotope Stage 11 as an analogue for the present interglacial. *Global and Planetary Change* **36**, 205–213.

Louys J, Curnoe D, and Tong H (2007) Characteristics of Pleistocene megafauna extinctions in southeast Asia. *Palaeogeography, Palaeoclimatology, Palaeoecology* **243**, 152–173.

Lozano-Garcia MdS, Caballero M, Ortega B, Rodriguez A, and Sosa S (2007) Tracing the effects of the Little Ice Age in the tropical lowlands of eastern Mesoemerica. *Proceedings of the National Academy of Sciences USA* **104**, 16200–16203.

Lund D, Lynch-Stieglitz J, and Curry WB (2006) Gulf Stream density structure and transport during the past millennium. *Nature* **444**, 601–604.

Lusk CH (2002) Does photoinhibition avoidance explain divarication in the New Zealand flora? *Functional Ecology* **16**, 858–860.

Lydekker R (1890) On the remains of some large extinct birds from the cavern-deposits of Malta. *Proceedings of the Zoological Society of London* **1890**, 403–411.

Lydekker R (1891) *Catalogue of the Fossil Birds in the British Museum (Natural History)*. Trustees of the British Museum (Natural History), London.

Lyman RL (1984) A model of large freshwater clam exploitation in the prehistoric southern Columbia Plateau culture area. *Northwest Anthropological Research Notes* **18**, 97–107.

Lyman RL (1994) Relative abundances of skeletal specimens and taphonomic analysis of vertebrate remains. *Palaios* **9**, 288–298.

Lyman RL (2006) Paleozoology in the service of conservation biology. *Evolutionary Anthropology* **15**, 11–19.

Lysaght A (1956) A note on the Polynesian black or sooty rail *Porzana nigra* (Miller) 1784. *Bulletin of the British Ornithological Club* **76**, 97–98.

Ma A and Tang H (1992) On discovery and significance of a Holocene *Ailuropoda-Stegodon* fauna from Jinhua, Zhejiang. *Vertebrata PalAsiatica* **30**, 295–312.

MacArthur R and Wilson EO (1967) *The Theory of Island Biogeography*. Princeton University Press, Princeton.

MacCall AD (1990) *Dynamic Geography of Marine Fish Populations*. University of Washington, Seattle.

Mace G, Masundire H, Baillie J, Ricketts T, and Brooks T (2005) Biodiversity. In Hassan R, Scholes R, and Ash N, eds, *Ecosystems and Human Well-Being: Current State and Trends. Findings of the Condition and Trends Working Group*, pp. 77–122. Island Press, Washington DC.

Mace PM (2004) In defence of fisheries scientists, single-species models and other scapegoats: confronting the real problems. *Marine Ecology Progress Series* **274**, 285–291.

Mack M and D'Antonio CM (1998) Impacts of biological invasions on disturbance regimes. *Trends in Ecology & Evolution* **13**, 195–198.

Mackay AW (2007) The paleoclimatology of Lake Baikal: a diatom synthesis and prospectus. *Earth-Science Reviews* **82**, 181–215.

Mackay AW, Battarbee RW, Birks HJB, and Oldfield F (eds) (2003a) *Global Change in the Holocene*. Arnold, London.

Mackay AW, Jones VJ, and Battarbee RW (2003b) Approaches to Holocene climate reconstruction using diatoms. In Mackay AW, Battarbee RW, Birks HJB, and Oldfield F, eds, *Global Change in the Holocene*, pp. 294–309. Arnold, London.

MacPhee RDE and Fleagle JG (1991) Postcranial remains of *Xenothrix mcgregori* (Primates, Xenotrichidae) and other late Quaternary mammals from Long Mile Cave, Jamaica. *Bulletin of the American Museum of Natural History* **206**, 287–321.

MacPhee RDE and Flemming C (1999) Requiem æternam: the last five hundred years of mammalian species extinctions. In MacPhee RDE, ed., *Extinctions in Near*

Time: Causes, Contexts, and Consequences, pp. 333–371. Kluwer Academic/Plenum, New York.

MacPhee RDE and Flemming C (2003) A possible heptaxodontine and other caviidan rodents from the Quaternary of Jamaica. *American Museum Novitates* **3422**, 1–42.

MacPhee RDE and Meldrum J (2006) Postcranial remains of the extinct monkeys of the Greater Antilles, with evidence for semiterrestriality in *Paralouatta*. *American Museum Novitates* **3516**, 1–65.

MacPhee RDE, Flemming C, and Lunde DP (1999) 'Last occurrence' of the Antillean insectivore *Nesophontes*: new radiometric dates and their interpretation. *American Museum Novitates* **3261**, 1–20.

MacPhee RDE, White JL, and Woods CA (2000) New megalonychid sloths (Phyllophaga, Xenarthra) from the Quaternary of Hispaniola. *American Museum Novitates* **3303**, 1–32.

MacPhee RDE, Tikhonov AN, Mol D, de Marliave C, van der Plicht H, Greenwood AD, Flemming C, and Agenbroad L (2002) Radiocarbon chronologies and extinction dynamics of the late Quaternary mammalian megafauna of the Taimyr Peninsula, Russian Federation. *Journal of Archaeological Science* **29**, 1017–1042.

MacPhee RDE, Tikhonov AN, Mol D, and Greenwood AD (2005) Late Quaternary loss of genetic diversity in muskox (*Ovibos*). *BMC Evolutionary Biology* **5**, 49.

MacPhee RDE, Iturralde-Vinent M, and Jiménez-Vásquez O (2007) Prehistoric sloth extinctions in Cuba: implications of a new "last" appearance date. *Caribbean Journal of Science* **43**, 94–98.

MacPherson AH (1965) The origin of diversity in mammals of the Canadian Arctic tundra. *Systematic Zoology* **14**, 153–173.

Magallon S and Sanderson MJ (2001) Absolute diversification rates in angiosperms. *Evolution* **55**, 1762–1780.

Mäki-Petäys H and Breen J (2007) Genetic vulnerability of a remnant ant population. *Conservation Genetics* **8**, 427–435.

Malakoff D (1998). Death by suffocation in the Gulf of Mexico. *Science* **281**, 190–192.

Malcomson RO (1937) Two new Mallophaga. *Annals of the Entomological Society of America* **30**, 53–56.

Mallory ML (2006) The northern fulmar (*Fulmarus glacialis*) in Arctic Canada: ecology, threats, and what it tells us about marine environmental conditions. *Environmental Reviews* **14**, 187–216.

Mancina CA and García-Rivera L (2005) New genus and species of fossil bat (Chiroptera: Phyllostomidae) from Cuba. *Caribbean Journal of Science* **41**, 22–27.

Mann CC (2005) *1491: New Revelations of the Americas Before Columbus*. Knopf Press, New York.

Manne LL, Brooks TM, and Pimm SL (1999) Relative risk of extinction of passerine birds on continents and islands. *Nature* **399**, 258–261.

Marchant RA (2007) Late Holocene environmental change and cultural response in south-western Uganda. In Lille M and Ellis S, eds, *Wetlands Archaeology and Environments*, pp. 275–288. Oxbow Books, Oxford.

Marchant RA and Hooghiemstra H (2001) A response to "Climate of East Africa 6000 ^{14}C yr B.P. as inferred from pollen data" by Peyron *et al.* (2000). *Quaternary Research* **56**, 133–135.

Marchant RA and Hooghiemstra H (2004) Rapid environmental change in tropical Africa and Latin America about 4000 years before present: a review. *Earth-Science Reviews* **66**, 217–260.

Marchant R, Hooghiemstra H, and Islebe G (2004) The rise and fall of Peruvian and Central American civilisations: interconnections with Holocene climatic change—a necessarily complex model. In Yasuda Y and Shinde V, eds, *Monsoon and Civilization*, pp. 351–376. Roli Books, Delhi.

Marchetti MP, Light TS, Feliciano J, Armstrong TW, Hogan Z, and Moyle PB (2001) Physical homogenization and biotic homogenization in aquatic systems. In JL Lockwood and ML McKinney, eds, *Biotic Homogenization*, pp. 259–278. Kluwer/Academic Press, New York.

Marchetti MP, Lockwood JL, and Light T (2006) Effects of urbanization on California's fish diversity: differentiation, homogenization and the influence of spatial scale. *Biological Conservation* **127**, 310–318.

Marinho-Filho J and Verissimo EW (1997) The rediscovery of *Callicebus personatus barbarabrownae* in northeastern Brazil with a new western limit for its distribution. *Primates* **38**, 429–433.

Markova AK, Smirnov NG, Kozharinov AV, Kazantseva NE, Simakova AN, and Kitaev LM (1995) Late Pleistocene distribution and diversity of mammals in Northern Eurasia. *Paleontologia i Evolucio* **28–29**, 5–143.

Marmontel M, Humphrey SR, and O'Shea TJ (1997) Population viability analysis of the Florida manatee (*Trichechus manatus latirostris*), 1976–1991. *Conservation Biology* **11**, 467–481.

Marples BJ (1946) Notes on some neognathous bird bones from the Early Tertiary of New Zealand. *Transactions of the Royal Society of New Zealand* **76**, 132–134.

Marsh H, De'ath G, Gribble N, and Lane B (2005) Historical marine population estimates: triggers or targets for conservation? The dugong case study. *Ecological Applications* **15**, 481–492.

Marsh RE (1985) More about the Bajan mouse. *Journal of the Barbados Museum and Historical Society* **37**, 310.

Marshall CR (1994) Confidence intervals on stratigraphic ranges: partial relaxation of the assumption of randomly distributed fossil horizons. *Paleobiology* **20**, 459–469.

Marshall CR (1997) Confidence intervals on stratigraphic ranges with non-random distributions of fossil horizons. *Paleobiology* **23**, 165–173.

Marshall K and Edwards-Jones G (1998) Reintroducing capercaillie (*Tetrao urogallus*) into southern Scotland: identification of minimum viable populations at potential release sites. *Biodiversity and Conservation* **7**, 275–296.

Martin J-L, Thibault J-C, and Bretagnolle V (2000) Black rats, island characteristics, and colonial nesting birds in the Mediterranean: consequences of an ancient introduction. *Conservation Biology* **14**, 1452–1466.

Martin PS (1984) Prehistoric overkill: the global model. In Martin PS and Klein RG, eds, *Quaternary Extinctions: a Prehistoric Revolution*, pp. 354–403. Arizona University Press, Tucson.

Martin PS and Steadman DW (1999) Prehistoric extinctions on islands and continents. In MacPhee RDE, ed., *Extinctions in Near Time: Causes, Contexts, and Consequences*, pp. 17–55. Kluwer Academic/Plenum, New York.

Martinez J (1987) Un nouveau cas probable d'endemisme insulaire: le canard de l'île Amsterdam. *Documents des Laboratoires de Gèologie de la Faculté des Sciences de Lyon* **99**, 211–219.

Maslin M, Seidov D, and Lowe J (2001) Synthesis of the nature and causes of rapid climate transitions during the Quaternary. In Seidov D, Haupt BJ, and Maslin MA, eds, *The Oceans and Rapid Climate Change: Past, Present and Future*, pp. 9–52. Geophysical Monograph Series 126. American Geophysical Union, Washington DC.

Massetti M (2001) Did endemic dwarf elephants survive on Mediterranean islands up to protohistorical times? In Cavaretta G, Gioia P, Mussi M, and Palombo MR, eds, *The World of Elephants: Proceedings of the 1st International Congress*, Rome, 16–20 October 2001, pp. 402–406. Consiglio Nazionale delle Ricerche, Rome.

Massuti E and Moranta J (2003) Demersal assemblages and depth distribution of elasmobranchs from the continental shelf and slope off the Balearic Islands (western Mediterranean). *ICES Journal of Marine Science* **60**, 753–766.

Mathews S and Donoghue MJ (1999) The root of angiosperm phylogeny inferred from duplicate phytochrome genes. *Science* **286**, 947–950.

Matisoo-Smith E, Roberts RM, Irwin GJ, Allen JS, Penny D, and Lambert DM (1998) Patterns of prehistoric human mobility in Polynesia indicated by mtDNA from the Pacific rat. *Proceedings of the National Academy of Sciences USA* **95**, 15145–15150.

Matsuoka H (2000) The Late Pleistocene fossil birds of the central and southern Ryukyu Islands, and their zoogeographical implications for the recent avifauna of the Archipelago. *Tropics* **10**, 165–188.

Matsuoka H, Oshiro I, Yamauchi T, Ono K, and Hasegawa Y (2002) Seabird—wood pigeon paleoavifauna of the Kita-Daito Island: fossil assemblage from the cave deposit and its implication. *Bulletin of Gunma Museum of Natural History* **6**, 1–14.

Matteson MR (1959) An analysis of the shells of freshwater mussels gathered by Indians in southwestern Illinois. *Transactions of the Illinois Academy of Science* **52**, 52–58.

Matteson MA (1960) Reconstruction of pre-Columbian environments through the analysis of molluscan collections from shell middens. *American Antiquity* **26**, 117–120.

Maxwell DL and Jennings S (2005) Power of monitoring programmes to detect decline and recovery of rare and vulnerable fish. *Journal of Applied Ecology* **42**, 25–37.

May RM (1988) How many species are there on Earth? *Science* **241**, 1441–1449.

Mayewski PA, Rohling EE, Stager JC, Karlén W, Maasch KA, Meeker LD, Meyerson EA, Gasse F, van Kreveld S, Holmgren K *et al.* (2004) Holocene climate variability. *Quaternary Research* **62**, 243–255.

Mayr E and Diamond JD (2001) *The Birds of Northern Melanesia: Speciation, Ecology and Biogeography*. Oxford University Press, New York.

Mayr G, Peters DS, and Rietschel S (2002) Petrel-like birds with a peculiar foot morphology from the Oligocene of Germany and Belgium (Aves: Procellariiformes). *Journal of Vertebrate Paleontology* **22**, 667–676.

Mazzanti DL and Quintana CA (1997) Asociación cultural con fauna extinguida en el sitio arqueológico Cueva Tixi, provincial de Buenos Aires, Argentina. *Revista Española de Antropología Americana* **27**, 11–21.

McAndrews J (1988) Human disturbance of North American forests and grasslands: the fossil pollen record. In Huntley B and Webb III T, eds, *Handbook of Vegetation Science. Volume VII. Vegetation History*, pp. 673–697. Kluwer Academic Publications, Dordrecht.

McClanahan TR (1995) A coral reef ecosystem-fisheries model: impacts of fishing intensity and catch selection on reef structure and processes. *Ecological Modelling* **80**, 1–19.

McClenachan L and Cooper AB (2008) Extinction rate, historical population structure and ecological role of the Caribbean monk seal. *Proceedings of the Royal Society of London Series B Biological Sciences* **275**, 1351–1358.

McCook LJ, Jompa J, and Diaz-Pulido G (2001) Competition between corals and algae on coral reefs: a review of the evidence and mechanisms. *Coral Reefs* **19**, 400–417.

McCullagh WH, Williams JD, McGregor SW, Pierson JM, and Lydeard C (2002) The unionid (Bivalvia) fauna of the Sipsey River northwestern Alabama, an aquatic hotspot. *American Malacological Bulletin* **17**, 1–15.

McDougall I, Brown FH, and Fleagle JG (2005) Stratigraphic placement and age of modern humans from Kibish, Ethiopia. *Nature* **433**, 733–736.

McDowell RM (1996) Threatened fishes of the world: *Prototoctes oxyrhynchus* Gunther, 1870 (Prototrocidae). *Environmental Biology of Fishes* **46**, 60.

McFarlane DA (1999a) A comparison of methods for the probabilistic determiniation of vertebrate extinction chronologies. In MacPhee R, ed., *Extinctions in Near Time: Causes, Contexts, and Consequences*, pp. 95–103. Kluwer Academic/Plenum, New York.

McFarlane DA (1999b) A note on sexual dimorphism in *Nesophontes edithae* (Mammalia: Insectivora), an extinct island-shrew from Puerto Rico. *Caribbean Journal of Science* **35**, 142–143.

McFarlane DA and Lundberg J (2002a) A Middle Pleistocene age and biogeography for the extinct rodent *Megalomys curazensis* from Curaçao, Netherlands Antilles. *Caribbean Journal of Science* **38**, 278–281.

McFarlane DA and Lundberg J (2002b) A new fossil rodent from the Holocene of Bonaire, Netherlands Antilles. *Geological Society of America Annual Meeting Abstract* 90.

McFarlane DA, MacPhee RDE, and Ford D (1998) Body size variability and a Sangamonian extinction model for *Amblyrhiza*, a West Indian megafaunal rodent. *Quaternary Research* **50**, 80–89.

McFarlane DA, Vale A, Christenson K, Lundberg J, Atilles G, and Lauritzen S-E (2000) New specimens of Late Quaternary extinct mammals from caves in Sanchez Ramirez Province, Dominican Republic. *Caribbean Journal of Science* **36**, 163–166.

McFarlane DA, Lundberg J, and Fincham AG (2002) A late Quaternary paleoecological record from caves of southern Jamaica, West Indies. *Journal of Cave and Karst Studies* **64**, 117–125.

McIntyre TM and Hutchings JA (2004) Small-scale temporal and spatial variation in Atlantic cod (*Gadus morhua*) life history. *Canadian Journal of Fisheries and Aquatic Sciences* **60**, 1111–1121.

McKean JL (1973) The bats of Lord Howe Island with the description of a new nyctophiline bat. *Australian Mammalogy* **1**, 329–332.

McKechnie S (2006) Biopedturbation by an island ecosystem engineer: burrowing volumes and litter deposition by sooty shearwaters (*Puffinus griseus*). *New Zealand Journal of Zoology* **33**, 259–265.

McKey D (1989) Population biology of figs: applications for conservation. *Experientia* **45**, 661–673.

McKinney ML (1997) Extinction vulnerability and selectivity: combining ecological and paleontological views. *Annual Review of Ecology and Systematics* **28**, 495–516.

McKinney ML and Lockwood JL (1999) Biotic homogenization: a few winners replacing many losers in the next mass extinction. *Trends in Ecology & Evolution* **14**, 450–453.

McLauchlan K (2003) Plant cultivation and forest clearance by prehistoric North Americans: pollen evidence from Fort Ancient, Ohio, USA. *The Holocene* **13**, 557–566.

McMinn M, Jaume D, and Alcover JA (1990) *Puffinus olsoni* n.sp.: nova espècie de baldritja recentment extingida provinent de depòsits espeleològics de Fuerteventura i Lanzarote (Illes Canàries, Atlàntic oriental). *Endins* **16**, 63–71.

McMinn M, Palmer M, and Alcover JA (2005) A new species of rail (Aves: Rallidae) from the Upper Pleistocene and Holocene of Eivissa (Pityusic Islands, western Mediterranean). *Ibis* **147**, 706–716.

McNamara JA (1997) Some smaller macropodid fossils of South Australia. *Proceedings of the Linnean Society of New South Wales* **117**, 97–105.

Mead JI and Grady F (1996) *Ochotona* (Lagomorpha) from Late Quaternary cave deposits in eastern North America. *Quaternary Research* **45**, 93–101.

Mead JI, Spiess AE, and Sobolik KD (2000) Skeleton of extinct North American sea mink (*Mustela macrodon*). *Quaternary Research* **53**, 247–262.

Mead JI, Steadman DW, Bedford SH, Bell CJ, and Spriggs M (2002) New extinct mekosuchine crocodile from Vanuatu, South Pacific. *Copeia* **2002**, 632–641.

Medway DG (2001) Pigs and petrels on the Poor Knights islands. *New Zealand Natural Sciences* **26**, 87–90.

Medway DG (2002) Why were Providence petrels (*Pterodroma solandri*) nocturnal at Norfolk Island? *Notornis* **49**, 268–270.

Meehan HJ, McConkey KR, and Drake DR (2002) Potential disruptions to seed dispersal mutualisms in Tonga, Western Polynesia. *Journal of Biogeography* **29**, 695–712.

Megyesi JL and O'Daniel DL (1997) Bulwer's Petrel (*Bulweria bulwerii*). In Poole A, and Gill F, eds, *The Birds of North America*, no. 281. The Academy of Natural Sciences, Philadelphia, and The American Ornithologists' Union, Washington DC.

Meier G (2004) Success and disappointment while searching for hutia. *Species* **41**, 7–8.

Memmott J, Waser NM, and Price MV (2004) Tolerance of pollination networks to species extinctions. *Proceedings of the Royal Society of London Series B Biological Sciences* **271**, 2605–2611.

Mengel RM (1964) The probable history of species formation in some northern wood warblers (Parulidae). *Living Bird* **3**, 9–43.

Menzies JI (1977) Fossil and subfossil bats from the mountains of New Guinea. *Australian Journal of Zoology* **25**, 329–336.

Meredith CW (1991) Vertebrate fossil faunas from islands in Australasia and the southwest Pacific. In Vickers-Rich P, Monaghan JM, Baird RF, and Rich TH, eds, *Vertebrate Palaeontology of Australasia*, pp. 1345–1382. Monash University Publications Committee, Melbourne.

Messerli B and Winiger M (1992) Climate, environmental change, and resources of the African mountains from the Mediterranean to the equator. *Mountain Research and Development* **12**, 315–336.

Metcalfe-Smith JL, Staton SK, Mackie GL, and Lane NM (1998) Changes in the biodiversity of freshwater mussels in the Canadian waters of the lower Great Lakes drainage basin over the past 140 years. *Journal of Great Lakes Research* **24**, 845–858.

Michaux J, López-Martínez N, and Hernández-Pacheco JJ (1996) A ^{14}C dating of *Canariomys bravoi* (Mammalia Rodentia), the extinct giant rat from Tenerife (Canary Islands, Spain), and the recent history of the endemic mammals in the archipelago. *Vie et Milieu* **46**, 261–266.

Micheli F (1999) Eutrophication, fisheries, and consumer-resource dynamics in marine pelagic ecosystems. *Science* **285**, 1396–1398.

Miettinen JO, Simola H, Grönlund E, Lahtinen J, and Niinioja R (2005) Limnological effects of growth and cessation of agricultural land use in Ladoga Karelia: sedimentary pollen and diatom analyses. *Journal of Paleolimnology* **34**, 229–243.

Mikkola H (1983) *Owls of Europe.* T & AD Poyser, London.

Milberg P and Tyrberg T (1993) Naïve birds and noble savages—a review of man-made prehistoric extinctions of island birds. *Ecography* **16**, 229–250.

Millar AJK (2001) The world's first recorded extinction of a seaweed. In Chapman ARO, Anderson RJ, Vreeland VJ, and Davison IR, eds, *Proceedings of the 17th International Seaweed Symposium*, pp. 313–318. Oxford University Press, Cape Town.

Millener PR (1988) Contributions to New Zealand's late Quaternary avifauna. 1: *Pachyplichas*, a new genus of wren (Aves: Acanthisittidae) with two new species. *Journal of the Royal Society of New Zealand* **18**, 383–406.

Millener PR (1999) The history of the Chatham Islands' bird fauna of the last 7000 years—a chronicle of change and extinction. In Olson SL, ed., *Avian Paleontology at the Close of the 20th Century: Proceedings of the 4th International Meeting of the Society of Avian Paleontology and Evolution*, Washington DC., 4–7 June 1996. *Smithsonian Contributions to Paleobiology* **89**, 85–109.

Millener PR and Worthy TH (1991) Contributions to New Zealand's late Quaternary avifauna. II: *Dendroscansor decurvirostris*, a new genus and species of wren (Aves: Acanthisittidae). *Journal of the Royal Society of New Zealand* **21**, 179–200.

Millennium Ecosystem Assessment (2005) *Ecosystems and Human Well-being: Biodiversity Synthesis.* World Resources Institute, Washington DC.

Miller AC, Rhodes L, and Tippit R (1984) Changes in the naiad fauna of the Cumberland River below Lake Cumberland in central Kentucky. *The Nautilus* **98**, 107–110.

Miller GH, Fogel ML, Magee JW, Gagan MK, Clarke SJ, and Johnson BJ (2005) Ecosystem collapse in Pleistocene Australia and a human role in megafaunal extinction. *Science* **309**, 287–290.

Miller Jr GS (1929) Mammals eaten by Indians, owls, and Spaniards in the coast region of the Dominican Republic. *Smithsonian Miscellaneous Collections* **82**, 1–16.

Miller LH (1925) *Chendytes*, a diving goose from the California Pleistocene. *The Condor* **27**, 145–149.

Milne-Edwards A and Grandidier A (1866) *Recherches sur la faune ornithologique éteinte des Iles Mascareignes et de Madagascar.* Paris.

Milne-Edwards A and Grandidier A (1894) Observations sur les *Aepyornis* de Madagascar. *Comptes Rendus Hebdomadaires des Séances de l'Académie des Sciences (Paris)* **118**, 122–127.

Milne-Edwards A and Grandidier A (1895) Sur les ossements d'oiseaux provenant des terrains récents de Madagascar. *Bulletin du Muséum d'Histoire Naturelle, Paris* **1**, 9–11.

Miskelly C, Timlin G, and Cotter R (2004) Common diving petrels (*Pelecanoides urinatrix*) recolonise Mana Island. *Notornis* **51**, 245–246.

Miskelly CM, Bester AJ, and Bell M (2006) Additions to the Chatham Islands' bird list, with further records of vagrant and colonising bird species. *Notornis* **53**, 215–230.

Mitchell FJG (2004) How open were European primeval forests? Hypothesis testing using palaeoecological data. *Journal of Ecology* **93**, 168–177.

Mithen S (1993) Simulating mammoth hunting and extinction: implications for the Late Pleistocene of the Central Russian Plain. *Archeological Papers of the American Anthropological Association* **4**, 163–178.

Moen RA, Pastor J, and Cohen Y (1999) Antler growth and extinction of Irish elk. *Evolutionary Ecology Research* **1**, 235–249.

Mohd-Azlan J and Sanderson J (2007) Geographic distribution and conservation status of the bay cat *Catopuma badia*, a Bornean endemic. *Oryx* **41**, 394–397.

Monnier L (1913) Paléontologie de Madagascar. VII. Les *Aepyornis*. *Annales de Paléontologie* **8**, 125–172.

Mooers AØ and Atkins R (2003) Indonesia's threatened birds: over 500 million years of evolutionary heritage at risk. *Animal Conservation* **6**, 183–188.

Mooers AØ, Harmon LJ, Blum MGB, Wong DHJ, and Heard SB (2007) Some models of phylogenetic tree shape. In Gascuel O and Steel M, eds, *Reconstructing Evolution: New Mathematical and Computational Advances*, pp. 149–170. Oxford University Press, Oxford.

Moojen J (1965) Nôvo gênero de Cricetidae do Brasil Central (Glires, Mammalia). *Revista Brasileira de Biología* **25**, 281–285.

Moore PD (2005) Down to the woods yesterday. *Nature* **433**, 588–589.

Morales A, Rosello E, and Canas JM (1994) Cueva de Nerja (prov. Malaga): a close look at a twelve thousand year ichthyofaunal sequence from southern Spain. In van Neer W, ed., *Fish Exploitation in the Past*, pp. 253–262. Annales du Musee Royale pour L'Afrique Centrale, Tervuren, Belgium.

Morato T, Cheung WWL, and Pitcher TJ (2006a) Vulnerability of seamount fish to fishing: fuzzy analysis of life history attributes. *Journal of Fish Biology* **68**, 209–221.

Morato T, Watson R, Pitcher TJ, and Pauly D (2006b) Fishing down the deep. *Fish and Fisheries* **7**, 24–34.

Morejohn GV (1976) Evidence of the survival to recent times of the extinct flightless duck *Chendytes lawi* Miller. *Smithsonian Contributions to Paleobiology* **27**, 207–211.

Morey DF and Crothers GM (1998) Clearing up clouded waters: palaeoenvironmental analysis of freshwater mussel assemblages from the Green River shell middens, western Kentucky. *Journal of Archaeological Science* **25**, 907–926.

Morgan GS (1989a) *Geocapromys thoracatus*. *Mammalian Species* **341**, 1–5.

Morgan GS (1989b) Fossil Chiroptera and Rodentia from the Bahamas, and the historical biogeography of the Bahamian mammal fauna. In Woods CA, ed., *Biogeography of the West Indies: Past, Present and Future*, pp. 685–740. Sandhill Crane Press, Gainesville.

Morgan GS (1993) Quaternary land mammals of Jamaica. In Wright RM and Robinson E, eds, *Biostratigraphy of Jamaica*, pp. 417–442. Geological Society of America, Denver.

Morgan GS (1994) Late Quaternary fossil vertebrates from the Cayman Islands. In Brunt MA and Davies JE, eds, *The Cayman Islands: Natural History and Biogeography*, pp. 465–508. Kluwer, Dordrecht.

Morgan GS (2001) Patterns of extinction in West Indian bats. In Woods CA and Sergile FE, eds, *Biogeography of the West Indies: Patterns and Perspectives*, pp. 369–406. CRC Press, Boca Raton.

Morgan GS and Wilkins L (2003) The extinct rodent *Clidomys* (Heptaxodontidae) from a Late Quaternary cave deposit in Jamaica. *Caribbean Journal of Science* **39**, 34–41.

Moritz C and Cicero C (2004) DNA barcoding: promises and pitfalls. *PLoS Biology* **2**, e354.

Morley SA and Karr JR (2002) Assessing and restoring the health of urban streams in the Puget Sound basin. *Conservation Biology* **16**, 1498–1509.

Morrison JC, Sechrest W, Dinerstein E, Wilcove DS, and Lamoreux JF (2007) Persistence of large mammal faunas as indicators of global human impacts. *Journal of Mammalogy* **88**, 1363–1380.

Morrison JPE (1942) Preliminary report on mollusks found in shell mounds of the Pickwick Landing Basin in the Tennessee River Valley. *Bureau of American Ethnology Bulletin* **129**, 339–392.

Mourer-Chauviré C and Weesie PDM (1986) *Bubo insularis* n. sp., forme endémique insulaire de grand-duc (Aves, Strigiformes) du pléistocène de Sardaigne et de Corse. *Revue de Paléobiologie* **5**, 197–205.

Mourer-Chauviré C and Moutou F (1987) Découverte d'une forme récemment éteinte d'ibis endémique insulaire de l'ile de la Réunion, *Borbonibis latipes* nov. gen. nov. sp. *Comptes Rendus de l'Academie des Sciences (Paris), Séries 2a* **305**, 419–423.

Mourer-Chauviré C and Antunes MT (2000) L'avifaune Pléistocène et Holocène de Gruta da Figueira Brava (Arrábida, Portugal). Last Neanderthals in Portugal. Odontologic and other evidence. *Memórias da Academia das Ciências de Lisboa, Classe de Ciências* **38**, 129–161.

Mourer-Chauviré C and Balouet JC (2005) Description of the skull of the genus *Sylviornis* Poplin, 1980 (Aves, Galliformes, Sylviornithidae new family), a giant extinct bird from the Holocene of New Caledonia. *Monografies de la Societat d'Història Natural de les Balears* **12**, 205–218.

Mourer-Chauviré C, Bour R, Moutou F, and Ribes S (1994) *Mascarenotus* nov. gen. (Aves, Strigiformes), genre endémique éteintdes Mascareignes et *M. grucheti* n. sp., espèce éteinte de La Réunion. *Comptes Rendus de l'Academie des Sciences (Paris), Séries 2a* **318**, 1699–1706.

Mourer-Chauviré C, Bour R, and Ribes S (1995a) Was the solitaire of Réunion an ibis? *Nature* **373**, 568.

Mourer-Chauviré C, Bour R, and Ribes S (1995b) Position systématique du Solitaire de la Réunion: nouvelle interprétation basée sur les restes fossiles et les récits des anciens voyageurs. *Comptes Rendus de l'Academie des Sciences (Paris), Séries 2a* **320**, 1125–1131.

Mourer-Chauviré C, Salotti M, Pereira M, Quinif Y, Courtois J-Y, Dubois J-N, and La Milza JC (1997) *Athene angelis* n. sp. (Aves, Strigiformes) nouvelle espèce endémique insulaire éteinte du Pléistocène moyen et supérieur de Corse (France). *Comptes Rendus de l'Academie des Sciences (Paris), Séries 2a* **324**, 677–684.

Mourer-Chauviré C, Bour R, Ribes S, and Moutou F (1999) The avifauna of Réunion Island (Mascarene Islands) at the time of the arrival of the first Europeans. In Olson SL, ed., *Avian Paleontology at the Close of the 20th Century: Proceedings of the 4th International Meeting of the Society of Avian Paleontology and Evolution*, Washington DC, 4–7 June 1996. *Smithsonian Contributions to Paleobiology* **89**, 1–38.

Mulder CPH and Keall SN (2001) Burrowing seabirds and reptiles: impacts on seeds, seedlings and soils in an island forest in New Zealand. *Oecologia* **127**, 350–360.

Mullon C, Freon P, and Cury P (2005) The dynamics of collapse in world fisheries. *Fish and Fisheries* **6**, 111–120.

Mumby PJ, Edwards AJ, Arias-Gonzalez JE, Lindeman KC, Blackwell PG, Gall A, Gorczynska MI, Harborne AR, Pescod CL, Renken H *et al.* (2004) Mangroves enhance the biomass of coral reef fish communities in the Caribbean. *Nature* **427**, 533–536.

Munday PL (2004) Habitat loss, resource specialization, and extinction on coral reefs. *Global Change Biology* **10**, 1–6.

Murphy GI (1981) Guano and the anchovetta fishery. *Resource Management and Environmental Uncertainty* **11**, 81–106.

Murphy RC and Snyder JP (1952) The "Pealea" phenomenon and other notes on storm petrels. *American Museum Novitates* **1596**, 1–16.

Murray JW (2002) Introduction to benthic foraminifera. In Haslett SK, ed., *Quaternary Environmental Micropalaeontology*, pp. 5–13. Arnold, London.

Murray P (1984) Extinctions downunder: a bestiary of extinct Australian late Pleistocene monotremes and marsupials. In Martin PS and Klein RG, eds, *Quaternary Extinctions: a Prehistoric Revolution*, pp. 600–628. The University of Arizona Press, Tucson and London.

Musick JA, Harbin MM, Berkeley SA, Burgess GH, Eklund AM, Findley L, Gilmore RG, Golden JT, Ha DS, Huntsman GR *et al.* (2000) Marine, estuarine, and diadromous fish stocks at risk of extinction in North America (exclusive of Pacific salmonids). *Fisheries* **25**, 6–30.

Musil R (1985) Paleobiogeography of terrestrial communities in Europe during the Last Glacial. *Acta Musei Nationalis Pragae XLI B* **1–2**, 1–83.

Musser GG (1981) The giant rat of Flores and its relatives east of Borneo and Bali. *Bulletin of the American Museum of Natural History* **169**, 71–175.

Musser GG and Holden ME (1991) Sulawesi rodents (Muridae: Murinae): morphological and geographical boundaries of species in the *Rattus hoffmanni* group and a new species from Pulau Peleng. *Bulletin of the American Museum of Natural History* **206**, 322–413.

Muzzolini A (1986) L'art rupestre préhistorique des massifs centraux sahariens. *Alfred Muzzolini BAR International Series* **318**, 1–356.

Myers RA and Worm B (2003) Rapid worldwide depletion of predatory fish communities. *Nature* **423**, 280–283.

Myers RA, Hutchings JA, and Barrowman NJ (1996) Hypotheses for the decline of cod in the North Atlantic. *Marine Ecology Progress Series* **138**, 293–308.

Myers RA, Baum JK, Shepherd TD, Powers SP, and Peterson CH (2007) Cascading effects of the loss of apex predatory sharks from a coastal ocean. *Science* **315**, 1846–1850.

Nadel H, Frank JH, and Knight JRJ (1992) Escapees and accomplices: the naturalization of exotic *Ficus* and their associated faunas in Florida. *Florida Entomologist* **75**, 29–38.

Naylor RL, Gloldburg RL, Primavera JH, Kautsky N, Beveridge MCM, Clay J, Folke C, Lubchenco J, Mooney H, and Troell M (2000) Effect of aquaculture on world fish supplies. *Nature* **405**, 1017–1024.

Neas JF and Jenkinson MA (1986) Type and figured specimens of fossil bertebrates in the collection of the University of Kansas Museum of Natural History. Part III. Fossil birds. *University of Kansas Museum of Natural History, Miscellaneous Publication* **78**, 1–13.

Nee S and May RM (1997) Extinction and the loss of evolutionary history. *Science* **278**, 692–694.

Neel JK and Allen WR (1964) The mussel fauna of the Upper Cumberland River basin before impoundment. *Malacologia* **1**, 427–459.

Nesje A, Jansen E, Birks HJB, Bjune AE, Bakke J, Andersson C, Dahl SO, Kristensen DK, Lauritzen S-E, Lie Ø *et al.* (2005) Holocene climate variability in the northern North Atlantic region: a review of terrestrial and marine evidence. In: *The Nordic Seas: an Integrated Perspective*, pp. 289–322. Geophysical Monograph Series 158. American Geophysical Union, Washington DC.

Neumann K (2005) The romance of farming: plant cultivation and domestication in Africa. In Stahl AB, ed., *African Archaeology*, pp. 249–275. Blackwell Publishing, Oxford.

Neves RJ, Bogan AE, Williams JD, Ahlstedt SA, and Hartfield PD (1997) Status of aquatic mollusks in the southeastern United States: a downward spiral of diversity. In Benz GW and Collins DE, eds, *Aquatic Fauna in Peril: the Southeastern Perspective*, pp. 43–86. Lenz Design & Communications, Decatur, Georgia.

Neves WA, González-José R, Hubbe M, Kipnis R, Araujo AGM, and Blasi O (2004) Early Holocene human skeletal remains from Cerca Grande, Lagoa Santa, Central Brazil, and the origins of the first Americans. *World Archaeology* **36**, 479–501.

Newsom LA and Wing ES (2004) *On Land and Sea. Native American Uses of Biological Resources in the West Indies.* University of Alabama Press, Tuscaloosa & London.

Newton A (1869) On a picture supposed to represent the didine bird of the island of Bourbon (Réunion). *Transactions of the Zoological Society of London* **6**, 373–376.

Newton K, Côté IM, Pilling GM, Jennings S, and Dulvy NK (2007) Current and future sustainability of island coral reef fisheries. *Current Biology* **17**, 655–658.

Nicholls GK and Jones M (2001) Radiocarbon dating with temporal order constraints. *Journal of the Royal Statistical Society, Series C* **50**, 503–521.

Nichols RA (1943) The breeding birds of St. Thomas and St. John, Virgin Islands. *Memoirs of the Society of Cuban Natural History "Felipe Poey"* **17**, 23–37.

Nicoll K (2001) Radiocarbon chronologies for prehistoric human occupation and hydroclimatic change in Egypt and northern Sudan. *Geoarchaeology* **16**, 47–64.

Nicoll K (2004) Recent environmental change and prehistoric human activity in Egypt and northern Sudan. *Quaternary Science Reviews* **23**, 561–580.

Niemi TM and Smith II AM (1999) Initial results of the southeastern Wadi Araba, Jordan geoarchaeological study: implications for shifts in Late Quaternary aridity. *Geoarchaeology* **14**, 791–820.

Nieves-Rivera ÁM and McFarlane DA (2001) In search of the extinct hutia in cave deposits of Isla de Mona, P.R. *NSS News* **59**, 92–95.

Nixon SW (1981) Remineralisation and nutrient recycling in coastal marine ecosystems. In Neilson BJ and Cronin LE, eds, *Estuaries and Nutrients*, pp. 111–138. Humana Press, Clifton.

Nogales M, Martín A, Tershie BR, Donlan CJ, Veitch D, Puerta N, Wood B, and Alonso J (2004) A review of feral cat eradication on islands. *Conservation Biology* **18**, 310–319.

Norse EA (1993) *Global Marine Biodiversity: a Strategy for Building Conservation into Decision Making.* Island Press, Washington DC.

Northcote EM (1982a) The extinct Maltese crane *Grus melitensis*. *Ibis* **124**, 76–80.

Northcote EM (1982b) Size, form and habit of the extinct Maltese swan *Cygnus falconeri*. *Ibis* **124**, 148–158.

Northcote EM (1984) Crane *Grus* fossils from the Maltese Pleistocene. *Palaeontology* **27**, 729–735.

Northcote EM (1988) An extinct 'swan-goose' from the Pleistocene of Malta. *Palaeontology* **31**, 725–740.

Northcote EM (1992) Swans (*Cygnus*) and cranes (*Grus*) from the Maltese Pleistocene. *Natural History Museum of Los Angeles County, Science Series* **36**, 285–292.

Northcote EM and Mourer-Chauviré C (1985) The distinction between the extinct pleistocene European crane *Grus primigenia*, and the extant Asian sarus crane, *G. antigone*. *Geobios* **18**, 877–881.

Northcote EM and Mourer-Chauviré C (1988) The extinct crane *Grus primigenia* MILNE-EDWARDS, in Majorca (Spain). *Geobios* **21**, 201–208.

Novotny V, Drozd P, Miller SE, Kulfan M, Janda M, Basset Y, and Weiblen GD (2006) Why are there so many species of herbivorous insects in tropical rainforests? *Science* **313**, 1115–1118.

Nowell K and Jackson P (eds) (1996) *Wild Cats: Status Survey and Conservation Action Plan.* IUCN, Gland.

Nunes Amaral LA and Meyer M (1998) Environmental changes, coextinction, and patterns in the fossil record. *Physical Review Letters* **82**, 652–655.

Núñez L, Grosjean M, and Cartajena I (2002) Human occupations and climate change in the Puna de Atacama, Chile. *Science* **298**, 821–824.

Nurse D (1997) The contributions of linguistics to the study of history in Africa. *Journal of African History* **38**, 359–391.

O'Connor S and Aplin K (2007) A matter of balance: an overview of Pleistocene occupational history and the impact of the Last Glacial Phase in east Timor and the Aru Islands, eastern Indonesia. *Archaeology in Oceania* **42**, 82–90.

O'Connor TP (1993) Birds and the scavenger niche. *Archaeofauna* **2**, 155–162.

O'Donnell CFJ and Phillipson SM (1996) Predicting the incidence of mohua predation from the seedfall, mouse, and predator fluctuations in beech forests. *New Zealand Journal of Zoology* **23**, 287–293.

Oka N (2004) The distribution of streaked shearwater (*Calonectris leucomelas*) colonies, with special attention to population size, area of sea where located and surface water temperature. *Journal of the Yamashina Institute for Ornithology* **35**, 164–188.

Olden JD (2006) Biotic homogenization: a new research agenda for conservation biogeography. *Journal of Biogeography* **33**, 2027–2039.

Olden JD and Poff NL (2003) Toward a mechanistic understanding and prediction of biotic homogenization. *American Naturalist* **162**, 442–460.

Olden JD and Poff NL (2004) Ecological processes driving biotic homogenization: testing a mechanistic model using fish faunas. *Ecology* **85**, 1867–1875.

Olden JD, Poff NL, Douglas MR, Douglas ME, and Fausch KD (2004) Ecological and evolutionary consequences of biotic homogenization. *Trends in Ecology & Evolution* **19**, 18–24.

Olden JD, Douglas ME, and Douglas MR (2005) The human dimensions of biotic homogenization. *Conservation Biology* **19**, 2036–2038.

Olden JD, Poff NL, and McKinney ML (2006) Forecasting faunal and floral homogenization associated with human population geography in North America. *Biological Conservation* **127**, 261–271.

Oldfield F (2005) *Environmental Change: Key Issues and Alternative Approaches*. Cambridge University Press, Cambridge.

Oliver WRB (1949) *The Moas of New Zealand and Australia*. Dominion Museum Bulletin 15. Dominion Museum, Wellington.

Oliver WRB (1955) *New Zealand Birds*, 2nd edn. AH and AW Reed, Wellington (1974 reprint).

Olsen EM, Heino M, Lilly GR, Morgan MJ, Brattey J, Ernande B, and Dieckamm J (2004) Maturation trends indicative of rapid evolution preceded the collapse of northern cod. *Nature* **428**, 932–935.

Olson SL (1973) Evolution of the rails of the South Atlantic Islands (Aves: Rallidae). *Smithsonian Contributions to Zoology* **152**, 1–53.

Olson SL (1974) A new species of *Nesotrochis* from Hispaniola, with notes on other fossil rails from the West Indies (Aves: Rallidae). *Proceedings of the Biological Society of Washington* **87**, 439–450.

Olson SL (1975) Paleornithology of St. Helena Island, South Atlantic Ocean. *Smithsonian Contributions to Paleobiology* **23**, 1–49.

Olson SL (1976a) A new species of *Milvago* from Hispaniola, with notes on other fossil caracaras from the West Indies (Aves: Falconidae). *Proceedings of the Biological Society of Washington* **88**, 355–366.

Olson SL (1976b) Fossil woodcocks: an extinct species from Puerto Rico and an invalid species from Malta (Aves: Scolopacidae: *Scolopax*). *Proceedings of the Biological Society of Washington* **89**, 265–74.

Olson SL (1977a) Notes on subfossil Anatidae from New Zealand, including a new species of pink-eared duck (*Malacorhynchus*). *Emu* **77**, 132–135.

Olson SL (1977b) A synopsis of the fossil Rallidae. In Ripley SD, *Rails of the World: a Monograph of the Family Rallidae*, pp. 339–373. David R. Godine, Boston.

Olson SL (1977c) Additional notes on subfossil bird remains from Ascension Island. *Ibis* **119**, 37–43.

Olson SL (1982a) A new species of palm swift (*Tachornis*: Apodidae) from the Pleistocene of Puerto Rico. *The Auk* **99**, 230–235.

Olson SL (1982b) Natural history of vertebrates on the Brazilian islands of the mid South Atlantic. *National Geographic Society Research Reports* **13**, 481–492.

Olson SL (1985a) A new species of *Siphonorhis* from Quaternary cave deposits in Cuba (Aves: Caprimulgidae). *Proceedings of the Biological Society of Washington* **98**, 526–532.

Olson SL (1985b) Pleistocene birds of Puerto Rico. *National Geographic Society Research Reports* **18**, 563–566.

Olson SL (1985c) Early Pliocene Procellariiformes (Aves) from Langebaanweg, south-western Cape Province, South Africa. *Annals of the South African Museum* **95**, 123–145.

Olson SL (1986a) Emendation of the name of the fossil rail *Rallus hodgeni* Scarlett. *Notornis* **33**, 32.

Olson SL (1986b) *Gallirallus sharpei* (Büttikofer) nov. comb. A valid species of rail (Rallidae) of unknown origin. *Le Gerfaut* **76**, 263–269.

Olson SL (1986c) An early account of some birds from Mauke, Cook Islands, and the origin of the "mysterious starling" *Aplonis mavornata* Buller. *Notornis* **33**, 197–208.

Olson SL (1991) The fossil record of the genus *Mycteria* (Ciconiidae) in North America. *The Condor* **93**, 1004–1006.

Olson SL (2004) Taxonomic review of the fossil Procellariidae (Aves: Procellariiformes) described from Bermuda by R. W. Shufeldt. *Proceedings of the Biological Society of Washington* **117**, 575–581.

Olson SL (2005) Refutation of the historical evidence for a Hispaniolan macaw (Aves: Psittacidae: *Ara*). *Caribbean Journal of Science* **41**, 319–323.

Olson SL (2006) Birds, including extinct species, encountered by the Malaspina Expedition on Vava'u, Tonga, in 1793. *Archives of Natural History* **33**, 42–52.

Olson SL (2007) The "walking eagle" *Wetmoregyps daggetti* Miller: a scaled-up version of the savanna hawk (*Buteogallus meridionalis*). *Ornithological Monographs* **63**, 110–114.

Olson SL and Wetmore A (1976) Preliminary diagnoses of two extraordinary new genera of birds from Pleistocene deposits in the Hawaiian Islands. *Proceedings of the Biological Society of Washington* **89**, 247–258.

Olson SL and Steadman DW (1977) A new genus of flightless ibis (Threskiornithidae) and other fossil birds from cave deposits in Jamaica. *Proceedings of the Biological Society of Washington* **90**, 447–457.

Olson SL and Steadman DW (1979) The humerus of *Xenicibis*, the extinct flightless ibis of Jamaica.

Proceedings of the Biological Society of Washington **92**, 23–27.

Olson SL and Hilgartner WB (1982) Fossil and subfossil birds from the Bahamas. *Smithsonian Contributions to Paleobiology* **48**, 22–56.

Olson SL and McKitrick MC (1982) A new genus and species of emberizine finch from Pleistocene cave deposits in Puerto Rico (Aves: Passeriformes). *Journal of Vertebrate Paleontology* **1**, 279–283.

Olson SL and James HF (1982) Prodromus of the fossil avifauna of the Hawaiian Islands. *Smithsonian Contributions to Zoology* **365**, 1–59.

Olson SL and James HF (1984) The role of Polynesians in the extinction of the avifauna of the Hawaiian islands. In Martin PS and Klein RG, eds, *Quaternary Extinctions: a Prehistoric Revolution*, pp. 768–780. University of Arizona Press, Tucson.

Olson SL and Kurochkin EN (1987) Fossil evidence of a tapaculo in the Quaternary of Cuba (Aves: Passeriformes: Scytalopodidae). *Proceedings of the Biological Society of Washington* **100**, 353–357.

Olson SL and Parris DC (1987) The Cretaceous birds of New Jersey. *Smithsonian Contributions to Paleobiology* **63**, 1–22.

Olson SL and James HF (1991) Descriptions of thirty-two new species of birds from the Hawaiian Islands: Part I. Non-Passeriformes. *Ornithological Monographs* **45**, 1–88.

Olson SL and Ziegler AC (1995) Remains of land birds from Lisianski Island, with observations on the terrestrial avifauna of the northwestern Hawaiian Islands. *Pacific Science* **49**, 111–125.

Olson SL and Jouventin P (1996) A new species of small flightless duck from Amsterdam Island, southern Indian Ocean (Anatidae: *Anas*). *The Condor* **98**, 1–9.

Olson SL and Rasmussen PD (2001) Miocene and Pliocene birds from the Lee Creek Mine, North Carolina. *Smithsonian Contributions to Paleobiology* **90**, 233–365.

Olson SL and Hearty PJ (2003) Probable extirpation of a breeding colony of short-tailed Albatross (*Phoebastria albatrus*) on Bermuda by Pleistocene sea-level rise. *Proceedings of the National Academy of Sciences USA* **100**, 12825–12829.

Olson SL and Wingate DB (2006) A new species of night-heron (Ardeidae: *Nyctanassa*) from Quaternary deposits on Bermuda. *Proceedings of the Biological Society of Washington* **119**, 326–337.

Olson SL, Fleischer RC, Fisher CT, and Bermingham E (2005a) Expunging the 'Mascarene starling' *Necropsar leguati*: archives, morphology and molecules topple a myth. *Bulletin of the British Ornithologists' Club* **125**, 31–42.

Olson SL, Wingate DB, Hearty PJ, and Grady FV (2005b) Prodromus of vertebrate paleontology and geochronology of Bermuda. *Monografies de la Societat d'Història Natural de les Balears* **12**, 219–232.

Onley D and Scofield RP (2007) *Albatrosses, Petrels and Shearwaters of the World*. Chistopher Helm, London.

Opler PA (1978) Insects of American chestnut: possible importance and conservation concern. In McDonald J, ed., *The American Chestnut Symposium*, pp. 83–85. West Virginia University Press, Morgantown.

Orlando L, Darlu P, Toussaint D, Bonjean D, Otte M, and Hänni C (2006a) Revisiting Neanderthal diversity with a 100,000 year old mtDNA sequence. *Current Biology* **16**, 400–401.

Orlando L, Mashkour M, Burke A, Douady CJ, Eisenmann V, and Hänni C (2006b) Geographic distribution of an extinct equid (*Equus hydruntinus*: Mammalia, Equidae) revealed by morphological and genetical analyses of fossils. *Molecular Ecology* **15**, 2083–2093.

Orliac C (2000) The woody vegetation of Easter Island between the early 14th and the mid-17th centuries A.D. In Stevenson C and Ayres W, eds, *Easter Island Archaeology: Research on Early Rapanui Culture*, pp. 199–207. Easter Island Foundation, Los Osos.

Oro D, Aguilar JS, Igual JM, and Louzao M (2004) Modelling demography and extinction risk in the endangered Balearic shearwater. *Biological Conservation* **116**, 93–102.

Ortmann AE (1909) Unionidae from an Indian garbage heap. *The Nautilus* **23**, 11–15.

Ortmann AE (1918) The nayades (freshwater mussels) of the Upper Tennessee drainage. With notes on synonymy and distribution. *Proceedings of the American Philosophical Society* **57**, 521–626.

Ortmann AE (1924) The naiad-fauna of Duck River in Tennessee. *The American Midland Naturalist* **9**, 18–62.

Ortmann AE (1925) The naiad-fauna of the Tennessee River system below Walden Gorge. *The American Midland Naturalist* **9**, 321–372.

Ortmann AE (1926) The naides of the Green River drainage in Kentucky. *Annals of the Carnegie Museum* **17**, 167–189.

Otto-Bliesner BL, Marshall SJ, Overpeck JT, Miller GH, Hu A, and CAPE Last Interglacial Project Members (2006) Simulating Arctic climate warmth and icefield retreat in the last interglaciation. *Science* **311**, 1751–1753.

Overpeck JT, Webb RS, and Webb III T (1992) Mapping eastern North American vegetation changes of the 18 ka: no-analogs and the future. *Geology* **20**, 1071–1074.

Owens IPF and Bennett PM (2000) Ecological basis of extinction risk in birds: habitat loss versus human persecution and introduced predators. *Proceedings of the National Academy of Sciences USA* **97**, 12144–12148.

Owen-Smith N (1987) Pleistocene extinctions: the pivotal role of megaherbivores. *Paleobiology* **13**, 351–362.

Pace ML, Cole JJ, Carpenter SR, and Kitchell JF (1999) Trophic cascades revealed in diverse ecosystems. *Trends in Ecology & Evolution* **14**, 483–488.

Pachur H-J and Roper H-P (1984) The Libyan (Western) Desert and northern Sudan during the Late Pleistocene and Holocene. In Klitzsch E, Said S, and Schrank E, eds, *Research in Egypt and Sudan: results of the Special Research Project Arid Areas 1981–1984*, pp. 249–284. D. Reimer Publishers, Berlin.

Paddle R (2000) *The Last Tasmanian Tiger: the History and Extinction of the Thylacine*. Cambridge University Press, Cambridge.

Paine RT (1980) Food webs: linkage, interaction strength and community infrastructure. *Journal of Animal Ecology* **49**, 667–685.

Palmer M, Pons GX, Cambefort I, and Alcover JA (1999) Historical processes and environmental factors as determinants of inter-island differences in endemic faunas: the case of the Balearic Islands. *Journal of Biogeography* **26**, 813–823.

Pandolfi JM and Jackson JBC (2006) Ecological persistence interrupted in the Caribbean. *Ecology Letters* **9**, 818–826.

Pandolfi JM, Bradbury RH, Sala E, Hughes TP, Bjorndal KA, Cooke RG, McArdle D, McClenachan L, Newman MJH, Paredes G *et al.* (2003) Global trajectories of the long-term decline of coral reef ecosystems. *Science* **301**, 955–958.

Panteleyev AV and Nessov LA (1987) A small tubinare (Aves: Procellariiformes) from the Eocene of Middle Asia. *Trudy Zoologicheskogo Instituta* **252**, 95–103.

Paquet PC, Carroll C, Noss RF, and Schumaker NH (2004) Extinction debt of protected areas in developing landscapes. *Conservation Biology* **18**, 1110–1120.

Pardiñas UFJ and Tonni EP (2000) A giant vampire (Mammalia, Chiroptera) in the Late Holocene from the Argentinian pampas: paleoenvironmental significance. *Palaeogeography, Palaeoclimatology, Palaeoecology* **160**, 213–221.

Parker SA (1984) The extinct Kangaroo Island emu, a hitherto-unrecognized species. *Bulletin of the British Ornithologists' Club* **104**, 19–22.

Parker WK (1865) Preliminary notes on some fossil birds from the Zebbug Cave, Malta. *Proceedings of the Zoological Society of London* **1865**, 752–753.

Parker WK (1869) On some fossil birds from the Zebbug Cave, Malta. *Transactions of the Zoological Society of London* **6**, 119–24.

Parmalee PW (1956) A comparison of past and present populations of fresh-water mussels in southern Illinois. *Illinois Academy of Science Transactions* **49**, 184–192.

Parmalee PW (1967) The fresh-water mussels of Illinois. *Illinois State Museum Popular Science Series* **8**, 1–108.

Parmalee PW (1969) Animal remains from the archaic Riverton, Swan, Island, and Robeson Hills sites, Illinois. In Winters HD, ed., *The Riverton Culture*, pp. 104–113. Illinois State Museum Reports of Investigations 13 and Illinois State Archaeological Survey Monograph 1. Illinois State Museum & Illinois State Archaeological Survey, Springfield.

Parmalee PW (1994) Freshwater mussels from Dust and Smith Bottom caves, Alabama. *Journal of Alabama Archaeology* **40**, 135–162.

Parmalee PW and Bogan AE (1986) Molluscan remains from aboriginal middens at the Clinch River breeder reactor plant site, Roane County, Tennessee. *American Malacological Bulletin* **4**, 25–37.

Parmalee PW and Klippel WE (1974) Freshwater mussels as a pre-Columbian food source. *American Antiquity* **39**, 421–434.

Parmalee PW and Klippel WE (1982) A relic population of *Obovaria retusa* in the Middle Cumberland River, Tennessee. *The Nautilus* **96**, 30–32.

Parmalee PW and Klippel WE (1986) A pre-Columbian aboriginal freshwater mussel assemblage from the Duck River in Middle Tennessee. *The Nautilus* **100**, 134–140.

Parmalee PW and Polhemus RR (2004) Pre-Columbian and pre-impoundment populations of freshwater mussels (Bivalvia: Unionidae) in the South Fork Holston River, Tennessee. *Southeastern Naturalist* **3**, 231–240.

Parmalee PW, Klippel WE, and Bogan AE (1980) Notes on the pre-Columbian and present status of the naiad fauna of the middle Cumberland River, Smith County, Tennessee. *The Nautilus* **94**, 93–105.

Parmalee PW, Klippel WE, and Bogan AE (1982) Aboriginal and modern freshwater mussel assemblages (Pelecypoda: Unionidae) from the Chickamauga Reservoir, Tennessee. *Brimleyana* **8**, 75–90.

Parshall T and Foster DR (2002) Fire on the New England landscape: regional and temporal variation, cultural and environmental controls. *Journal of Biogeography* **29**, 1309–1317.

Partridge JH (1967) A 3,300 year old thylacine (Marsupialia: Thylacinidae) from the Nullarbor Plain, Western Australia. *Journal of the Royal Society of Western Australia* **50**, 57–59.

Pascal M (1980) Structure et dynamique de la population de chats harets de l'archipel des Kerguelen. *Mammalia* **44**, 161–182.

Pascal M and Lorvelec O (2005) Holocene turnover of the French vertebrate fauna. *Biological Invasions* **7**, 99–106.

Pasveer JM (1998) Kria Cave: an 8000-year occupation sequence from the Bird's Head of Irian Jaya. In Bartstra

G-J, ed., *Bird's Head approaches: Irian Jaya studies—a programme for interdisciplinary research*, pp. 67–89. A.A. Balkema, Rotterdam.

Patel SR (2006) The value of small things. Why save what we love to kill? *SEED* August, 8.

Patterson III WA and Backman AE (1988) Fire and disease history of forests. In Huntley B and Webb III T, eds, *Handbook of Vegetation Science. Volume VII. Vegetation History*, pp. 673–697. Kluwer Academic Publications, Dordrecht.

Pauly D (2004) Much rowing for fish. *Nature* **432**, 813–814.

Pauly D and Alder J (2006) Marine fisheries systems. In Hassan R, Scholes R, and Ash N, eds, *Ecosystems and Human Well-being: Current Trends and Status*, pp. 477–511. Island Press, Washington DC.

Pauly D, Christensen V, Dalsgaard J, Froese R, and Torres Jr F (1998) Fishing down marine food webs. *Science* **279**, 860–863.

Pauly D, Alder J, Bennett E, Christensen V, Tyedmers P, and Watson R (2003) The future for fisheries. *Science* **302**, 1359–1361.

Pauly D, Watson R, and Alder J (2005) Global trends in world fisheries: impacts on marine ecosystems and food security. *Philosophical Transactions of the Royal Society of London B* **360**, 5–12.

Pavia M (2000) *Le avifaune pleistoceniche dell'Italia meridionale.* Tesi Università di Torino Scienze de la Terra, Torino.

Paxinos EE, James H, Olson SL, Sorensen MD, Jackson J, and Fleischer R (2002) mtDNA from fossils reveals a radiation of Hawaiian geese recently derived from the Canada goose (*Branta canadensis*). *Proceedings of the National Academy of Sciences USA* **99**, 1399–1404.

Peacock E (1998) Historical and applied perspectives on prehistoric land use in eastern North America. *Environment and History* **4**, 1–29.

Peacock E (2000a) Assessing bias in archaeological shell assemblages. *Journal of Field Archaeology* **27**, 183–196.

Peacock E (2000b) Molluscan analysis. In Walling R, Alexander L, and Peacock E, ed., *The Jefferson Street Bridge Project: Archaeological Investigations at the East Nashville Mounds Site (40Dv4) and the French Lick/Sulphur Dell site (40Dv5) in Nashville, Davidson County, Tennessee, Vol. II*, pp. 391–419. Publications in Archaeology 7. Tennessee Department of Transportation, Office of Environmental Planning and Permits, Nashville.

Peacock E and James TR (2002) A pre-Columbian unionid assemblage from the Big Black River drainage in Hinds County, Mississippi. *Journal of the Mississippi Academy of Sciences* **47**, 119–123.

Peacock E, Haag WR, and Warren Jr ML (2005) Pre-Columbian decline in freshwater mussels coincident with the advent of maize agriculture. *Conservation Biology* **19**, 547–551.

Pearce F (1999) Sungbo Erdo of Nigeria, a neglected past. *New Scientist* 11 September, 39–41.

Pearl FB and Dickson DB (2004) Geoarchaeology and prehistory of the Kipsing and Tol River watersheds in the Mukogodo Hills region of central Kenya. *Geoarchaeology* **19**, 565–582.

Peebles CS (1978) Determinants of settlement size and location in the Moundville phase. In Smith BD, ed., *Mississippian Settlement Patterns*, pp. 369–416. Academic Press, New York.

Pereira E, Ottaviani-Spella M-M, Salotti M, Louchart A, and Quinif Y (2006) Tentative de reconstitution paléoenvironmentale de deux dépôts Quaternaires Corses. *Geologica Belgica* **9**, 267–273.

Perez JM and Palma RL (2001) A new species of *Felicola* (Phthiraptera: Trichodectidae) from the endangered Iberian lynx: another reason to ensure its survival. *Biodiversity and Conservation* **10**, 929–937.

Perez VR, Godfrey LR, Nowak-Kemp M, Burney DA, Ratsimbazafy J, and Vasey N (2005) Evidence of early butchery of giant lemurs in Madagascar. *Journal of Human Evolution* **49**, 722–742.

Perrings C, Dehnen-Schmutz K, Touza J, and Williamson M (2005) How to manage biological invasions under globalization. *Trends in Ecology & Evolution* **20**, 212–215.

Peterken GF (1996) *Natural Woodland: Ecology and Conservation in Northern Temperate Regions.* Cambridge University Press, Cambridge.

Peterman RM and M'Gonigle M (1992) Statistical power analysis and the precautionary principle. *Marine Pollution Bulletin* **24**, 231–234.

Peters J (1992) Late Quaternary mammalian remains from central and eastern Sudan and their palaeoenvironmental significance. *Palaeoecology of Africa* **23**, 91–115.

Peters J (1998) *Camelus thomasi* Pomel, 1893, a possible ancestor of the one-humped camel? *Zeitschrift für Säugetierkunde* **63**, 372–376.

Peters J, Gautier A, Brink JS, and Haenen W (1994) Late Quaternary extinction of ungulates in sub-Saharan Africa: a reductionist's approach. *Journal of Archaeological Science* **21**, 17–28.

Peters SE and Foote M (2002) Determinants of extinction in the fossil record. *Nature* **416**, 420–424.

Petersen JB (1996) Archaeology of Trants, Montserrat. Part 3. Chronological and settlement data. *Annals of Carnegie Museum* **65**, 323–361.

Petts GE (1994) Rivers: dynamic components of catchment ecosystems. In Calow P, and Petts GE, eds, *The Rivers Handbook, Volume 2*, pp. 3–22. Blackwell Science, Oxford.

Pfenninger M, Posada D, and Magnin F (2003) Evidence for the survival of Pleistocene climatic changes in Northern refugia by the land snail *Trochoidea geyeri* (Soós 1926) (Helicellinae, Stylommatophora). *BMC Evolutionary Biology* **3**, 8.

Phillimore AB, Freckleton RP, Orme CDL, and Owens IPF (2006) Ecology predicts large-scale patterns of phylogenetic diversification in birds. *American Naturalist,* **168**, 220–229.

Phillipps WJ (1959) The last (?) occurrence of *Notornis* in the North Island. *Notornis* **8**, 93–94.

Phillips RA, Petersen MK, Lilliendahl K, Solmundsson J, Hamer KC, Camphuysen CJ, and Zonfrillo B (1999) Diet of the northern fulmar *Fulmarus glacialis*: reliance on commercial fisheries? *Marine Biology* **135**, 159–170.

Pickering J and Norris CA (1996) New evidence concerning the extinction of the endemic murid *Rattus macleari* from Christmas Island, Indian Ocean. *Australian Mammalogy* **19**, 19–25.

Piet GJ and Rice JC (2004) Performance of precautionary reference points in providing management advice on North Sea fish stocks. *ICES Journal of Marine Science* **61**, 1305–1312.

Pieper H (1984) A new species of *Mesocricetus* (Mammalia: Cricetidae) from the Greek island Armathia. *Stuttgarter Beiträge zur Naturkunde, Serie B (Geologie und Paläontologie)* **107**, 1–9.

Pieper H (1985) The fossil land birds of Madeira and Porto Santo. *Bocagiana* **88**, 1–6.

Pikitch EK, Santora C, Babcock EA, Bakun A, Bonfil R, Conover DO, Dayton PK, Doukakis P, Fluharty D, Heneman B *et al.* (2004) Ecosystem-based fishery management. *Science* **305**, 346–347.

Pimm SL (1996) Lessons from the kill. *Biodiversity and Conservation* **5**, 1059–1067.

Pimm SL and Askins RA (1995) Forest losses predict bird extinctions in eastern North America. *Proceedings of the National Academy of Sciences USA* **92**, 9343–9347.

Pimm SL, Jones HL, and Diamond J (1988) On the risk of extinction. *American Naturalist* **132**, 757–785.

Pimm SL, Moulton MP, and Justice LJ (1994) Bird extinctions in the central Pacific. *Philosophical Transactions of the Royal Society of London B* **344**, 27–33.

Pimm SL, Moulton MP, and Justice LJ (1995) Bird extinctions in the central Pacific. In Lawton JH and May RM, eds, *Extinction Rates*, pp. 75–87. Oxford University Press, Oxford.

Pimm SL, Raven P, Peterson A, Sekercioglu ÇH, and Ehrlich PR (2006) Human impacts on the rates of recent, present and future bird extinctions. *Proceedings of the National Academy of Sciences USA* **103**, 10941–10946.

Pinnegar JK and Engelhard GH (2008) The 'shifting baseline' phenomenon: a global perspective. *Reviews in Fish Biology and Fisheries* **18**, 1–16.

Pinnegar JK, Polunin NVC, Francour P, Badalamenti F, Chemello R, Harmelin-Vivien M-L, Hereu B, Milazzo M, Zabala M, D'Anna G, and Pipitone C (2000) Trophic cascades in benthic marine ecosystems: lessons for fisheries and protected-area management. *Environmental Conservation* **27**, 179–200.

Piperno DR and Pearsall DM (2002) *The Origins of Agriculture in the Lowland Neotropics.* Academic Press, San Diego.

Pitcher TJ (1998) A cover story: fisheries may drive stocks to extinction. *Reviews in Fish Biology and Fisheries* **8**, 367–370.

Plane M (1976) The occurrence of *Thylacinus* in Tertiary rocks from Papua New Guinea. *BMR Journal of Australian Geology and Geophysics* **1**, 78–79.

Podolsky RH and Kress SW (1989) Factors affecting colony formation in Leach's storm-petrel. *The Auk* **106**, 332–336.

Podolsky R and Kress SW (1992) Attraction of the endangered dark-rumped petrel to recorded vocalizations in the Galapagos Islands. *The Condor* **94**, 448–453.

Polis GA, Anderson WB, and Holt RD (1997) Toward an integration of landscape and food web ecology: the dynamics of spatially subsidized food webs. *Annual Review of Ecology and Systematics* **28**, 289–316.

Pope KO, Pohl MED, Jones JG, Lentz DL, von Nagy C, Vega FJ, and Quitmyer IR (2001) Origin and environmental setting of ancient agriculture in the lowlands of Mesoamerica. *Science* **292**, 1370–1373.

Poplin F (1980) *Sylviornis neocaledoniae* n. g., n. sp. (Aves), ratite eteint de la Nouvelle-Caledonie. *Comptes Rendus des Séances de l'Academie des Sciences, Série D: Sciences Naturelles* **290**, 691–694.

Poplin F and Mourer-Chauviré C (1985) *Sylviornis neocaledoniae* (Aves, Galliformes, Megapodiidae), oiseau géant éteint de l'Ile des Pins (Nouvelle-Calédonie). *Geobios* **18**, 73–97.

Poplin F, Mourer-Chauviré C, and Evin J (1983) Position systématique et datation de *Sylviornis neocaledoniae,* mégapode géant (Aves, Galliformes, Megapodiidae) éteint de la Nouvelle-Calédonie. *Comptes Rendus de l'Academie des Sciences (Paris), Séries 2a* **297**, 301–304.

Potts TH (1871) On the birds of New Zealand, part II. *Transactions and Proceedings of the New Zealand Institute* **3**, 59–106.

Poulin R and Morand S (2000) The diversity of parasites. *The Quarterly Review of Biology* **75**, 277–293.

Powell DC, Aulerich R, Stromborg KL, and Bursian SJ (1996) Effects of 3,3′,4,4′-tetrachlorobiphenyl, 2,3,3′,4,4′-

pentachlorobiphenyl, and 3,3′,4,4′,5-pentachlorobiphenyl on the developing chicken embryo when injected prior to incubation. *Journal of Toxicology and Environmental Health* **49**, 319–38.

Preece RC (1997) The spatial response of non-marine Mollusca to past climate change. In Huntley B, Cramer W, Prentice AV, and Allen JRM, eds, *Past and Future Rapid Environmental Change: the Spatial and Evolutionary Responses of Terrestrial Biota*. Springer, Berlin.

Preece RC (1998) Impact of early Polynesian occupation on the land snail fauna of Henderson Island, Pitcairn Group (south Pacific). *Philosophical Transactions of the Royal Society of London B* **353**, 347–368.

Preece RC and Bridgland DR (1998) *Late Quaternary Environmental Change in North-west Europe: Excavations at Holywell Coombe, South-east England*. Chapman and Hall, London.

Pregill G (1981) Late Pleistocene herpetofaunas from Puerto Rico. *University of Kansas Museum of Natural History, Miscellaneous Publication* **71**, 1–72.

Pregill GK and Olson SL (1981) Zoogeography of West Indian vertebrates in relation to Pleistocene climatic cycles. *Annual Review of Ecology and Systematics* **12**, 75–98.

Pregill GK, Steadman DW, and Watters DR (1994) Late Quaternary vertebrate faunas of the Lesser Antilles: historical components of Caribbean biogeography. *Bulletin of Carnegie Museum of Natural History* **30**, 1–51.

Prentice IC (1986) Vegetation response to past climate changes. *Vegetatio* **67**, 131–141.

Prentice IC, Jolly D, and BIOME 6000 participants (2000) Mid-Holocene and glacial-maximum vegetation geography of the northern continents and Africa. *Journal of Biogeography* **27**, 507–519.

Prevosti FJ and Vizcaíno SF (2006) Paleoecology of the large carnivore guild from the late Pleistocene of Argentina. *Acta Palaeontologica Polonica* **51**, 407–422.

Price PW (1980) *Evolutionary Biology of Parasites*. Princeton University Press, Princeton.

Price RD, Clayton DH, and Adams RJ (2000) Pigeon lice down under: taxonomy of Australian *Campanulotes* (Phthiraptera: Philopteridae), with a description of *C. durdeni* n. sp. *Journal of Parasitology* **86**, 948–950.

Price RD, Hellenthal RA, Palma RL, Johnson KP, and Clayton DH (2003) *Chewing Lice: World Checklist and Biological Overview*. Special Publication 24. Illinois Natural History Survey, Champaign.

Priddel D, Carlile N, and Wheeler R (2006) Establishment of a new breeding colony of Gould's petrel (*Pterodroma leucoptera leucoptera*) through the creation of artificial nesting habitat and the translocation of nestlings. *Biological Conservation* **128**, 553–563.

Pringle H (1998) The slow birth of agriculture. *Science* **282**, 1446–1450.

Pritchard P (1982) Nesting of leatherback turtle *Dermochelys coriacea* in Pacific Mexico, with a new estimate of the world population status. *Copeia* **4**, 741–747.

Pruett CL and Winker K (2005) Biological impacts of climate change on a Beringian endemic: cryptic refugia in the establishment and differentiation of the rock sandpiper (*Calidris ptilocnemis*). *Climatic Change* **68**, 219–240.

Pruett CL and Winker K (2008) Evidence for cryptic northern refugia among high- and temperate-latitude species in Beringia. A response to Stewart and Dalen (2008). *Climatic Change* **86**, 23–27.

Prummel W (2005) The avifauna of the Hellenistic town of New Halos, Thessaly, Greece. In Grupe G and Peters J, eds, *Feathers, Grit and Symbolism: Birds and Humans in the Ancient Old and New World. Proceedings of the Fifth Meeting of the ICAZ Bird Working Group. Documenta Archaeobiologiae* **3**, 349–360.

Purvis A and Hector A (2000) Getting the measure of biodiversity. *Nature* **405**, 212–219.

Purvis AP, Agapow P-M, Gittleman JL, and Mace GM (2000a) Nonrandom extinction and the loss of evolutionary history. *Science* **288**, 328–330.

Purvis A, Jones KE, and Mace GM (2000b) Extinction. *BioEssays* **22**, 1123–1133.

Purvis A, Katzourakis A, and Agapow P-M (2002) Evaluating phylogenetic tree shape: two modifications to Fusco and Cronk's method. *Journal of Theoretical Biology* **214**, 99–103.

Quitmyer IR (2003) Zooarchaeology of Cinnamon Bay, St. John, U.S. Virgin Islands: pre-Columbian overexploitation of animal resources. *Bulletin of the Florida Museum of Natural History* **44**, 131–158.

Raab LM (1992) An optimal foraging analysis of pre-Columbian shellfish collecting on San Clemente Island, California. *Journal of Ethnobiology* **12**, 63–80.

Rackham O (1980) *Ancient Woodland: its History, Vegetation and Uses in England*. Edward Arnold, London.

Raffaele HA (1979) The status of some endangered species in Puerto Rico with particular emphasis on *Isolobodon* (Rodentia). In *Tercer simposio Departamento de Recursos Naturales*, pp. 100–104. Departamento de Recursos Naturales, San Juan.

Rahmstorf S and Schellnhuber HJ (2007) *Der Klimawandel*. Verlag C.H. Beck, Munchen.

Ramis D and Alcover JA (2005) Holocene extinction of endemic mammals of the Mediterranean Islands: some methodological questions and an update. *Monografies de la Societat d'Història Natural de les Balears* **12**, 309–318.

Rando JC (2002) New data of fossil birds from El Hierro (Canary Islands): probable causes of extinction and some biogeographical considerations. *Ardeola* **49**, 39–49.

Rando JC and Perera MA (1994) Primeros datos de ornitofagia entre los aborigenes de Fuerteventura (Islas Canarias). *Archaeofauna* **3**, 13–19.

Rando JC and Lopez M (1996) Un nuevo yacimiento de vertebrados fosiles en Tenerife (Islas Canarias). *Proceedings 7th International Symposium on Vulcanospeleology* **1**, 171–173.

Rando JC and Alcover JA (2007) Evidence for a second western Palaearctic seabird extinction during the last Millennium: the lava shearwater *Puffinus olsoni*. *Ibis* **150**, 188–192.

Rando JC, Lopez M, and Jimenez MC (1997) Bird remains from the archaeological site of Guinea (El Hierro, Canary Islands). *International Journal of Osteoarchaeology* **7**, 298–302.

Rando JC, Lopez M, and Seguí B (1999) A new species of extinct flightless passerine (Emberizidae: *Emberiza*) from the Canary Islands. *The Condor* **101**, 1–13.

Rasmussen PC and King BF (1998) The rediscovery of the forest owlet *Athene (Heteroglaux) blewitti*. *Forktail* **14**, 53–55.

Rasmussen PC and Prŷs-Jones RP (2003) History vs mystery: the reliability of museum specimen data. *Bulletin of the British Ornithologists' Club* **123A**, 66–94.

Rasoloarison RM, Goodman SM, and Ganzhorn JU (2000) Taxonomic revisions of mouse lemurs (*Microcebus*) in the western portions of Madagascar. *International Journal of Primatology* **21**, 963–1019.

Rauch EM and Bar-Yam Y (2004) Theory predicts the uneven distribution of genetic diversity within species. *Nature* **43**, 449–452.

Ravier C and Fromentin J-M (2001) Long-term fluctuations in the eastern Atlantic and Mediterranean bluefin tuna population. *ICES Journal of Marine Science* **58**, 1299–1317.

Ray CE (1962) *Oryzomyine Rodents of the Antillean Subregion*. PhD thesis, Harvard University.

Raynaud D, Barnola J-M, Chappellaz J, Blunier T, Indermühle A, and Stauffer B (2000) The ice record of greenhouse gases: a view in the context of future changes. *Quaternary Science Reviews* **19**, 9–17.

Raynal M (2002) Une répresentation picturale de l'oiseau mystérieux d'Hiva-Oa. *Cryptozoologia* **47**, 3–10.

Rayner MJ, Hauber ME, Imber MJ, Stamp RK, and Clout MN (2007) Spatial heterogeneity of mesopredator release within an oceanic island system. *Proceedings of the National Academy of Sciences USA* **104**, 20862–20865.

Rea AM (1980) Late Pleistocene and Holocene turkeys in the Southwest. *Contributions in Science, Natural History Museum of Los Angeles County* **330**, 209–224.

Reaka-Kudla ML (1997) The global biodiversity of coral reefs: a comparison with rainforests. In Reaka-Kudla ML, Wilson D, and Wilson EO, eds, *Biodiversity II*, pp. 83–108. Joseph Henry Press, Washington DC.

Reed DL, Smith VS, Hammond SL, Rogers AR, and Clayton DH (2004) Genetic analysis of lice supports direct contact between modern and archaic humans. *PloS Biology* **2**, 1972–1983.

Reed JM (1996) Using statistical probability to increase confidence of inferring species extinction. *Conservation Biology* **10**, 1283–1285.

Reese DS, Belluomini G, and Ikeya, M (1996) Absolute dates for the Pleistocene fauna of Crete. In Reese DS, ed., *Pleistocene and Holocene Fauna of Crete and its First Settlers*, pp. 47–51. Prehistory Press, Madison.

Reimer PJ, Baillie MGL, Bard E, Bayliss A, Beck J, Bertrand CJH, Blackwell PG, Buck CE, Burr GS, Cutler KB *et al.* (2004) IntCal04 terrestrial radiocarbon age calibration, 0–26 cal kyr BP. *Radiocarbon* **46**, 1029–1058.

Reis KR and Steadman DW (1999) Archaeology of Trants, Montserrat. Part 5: prehistoric avifauna. *Annals of the Carnegie Museum* **68**, 275–287.

Reise K (2005) Coast of change: habitat loss and transformations in the Wadden Sea. *Helgoland Marine Research* **59**, 9–21.

Reitz EJ and McEwan BG (1995) Animals, environment, and the Spanish diet at Puerto Real. In Deagan K, ed., *Puerto Real: the Archaeology of a Sixteenth-Century Spanish Town in Hispaniola*, pp. 287–334. University Press of Florida, Gainesville.

Reyes-Arriagada R, Campos-Ellwanger P, Schlatter R, and Baduini C (2007) Sooty Shearwater (*Puffinus griseus*) on Guafo Island: the largest seabird colony in the world? *Biodiversity and Conservation* **16**, 913–930.

Reynolds JD, Dulvy NK, Goodwin NB, and Hutchings JA (2005) Biology of extinction risk in marine fishes. *Proceedings of the Royal Society of London Series B Biological Sciences* **272**, 2337–2344.

Rezende E, Lavabre JE, Guimaraes PR, Jordano P, and Bascompte J (2007) Non-random coextinctions in phylogenetically structured mutualistic networks. *Nature* **448**, 925–928.

Ricciardi A (2007) Are modern biological invasions an unprecedented form of global change? *Conservation Biology* **21**, 329–336.

Rice DS and Rice PM (1984) Collapse intact. Postclassical archaeology of the Peten Maya. *Archaeology* **37**, 46–51.

Rice DW (1998) *Marine Mammals of the World: Systematics and Distribution*. Society for Marine Mammalogy, Lawrence.

Rice DW and Kenyon KW (1962) Breeding distribution, history and populations of North Pacific albatrosses. *The Auk* **79**, 365–386.

Richardson DM, Allsopp N, D'Antonio C, Milton SJ, and Rejmanek M (2000) Plant invasions: the role of mutualisms. *Biological Reviews* **75**, 65–93.

Rick TC, Erlandson JM, and Vellanoweth RL (2001) Paleocoastal marine fishing on the Pacific coast of the Americas: perspectives from Daisy Cave, California. *American Antiquity* **66**, 595–613.

Ricklefs RE (2003) Global diversification rates of passerine birds. *Proceedings of the Royal Society of London Series B Biological Sciences* **270**, 2285–2291.

Riera N, Traveset A, and García O (2002) Breakage of mutualisms by exotic species: the case of *Cneorum tricoccon* L. in the Balearic Islands (Western Mediterranean Sea). *Journal of Biogeography* **29**, 713–719.

Rijsdijk KF, Hume JP, Bunnik F, Florens V, Baider C, Shapiro B, van der Plicht H, Janoo A, Griffiths O, van den Hoek Ostende LW *et al.* (2009) Prehuman middle Holocene Concentration-Lagerstätten on an oceanic volcanic island: Mare aux Songes, Mauritius. *Quaternary Science Reviews* (in press).

Rímoli R (1977) Une nueva especie de monos (Cebidae: Saiminae: *Saimiri*) de la Hispaniola. *Cuadernos del CENDIA, Universidad Autonoma de Santo Domingo* **242**, 5–14.

Ritchie JC (1984) *Past and Present Vegetation of Far Northwest Canada.* University of Toronto Press, Toronto.

Roark EB, Guilderson TP, Dunbar RB, and Ingram BL (2006) Radiocarbon-based ages and growth rates of Hawaiian deep-sea corals. *Marine Ecology Progress Series* **327**, 1–14.

Roberts CM and Hawkins JP (1999) Extinction risk in the sea. *Trends in Ecology & Evolution* **14**, 241–246.

Roberts C, McClean C, Allen G, Hawkins J, McAllister D, Mittermeier C, Schueler F, Spalding M, Veron E, Wells F *et al.* (2002) Biodiversity hotspots and conservation priorities in the sea. *Science* **295**, 1280–1284.

Roberts DL (2006) Extinct or possibly extinct? *Science* **312**, 997.

Roberts DL and Solow AR (2003) When did the dodo become extinct? *Nature* **426**, 245.

Roberts DL and Saltmarsh A (2006) How confident are we that a species is extinct? Quantitative inference of extinction from biological records. *Bulletin of the British Ornithologists' Club* **126** (suppl), 55–58.

Roberts JM, Wheeler AJ, and Freiwald A (2006) Reefs of the deep: the biology and geology of cold-water coral ecosystems. *Science* **312**, 543–547.

Roberts N (1998) *The Holocene: an Environmental History*, 2nd edn. Blackwell, Oxford.

Roberts N, Stevenson AC, Davis B, Cheddadi R, Brewer S, and Rosen A (2004) Holocene climate, environment and cultural change in the circum-Mediterranean region. In Battarbee RW, Gasse F, and Stickley CE, eds, *Past Climate Variability Through Europe and Africa*, pp. 343–362. Kluwer Academic Publishers, Dordrecht.

Roberts RG, Flannery TF, Ayliffe LK, Yoshida H, Olley JM, Prideaux GJ, Laslett GM, Baynes A, Smith MA, Jones R, and Smith BL (2001) New ages for the last Australian megafauna: continent-wide extinction about 46,000 years ago. *Science* **292**, 1888–1892.

Robertson C (1929) *Flowers and Insects: Lists of Visitors to Four Hundred and Fifty-three Flowers.* C. Robertson, Carlinville.

Robins JH, Ross HA, Allen MS, and Matisoo-Smith E (2006) Taxonomy: *Sus bucculentus* revisited. *Nature* **440**, E7.

Robinson AC and Young MC (1983) *The Toolache Wallaby.* Report no. 2. South Australian Department of Environment and Planning, Adelaide.

Robinson RA and Sutherland WJ (2002) Post-war changes in arable farming and biodiversity in Great Britain. *Journal of Applied Ecology* **39**, 157–176.

Robison ND (1983) Archaeological records of naiad mussels along the Tennessee-Tombigbee Waterway. In Miller AC, ed., *Report of Freshwater Mussels Workshop*, 26–27 October 1982, pp. 115–129. U.S. Army Engineer Waterways Experiment Station, Vicksburg.

Robock A (2000) Volcanic eruptions and climate. *Reviews of Geophysics* **38**, 191–219.

Robson DS and Whitlock JH (1964) Estimation of a truncation point. *Biometrika* **51**, 33–39.

Roca AL, Georgiadis N, Pecon-Slattery J, and O'Brien SJ (2001) Genetic evidence for two species of elephant in Africa. *Science* **293**, 1473–1477.

Rocha-Camarero G and Hildago de Truios S (2002) The spread of the collared dove *Streptopelia decaocto* in Europe: colonization patterns in the west of the Iberian Peninsula. *Bird Study* **49**, 11–16.

Rodriguez JP (2002) Range contraction in declining North American bird populations. *Ecological Applications* **12**, 238–248.

Roff DA and Roff RJ (2003) Of rats and Maoris: a novel method for the analysis of patterns of extinction in the New Zealand avifauna before human contact. *Evolutionary Ecology Research* **5**, 759–779.

Rohland N, Pollack JL, Nagel D, Beauval C, Airvaux J, Pääbo S, and Hofreiter M (2005) The population history of extant and extinct hyenas. *Molecular Biology and Evolution* **22**, 2435–2443.

Rohling EJ, Grant K, Hemleben C, Siddall M, Hoogakker BAA, Bolshaw M, and Kucera M (2008) High rates of sea-level rise during the last interglacial period. *Nature Geoscience* **1**, 38–42.

Rojas-Bracho L, Reeves RR, and Jaramillo-Legorreta A (2006) Conservation of the vaquita *Phocoena sinus*. *Mammal Review* **36**, 179–216.

Rolett B and Diamond J (2004) Environmental predictors of pre-European deforestation on Pacific islands. *Nature* **431**, 443–446.

Roman J and Palumbi SR (2003) Whales before whaling in the North Atlantic. *Science* **301**, 508–510.

Roosevelt AC, Lima da Costa M, Lopes Mochado G, Michnab M, Mercier N, Valladas H, Feathers J, Barnett W, Imazio da Silveira M, Henderson A *et al.* (1996) Paleoindian cave dwellers in the Amazon: the peopling of the Americas. *Science* **272**, 373–383.

Rose GA (2004) Reconciling overfishing and climate change with stock dynamics of Atlantic cod (*Gadus morhua*) over 500 years. *Canadian Journal of Fisheries and Aquatic Science* **61**, 1553–1557.

Rosen B (1994) Mammoths in ancient Egypt? *Nature* **369**, 364.

Rosenzweig ML (2001) The four questions: what does the introduction of exotic species do to diversity? *Evolutionary Ecology Research* **3**, 361–367.

Rothschild W (1893–1900) *Avifauna of Laysan and the Neighbouring Islands with a Complete History to Date of the Birds of the Hawaiian Possession*. RH Porter, London.

Rothschild W (1907) *Extinct Birds*. Hutchinson & Co., London.

Rounsevell DE and Smith SJ (1982) Recent alleged sightings of the thylacine (Marsupialia, Thylacinidae) in Tasmania. In Archer M, ed., *Carnivorous Marsupials*, pp. 233–236. Royal Zoological Society of New South Wales, Mosman.

Rouse I (1992) *The Tainos: Rise and Decline of the People who Greeted Columbus*. Yale University Press, New Haven.

Roy K, Jablonski D, and Valentine JW (1995) Thermally anomalous assemblages revisited: patterns in the extraprovincial latitudinal range shift of Pleistocene marine mollusks. *Geology* **23**, 1071–1074.

Roy K, Valentine JW, Jablonski D, and Kidwell SM (1996) Scales of climatic variability and time averaging in Pleistocene biotas: implications for ecology and evolution. *Trends in Ecology & Evolution* **11**, 458–463.

Ruddiman WF (2003) The anthropogenic greenhouse era began thousands of years ago. *Climatic Change* **61**, 261–293.

Russell EWB (1983) Indian-set fires in the forests of the northeastern United States. *Ecology* **64**, 78–88.

Russell EWB (1997) *People and the Land Through Time*. Yale University Press, New Haven.

Russell EWB, Davis RB, Anderson RS, Rhodes TE, and Anderson DS (1993) Recent centuries of vegetation change in the glaciated north-eastern United States. *Journal of Ecology* **81**, 647–664.

Russell GJ, Brooks TJ, McKinney ML, and Anderson CG (1998) Present and future taxonomic selectivity in bird and mammal extinctions. *Conservation Biology* **12**, 1365–1376.

Ruzzante DE, Wroblewski JS, Taggart CT, Smedbol RK, Cook D, and Goddard SV (2000) Bay-scale population structure in coastal Atlantic cod in Labrador and Newfoundland, Canada. *Journal of Fish Biology* **56**, 431–447.

Saavedra B and Simonetti JA (1998) Small mammal taphonomy: intraspecific bone assemblage comparison between South and North American barn owl, *Tyto alba*, populations. *Journal of Archaeological Science* **25**, 165–170.

Sadovy Y (2005) Trouble on the reef: the imperative for managing vulnerable and valuable fisheries. *Fish and Fisheries* **6**, 167–185.

Sadovy Y and Cheung WL (2003) Near extinction of a highly fecund fish: the one that nearly got away. *Fish and Fisheries* **4**, 86–99.

Sainsbury K and Sumaila UR (2003) Incorporating ecosystem objectives into management of sustainable marine fisheries, including 'best practice' reference points and use of marine protected areas. In Sinclair M and Valdimarsson G, eds, *Responsible Fisheries in the Marine Ecosystem*, pp. 343–361. CAB International, Wallingford.

Salomonsen F (1965) The geographical variation of the fulmar (*Fulmarus glacialis*) and the zones of marine environment in the North Atlantic. *The Auk* **85**, 327–355.

Salzmann U and Waller M (1998) Holocene vegetation history of the Nigerian Sahel based on multiple pollen profiles. *Review of Palaeobotany and Palynology* **100**, 39–72.

Sánchez Marco A (2004) Avian zoogeographical patterns during the Quaternary in the Mediterranean region and paleoclimatic interpretation. *Ardeola* **51**, 91–132.

Sanchez-Pinero F and Polis GA (2000) Bottom-up dynamics of allochthonous input: direct and indirect effects of seabirds on islands. *Ecology Letters* **81**, 3117–3132.

Sandweiss DH (2003) Terminal Pleistocene through mid-Holocene archaeological sites as paleoclimatic archives for the Peruvian coast. *Palaeogeography, Palaeoclimatology, Palaeoecology* **194**, 23–40.

Sanfilippo R (1998) Spirorbid polychaetes as boreal guests in the Mediterranean Pleistocene. *Rivista Italiano di Paleontologia e Stratigraphia* **104**, 279–286.

Sangster G, Collinson JM, Helbig AJ, Knox AG, and Parkin DT (2004) Taxonomic recommendations for British birds: second report. *Ibis* **146**, 153–157.

Santley RS, Killion TW, and Lycett MT (1986) On the Maya collapse. *Journal of Anthropological Research* **42**, 123–159.

Savidge JA (1987) Extinction of an island avifauna by an introduced snake. *Ecology* **68**, 660–668.

Sax DF and Gaines SD (2003) Species diversity: from global decreases to local increases. *Trends in Ecology & Evolution* **18**, 561–566.

Sax DF, Gaines SD, and Brown JH (2002) Species invasions exceed extinctions on islands worldwide: a comparative study of plants and birds. *American Naturalist* **160**, 766–783.

Schodde R, Fullagar R, and Hermes N (1983) *A Review of Norfolk Island Birds: Past and Present*. Special Publication 8. Australian National Parks and Wildlife Service, Canberra.

Schoenbrun DL (1993) We are what we eat: ancient agriculture between the Great Lakes. *Journal of African History* **34**, 1–31.

Schomburgk RH (1848) *The History of Barbados*. Longman, Brown, Green & Longmans, London.

Schorger AW (1955) *The Passenger Pigeon: its Natural History and Extinction*. University of Wisconsin Press, Madison.

Schroering GB (1995) Swamp deer resurfaces. *Wildlife Conservation* **98(6)**, 22.

Schubert BW, Mead JI, and Graham RW (eds) (2003) *Ice Age Cave Faunas of North America*. Indiana University Press, Bloomington.

Schwartz D (1992) Assèchement climatique vers 3000 BP et expansion Bantu en Afrique centrale atlantique: quelques réflexions. *Bulletin de la Societé Géologique de France* **136**, 353–361.

Scofield P, Worthy TH, and Schlumpf H (2003) What birds were New Zealand's first people eating? Wairau Bar's avian remains re-examined. *Records of the Canterbury Museum* **17**, 17–35.

Scofield RP, Hiller N, and Mannering AA (2006) A fossil diving petrel (Aves: Pelecanoididae) from the mid-Miocene of North Canterbury, New Zealand. *Records of the Canterbury Museum* **20**, 65–71.

Scotland RW and Sanderson MJ (2003) The significance of few versus many in the tree of life. *Science* **303**, 643.

Scruggs Jr GD (1960) *Status of Freshwater Mussel Stocks in the Tennessee River*. Special Scientific Report, Fisheries, 370. US Fish and Wildlife Service, Washington DC.

Sealfon RA (2007) Dental divergence supports species status of the extinct sea mink (Carnivora: Mustelidae: *Neovison macrodon*). *Journal of Mammalogy* **88**, 371–383.

Sechrest W, Brooks TM, da Fonseca GAB, Konstant WR, Mittermeier RA, Purvis A, Rylands AB, and Gittleman JL (2002) Hotspots and the conservation of evolutionary history. *Proceedings of the National Academy of Sciences USA* **99**, 2067–2071.

Seguí B (1998) *Els ocells fossils de Mallorca i de Menorca. Succesio estratigrafica d'aus en els rebliments carstics de les Gimnesies*. Doctoral thesis, Universitat de les Illes Balears.

Seguí B and Alcover JA (1999) Comparison of paleoecological patterns in insular bird faunas: a case study from the eestern Mediterranean and Hawaii. In Olson SL, ed., *Avian Paleontology at the Close of the 20th Century: Proceedings of the 4th International Meeting of the Society of Avian Paleontology and Evolution*, Washington DC, 4–7 June 1996. *Smithsonian Contributions to Paleobiology* **89**, 67–73.

Seguí B, Quintana J, Fornós JJ, and Alcover JA (2001) A new fulmarine petrel (Aves: Procellariiformes) from the upper Miocene of the western Mediterranean. *Palaeontology* **44**, 933–948.

Sekercioglu ÇH (2006) Increasing awareness of avian ecological function. *Trends in Ecology & Evolution* **21**, 464–469.

Sekercioglu ÇH, Daily GC, and Ehrlich PR (2004) Ecosystem consequences of bird declines. *Proceedings of the National Academy of Sciences USA* **101**, 18042–18047.

Semken HA, Stafford TH, and Graham RW (1998) Contemporaneity of megamammal extinctions and the reorganisation of non-analogue micromammal associations during the Late Pleistocene of North America. *Final Program and Abstracts of the 8th International Congress of the Internatuonal Council for Archaeology* 257.

Seppä H and Birks HJB (2001) July mean temperature and annual precipitation trends during the Holocene in Fennoscandian tree-line area: pollen based climate reconstructions. *The Holocene* **11**, 527–539.

Seppä H and Bennett KD (2003) Quaternary pollen analysis: recent progress in palaeoecology and palaeoclimatology. *Progress in Physical Geography* **27**, 548–579.

Serb JM (2006) Discovery of genetically distinct sympatric lineages in the freshwater mussel *Cyprogenia aberti* (Bivalvia: Unionidae). *Journal of Molluscan Studies* **72**, 425–434.

Serjeantson D (2005) Archaeological records of a gadfly petrel *Pterodroma* sp. from Scotland in the first millennium A.D. In Grupe G and Peters J, eds, *Feathers, Grit and Symbolism: Birds and Humans in the Ancient Old and New World. Proceedings of the Fifth Meeting of the ICAZ Bird Working Group. Documenta Archaeobiologiae* **3**, 235–246.

Seto NWH (2001) Christmas shearwater (*Puffinus nativitatis*). In Poole A and Gill F, eds, *The Birds of North America*, no. 561. The Academy of Natural Sciences, Philadelphia, and The American Ornithologists' Union, Washington DC.

Shackleton NJ, Sánchez-Goñi MF, Pailler D, and Lancelot Y (2003) Marine isotope stage 5e and the Eemian interglacial. *Global and Planetary Change* **36**, 151–155.

Shapiro B, Drummond AJ, Rambaut A, Wilson MC, Matheus PE, Sher AV, Pybus OG, Gilbert MTP, Barnes I, Binladen J *et al.* (2004) Rise and fall of the Beringian steppe bison. *Science* **306**, 1561–1565.

Sharland MSR (1940) In search of the thylacine. Society's interest in the preservation of a unique marsupial. *Proceedings of the Royal Zoological Society of New South Wales* 1939–1940, 20–38.

Sheldon PR (1987) Parallel gradualistic evolution of Ordovician trilobites. *Nature* **330**, 561–563.

Shennan I, Lambeck K, Flather R, Horton B, McArthur J, Innes J, Lloyd J, Rutherford M, and Wingfield R (2000) Modelling western North Sea palaeogeographies and tidal changes during the Holocene. In Shennan I and Andrews J, eds, *Holocene Land-Ocean Interaction and Environmental Change Around the North Sea*, pp. 299–319. Special Publication 166. Geological Society, London.

Sheppard CRC (2003) Predicted recurrences of coral mortality in the Indian Ocean. *Nature* **425**, 294–297.

Sherratt A (1997) Climatic cycles and behavioural revolutions: the emergence of modern humans and the beginning of farming. *Antiquity* **71**, 271–281.

Shindell DT, Schmidt GA, Mann ME, Rind D, and Waple A (2001) Solar forcing of regional climate change during the Maunder Minimum. *Science* **294**, 2149–2152.

Shufeldt RW (1916) The bird-caves of the Bermudas and their former inhabitants. *Ibis* **(10)4**, 623–635.

Shuman BN, Bartlein PJ, Logar N, Newby P, and Webb III T (2002) Parallel vegetation and climate responses to the early-Holocene collapse of the Laurentide Ice Sheet. *Quaternary Science Reviews* **21**, 1793–1805.

Shuman BN, Bartlein PJ, and Webb III T (2005) The relative magnitude of millennial- and orbital-scale climatic change in eastern North America during the Late-Quaternary. *Quaternary Science Reviews* **24**, 2194–2206.

Shurin JB and Seabloom EW (2005) The strength of trophic cascades across ecosystems: predictions from allometry and energetics. *Journal of Animal Ecology* **74**, 1029–1038.

Siegenthaler U, Stocker TF, Monnin E, Lüthi D, Schwander J, Stauffer B, Raynaud D, Barnola J-M, Fischer H, Masson-Delmotte V, and Jouzel J (2005) Stable carbon cycle-climate relationship during the late Pleistocene. *Science* **310**, 1313–1317.

Sievert PR and Sileo L (1993) The effects of ingested plastic on growth and survival of albatross chicks. In Vermeer K, Briggs KT, Morgan KH, and Siegel-Causey D, eds, *The Status, Ecology, and Conservation of Marine Birds of the North Pacific*, pp. 212–217. Canadian Wildlife Service Special Publication, Ottawa.

Signor PW III and Lipps JH (1982) Sampling bias, gradual extinction patterns, and catastrophes in the fossil record. In Silver LT and Schultz, eds, *Geological Implications of Impacts of Large Asteroids and Comets on the Earth*, pp. 291–296. Special Publication 190. Geological Society of America, Boulder.

Sikes DS and Raithel CJ (2002) A review of the hypotheses of decline of the endangered American burying beetle (Silphidae: *Nicrophorus americanus* Olivier). *Journal of Insect Conservation* **6**, 103–113.

Sikes EL, Burgess SN, Grandpre R, and Guilderson TP (2008) Assessing modern deep-water ages in the New Zealand region using deep-water corals. *Deep Sea Research Part I: Oceanographic Research Papers* **55**, 38–49.

Silva AL, Tamashiro J, and Begossi A (2007) Ethnobotany of riverine populations from the Rio Negro, Amazonia. *Journal of Ethnobiology* **27**, 46–72.

Silvano RAM, MacCord PFL, Lima RV, and Begossi A (2006) When does this fish spawn? Fishermen's local knowledge of migration and reproduction of Brazilian coastal fishes. *Environmental Biology of Fishes* **76**, 371–386.

Simberloff D (2000) Extinction-proneness of island species—causes and management implications. *Raffles Bulletin of Zoology* **48**, 1–9.

Simberloff D and Von Holle B (1999) Positive interactions of nonindigenous species: invasional meltdown? *Biological Invasions* **1**, 21–32.

Simenstad CA, Estes JA, and Kenyon KW (1978) Aleuts, sea otters, and alternate stable-state communities. *Science* **200**, 403–411.

Simmons AH, ed., (1999) *Faunal Extinction in an Island Society: Pygmy Hippopotamus Hunters of Cyprus*. Kluwer Academic/Plenum, New York.

Simms MJ (1994) Emplacement and preservation of vertebrates in caves and fissures. *Zoological Journal of the Linnean Society* **112**, 261–283.

Simon JE, Frank KT, and Kulka DW (2002) *Distribution and Abundance of Barndoor Skate Dipturus laevis in the Canadian Atlantic Based upon Research Vessel Surveys and Industry/Science Surveys*. Canadian Science Advisory Secretariat Research Document. Department of Fisheries and Oceans, Ottawa.

Simons EL (1997) Lemurs: old and new. In Goodman SM and Patterson BD, eds, *Natural Change and Human Impact in Madagascar*, pp. 142–166. Smithsonian Institution Press, Washington.

Simons TR (1983) *Biology and Conservation of the Endangered Dark-rumped Petrel (Pterodroma phaeopygia sandwichensis)*. CPSU/UW83–2. National Park Service, Cooperative Park Studies Unit, University of Washington, Seattle.

Sinha A, Datta A, Madhusudan MD, and Mishra C (2005) The Arunachal macaque *Macaca munzala*: a new species from western Arunachal Pradesh, northeastern India. *International Journal of Primatology* **26**, 977–989.

Slikas B, Olson SL, and Fleischer RC (2002) Rapid independent evolution of flightlessness in four species of Pacific island rails (Rallidae): an analysis based on mitochondrial sequence data. *Journal of Avian Biology* **33**, 5–14.

Smedbol RK, McPherson A, Hansen MM, and Kenchington E (2002) Myths and moderation in marine 'metapopulations'. *Fish and Fisheries* **3**, 20–35.

Smith DR (2006) Survey design for detecting rare freshwater mussels. *Journal of the North American Benthological Society* **25**, 701–711.

Smith FA and Betancourt JL (2006) Predicting woodrat (*Neotoma*) responses to anthropogenic warming from studies of the palaeomidden record. *Journal of Biogeography* **33**, 2061–2076.

Smith FA, Bestelmeyer BT, Biardi J, and Strong M (1993) Anthropogenic extinction of the endemic woodrat, *Neotoma bunkeri* Burt. *Biodiversity Letters* **1**, 149–155.

Smith RL and Weissman I (1985) Maximum likelihood estimation of the lower tail of a probability distribution. *Journal of the Royal Statistical Society, Series B* **47**, 285–298.

Smith TB, Freed LA, Kaimanu Lepson J, and Carothers JA (1995) Evolutionary consequences of extinctions in populations of a Hawaiian honeycreeper. *Conservation Biology* **9**, 107–113.

Smith VR (1978) Animal-plant-soil nutrient relationships on Marion Island (Subantarctic). *Oecologia* **32**, 239–253.

Smol JP, Birks HJB, and Last WM (eds) (2001) *Tracking Environmental Change using Lake Sediments: Terrestrial, Algal, and Siliceous Indicators*. Kluwer, Dordrecht.

Snyder NFR (2007) An alternative hypothesis for the cause of the ivory-billed woodpecker's decline. *Monographs of the Western Foundation of Vertebrate Zoology* **2**, 1–58.

Snyder NFR and Snyder H (2000) *The California Condor: a Saga of Natural History and Conservation*. Academic Press, San Diego.

Solow AR (1993) Inferring extinction in a declining population. *Journal of Mathematical Biology* **32**, 79–82.

Solow AR (2003) Estimation of stratigraphic ranges when fossil finds are not randomly distributed. *Palaeobiology* **29**, 181–185.

Solow AR (2005) Inferring extinction from a sighting record. *Mathematical Biosciences* **195**, 47–55.

Solow AR and Roberts DL (2003) A nonparametric test for extinction based on a sightings record. *Ecology* **84**, 1329–1332.

Solow AR, Roberts DL, and Robbirt KM (2006) On the Pleistocene extinctions of Alaskan mammoths and horses. *Proceedings of the National Academy of Sciences USA* **103**, 7351–7353.

Sommer R and Benecke N (2004) Late- and post-glacial history of the Mustelidae in Europe. *Mammal Review* **34**, 249–284.

Sommer RS and Benecke N (2006) Late Pleistocene and Holocene development of the felid fauna (Felidae) of Europe: a review. *Journal of Zoology* **269**, 7–19.

Soper RC (1971) A general review of the Early Iron Age for the southern half of Africa. *Azania* **6**, 5–38.

Spalding MD, Blasco F, and Fields CD (1997) *World Mangrove Atlas*. The International Society for Mangrove Ecosystems, Okinawa.

Spector JD (1993) *What this Awl Means*. Minnesota Historical Society Press, St Paul.

Spennemann DHR (1997) Distribution of rat species (*Rattus*) on the atolls of the Marshall Islands: past and present dispersal. *Atoll Research Bulletin* **445**, 1–16.

Spotila JR, Dunham AE, Leslie AJ, Steyermark AC, Plotkin PT, and Paladino FV (1996) Worldwide population decline of *Dermochelys coriacea*: are leatherback turtles going extinct? *Chelonian Conservation and Biology* **2**, 209–222.

Springer AM, Estes JA, van Vliet GB, Williams TM, Doak DF, Danner EM, Forney KA, and Pfister B (2003) Sequential megafaunal collapse in the North Pacific Ocean: an ongoing legacy of industrial whaling? *Proceedings of the National Academy of Sciences USA* **100**, 12223–12228.

Springer M and Lilje A (1988) Biostratigraphy and gap analysis: the expected sequence of biostratigraphic events. *Journal of Geology* **96**, 228–236.

Srinivasan UT, Dunne JA, Harte J, and Martinez ND (2007) Response of complex food webs to realistic extinction sequences. *Ecology* **88**, 671–82.

Stafford Jr TM, Semken Jr HA, Graham RW, Klipel WF, Markova A, Smirnov N, and Southon J (1999) First accelerator mass spectrometry ^{14}C dates documenting contemporaneity of nonanalog species in late Pleistocene mammal communities. *Geology* **27**, 903–906.

Stager JC, Ryves D, Cumming BF, Meeker LD, and Beer J (2005) Solar variability and the levels of Lake Victoria, east Africa, during the last millennium. *Journal of Paleolimnology* **33**, 243–251.

Stahl AB (1986) Early food production in West Africa: rethinking the role of the Kintampo Culture. *Current Anthropology* **27**, 532–536.

Stanley J-D, Krom MD, Cliff RA, and Woodward JC (2003) Nile Flow failure at the end of the Old Kingdom, Egypt: strontium isotopic evidence and pedological evidence. *Geoarchaeology* **18**, 395–402.

Stansbery DH (1964) The mussel (muscle) shoals of the Tennessee River revisited. (Abstract). *American Malacological Union, Inc. Annual Reports* **1964**, 25–28.

Stansbery DH (1970) 2. Eastern freshwater mollusks. (I.) The Mississippi and St. Lawrence River systems.

Proceedings of the American Malcological Union Symposium on Rare and Endangered Mollusks. *Malacologia* **10**, 9–22.

Stansbery DH (1971) Rare and endangered freshwater mollusks in eastern United States. In Jorgenson SE and Sharp RW, eds, *Proceedings of a Symposium on Rare and Endangered Mollusks (Naiads) of the U.S.*, pp. 5–18. US Fish and Wildlife Service, Twin Cities, MN.

Stansbery DH (1976) *The Status of Endangered Fluviatile Mollusks in Central North America. Epioblasma turgidula (Lea, 1858)* Unpublished report to US Department of the Interior, Fish and Wildlife, Bureau of Sport Fisheries and Wildlife, Washington DC. Ohio State University Museum of Zoology, Columbus.

Stapp P (2002) Stable isotopes reveal evidence of predation by ship rats on seabirds on the Shiant Islands, Scotland. *Journal of Applied Ecology* **39**, 831–840.

Starrett WC (1971) A survey of the mussels (Unionacea) of the Illinois River: a polluted stream. *Illinois Natural History Survey Bulletin* **30**, 267–403.

Staubwasser M and Weiss H (2006) Holocene climate and cultural evolution in late prehistoric-early historic West Asia. *Quaternary Research* **66**, 372–387.

Staunton Smith J and Johnson CR (1995) Nutrient inputs from seabirds and humans on a populated coral cay. *Marine Ecology, Progress Series* **124**, 189–200.

Steadman DW (1987) Two new species of rails (Aves: Rallidae) from Mangaia, Southern Cook Islands. *Pacific Science* **40**, 27–43.

Steadman DW (1988) A new species of *Porphyrio* (Aves: Rallidae) from archaeological sites in the Marquesas Islands. *Proceedings of the Biological Society of Washington* **101**, 162–170.

Steadman DW (1989a) A new species of starling (Sturnidae, *Aplonis*) from an archaeological site on Huahine, Society Islands. *Notornis* **36**, 161–169.

Steadman DW (1989b) New species and records of birds (Aves: Megapodiidae, Columbidae) from an archaeological site on Lifuka, Tonga. *Proceedings of the Biological Society of Washington* **102**, 537–552.

Steadman DW (1990) Archaeological bird bones from Ofu, Manu'a, American Samoa: extirpation of shearwaters and petrels. *Archaeology in Oceania* **25**, 14–15.

Steadman DW (1991) Extinct and extirpated birds from Aitutaki and Atiu, southern Cook Islands. *Pacific Science* **45**, 325–347.

Steadman DW (1992) New species of *Gallicolumba* and *Macropygia* (Aves: Columbidae) from archaeological sites in Polynesia. *Natural History Museum of Los Angeles County, Science Series* **36**, 329–350.

Steadman DW (1994) Bird bones from the To'aga site, Ofu, American Samoa: prehistoric loss of seabirds and megapodes. *University of California Archaeological Research Facility Contributions* **51**, 217–228.

Steadman DW (1995) Prehistoric extinctions of Pacific Island birds: biodiversity meets zooarchaeology. *Science* **267**, 1123–1131.

Steadman DW (1997a) The historic biogeography and community ecology of Polynesian pigeons and doves. *Journal of Biogeography* **24**, 737–753.

Steadman DW (1997b) A re-examination of the bird bones excavated on New Caledonia by E. W. Gifford in 1952. *Kroeber Anthropological Society Papers* **82**, 38–48.

Steadman DW (1999) The prehistory of vertebrates, especially birds, on Tinian, Aguiguan, and Rota, Northern Mariana Islands. *Micronesica* **31**, 319–345.

Steadman DW (2002a) A new species of gull (Laridae, *Larus*) from an archaeological site on Huahine, Society Islands. *Proceedings of the Biological Society of Washington* **115**, 1–17.

Steadman DW (2002b) A new species of swiftlet (Aves, Apodidae) from the late Quaternary of Mangaia, Cook Islands, Oceania. *Journal of Vertebrate Paleontology* **22**, 326–331.

Steadman DW (2005) A new species of extinct parrot (Psittacidae: *Eclectus*) from Tonga and Vanuatu, South Pacific. *Pacific Science* **60**, 137–145.

Steadman DW (2006a) An extinct species of tooth-billed pigeon (*Didunculus*) from the Kingdom of Tonga, and the concept of endemism in insular landbirds. *Journal of Zoology* **268**, 233–241.

Steadman DW (2006b) *Extinction and Biogeography of Tropical Pacific Birds*. University of Chicago Press, Chicago.

Steadman DW and Morgan GS (1985) A new species of bullfinch (Aves, Emberizidae) from a late Quaternary cave deposit on Cayman Brac, West Indies. *Proceedings of the Biological Society of Washington* **98**, 544–553.

Steadman DW, Pregill GK, Olson SL (1984) Fossil vertebrates from Antigua, Lesser Antilles: evidence for late Holocene human-caused extinctions in the West Indies. *Proceedings of the National Academy of Sciences USA* **81**, 4448–4451.

Steadman DW and Olson SL (1985) Bird remains from an archaeological site on Henderson Island, South Pacific: man-caused extinctions on an "uninhabited" island. *Proceedings of the National Academy of Sciences USA* **82**, 6191–6195.

Steadman DW and Zarriello MC (1987) Two new species of parrots (Aves: Psittacidae) from archeological sites in the Marquesas Islands. *Proceedings of the Biological Society of Washington* **100**, 518–528.

Steadman DW and Kirch PV (1990) Prehistoric extinction of birds on Mangaia, Cook Islands, Polynesia.

Proceedings of the National Academy of Sciences USA **87**, 9605–9609.

Steadman DW and Rolett B (1996) A chronostratigraphic analysis of the extinction of landbirds on Tahuata, Marquesas Islands. *Journal of Archaeological Science* **23**, 81–94.

Steadman DW and Justice LJ (1998) Prehistoric exploitation of birds on Mangareva, Gambier Islands, French Polynesia. *Man and Culture in Oceania* **14**, 81–98.

Steadman DW and Hilgartner WB (1999) A new species of extinct barn owl (Aves: *Tyto*) from Barbuda, Lesser Antilles. In Olson SL, ed., *Avian Paleontology at the Close of the 20th Century: Proceedings of the 4th International Meeting of the Society of Avian Paleontology and Evolution*, Washington DC, 4–7 June 1996. *Smithsonian Contributions to Paleobiology* **89**, 75–84.

Steadman DW and Martin PS (2003) The late Quaternary extinction and future resurrection of birds on Pacific islands. *Earth-Science Reviews* **61**, 133–147.

Steadman DW, Stafford Jr TW, Donahue DJ, and Jull AJT (1991) Chronology of Holocene vertebrate extinction in the Galápagos Islands. *Quaternary Research* **35**, 126–133.

Steadman DW, Vargas CP, and Cristino FC (1994) Stratigraphy, chronology, and cultural context of an early faunal assemblage from Easter Island. *Asian Perspectives* **33**, 79–96.

Steadman DW, White JP, and Allen J (1999) Prehistoric birds from New Ireland, Papua New Guinea: extinctions on a large Melanesian island. *Proceedings of the National Academy of Sciences USA* **96**, 2563–2568.

Steadman DW, Worthy TH, Anderson AJ, and Walter R (2000) New species and records of birds from prehistoric sites on Niue, Southwest Pacific. *Wilson Bulletin* **112**, 165–186.

Steadman DW, Plourde A, and Burley DV (2002a) Prehistoric butchery and consumption of birds in the Kingdom of Tonga, South Pacific. *Journal of Archaeological Science* **29**, 571–584.

Steadman DW, Pregill GK, and Burley DV (2002b) Rapid prehistoric extinction of iguanas and birds in Polynesia. *Proceedings of the National Academy of Sciences USA* **99**, 3673–3677.

Steadman DW, Tellkamp MP, and Wake TA (2003) Prehistoric exploitation of birds on the Pacific coast of Chiapas, Mexico. *The Condor* **105**, 572–579.

Steadman DW, Martin PS, MacPhee RDE, Jull AJT, McDonald HG, Woods CA, Iturralde-Vinent M, and Hodgins GWL (2005) Asynchronous extinction of late Quaternary sloths on islands and continents. *Proceedings of the National Academy of Sciences USA* **102**, 11763–11768.

Steadman DW, Franz R, Morgan GS, Albury NA, Kakuk B, Broad K, Franz SE, Tinker K, Pateman MP, Lott TA *et al.* (2007) Exceptionally well preserved late Quaternary plant and vertebrate fossils from a blue hole on Abaco, The Bahamas. *Proceedings of the National Academy of Sciences USA* **104**, 19897–19902.

Stebich M, Brüchmann C, Kulbe T, Schettler G, and Negendank JFW (2005) Vegetation history, human impact and climate change during the last 700 years recorded in annually laminated sediments of Lac Pavin, France. *Review of Palaeobotany and Palynology* **133**, 115–133.

Stejneger L (1887) How the great northern sea-cow (*Rytina*) became exterminated. *American Naturalist* **21**, 1047–1054.

Steneck RS, Graham MH, Bourque BJ, Corbett D, Erlandson JM, Estes JA, and Tegner MJ (2003) Kelp forest ecosystems: biodiversity, stability, resilience and future. *Environmental Conservation* **29**, 436–459.

Steponaitis VP (1986) Prehistoric archaeology in the southeastern United States, 1970–1985. *Annual Review of Anthropology* **15**, 363–404.

Stevens JD, Bonfil R, Dulvy NK, and Walker P (2000) The effects of fishing on sharks, rays and chimaeras (chondrichthyans), and the implications for marine ecosystems. *ICES Journal of Marine Science* **57**, 476–494.

Stewart JR (1999) Intraspecific variation in modern and Quaternary European *Lagopus*. *Smithsonian Contributions to Paleobiology* **89**, 159–168.

Stewart JR (2001) Wetland birds in the archaeological and recent palaeontological record of Britain and Europe. In Coles B and Bull DE, eds, *Heritage Management of Wetlands*, pp. 141–148. Europae Archaeologiae Consilium.

Stewart JR (2003) Comment on "Buffered tree population changes in a Quaternary refugium: evolutionary implications". *Science* **299**, 825a.

Stewart JR (2004) Wetland birds in the recent fossil record of Britain and north-west Europe. *British Birds* **97**, 33–43.

Stewart JR (2005) The ecology and adaptation of Neanderthals during the non-analogue environment of Oxygen Isotope Stage 3. *Quaternary International* **137**, 35–46.

Stewart JR (2007a) The fossil and archaeological record of the eagle owl in Britain. *British Birds* **100**, 481–486.

Stewart JR (2007b) *An Evolutionary Study of Some Archaeologically Significant Avian Taxa in the Quaternary of the Western Palaearctic*. BAR International Series 405, Oxford.

Stewart JR (2008) The progressive independent response to Quaternary climate change. *Quaternary Science Reviews* **27**, 2499–2508.

Stewart JR and Lister AM (2001) Cryptic northern refugia and the origins of the modern biota. *Trends in Ecology & Evolution* **16**, 608–613.

Stewart JR and Dalén L (2008) Is the glacial refugium concept relevant for northern species? A comment on Pruett and Winker 2005. *Climatic Change* **86**, 19–22.

Stewart JR, van Kolfschoten M, Markova A, and Musil R (2003) The mammalian faunas of Europe during Oxygen Isotope Stage Three. In van Andel TH and Davies SW, eds, *Neanderthals and Modern Humans in the European Landscape During the Last Glaciation, 60,000 to 20,000 years ago: Archaeological Results of the Stage 3 Project*. MacDonald Institute Monograph Series. MacDonald Institute for Archaeological Research, Cambridge.

Stine JK (1993) *Mixing the Waters: Environment, Politics, and the Building of the Tennessee-Tombigbee Waterway*. University of Akron Press, Akron.

Stork NE (1988) Insect diversity: facts, fiction and speculation. *Biological Journal of the Linnean Society* **353**, 321–337.

Stork NE and Lyal CHC (1993) Extinction or coextinction rates. *Nature* **366**, 307.

Strahan R, ed. (1995) *The Mammals of Australia*, 2nd edn. Australian Museum/Reed New Holland, Sydney.

Strauss D and Sadler PM (1989) Classical confidence intervals and Bayesian probability estimates ends of local taxon ranges. *Mathematical Geology* **21**, 411–427.

Strayer DL, Downing JA, Haag WR, King TL, Layzer JB, Newton TJ, and Nichols SJ (2004) Changing perspectives on pearly mussels, America's most imperiled animals. *BioScience* **54**, 429–439.

Strickland HE and Melville AG (1848) *The Dodo and its Kindred*. Reeve, Benham & Reeve, London.

Stuart AJ (1991) Mammalian extinctions in the Late Pleistocene of northern Eurasia and North America. *Biological Reviews* **66**, 453–562.

Stuart AJ (1999) Late Pleistocene megafaunal extinctions: a European perspective. In MacPhee RDE, ed., *Extinctions in Near Time: Causes, Contexts, and Consequences*, pp. 257–269. Kluwer Academic/Plenum, New York.

Stuart AJ, Sulerzhitsky LD, Orlova LA, Kuzmin YV, and Lister AM (2002) The latest woolly mammoths (*Mammuthus primigenius* Blumenbach) in Europe and Asia: a review of the current evidence. *Quaternary Science Reviews* **21**, 1559–1569.

Stuart AJ, Kosintsev PA, Higham TFG, and Lister AM (2004) Pleistocene to Holocene extinction dynamics in giant deer and woolly mammoth. *Nature* **431**, 684–689.

Suárez W (2000a) Contribucion al conocimiento del estatus generico del condor extinto (Ciconiiformes: Vulturidae) del Cuaternario cubano. *Ornitologia Neotropical* **11**, 109–122.

Suárez W (2000b) Fossil evidence for the occurrence of the Cuban poorwill *Siphonorhis daiquiri* in western Cuba. *Cotinga* **14**, 66–68.

Suárez W (2001) A reevaluation of some fossils identified as vultures (Aves: Vulturidae) from Quaternary cave deposits of Cuba. *Caribbean Journal of Science* **37**, 110–111.

Suárez W (2004a) The identity of the fossil raptor of the genus *Amplibuteo* (Aves: Accipitridae) from the Quaternary of Cuba. *Caribbean Journal of Science* **40**, 120–125.

Suárez W (2004b) The enigmatic snipe *Capella* sp. (Aves: Scolopacidae) in the fossil record of Cuba. *Caribbean Journal of Science* **40**, 155–157.

Suárez W (2005) Taxonomic status of the Cuban vampire bat (Chiroptera: Phyllostomidae: Desmodontinae: *Desmodus*). *Caribbean Journal of Science* **41**, 761–767.

Suárez W and Olson SL (2001a) A remarkable new species of small falcon from the Quaternary of Cuba (Aves: Falconidae: *Falco*). *Proceedings of the Biological Society of Washington* **114**, 34–41.

Suárez W and Olson SL (2001b) Further characterization of *Caracara creightoni* Brodkorb based on fossils from the Quaternary of Cuba (Aves: Falconidae). *Proceedings of the Biological Society of Washington* **114**, 501–508.

Suárez W and Díaz-Franco S (2003) A new fossil bat (Chiroptera: Phyllostomidae) from a Quaternary cave deposit in Cuba. *Caribbean Journal of Science* **39**, 371–377.

Suárez W and Emslie SD (2003) New fossil material with a redescription of the extinct condor *Gymnogyps varonai* (Arredondo 1971) (Aves: Vulturidae) from the Quaternary of Cuba. *Proceedings of the Biological Society of Washington* **116**, 29–37.

Suárez W and Olson SL (2003a) A new species of caracara (*Milvago*) from Quaternary asphalt deposits in Cuba, with notes on new material of *Caracara creightoni* Brodkorb (Aves: Falconidae). *Proceedings of the Biological Society of Washington* **116**, 301–307.

Suárez W and Olson SL (2003b) New records of storks (Ciconiidae) from Quaternary asphalt deposits in Cuba. *The Condor* **105**, 150–153.

Suárez W and Olson SL (2007) The Cuban fossil eagle *Aquila borrasi* Arredondo: a scaled-up version of the great black-hawk *Buteogallus urubitinga* (Gmelin). *Journal of Raptor Research* **41**, 288–298.

Sullivan J, Arellano E, and Rogers DS (2000) Comparative phylogeography of Mesoamerican highland rodents: concerted versus independent response to past climatic fluctuations. *American Naturalist* **155**, 755–768.

Sutherland WJ (2002) Openness in management. *Nature* **418**, 834–835.

Sutherland WJ, Pullin AS, Dolman PM, and Knight TM (2004) The need for evidence-based conservation. *Trends in Ecology & Evolution* **19**, 305–308.

Svenning J-C (2002) A review of natural vegetation openness in north-western Europe. *Biological Conservation* **104**, 133–148.

Symeonidis NK, Bachmayer F, and Zapfe H (1973) Grabungen in der Zwergelefanten-Höhle 'Charkadio' auf der Insel Tilos (Dodekanes, Griechenland). *Annalen des Naturhistorischen Museums Wien* **77**, 133–139.

Symonds MRE (2005) Phylogeny and life histories of the 'Insectivora': controversies and consequences. *Biological Reviews* **80**, 93–128.

Taberlet P, Fumagalli L, Wust-Saucy A, and Cosson J (1998) Comparative phylogeography and postglacial colonization routes in Europe. *Molecular Ecology* **8**, 1923–1934.

Tankersley KB (1999) Sheriden: a stratified Pleistocene-Holocene cave site in the Great Lakes Region of North America. In Driver JC, ed., *Zooarchaeology of the Pleistocene/Holocene Boundary. Proceedings of a Symposium held at the International Council for Archaeozoology (ICAZ)*, Victoria, British Columbia, Canada, August 1998, pp. 67–75. BAR International Series 800.

Tarasov L and Peltier WR (2005) Arctic freshwater forcing of the Younger Dryas cold reversal. *Nature* **435**, 662–665.

Tasker ML, Camphuysen CJ, Cooper J, Garthe S, Montevecchi WA, and Blaber SJM (2000) The impacts of fishing on marine birds. *ICES Journal of Marine Science* **57**, 531–547.

Taylor DM (1990) Late Quaternary pollen records from two Ugandan mires: evidence for environmental change in the Rukiga Highlands of southwest Uganda. *Palaeogeography, Palaeoclimatology, Palaeoecology* **80**, 283–300.

Taylor DM, Robetshaw P, and Marchant RA (2000) Environmental change and political upheaval in pre-colonial western Uganda. *The Holocene* **10**, 527–536.

Taylor GA (2000) *Threatened Seabirds Recovery Plan*. Department of Conservation, Wellington.

Taylor RW (1989) Changes in freshwater mussel populations of the Ohio River: 1,000 BP to recent times. *Ohio Journal of Science* **89**, 188–191.

Teilhard de Chardin P, and Young CC (1936) On the mammalian remains from the archæological site of Anyang. *Palæontologia Sinica, Series C* **7**, 5–61.

Telford RJ, Heegaard E, and Birks HJB (2004) The intercept is a poor estimate of a calibrated radiocarbon age. *The Holocene* **14**, 296–298.

Teller T, Leverington DW, and Mann JD (2002) Freshwater outbursts to the oceans from glacial Lake Agassiz and their role in climate change during the last deglaciation. *Quaternary Science Reviews* **21**, 879–887.

Temple SA (1977) Plant-animal mutualism: co-evolution with dodo leads to near extinction of plant. *Science* **197**, 885–886.

Temple SA (1979) The dodo and the tambalacoque tree. *Science* **203**, 1364–1364.

Tendeiro J (1969) Estudos sobre os Goniodideos (Mallophaga, Ischnocera) dos Columbiformes. IV—Genero *Campanulotes* Keler, 1939. *Revista de Ciencias Veterinarias, Universidade de Lourenco Marques (Serie A)* **2**, 365–466.

Tennessee-Tombigbee Waterway Development Authority (2007) *About the Tenn-Tom Waterway: Waterway Construction*. Tennessee-Tombigbee Waterway Development Authority, Columbus. www.tenntom.org/about/ttwconstruction.htm.

Tennyson AJD and Martinson P (2006) *Extinct Birds of New Zealand*. Te Papa Press, Wellington.

Tennyson AJD and Millener PR (1994) Bird extinctions and fossil bones from Mangere Island, Chatham Islands. *Notornis* **41**(suppl.), 165–178.

Tennyson AJD, Palma RL, Robertson HA, Worthy TH, and Gill BJ (2003) A new species of kiwi (Aves, Apterygiformes) from Okarito, New Zealand. *Records of the Auckland Museum* **40**, 55–64.

Theler JL (1987a) Pre-Columbian freshwater mussel assemblages of the Mississippi River in southwestern Wisconsin. *The Nautilus* **101**, 143–150.

Theler JL (1987b) The pre-Columbian freshwater mussels (*Naiades*) from Brogley Rockshelter in southwestern Wisconsin. *American Malacological Bulletin* **5**, 165–171.

Theler JL (1991) Aboriginal utilization of freshwater mussels at the Azatlan site, Wisconsin. In Purdue JR, Klippel WE, and Styles BW, eds, *Beamers, Bobwhites, and Blue-points. Tributes to the Career of Paul W. Parmalee*, pp. 315–332. Illinois State Museum Scientific Papers, 23. Illinois State Museum, Springfield.

Theler JL and Boszhardt RF (2006) Collapse of crucial resources and culture change: a model for the Woodland to Oneota transformation in the Upper Midwest. *American Antiquity* **71**, 433–472.

Thibault JC and Bretagnolle V (1998) A Mediterranean breeding colony of Cory's shearwater *Calonectris diomedea* in which individuals show behavioural and biometric characters of the Atlantic subspecies. *Ibis* **140**, 523–529.

Thomas CD, Cameron A, Green RE, Bakkenes M, Beaumont LJ, Collingham YC, Erasmus BFN, de Siqueira MF, Grainger A, Hannah L *et al.* (2004) Extinction risk from climate change. *Nature* **427**, 145–148.

Thomas O (1910) A collection of mammals from eastern Buenos Ayres, with descriptions of related new mammals from other localities. *Annals and Magazine of Natural History* **8(5)**, 239–247.

Thompson LG, Davis ME, Mosely-Thompson E, and Liu K-B (1988) Pre-Incan agricultural activity recorded in dust layers in two tropical ice cores. *Nature* **336**, 763–765.

Thompson WL (2004) *Sampling Rare or Elusive Species: Concepts, Designs, and Techniques for Estimating Population Parameters*. Island Press, Washington DC.

Thomson GM (1922) *The Naturalisation of Plants and Animals in New Zealand*. Cambridge University Press, Cambridge.

Thoreau HD (1951) *Cape Cod*. Bramhall House, New York.

Thuiller W, Brotons L, Araújo MB, and Lavorel S (2004) Effects of restricting environmental range of data to project current and future species distributions. *Ecography* **27**, 165–172.

Tilman D, May RM, Lehman CL, and Nowak MA (1994) Habitat destruction and the extinction debt. *Nature* **371**, 65–66.

Timm RM, Salazar RM, and Peterson AT (1997) Historical distribution of the extinct tropical seal *Monachus tropicalis* (Carnivora: Phocidae). *Conservation Biology* **11**, 549–551.

Tipping R, Buchanan J, Davies A, and Tisdell E (1999) Woodland biodiversity, palaeo-human ecology and some implications for conservation management. *Journal of Biogeography* **26**, 33–43.

Tittensor DP, Worm B, and Myers RA (2008) Macroecological changes in exploited marine systems. In Witman JD and Roy K, eds, *Marine Macroecology* (in press). University of Chicago Press, Chicago.

Tobias JA and Ekstrom JMM (2002) The New Caledonian owlet-nightjar (*Aegotheles savesi*) rediscovered? *Bulletin of the British Ornithologists' Club* **122**, 282–283.

Tong H and Liu J (2004) The Pleistocene-Holocene extinctions of mammals in China. In Dong W, ed., *Proceedings of the Ninth Annual Symposium of the Chinese Society of Vertebrate Paleontology*, pp. 111–119. China Ocean Press, Beijing.

Tonni EP and Politis G (1982) Un gran cánido del Holoceno de la provincial de Buenos Aires y el registro prehispánico de *Canis* (*Canis*) *familiaris* en la areas pampeanas y patagónica. *Ameghiniana* **18**, 251–265.

Towns DR and Broome KG (2003) From small Maria to massive Campbell: forty years of rat eradications from New Zealand islands. *New Zealand Journal of Ecology* **30**, 377–398.

Tremblay NO and Schoen DJ (1999) Molecular phylogeography of *Dryas integrifolia*: glacial refugia and post-glacial recolonization. *Molecular Ecology* **8**, 1187–1198.

Trewick SA (1997) Flightlessness and phylogeny amongst endemic rails (Aves: Rallidae) of the New Zealand region. *Philosophical Transactions of the Royal Society of London B* **352**, 429–446.

Trewick SA, Morgan-Richards M, Russell SJ, Henderson S, Rumsey FJ, Pínter I, Barrett JA, Gibby M, and Vogel JC (2002) Polyploidy, phylogeography and Pleistocene refugia of rockfern *Asplenum ceterah*: evidence from chloroplast DNA. *Molecular Ecology* **11**, 2003–2012.

Trotter MM and McCulloch B (1984) Moas, men and middens. In Martin PS and Klein RG, eds, *Quaternary Extinctions: a Prehistoric Revolution*, pp. 708–727. Arizona University Press, Tucson.

Tuck GN, Polacheck T, Croxall JP, and Weimerskirch H (2001) Modelling the impact of fishery by-catches on albatross populations. *Journal of Applied Ecology* **38**, 1182–1196.

Turgeon DD, Quinn Jr JF, Bogan AE, Coan EV, Hochberg RJ, Lyons WG, Mikkelsen PM, Neves RJ, Roper CFE, Rosenberg G *et al.* (1998) *Common and Scientific Names of Aquatic Invertebrates from the United States and Canada: Mollusks*, 2nd edn. Special Publication 26. American Fisheries Society, Bethesda.

Turk T (2004) The eagle owl in Britain 2004. Has the native returned? *Tyto* **9(3)**, 9–20.

Turley CM, Roberts JM, and Guinotte JM (2007) Corals in deep water: will the unseen hand of ocean acidification destroy cold-water ecosystems? *Coral Reefs* **26**, 445–448.

Turvey ST and Risley CL (2006) Modelling the extinction of Steller's sea cow. *Biology Letters* **2**, 94–97.

Turvey ST, Green OR, and Holdaway RN (2005) Cortical growth marks reveal extended juvenile development in New Zealand moa. *Nature* **435**, 940–943.

Turvey ST, Oliver JR, Narganes Storde YM, and Rye P (2007a) Late Holocene extinction of Puerto Rican native land mammals. *Biology Letters* **3**, 193–196.

Turvey ST, Pitman RL, Taylor BL, Barlow J, Akamatsu T, Barrett LA, Zhao X, Reeves RR, Stewart BS, Pusser LT *et al.* (2007b) First human-caused extinction of a cetacean species? *Biology Letters* **3**, 537–540.

Tyler T (2002a) Geographical distribution of allozyme variation in relation to post-glacial history in *Carex digitata*, a widespread European woodland sedge. *Journal of Biogeography* **29**, 919–930.

Tyler T (2002b) Large-scale geographic patterns of genetic variation in *Melica nutans*, a widespread Eurasian woodland grass. *Plant Systematics and Evolution* **236**, 73–87.

Tyrberg T (1991a) Crossbill (Genus *Loxia*) evolution in the West Palearctic—a look at the fossil evidence. *Ornis Svecica* **1**, 3–10.

Tyrberg T (1991b) Arctic, montane and steppe birds as glacial relicts in the West Palearctic. *Ornithologische Verhandlungen* **25**, 29–49.

Tyrberg T (1998) *Pleistocene Birds of the Palearctic: a Catalogue.* Nuttall Ornithological Club, Cambridge, MA.

Umbanhower Jr CE (2004) Interaction of fire, climate and vegetation change at a large landscape scale in the Big Woods of Minnesota, USA. *The Holocene* **14**, 661–676.

Ursenbacher S, Carlsson M, Helfer V, Tegelström H, and Fumagalli L (2006) Phylogeography and Pleistocene refugia of the adder (*Vipera berus*) as inferred from mitochondrial DNA sequence data. *Molecular Ecology* **15**, 3425–3437.

US Fish and Wildlife Service (1983) *Green Blossom Pearly Mussel Recovery Plan.* US Fish and Wildlife Service, Atlanta.

US Fish and Wildlife Service (1991) *Speckled Pocketbook Mussel (Lampsilis streckeri) Recovery Plan.* US Fish and Wildlife Service, Jackson.

US Fish and Wildlife Service (2000) *Mobile River Basin Aquatic Ecosystem Recovery Plan.* US Fish and Wildlife Service, Atlanta.

US Fish and Wildlife Service (2003) *Recovery Plan for Endangered Fat Threeridge (Amblema neislerii), Shinyrayed Pocketbook (Lampsilis subangulata), Gulf Moccasinshell (Medionidus penicillatus), Ochlockonee Moccasinshell (Medionidus simpsonianus), and Oval Pigtoe (Pleurobema pyriforme); and Threatened Chipola Slabshell (Elliptio chipolaensis) and Purple Bankclimber (Elliptoideus sloatianus).* US Fish and Wildlife Service, Atlanta.

Valdes P (2003) An introduction to climate modelling of the Holocene. In Mackay AW, Battarbee RW, Birks HJB, and Oldfield F, eds, *Global Change in the Holocene*, pp. 20–35. Arnold, London.

Valdiosera CE, García N, Anderung C, Dalén L, Crégut-Bonnoure E, Kahlke R-D, Stiller M, Brandström M, Thomas MG, Arsuaga JL *et al.* (2007) Staying out in the cold: glacial refugia and mitochondrial DNA Phylogeography in ancient European bears. *Molecular Ecology* **24**, 5140–5148.

van Aarde RJ (1980) The diet and feeding behavior of feral cats, *Felis catus*, at Marion Island. *South African Journal of Wildlife Research* **10**, 123–128.

van der Ree R and McCarthy MA (2005) Inferring persistence of indigenous mammals in response to urbanisation. *Animal Conservation* **8**, 309–319.

van der Schalie H (1981) Perspective on North American malacology I. Mollusks in the Alabama River drainage; past and present. *Sterkiana* **71**, 24–40.

van Deusen HM (1963) First New Guinea record of *Thylacinus*. *Journal of Mammalogy* **44**, 279–280.

van Dyck S and Strahan R (2008) *The Mammals of Australia.* 3rd edn. Reed New Holland, Sydney.

Vane-Wright RI, Humphries CJ, and Williams PH (1991) What to protect—systematics and the agony of choice. *Biological Conservation* **55**, 235–254.

van Hoof TB, Bunnik FPM, Waucomont JGM, Kurschner WM, and Visscher H (2005) Forest re-growth on medieval farmland after the Black Death pandemic—implications for atmospheric CO_2 levels. *Palaeogeography, Palaeoclimatology, Palaeoecology* **237**, 396–411.

van Kolschoten T (1990) The evolution of the mammal fauna in the Netherlands and the Middle Rhine area (West Germany) during the late Middle Pleistocene. *Mededelingen Rijks Geologische Dienst* **43**, 1–69.

Vanni MJ, DeAngelis DL, Schindler DE, and Huxel GR (2004) Overview: cross-habitat flux of nutrients and detritus. In Polis GA, Power ME, and Huxel GR, eds, *Food Webs at the Landscape Level*, pp. 3–12. University of Chicago Press, Chicago.

Vansina J (1999) Linguistic evidence and historical reconstruction. *Journal of African History* **40**, 469–473.

Van Tets GF (1994) An extinct new species of cormorant (Phalacrocoracidae, Aves) from a Western Australian peat swamp. *Records of the South Australian Museum* **27**, 135–138.

Van Tets GF and O'Connor S (1983) The Hunter Island penguin, an extinct new genus and species from a Tasmanian midden. *Records of the Queen Victoria Museum of Launceston* **81**, 1–13.

van Vuure C (2005) *Retracing the Aurochs: History, Morphology and Ecology of an Extinct Wild Ox.* Pensoft, Sofia and Moscow.

Vartanyan SL, Garutt VE, and Sher AV (1993) Holocene dwarf mammoths from Wrangel Island in the Siberian Arctic. *Nature* **362**, 337–340.

Vazquez DP and Gittleman JL (1998) Biodiversity conservation: does phylogeny matter? *Current Biology* **8**, R379–R381.

Veitch CR (1999) The eradication of feral cats (*Felis catus*) from Little Barrier Island, New Zealand. *New Zealand Journal of Zoology* **28**, 1–12.

Veitch CR, Miskelly CM, Harper GA, Taylor GA, and Tennyson AJD (2004) Birds of the Kermadec Islands, south-west Pacific. *Notornis* **51**, 61–90.

Vélez-Juarbe J and Miller TE (2007) First report of a Quaternary crocodilian from a cave deposit in northern Puerto Rico. *Caribbean Journal of Science* **43**, 273–277.

Vera FWM (2000) *Grazing Ecology and Forest History.* CABI Publishing, Wallingford.

Vermeij GJ (1993) Biogeography of recently extinct marine species—implications for conservation. *Conservation Biology* **7**, 391–397.

Verschuren D, Laird KR, and Cumming BR (2000) Rainfall and drought in equatorial east Africa during the past 1100 years. *Nature* **403**, 410–414.

Verschuren D, Johnson TC, Kling HJ, Edgington DN, Leavitt PR, Brown ET, Talbot MR, and Hecky RE (2002) History and timing of human impact on Lake Victoria, East Africa. *Proceedings of the Royal Society of London Series B Biological Sciences* **269**, 289–294.

Vigne J-D and Valladas H (1996) Small mammal fossil assemblages as indicators of environmental change in northern Corsica during the last 2500 years. *Journal of Archaeological Science* **23**, 199–215.

Vigne J-D, Bailon S, and Cuisin J (1997) Biostratigraphy of amphibians, reptiles, birds and mammals in Corsica and the role of man in the Holocene faunal turnover. *Anthropozoologica* **25–26**, 587–604.

Vila E (2006) Data on equids from late fourth and third millenium sites in northern Syria. In Mashkour M, ed., *Equids in Time and Space*, pp. 101–123. Oxbow Books, Oxford.

Vitousek PM (2002) Oceanic islands as model systems for ecological studies. *Journal of Biogeography* **29**, 573–582.

Vitousek PM (2004) *Nutrient Cycling and Limitation: Hawai'i as a Model System*. Princeton University Press, Princeton.

Voightlander CW and Poppe WL (1989) The Tennessee River. In Dodge DP, ed., *Proceedings of the International Large River Symposium*, pp. 372–384. Canadian Special Publication of Fisheries and Aquatic Sciences 106, Ottawa.

Volman TP (1978) Early archaeological evidence for shellfish collecting. *Science* **201**, 911–913.

von den Driesch A (1999) The crane, *Grus grus*, in prehistoric Europe and its relation to the Pleistocene *Grus primigenia* [*primigenia*]. In Benecke N, ed., *The Holocene History of the European Vertebrate Fauna: Modern Aspects of Research. Archäologie in Eurasien* **6**, 201–207.

von Euler F (2001) Selective extinction and rapid loss of evolutionary history in the bird fauna. *Proceedings of the Royal Society of London Series B Biological Sciences* **268**, 127.

Vrba ES (1984) Evolutionary pattern and process in the sister-group Alcelaphini-Aepycerotini (Mammalia: Bovidae). In Eldgredge N and Stanley SM, eds, *Living Fossils*, pp. 62–79. Springer-Verlag, New York.

Wait DA, Aubrey DP, and Anderson WB (2005) Seabird guano influences on desert islands: soil chemistry and herbaceous species richness and productivity. *Journal of Arid Environments* **60**, 681–695.

Walker CA, Wragg GM, and Harrison CJO (1990) A new shearwater from the Pleistocene of the Canary Islands and its bearings on the evolution of certain *Puffinus* shearwaters. *Historical Biology* **3**, 203–224.

Walker P (1980) Archaeological evidence for the recent extinction of three terrestrial mammals on San Miguel Island. In Power DM, ed., *The California Islands: Proceedings of a Multidisciplinary Symposium*, pp. 703–717. Santa Barbara Museum of Natural History, Santa Barbara.

Walters M (1988) Probable validity of *Rallus nigra* Miller, an extinct species from Tahiti. *Notornis* **35**, 260–269.

Wanless RM, Angel A, Cuthbert RJ, Hilton GM, and Ryan PG (2007) Can predation by invasive mice drive seabird extinctions? *Biology Letters* **3**, 241–244.

Warheit KI (2002) The seabird fossil record and the role of paleontology in understanding seabird community structure. In Schreiber EA and Burger EA, eds, *Biology of Marine Birds*, pp. 17–55. CRC Press, Boca Raton.

Warren Jr ML and Haag WR (2005) Spatio-temporal patterns of the decline of freshwater mussels in the Little South Fork Cumberland River, USA. *Biodiversity and Conservation* **14**, 1383–1400.

Warren RE (1975) *Pre-Columbian Unionacean (Freshwater Mussel) Utilization at the Widows Creek site (1JA305), Northeast Alabama*. MA thesis, University of Nebraska.

Warren RE (1995) Premodern *Pleurobema rubrum* (Rafinesque 1820) from the Illinois River. *Transactions of the Illinois State Academy of Science* **88**, 5–12.

Waser NM, Chittka L, Price MV, Williams NM, and Ollerton J (1996) Generalization in pollination systems, and why it matters. *Ecology* **77**, 1043–1060.

Waters MR and Stafford Jr TW (2007).Redefining the age of Clovis: implications for the peopling of the Americas. *Science* **315**, 1122–1126.

Waterton C (1825) *Wanderings in South America*. J. Mawman, London.

Watters GT (1994) *An Annotated Bibliography of the Reproduction and Propagation of the Unionoidea (Primarily of North America)*. Ohio Biological Survey Miscellaneous Contributions 1. Ohio Biological Survey, Columbus.

Watts D (1987) *The West Indies: Patterns of Development, Culture and Environmental Change Since 1492*. Cambridge University Press, Cambridge.

Wcsislo WT and Cane JH (1996) Floral resource utilization by solitary bees (Hymenoptera: Apoidea) and exploitation of their stored food by natural enemies. *Annual Review of Entomology* **41**, 257–286.

Weaver ME and Ingram DL (1969) Morphological changes in swine associated with environmental temperature. *Ecology* **50**, 710–713.

Webb SL (1986) Potential role of passenger pigeons and other vertebrates in the rapid Holocene migrations of nut trees. *Quaternary Research* **26**, 367–375.

Webb III T (1973) A comparison of modern and presettlement pollen in southern Michigan (U.S.A.). *Review of Palaeobotany and Palynology* **16**, 137–156.

Webb III T (1982) Temporal resolution in Holocene pollen data. *Third North American Paleontological Convention, Proceedings* **2**, 569–572.

Webb III T (1986) Is vegetation in equilibrium with climate? How to interpret late Quaternary pollen data. *Vegetatio* **67**, 75–91.

Webb III T (1988) Eastern North America. In Huntley B and Webb III T, eds, *Handbook of Vegetation Science. Volume VII. Vegetation History*, pp. 385–414. Kluwer Academic Publications, Dordrecht.

Webb TJ, Noble D, and Freckleton RP (2007) Abundance-occupancy dynamics in a human dominated environment: linking interspecific and intraspecific trends in British farmland and woodland birds. *Journal of Animal Ecology* **76**, 123–134.

Webb WS and DeJarnette DL (1942) *An Archeological Survey in Pickwick Basin in Adjacent Portions of the States of Alabama, Mississippi, and Tennessee*. Bulletin 129. Bureau of American Ethnology, Washington DC.

Weckström J, Korhola A, Erästö P, and Holmström L (2006) Temperature patterns over the past eight centuries in Northern Fennoscandia inferred from sedimentary diatoms. *Quaternary Research* **66**, 78–86.

Weesie PDM (1982) A Pleistocene endemic island form within the genus Athene: *Athene cretensis* n. sp. (Aves, Strigiformes) from Crete. *Proceedings of the Koninklijke Nederlandse Akademie van Wetenschappen, Series B* **85**, 323–336.

Weesie PDM (1988) The Quaternary avifauna of Crete, Greece. *Palaeovertebrata* **18**, 1–94.

Weisler MJ (1994) The settlement of marginal Polynesia: new evidence from Henderson Island. *Journal of World Archaeology* **21**, 83–102.

Weller AA (1999) On types of trochilids in the Natural History Museum, Tring II. Re-evaluation of *Erythronota* (?) *elegans* Gould 1860: a presumed extinct species of the genus *Chlorostilbon*. *Bulletin of the British Ornithological Club* **119**, 197–202.

Wendorf F, Close AE, Schild R, Wasylikowa K, Housley RA, Harlan JR, and Królik H (1992) Saharan exploitation of plants 8,000 years BP. *Nature* **359**, 721–724.

Weninger B, Alram-Stern E, Bauer E, Clare L, Danzeglocke U, Jöris O, Kubatzki C, Rollefson G, Todorova H, and van Andel T (2006) Climate forcing due to the 8200 cal yr BP event observed at Early Neolithic sites in the Eastern Mediterranean. *Quaternary Research* **66**, 401–420.

West RG (1980) Pleistocene forest history in East Anglia. *New Phytologist* **85**, 571–622.

Wetmore A (1918) Bones of birds collected by Theodoor de Booy from kitchen midden deposits in the islands of St. Thomas and St. Croix. *Proceedings of the United States National Museum* **54**, 513–522.

Wetmore A (1920) Five new species of birds from cave deposits in Porto Rico. *Proceedings of the Biological Society of Washington* **33**, 77–82.

Wetmore A (1922a) Remains of birds from caves in the Republic of Haiti. *Smithsonian Miscellaneous Collections* **74(4)**, 1–4.

Wetmore A (1922b) Bird remains from the caves of Porto Rico. *Bulletin of the American Museum of Natural History* **46**, 297–333.

Wetmore A (1923) An additional record for the extinct Porto Rican quail-dove. *The Auk* **40**, 324.

Wetmore A (1927) The birds of Porto Rico and the Virgin Islands. *New York Academy of Science, Scientific Survey of Porto Rico and the Virgin Islands* **9**, 243–406, 409–598.

Wetmore A (1937a) Ancient records of birds from the island of St. Croix with observations on extinct and living birds of Puerto Rico. *Journal of Agriculture of the University of Puerto Rico* **21**, 5–16.

Wetmore A (1937b) Bird remains from cave deposits on Great Exuma Island in the Bahamas. *Bulletin of the Museum of Comparative Zoology* **80**, 427–441.

Wetmore A (1963) An extinct rail from the island of St. Helena. *Ibis* **103b**, 379–381.

White PCL and King CM (2006) Predation on native birds in New Zealand beech forests: the role of functional relationships between stoats *Mustela erminea* and rodents. *Ibis* **148**, 765–771.

Whitlock R (1952) *Rare and Extinct Birds of Britain*. Phoenix House, London.

Wick L, Lemscke G, and Sturm M (2003) Evidence for Late Glacial and Holocene climatic change and human impact in eastern Anatolia: high-resolution pollen, charcoal, isotopic, and geochemical records from the laminated sediments of Lake Van, eastern Turkey. *The Holocene* **13**, 97–107.

Wignall PB and Benton MJ (1999) Lazarus taxa and fossil abundance at times of biotic crisis. *Journal of the Geological Society, London* **156**, 453–456.

Wilcove DS (2005) Rediscovery of the ivory-billed woodpecker. *Science* **308**, 1422–1423.

Wilcox C and Donlan CJ (2007) Compensatory mitigation as a solution to fisheries bycatch-biodiversity conservation conflicts. *Frontiers in Ecology and the Environment* **5**, 325–331.

Wilkinson C (2000) *Status of Coral Reefs of the World: 2000*. Australian Institute for Marine Science, Townsville.

Wilkinson HE (1969) Description of an Upper Miocene albatross from Beaumaris, Victoria, Australia, and a review of fossil Diomedeidae. *Memiors of the National Museum of Victoria* **29**, 41–51.

Will KW and Rubinoff D (2004) Myth of the molecule: DNA barcodes for species cannot replace morphology for identification and classification. *Cladistics* **20**, 47–55.

Wille M, Hooghiemstra H, Behling H, Van der Borg K, and Negret AJ (2001) Environmental change in the Colombian subandean forest belt from 8 pollen records: the last 50 kyr. *Vegetation History and Archaeobotany* **10**, 61–77.

Willerslev E, Hansen AJ, Brand T, Binladen J, Gilbert TMP, Shapiro BA, Wiuf C, Gilichinsky DA, and Cooper A (2003) Diverse plant and animal DNA from Holocene and Pleistocene sedimentary records. *Science* **300**, 791–795.

Willerslev E, Cappellini E, Boomsma W, Nielsen R, Hebsgaard MB, Brand TB, Hofreiter M, Bunce M, Poinar HN, Dahl-Jensen D *et al.* (2007) Ancient biomolecules from deep ice cores reveal a forested southern Greenland. *Science* **317**, 111–114.

Williams JD, Bogan AE, Garner JT (2008) *Freshwater mussels of Alabama and the Mobile Basin in Georgia, Mississippi, and Tennessee.* University of Alabama Press, Tuscaloosa.

Williams EE and Koopman KF (1951) A new fossil rodent from Puerto Rico. *American Museum Novitates* **1515**, 1–9.

Williams JD and Butler RS (1994) Class Bivalvia, Freshwater bivalves. In Deyrup M and Franz R, eds, *Rare and Endangered Biota of Florida. IV. Invertebrates*, pp. 53–128, 740–742. University Press of Florida, Gainesville.

Williams JD and Fradkin A (1999) *Fusconaia apalachicola*, a new species of freshwater mussel (Bivalvia: Unionidae) from pre-Columbian archeological sites in the Apalachicola basin of Alabama, Florida, and Georgia. *Tulane Studies in Zoology* **31**, 51–62.

Williams JD, Fuller SLH, and Grace R (1992) Effects of impoundments on freshwater mussels (Mollusca: Bivalvia: Unionidae) in the main channel of the Black Warrior and Tombigbee rivers in western Alabama. *Bulletin of the Alabama Museum of Natural History* **13**, 1–10.

Williams JD, Warren Jr ML, Cummings KS, Harris JL, and Neves RJ (1993) Conservation status of freshwater mussels of the United States and Canada. *Fisheries* **18**, 6–22.

Williams JW, Shuman BN, Webb III T, Bartlein PJ, and Leduc PL (2004) Late Quaternary vegetation dynamics in North America: scaling from taxa to biomes. *Ecological Monographs* **74**, 309–334.

Williams JW, Jackson ST, and Kutzbach JE (2007) Projected distributions of novel and disappearing climates by 2100 AD. *Proceedings of the National Academy of Sciences USA* **104**, 5738–5742.

Williams M (2000) Dark ages and dark areas: global deforestation in the deep past. *Journal of Historical Geography* **26**, 28–46.

Williams M (2003) *Deforesting the Earth: from Prehistory to Global Crisis.* University of Chicago Press, Chicago.

Williams MI and Steadman DW (2001) The historic and prehistoric distribution of parrots (Psittacidae) in the West Indies. In Woods CA and Sergile FE, eds, *Biogeography of the West Indies: Patterns and Perspectives*, pp. 175–189. CRC Press, Boca Raton.

Williams PH and Gaston KJ (1994) Do conservationists and molecular biologists value differences between organisms in the same way? *Biodiversity Letters* **2**, 67–78.

Willis KJ and Whittaker RJ (2000) Paleoecology: the refugial debate. *Science* **287**, 1406–1407.

Willis KJ and van Andel T (2004) Trees or no trees? The environment of central and eastern Europe during the Last Glaciation. *Quaternary Science Reviews* **23**, 2369–2387.

Willis KJ and Birks HJB (2006) What is Natural? The need for a long-term perspective in biodiversity conservation. *Nature* **314**, 1261–1265.

Willis KJ, Gillson L, and Brencic TM (2004) How 'virgin' in virgin rainforest. *Science* **304**, 402–403.

Willis KJ, Araújo MB, Bennett KD, Figueroa-Rangel B, Froyd CA, and Myers N (2007) How can a knowledge of the past help to conserve the future? Biodiversity conservation and the relevance of long-term ecological studies. *Philosophical Transactions of the Royal Society B* **362**, 175–186.

Wilmshurst JM and Higham TFG (2004) Using rat-gnawed seeds to independently date the arrival of Pacific rats and humans in New Zealand. *The Holocene* **14**, 801–806.

Wilson CB and Clark HW (1914) *The Mussels of the Cumberland River and its Tributaries.* Document no. 781, pp. 1–63. United States Bureau of Fisheries, Washington DC.

Wilson DE (1991) Mammals of the Tres Marías Islands. *Bulletin of the American Museum of Natural History* **206**, 214–250.

Wilson DE and Reeder DM (eds) (2005) *Mammal Species of the World*, 3rd edn. Johns Hopkins University Press, Baltimore.

Wilson EO (1984) *Biophilia.* Cambridge University Press, Cambridge.

Wilson SM (ed.) (1997) *The Indigenous People of the Caribbean.* University Press of Florida, Gainesville.

Wilson SM (2007) *The Archaeology of the Caribbean.* Cambridge University Press, Cambridge.

Windsor DA (1995) Equal rights for parasites. *Conservation Biology* **9**, 1–2.

Wing SR and Wing ES (2001) Prehistoric fisheries in the Caribbean. *Coral Reefs* **20**, 1–8.

Witmer MC and Cheke AS (1991) The dodo and the tambalacoque tree – an obligate mutualism reconsidered. *Oikos* **61**, 133–137.

Wójcik JM, Ratkiewicz M, and Searle JB (2002) Evolution of the common shrew *Sorex araneus*: chromosomal and molecular aspects. *Acta Theriologica* **47** (suppl 1), 139–167.

Wonham MJ and Pachepsky E (2006) A null model of temporal trends in biological invasion records. *Ecology Letters* **9**, 663–672.

Wood CA and Wetmore A (1926) A collection of birds from the Fiji Islands. Part III. Field observations. *Ibis* **56**, 91–136.

Woodman P, McCarthy M, and Monaghan N (1997) The Irish Quaternary fauna project. *Quaternary Science Reviews* **16**, 129–159.

Woods CA (1989) A new capromyid rodent from Haiti: the origin, evolution, and extinction of West Indian rodents and their bearing on the origin of New World hystricognaths. *Los Angeles County Museum, Science Series* **33**, 59–89.

Wolff WJ (2000a) Causes of extirpations in the Wadden Sea, an estuarine area in the Netherlands. *Conservation Biology* **14**, 876–885.

Wolff WJ (2000b) The south-eastern North Sea: losses of vertebrate fauna during the past 2000 years. *Biological Conservation* **95**, 209–217.

Worm B, and Myers RA (2003) Top-down versus bottom-up control in oceanic food webs: a meta-analysis of cod-shrimp interactions in the North Atlantic. *Ecology* **84**, 162–173.

Worm B, Barbier EB, Beaumont N, Duffy JE, Folke C, Halpern BS, Jackson JBC, Lotze HK, Micheli F, Palumbi SR *et al.* (2006) Impacts of biodiversity loss on ocean ecosystem services. *Science* **314**, 787–790.

Worthy TH (1989a) An analysis of moa bones (Aves: Dinornithiformes) from three lowland North Island swamp sites: Makirikiri, Riverlands and Takapau Road. *Journal of the Royal Society of New Zealand* **19**, 419–432.

Worthy TH (1989b) Validation of *Pachyornis australis* Oliver (Aves: Dinornithiformis), a medium sized moa from the South Island, New Zealand. *New Zealand Journal of Geology and Geophysics* **32**, 255–266.

Worthy TH (1990) An analysis of the distribution and relative abundance of moa species (Aves: Dinornithiformes). *New Zealand Journal of Zoology* **17**, 213–241.

Worthy TH (1998a) Quaternary fossil faunas of Otago, South Island, New Zealand. *Journal of the Royal Society of New Zealand* **28**, 421–521.

Worthy TH (1998b) The Quaternary fossil avifauna of Southland, South Island, New Zealand. *Journal of the Royal Society of New Zealand* **28**, 537–589.

Worthy TH (2000) The fossil megapodes (Aves: Megapodiidae) of Fiji with descriptions of a new genus and two new species. *Journal of the Royal Society of New Zealand* **30**, 337–364.

Worthy TH (2001) A giant flightless pigeon gen. et sp. nov. and a new species of *Ducula* (Aves: Columbidae), from Quaternary deposits in Fiji. *Journal of the Royal Society of New Zealand* **31**, 763–794.

Worthy TH (2003) A new extinct species of snipe *Coenocorypha* from Viti Levu, Fiji. *Bulletin of the British Ornithologists' Club* **123**, 90–103.

Worthy TH (2004a) Letter to the Editor. *Journal of the Royal Society of New Zealand* **34**, 105–106.

Worthy TH (2004b) The fossil rails (Aves: Rallidae) of Fiji with descriptions of a new genus and species. *Journal of the Royal Society of New Zealand* **34**, 295–314.

Worthy TH (2005a) A new species of *Oxyura* (Aves: Anatidae) from the New Zealand Holocene. *Memoirs of the Queensland Museum* **51**, 255–272.

Worthy TH (2005b) Rediscovery of the types of *Dinornis curtus* Owen and *Palapteryx geranoides* Owen, with a new synonymy (Aves: Dinornithiformes). *Tuhinga* **16**, 57–67.

Worthy TH and Holdaway RN (1994) Scraps from an owl's table—predator activity as a significant taphonomic process newly recognised from New Zealand Quaternary deposits. *Alcheringa* **18**, 229–245.

Worthy TH and Jouventin P (1999) The fossil avifauna of Amsterdam Island, Indian Ocean. In Olson SL, ed., *Avian Paleontology at the Close of the 20th Century: Proceedings of the 4th International Meeting of the Society of Avian Paleontology and Evolution*, Washington DC, 4–7 June 1996, pp. 39–66. Smithsonian University Press, Washington DC.

Worthy TH and Molnar RE (1999) Megafaunal expression in a land without mammals – the first fossil faunas from terrestrial deposits in Fiji (Vertebrata: Amphibia, Reptilia, Aves). *Senckenbergiana Biologica* **79**, 237–242.

Worthy TH and Holdaway RN (2002) *The Lost World of the Moa.* Indiana University Press, Bloomington.

Worthy TH and Olson SL (2002) Relationships, adaptations, and habits of the extinct duck *Euryanas finschi. Notornis* **49**, 1–17.

Worthy TH and Wragg GM (2003) A new species of *Gallicolumba*: Columbidae from Henderson Island, Pitcairn Group. *Journal of the Royal Society of New Zealand* **33**, 769–793.

Worthy TH and Wragg GM (2008) A new genus and species of pigeon (Aves: Columbidae) from Henderson Island, Pitcairn Group. *Terra Australis* **29**, 499–510.

Worthy TH and Tennyson AJD (2004) Avifaunal assemblages from the Nenega-iti and Onemea sites. In Kirch PV and Conte E, eds, *Archaeological Investigations in the Mangareva Islands (Gambier Archipelago), French Polynesia*, pp. 122–127. Archaeological Research Facility, University of California, Berkeley.

Worthy TH, Anderson AJ, and Molnar RE (1999) Megafaunal expression in a land without mammals: the first fossil faunas from terrestrial deposits in Fiji (Vertebrata: Amphibia, Reptilia, Aves). *Senckenbergiana Biologica* **79**, 237–242.

Worthy TH, Bunce M, Cooper A and Scofield P (2005) *Dinornis* – an insular oddity, a taxonomic conundrum reviewed. *Monografies de la Societat d'Història Natural de les Balears* **12**, 377–390.

Worthy TH, Tennyson AJD, Jones C, McNamara JA, and Douglas BJ (2007) Miocene waterfowl and other birds from Central Otago, New Zealand. *Journal of Systematic Palaeontology* **5**, 1–39.

Wragg GM (1995) The fossil birds of Henderson Island, Pitcairn Group: natural turnover and human impact, a synopsis. *Biological Journal of the Linnean Society* **56**, 405–414.

Wragg GM and Weisler MI (1994) Extinctions and new records of birds from Henderson Island, Pitcairn Group, South Pacific Ocean. *Notornis* **41**, 61–70.

Wragg GM and Worthy TH (2006) A new species of extinct imperial pigeon (*Ducula*: Columbidae) from Henderson Island, Pitcairn Group. *Historical Biology* **18**, 127–140.

Wroe S, Myers TJ, Wells RT, and Gillespie A (1999) Estimating the weight of the Pleistocene marsupial lion, *Thylacoleo carnifex* (Thylacoleonidae: Marsupialia): implications for the ecomorphology of a marsupial super-predator and hypotheses of impoverishment of Australian marsupial carnivore faunas. *Australian Journal of Zoology* **47**, 489–498.

Wroe S, Field J, Fullagar R, Jermiin LS (2004) Megafaunal extinction in the late Quaternary and the global overkill hypothesis. *Alcheringa* **28**, 291–331.

Wroe S, Clausen P, McHenry C, Moreno K, and Cunningham E (2007) Computer simulation of feeding behaviour in the thylacine and dingo as a novel test for convergence and niche overlap. *Proceedings of the Royal Society of London Series B Biological Sciences* **274**, 2819–2828.

Wunderlich J (1989) *Untersuchungen zur Entwicklung des westlichen Nildetas in Holozan.* Gedruckt bei Wenzel, Marburg.

Wynne-Edwards VC (1962) *Animal Dispersion in Relation to Social Behaviour.* Oliver & Boyd, Edinburgh.

Yalden D (1999) *The History of British Mammals.* T & AD Poyser, London.

Yalden D (2007) The older history of the white-tailed eagle in Britain. *British Birds* **100**, 471–480.

Yang DY, Liu L, Chen X, and Speller CF (2008) Wild or domesticated: DNA analysis of ancient water buffalo remains from north China. *Journal of Archaeological Science* **35**, 2778–2785.

Yoder AD, Rasoloarison RM, Goodman SM, Irwin JA, Atsalis S, Ravosa MJ, and Ganzhorn JU (2000) Remarkable species diversity in Malagasy mouse lemurs (Primates, *Microcebus*). *Proceedings of the National Academy of Sciences USA* **97**, 11325–11330.

Youngman PM (1989) The status of the sea mink. *Canadian Field-Naturalist* **103**, 299–299.

Youngman PM and Schueler FW (1991) *Martes nobilis* is a synonym of *Martes americana*, not an extinct Pleistocene-Holocene species. *Journal of Mammalogy* **72**, 567–577.

Zeller D, Booth S, Craig P, and Pauly P (2006) Reconstruction of coral reef fisheries catches in American Samoa, 1950–2002. *Coral Reefs* **25**, 144–152.

Zhang X, Wang D, Liu R, Wei Z, Hua Y, Wang Y, Chen Z, and Wang L (2003) The Yangtze River dolphin or baiji (*Lipotes vexillifer*): population status and conservation issues in the Yangtze River, China. *Aquatic Conservation of Marine and Freshwater Ecosystems* **13**, 51–64.

Ziegler AC (2002) *Hawaiian Natural History, Ecology, and Evolution.* University of Hawaii Press, Honolulu.

Zimov SA (2005) Pleistocene Park: return of the mammoth's ecosystem. *Science* **308**, 796–798.

Zinsmeister WJ (1974) A new interpretation of thermally anomalous molluscan assemblages of the California Pleistocene. *Journal of Paleontology* **48**, 74–94.

Zogning A, Giresse P, Maley J, and Gadel F (1997) The Late Holocene paleoenvironment in the Lake Njupi area, west Cameroon: implications regarding the history of Lake Nyos. *Journal of African Earth Sciences* **24**, 285–300.

Zolitschka B (2003) Dating based on freshwater- and marine-laminated sediments. In Mackay AW, Battarbee RW, Birks HJB, and Oldfield F, eds, *Global Change in the Holocene*, pp. 92–106. Arnold, London.

Zolitschka B, Behre K-E, and Schneider J (2003) Human and climatic impact on the environment as derived from colluvial, fluvial and lacustrine archives—examples from the Bronze Age to the Migration period, Germany. *Quaternary Science Reviews* **22**, 81–100.

Index

Printed and bound by CPI Group (UK) Ltd, Croydon, CR0 4YY